IEE POWER AND ENERGY SERIES 40

Series Editors: Professor A. T. Johns
D. F. Warne

Advances in High Voltage Engineering

Other volumes in this series:

Advances in High Voltage Engineering

Edited by
A. Haddad and D. F. Warne

The Institution of Electrical Engineers

Published by: The Institution of Electrical Engineers, London,
United Kingdom

The Institution of Electrical Engineers,
Michael Faraday House,
Six Hills Way, Stevenage,
Herts., SG1 2AY, United Kingdom

The author thanks the International Electrotechnical Commission (IEC) for permission to
reproduce information from its Technical Report IEC 60479-1. All such extracts are copyright of
IEC, Geneva, Switzerland. All rights reserved. Further information on the IEC is available from
www.iec.ch. IEC has no responsibility for the placement and context in which the extracts and
contents are reproduced by IEE; nor is IEC in any way responsible for the other content or
accuracy therein.

The following figures from Chapter 8: 'Variation of resistivity with salt (a), moisture (b) and
temperature (c), 'Diagonal voltage profile for a 16-mesh square grid with 9 rods in a two layer
soil' and 'Potential distribution above an earth grid with a potential ramp and without ramp'
have been reprinted with permission from IEEE STD 81 1983, "IEEE Guide for Measuring Earth
Resistivity, Ground Impedance and Earth Surface Potentials of a Ground System" 1983, and
IEEE STD 80 2000, "IEEE Guide for Safety in AC Substation Grounding", by IEEE. The IEEE
disclaims any responsibility or liability resulting from the placement and use in the described
manner.

British Library Cataloguing in Publication Data

Advances in high voltage engineering. – (Power & energy series)
    1.High voltages
    I.Haddad, Manu II.Warne, D. F. III.Institution of Electrical Engineers
    621.3'1913

ISBN 0 85296 158 8

Typeset in India by Newgen Imaging Systems (P) Ltd., Chennai, India
Printed in the UK by MPG Books Limited, Bodmin, Cornwall

Contents

Contributors

N.L. Allen
Department of Electrical Engineering
and Electronics, UMIST,
Manchester, England, UK

A.E. Baker
AREVA T&D Technology Centre,
Stafford, England, UK

J.E. Dolan
Optics & Laser Technology Department
Advanced Technology Centre – Sowerby
BAE SYSTEMS
P.O. Box 5, FPC 30 Filton
Bristol, UK

O. Farish
Institute for Energy and Environment,
Strathclyde University, Glasgow,
Scotland, UK

J.C. Fothergill
Department of Engineering,
University of Leicester,
Leicester, England, UK

D.M. German
Cardiff University, Cardiff, Wales, UK

H. Griffiths
Cardiff University, Cardiff, Wales, UK

A. Haddad
Cardiff University, Cardiff, Wales, UK

B.F. Hampton
Diagnostic Monitoring Systems Ltd,
Glasgow, Scotland, UK

R.N. Hampton
Borealis AB
Stenungsund,
Sweden, 444 86

G.R. Jones
Centre for Intelligent Monitoring
Systems, Department of Electrical
Engineering and Electronics,
University of Liverpool, England, UK

M.D. Judd
Institute for Energy and Environment,
Strathclyde University, Glasgow,
Scotland, UK

I.J. Kemp
Glasgow Caledonian University,
Glasgow, Scotland, UK

J.S. Pearson
Diagnostic Monitoring Systems Ltd,
Glasgow, Scotland, UK

N. Pilling
Cardiff University, Cardiff, Wales, UK

T.W. Preston

H.M. Ryan
McLaren Consulting and University of
Sunderland, England, UK

J.W. Spencer
Centre for Intelligent Monitoring
Systems, Department of Electrical
Engineering and Electronics,
University of Liverpool, England, UK

J.P. Sturgess
AREVA T&D Technology Centre,
Stafford, England, UK

D.A. Swift
Cardiff University, Cardiff,
Wales, UK

R.T. Waters
Cardiff University, Cardiff,
Wales, UK

J. Yan
Centre for Intelligent Monitoring
Systems, Department of Electrical
Engineering and Electronics,
University of Liverpool,
England, UK

Introduction

In 1984 a book entitled 'Les propriétés diélectriques de l'air et les trés hautes tensions' was published. It represented a collection of research studies undertaken at EdF and other major world utilities over a significant period, and it has been widely used by manufacturers and users of high voltage equipment and systems, and by academic groups working in the area.

Undeniably, there have been a large number of developments in the high voltage field over the intervening years, but no such reference work has since been produced.

There have been a number of key advances in materials. Polymeric insulators are still under major trials for transmission voltages but have become widely used in overhead lines at distribution levels. Cable insulation has moved from paper systems to various polymer-based materials. SF_6 has taken over as an efficient and reliable insulation medium in high-voltage switchgear, and gas-insulated substations are now frequently the preferred option for new substations especially in urban areas. All of these changes have not been direct substitutions; they have each required a reconsideration of equipment configuration, and new installation and operational techniques.

ZnO surge arresters are also widely used worldwide. Their excellent overvoltage protection characteristics have allowed design of modern compact systems to become more reliably achievable. The propagation of lightning and fault currents through earthing arrangements can now be modelled and predicted in much more reliable fashion. Much of this improvement in understanding has come from better use of numerical methods such as boundary elements, finite element modelling and others, which has in turn put pressure on the experimental derivation of key material and system properties. The understanding of basic physical processes has also developed. The nature of breakdown in air and in other materials is now better characterized than it was twenty years ago.

Improvements have been made in instrumentation and experimental techniques. Who would have predicted thirty years ago, for instance, that it would now be possible to measure the distribution of space charge in solid insulation? So, the case seemed compelling for a new book reviewing, once again, the advances that have been made in high voltage engineering.

This project has had some sense of urgency because of the demographic trends in the research field. The age profile of researchers and experts in the field is disturbingly skewed, a number of experts having retired in the past few years or being about to retire. In many of the key areas of research and development, there is no obvious succession of expertise and a lack of any ongoing group within which the accumulated knowledge might be stored, protected and further developed. It was important then to encapsulate the accumulated wisdom and experience resident in these experts.

It is hoped, therefore, that this work on the one hand meets a need to update the advances that have undoubtedly been made in high voltage engineering during the past twenty years and, on the other hand, captures succinctly an accumulated knowledge which might otherwise be rather difficult to unearth through individual papers dispersed through the literature.

Edited by
A. Haddad and D.F. Warne

Chapter 1

Mechanisms of air breakdown

N.L. Allen

1.1 Introduction

1.1.1 Beginnings

Studies of air breakdown began in the eighteenth century. Two names are pre-eminent: Franklin [1] and Lichtenberg [2], although contemporaries were active. Franklin's work grew out of his interest in lightning – a long spark – while Lichtenberg drew tree-like discharges, now called corona, across the surface of a large cake of resin. These two men defined two broad approaches to the study of breakdown which are perpetuated to this day in experimental and theoretical work. In the late 19th century, the emergence of modern physics, exemplified by the work of Townsend [3] and his successors, permitted knowledge of the process of ionisation to be applied to these phenomena. The two approaches were thus linked and another concept from the 18th century, the electric field, became established as paramount in all discussions of the subject.

Indeed, many of the quantities used in discussion of the processes in discharge physics, such as ionisation and attachment coefficients, electron and ion temperatures, diffusion coefficients and so on, have been measured and are quoted in terms of electric field. Usually, it is assumed that the electric fields being considered are uniform. In fact, this is very rarely true in practice and the use of these quantities is always subject to such modifications as are dictated by non-uniform electric field conditions. This was true in the times of Franklin and Lichtenberg, and it will be assumed in most of the discussion in this chapter since practical engineers rarely enjoy the luxury of simple, uniform field configurations.

The concept of breakdown will be assumed to signify the collapse of the dielectric strength of air between two electrodes, which is in practice defined by the collapse of the voltage that had previously been sustained between them.

1.1.2   Basic breakdown processes

Over many decades, research has identified concepts which contribute to the formation of a basic picture of breakdown in air.

1.1.2.1   Primary electrons

Free electrons exist only for very short times in air that is not subject to a high electric field; normally they are trapped, after creation by cosmic rays, background radiation and so on, to form negative ions. These have a density commonly of the order of a few hundred per cubic centimetre. However, electrons can be detached again from negative ions by acceleration and resulting collisions with neutral molecules in a strong electric field.

1.1.2.2   Ionisation

The electrons so liberated can themselves accelerate in the field, collide with neutral molecules and settle down to a constant average drift velocity in the direction of the field. When sufficiently energetic, the collisions may liberate a further electron, so leaving behind a positive ion. The process is cumulative, quantified initially by Townsend [3], and resulting in the formation of avalanches of electrons. The growth in numbers of electrons and positive ions imparts a small conductivity to the air, which does not lead immediately to a breakdown, that is, to a collapse of voltage.

1.1.2.3   Excitation

Where electrons are sufficiently energetic to cause ionisation, there is usually a plentiful supply with lower energies that can excite neutral atoms without liberating electrons. When returning to the ground state, these atoms emit radiation as visible or ultra-violet light. This property is widely used in research to indicate the presence of ionisation.

1.1.2.4   Other electron processes

The electrons created by the growth of ionisation may be trapped, as described above, and so removed from the ionisation process. This is the attachment process; a net growth of electron and ion population occurs only when the field is sufficiently high for the rate of ionisation to exceed the rate of attachment. Subsequent detachment of electrons from negative ions occurs at the same time, through collisions with neutrals, with free electrons or by interaction with photons. Recombination between electrons and positive ions and between positive and negative ions is a further element in the competing processes that are active in an ionised gas.

1.1.2.5   Regeneration

Initially, Townsend postulated that the positive ions could also ionise, a process now recognised as insignificant. Also, that they move towards the negative electrode to release further electrons by secondary emission, so that the ionisation process could be sustained and grow indefinitely until breakdown occurred. Experiment later showed that breakdown could occur much more quickly than this process would allow. The solution lay in postulating that the positive ions, created by ionisation, are sufficient to create an electric field which, when added to the applied field, intensifies the ionisation process [4, 5]. Additional initiatory electrons are assumed to be created by ultra-violet radiation from the excited molecules in the electron avalanches in which ionisation takes place [6]. They will also be created by photo-emission from the negative electrode. In a sufficiently intense field, these events are cumulative and can occur very rapidly [7, 8]. The current density rises, heating the gas and reducing its density, leading to a rapid increase in energy input and conductivity. This results in a discharge of very low impedance and causes voltage collapse.

1.1.2.6   Reduced electric field

Common sense suggests that the above processes, which all depend on an applied electric field, are determined by the energy that electrons and ions acquire between collisions. Thus the ratio E/N of electric field E V/cm to the gas density N mols/cm^3, known as the reduced field, is now widely used as the reference variable when measuring values of fundamental quantities. The unit of this ratio is the Townsend (Td), which has the numerical value 10^{-17} Vcm^2. Older work used the equivalent ratio E/p where E was in V/cm and p in torr.

They are related as follows:

$$\frac{E}{N} \text{ Td} = 3.03 \frac{E}{p} \text{ Vcm}^{-1} \text{torr}^{-1}$$

when temperature is not a variable. It will be noted that custom has determined that c.g.s. units are still used for these quantities.

The breakdown mechanisms are now examined in more detail. In most of what follows, the discussion will assume that the non-uniform field occurs at a positive polarity electrode (i.e. diverging lines of force), with only a brief description of ionisation processes at a negative (converging field) electrode. The reason is that the processes in a diverging field lead more readily to breakdown than those in a converging field, so that, in engineering practice, the dielectric strength of a gap is lower when the more sharply radiused electrode is positive rather than negative. Thus, for example, positive surge voltages are frequently more dangerous to a power system than negative surge voltages.

1.2   Physical mechanisms

Discussion will first be general, in which physical processes are described in relation to the electric field or E/N value which sustains them. Later, differences will be discussed when the ionisation growth originates in the field at either a positive or a negative electrode.

1.2.1   Avalanche development

Analytic treatment of ionisation by collision assumes a continuous process in which the number of electron–ion pairs created by an electron of a given average energy in a given electric field is proportional to the distance that it travels in that field. The number of new electrons dn created in distance dx is thus:

$$dn = \alpha\, dx \tag{1.1}$$

where α is a proportionality constant which is the number of electron–ion pairs created by the electron per unit of distance in the direction of the field. Including now the electron progeny created by the first pair, all of which are drifting in the field at the same average rate and with the same ionising efficiency, then at any point x along its path some number n of electrons will enter an element of length dx and the number of new electrons created is:

$$dn = \alpha n\, dx \tag{1.2}$$

Over a distance d, starting with one electron at the origin, the total number N of ion pairs created becomes

$$N_a = \exp \alpha d \tag{1.3}$$

Here N_a is the avalanche number. Whereas integration of Equation (1.1) shows that the originating single electron produces a number of ion pairs which increases linearly with its path length in the field, and which is always a relatively small number, use of Equation (1.2) shows that when the effect of its progeny is taken into account, the number of ion pairs increases to a very large number.

The quantity α embraces complex physical processes which include several types of collision, including electronic excitations of neutral molecules and the subsequent production of radiation which may itself aid the ionisation process. These in turn depend on the electric field in relation to the air density or pressure, E/N or E/p, for this is an indicator of the energy gained by electrons between collisions with gas molecules. Indeed, it determines the function governing the distribution of the energies of the electrons in their gas-kinetic-type drift motion in the electric field. In round terms, the average energy required for an electron to ionise is of the order of 20 electron volts (eV). At a given instant, the electrons in the distribution have only a very small probability of gaining sufficient energy to ionise. The mean energy is in fact much less than 20 eV in the typical case of the field required for breakdown, referred to above: here, a value in the order of 2 eV has been determined from the small amount of data that is available [9, 10].

Some gases, of which the most important in electrical engineering are air and SF_6, exhibit a strong affinity for electrons. This is the process of attachment to a neutral molecule to form a negative ion. It is described by an attachment coefficient, η, defined as the number of attachments per unit length per electron moving in the direction of the electric field. This is analogous in form to the definition of the ionisation coefficient, α, describing the rate of loss of free electrons per unit length, rather than the rate of increase.

Where attachment is significant, the reverse process of detachment can also occur. Two mechanisms are possible:

(i)   by collisions with neutral molecules in a high electric field
(ii) by interaction with radiation.

Process (i) [11] is the mechanism by which it is assumed that primary free electrons are liberated at the outset of the ionisation process (section 1.1.2.1). Since the time between collisions is of the order of a nanosecond, such a mechanism is able to produce free electrons either for the initiation of avalanches or to restore to an avalanche some of the electrons that have been trapped by attachment. A discussion of the process is given in References 10 and 11.

Photodetachment, mechanism (ii), can take place only when avalanches are forming, as a result of the excitations occurring at the same time as the ionisation. It has been proposed by Boylett and Williams [12] as a possible mechanism in the propagation of a corona discharge, to be discussed later in this chapter.

Where the processes of ionisation, attachment and detachment exist together, the basic ionisation growth Equation (1.1) becomes:

$$dn = (\alpha - \eta + \delta)n\,dx \tag{1.4}$$

where δ is the detachment coefficient, so that:

$$N_a = \exp(\alpha - \eta + \delta)d \tag{1.5}$$

Data on the values of $(\alpha - \eta + \delta)/N$ in air, as a function of E/N, is available in References 9 to 11, 13 and 14.

As an example of the use of Equations (1.4) and (1.5), we take a value of electric field needed to break down air at normal temperature and pressure. This is in the order of 3 MV m^{-1}, where E/N is about 121 Td and E/p is about 40 Vcm^{-1}torr^{-1}. Here, the value of $(\alpha - \eta + \delta)$ is about 1800 per metre. Thus, in traversing a gap between electrodes of 10^{-2} metre, one electron creates about 18 ion pairs (Equation (1.4)). This may be compared with the total number of collisions made by the electron in crossing the gap, which is in the order of 10^5. However, Equation (1.5) shows that when the similar ionising power of the initial electron's progeny is taken into account, the total number of ion pairs created is about 10^8. Strictly, the number of positive ions created is $\exp(\alpha - \eta + \delta)d - 1$, since the integration takes into account the fact that one electron exists before any ionising collision has occurred.

In this example, where the net value of $(\alpha - \eta + \delta)$ is about 1800 per electron per metre, the value of η for the same condition of E/N is about 1000 per electron per metre [10, 15]. The effect of electronegative attaching molecules in the gas is therefore considerable. Atmospheric air contains two elements that are electronegative, namely oxygen and water vapour. In the latter case, it is the complex $O_2(H_2O)_n$ where $1 < n < 5$ that is most active in attaching an electron. The affinity is of the order of 1 eV and the energy given up in the attachment process is released as radiation, given to a third body as kinetic energy, or produces dissociation of the host molecule. The effective reduction in the ionisation coefficient caused by attachment means that an electronegative gas tends to have a higher breakdown strength than one which does not show this property. The concentration of water vapour in air has significant consequences on the formation of corona and the sparkover stress, as will be discussed later.

1.2.2 Avalanche properties

In an electric field, electrons move at much higher velocities than ions, so that in the ionisation process outlined in section 1.2.1, the positive ions can, on the timescale of avalanche formation, be regarded as remaining stationary. As the concentration of positive ions increases, so the effect of the electric field due to their own space charge increases. The electrons at the head of a developing avalanche thus experience a reduced net field, being the vector sum of the applied and the space charge fields. Likewise, the resultant field is increased behind the avalanche head.

Some important avalanche properties are determined, however, by the diffusion of the electrons during their ionising progress. For many purposes, it is a satisfactory approximation to picture the electron cloud, while growing in numbers according to Equation (1.5), as having diffused to the same extent as the same number of electrons starting at the same time as the single electron that initiates the avalanche. Then, diffusion theory states that the mean square radius of the cloud, assumed spherical, at time t is given by:

$$r^2 = 6Dt \tag{1.6}$$

where D is the coefficient of free diffusion of the electrons.

In time t, the centre of the electron cloud has progressed a distance x, so that:

$$x = vt \tag{1.7}$$

where v is the mean drift velocity of the centre of the cloud.

Then:

$$r^2 = \frac{6Dx}{v} \tag{1.8}$$

and the volume traced out by avalanche growth is approximately a paraboloid.

As the avalanche develops, however, effects of the field due to the positive ion space charge become important. In the axial direction, the reduced field at the head of the avalanche and the enhanced field behind (Figure 1.1) affect the symmetry of the diffusion process that has hitherto been assumed. Estimation of these effects is difficult in any simple picture and a proper treatment requires computation using the continuity equations for both the electron and positive ion components. Lateral diffusion of the cloud is also affected, but here a simple model can utilise the fact that lateral diffusion of electrons will be reduced to the level of that of the positive ions when the potential energy of the electrons in the field of the ions is of the same order as their random kinetic energy. This equality is formalised, in the case of a plasma of roughly equal densities of electrons and ions, by the expression for the 'Debye

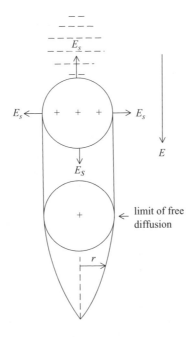

Figure 1.1   Outline of avalanche development, showing space charge field E_s in relation to applied field E. Also mean square radius r of avalanche electron cloud, with transition from free diffusion to diffusion limited by positive ion space charge

length' L_d over which the equality is achieved:

$$L_d = \left(\frac{\epsilon_0 kT}{ne^2} \right)^{1/2} \tag{1.9}$$

Here, n is the electron density, e is the electron charge, k is Boltzmann's constant and T is the electron temperature [9].

If the square of the Debye length is now equated with the mean square radius of the avalanche, the avalanche length x at which free diffusion ceases and is replaced by ambipolar diffusion is obtained from:

$$e \frac{\exp(\alpha - \eta + \delta)x}{4\pi\epsilon_0} = \left[\frac{\pi}{9} \left(\frac{D}{\mu} \right)^3 \frac{x}{E} \right]^{1/2} \tag{1.10}$$

where μ is the electron mobility and E the applied electric field. After this point, the insignificant rate of diffusion of the positive ions ensures that the radius of the avalanche remains almost constant thereafter (Figure 1.1). There is also a redistribution of electron density due to the internal field of the ion space charge; this will not be discussed here.

The drift of electrons in the space between electrodes is recorded in the external circuit by virtue of the displacement current existing between the moving electrons and the electrodes. The problem has been discussed by Shockley [16] for the case of a charge q between parallel plane electrodes. Here, the current I recorded due to charge q moving at velocity v is:

$$I = \frac{qv}{d} \tag{1.11}$$

where d is the distance between the electrodes. For the more general case of the charges due to electrons, positive and negative ions and including the effect of electron diffusion, the equation due to Sato [17] can be used. Ionic currents are of course small compared with those of the electrons, due to the disparity in velocities.

More rigorous modelling of the avalanche and its effect on external circuitry requires the solution of the continuity equations for the electrons and ions, with Poisson's equation. An example is the work of Morrow and Lowke [14], where the circuit current due to avalanche growth is computed taking into account the effects of electron diffusion and of photoionisation. For the conditions of electric field taken, the calculations show a typical current for a single avalanche rising to a few nanoamperes and lasting for a few tenths of a nanosecond.

1.2.3   The critical avalanche and the critical volume

The importance of the space charge field set up by the positive ions has already been mentioned. It reinforces the applied electric field, in which the avalanche has been created, behind its head of positive ions, that is, in the region from which the generating electrons have just come. It is thus able to extend the region of the applied field over which further avalanches might be initiated. Clearly, the extent to which this happens depends on the magnitude of the applied field itself, since this will determine both the size of the avalanche and the space charge field that it creates.

Experiment has shown, however, that breakdown occurs in air when the applied field reaches a value of about 3×10^6 Vm^{-1} implying that ionisation reaches a critical stage at this field. It is also known that the minimum field at which net ionisation can occur is about 2.6×10^6 Vm^{-1}, for at this lower value the sum of the coefficients α and δ in Equation (1.5) exactly balances the attachment coefficient η, so that there is no net gain or loss of electrons. Between these two field values, ionisation can occur, but not the development to breakdown, which requires the higher value.

This has led to the concept that the positive ion space charge field plays a critical role: if it reaches a certain value, it extends the volume around the electrode in which a successor avalanche can be started. Meek [7] proposed that the value of the space charge field at the boundary of the positive ions should be equal to that of the critical field of 3×10^6 Vm^{-1}. Thus, the avalanche number $\exp(\alpha - \eta + \delta)d$ of ions must be contained within the roughly spherical region created by the advancing electron cloud. Experiment, again, indicates that the critical avalanche number is of the order 10^8, that is, a charge of 16 picocoulombs, with a diameter of the order of 100 microns.

This model leads to the expectation that, once this critical avalanche size has been achieved, avalanches can form repeatedly in a direction away from the electrode. Before considering this, however, it is necessary to introduce a further concept; that of the critical volume of field around the electrode.

Obviously, ionisation occurs more efficiently the closer an initiating electron appears to the electrode surface. However, if it starts within only a few ionising free paths of the electrode, there is insufficient distance available for exponential growth to the critical avalanche number of positive ions. The electrons are quickly absorbed by the anode and the positive ions gradually disperse. Evidently, there can be defined a contour around the electrode within which no critical avalanche can be formed. The distance of this contour from the electrode will depend upon the voltage at the electrode and also on its radius.

Similarly, there is an outer contour at which the electric field falls to the critical value of 2.6×10^6 Vm^{-1}. Beyond this boundary, a free electron has insufficient net ionising power to initiate an avalanche. At any point between the outer and inner contours, a critical avalanche can be formed. The volume between the two contours is termed the critical volume. An example is the critical volume shape around the tip of a rod of radius 1 cm, which is of the order of a few cubic centimetres when the breakdown voltage for a rod–plane electrode gap of 1 m is applied. An example of such a critical volume is given in Figure 1.2 [18].

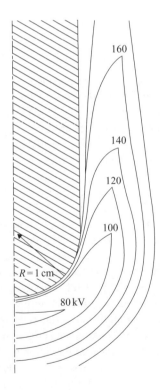

Figure 1.2   Right section showing the growth of the critical volume around the tip of a hemispherically ended rod of diameter 2 cm as voltage levels rise to 180 kV (Allen et al. [18])

Within this volume, an electron must be detached from a negative ion in order to start an avalanche. The concentration of ions has already been noted as a few hundred per cc. The probability of detachment per unit of time is not known with certainty. Where direct or alternating voltages are applied, the time required for a detachment to occur is of no importance, but where a rapidly rising impulse voltage is applied, the statistical uncertainty of the detachment process means that there is a variable statistical time lag before a free electron can start the avalanche, which is responsible for resultant statistical lags in corona formation with consequent effect on the time to the final breakdown. The effect of local atmospheric conditions on the statistical lag is discussed in Reference 18.

1.2.4   Streamer formation

It has already been noted that the space charge field of the critical avalanche is sufficient, when added to the applied field, to initiate a succeeding avalanche. If this development is considered to be along the axis of the system, in a simple case, then the successor is formed in a weaker applied field than the original. This raises two questions:

(i)   are there sufficient free electrons created in this direction to allow this process to proceed?
(ii)   what is the weakest applied field in which replication can occur?

Regarding case (i), the formation time of the avalanche is less than 1 ns, but the positive ion space charge, having a mobility of 10^{-4} metres per second per volt per metre of field, would have a velocity of only 300 ms^{-1} in a field of 3×10^6 Vm^{-1}. Thus, under the influence of this charge, a significant time is available for the detachment of electrons from negative ions. But if the first avalanche critical head occurs very close to the electrode surface, as might be expected, the additional critical volume which it adds to that already established by the voltage at the electrode is extremely small so that the probability of the observed rapid avalanche replication by this means (see section 1.2.5) is negligible.

For this reason, the alternative mechanism of photo-emission of electrons from neutral molecules within the avalanche critical volume is generally considered to be the most likely cause of further development. The photons are generated by excitation in the first avalanche and work reported on the number of photons created as a proportion of the number of ion pairs [6, 19] has allowed estimates to be made of the density of photoelectrons produced as a function of distance from the photon source. As an example, at a distance of 1 mm from the source in an avalanche, the number of photoelectrons generated is about 0.1 per ionising collision in the avalanche. Since the number of ionising collisions is very large, the number of photoelectrons produced in the critical volume of the avalanche space charge is also large.

The sequence of events is shown in Figure 1.3(*i*), taken from Reference 20, where the case of streamers emanating from a negative electrode is also dealt with (see section 1.2.9) [20]. Here, progression from the initiatory electron to the critical avalanche and subsequent streamer growth is illustrated in *a* to *e*.

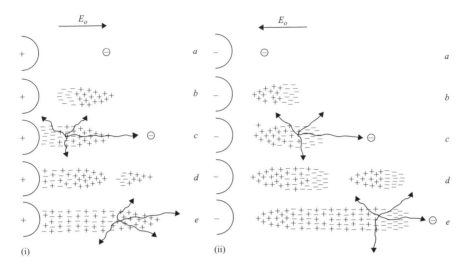

Figure 1.3   Models of streamer growth near a rod electrode

    (i)   positive streamer development from free electron *a*, avalanche *b*, *c*, streamer initiation *d* to growth *e*

   (ii) negative streamer development from free electron *a*, avalanche *b*, *c*, streamer and further avalanche *d*, to space stem *e* (see section 1.2.8) (courtesy of CIGRE's *Electra*, (74), pp. 67–216, Paris)

An alternative hypothesis is that the necessary free electrons are created by photodetachment [12] from the negative ions. This proposal has the attraction that the energy required is of the order of only 1 eV, whereas the energy required for photoionisation is of the order 10 eV so that a much larger flux of active photons is available. The hypothesis suffers from the same difficulty as that for collisional detachment, namely that the number of negative ions in the existing critical volume may be insufficient to maintain a rapid rate of replication. It is, however, consistent with several aspects of streamer propagation and branching.

As the number of successive avalanches increases, so the critical volume, by virtue of the potential at the end of the string of positive space charge heads, moves in the direction of propagation. Although the space charge head is of very small diameter, of the order 100 μm, the perturbation to the applied electric field, calculated in Reference 14 is significant, perpetuating to a great extent the field profile introduced by the rod itself (Figure 1.4). Thus the critical volume remains of appreciable size, comparable with that around the rod, but the known statistical lags associated with such a critical volume appear to rule out collisional detachment and photodetachment as a means of free electron generation, able to allow rapid replication of the heads. Photoionisation thus appears to be the most likely mechanism.

The concept of repeated formation of critical avalanches, leading to advancement of intense ionisation across the space between electrodes, led to the formulation of the

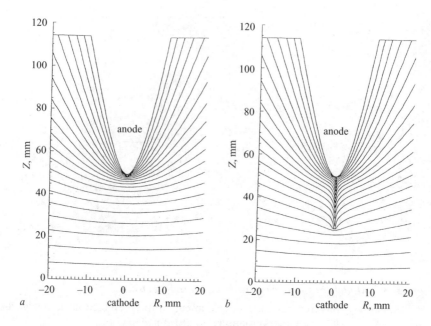

Figure 1.4   Potential distributions a before, b during development of a positive
streamer from an electrode. Anode at 20 kV, equipotential intervals
1 kV. Streamer has progressed ~23 mm in 54 ns (after Morrow and
Lowke [14])

streamer theory first clearly enunciated by Meek [7], although supported by several
earlier experimental observations, for example References 21 and 22, which indicated
some of the principles involved.

1.2.5   Streamer development

The picture so far presented is of successive avalanches progressing into regions of
decreasing electric field, thereby prompting the second of the two questions posed
earlier (page 10).

Meek propounded the criterion that for propagation the electric field at the boundary of an assumed spherical positive space charge is put equal to the applied field
at the electrode. The further criterion has been noted that the resultant field at the
boundary of the critical volume must be at least equal to that needed to sustain ionisation. More exactly, the sum of the applied and space charge fields in this latter
region must be sufficient to supply the rate of energy input for ionisation to produce
a critical avalanche.

This approach was adopted by Gallimberti [23], who considered the parameters
involved in the formation of a single avalanche which was, itself, the successor of
a previous avalanche in the electric field. It was concluded, from theoretical considerations, that an applied field of about $700 \, kVm^{-1}$ was necessary for continuous
propagation of critical avalanches. This conclusion was subjected to practical tests at

later dates, initially by Phelps and Griffiths [24], who initiated streamers at a point but propagated them in a uniform electric field, so removing any ambiguities due to the non-uniformity of field in the models discussed above. In this work, both the pressure (and therefore density) and the humidity of the air were varied, but the authors concluded that at the standard normal temperature and pressure (NTP) condition and at a standard humidity of 11 grams of moisture per cubic metre of air the minimum applied field needed to sustain streamer propagation was close to $500\,kVm^{-1}$.

The result has since been generally confirmed by other measurements [25–28] in a variety of experimental arrangements and the value of $500\,kVm^{-1}$ has been incorporated into the IEC Standard 60060-1 (1989) concerned with the effects of atmospheric changes on sparkover voltages in air (section 1.3.4).

The velocity of streamers, as a function of electric field, is an important parameter that has been investigated [28, 29]. At fields just above the minimum, the velocity is about 2×10^5 ms^{-1}, but this increases faster than linearly as the electric field increases and is commonly greater than 10^6 ms^{-1} at the start of propagation in the non-uniform field exemplified by the rod plane gap referred to above.

It is also worth noting here that at a given condition of air density and humidity, the streamer properties of minimum propagation field and of velocity of growth at higher fields are very precisely defined. It has been found that both of these quantities can be quoted with an uncertainty of less than one per cent. The statistical nature of sparkover measurements is thus due to other factors, but the precision of streamer growth has found application in the use of the rod–rod gap as a standard for measurement of direct voltage [30, 31].

1.2.6 Corona

Practical experience shows that streamers branch, after propagating short distances, to form what is generally termed corona which can then develop and extend to cause a sparkover of the gap. Branching is assumed to be caused by development of side avalanches in the field of an initial avalanche. An example is shown in Figure 1.5. The mechanism of branching is not fully understood. It might be thought that the flux of photons sufficient to cause rapid extension of the streamer by photoionisation would be sufficient to cause a much higher density of branches than is shown in Figure 1.5. The lack of a favourable combination of applied and space charge fields may be responsible, but it has also been postulated that the observed pattern may result from a reliance on photodetachment from surrounding, but relatively rare, negative ions in order to produce the necessary free electrons [12].

Many tests with various electrode configurations show that streamers, although originating in very high fields, can propagate into regions in which the field is much less than $500\,kVm^{-1}$. Figure 1.5 is an example where streamers in the corona progress to the plane, where the electric field is of the order $100\,kVm^{-1}$. Direct measurement has shown, however, that in such a case, the field at the plane, at the instant of arrival of the streamers, was about $450\,kVm^{-1}$ [26, 32]. The largest component of the field was thus due to the sum of the positive space charges at the heads of large numbers

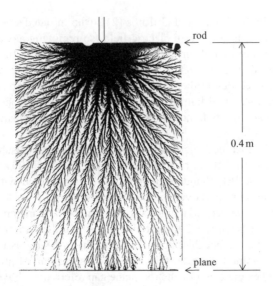

Figure 1.5   *Image of positive corona obtained by sheet film placed along the axis of a rod–plane gap 0.4 m long. Voltage ~170 kV. Note evidence of secondary emission when streamers reach the plane (after Hassanzahraee, PhD thesis, University of Leeds, 1989)*

of streamers which combined to allow propagation in a comparatively weak applied field. This phenomenon allows propagation to take place over distances of the order of metres.

1.2.7   The streamer trail

The electrons produced in each replicating avalanche pass through the positive space charge head left by the preceding avalanche and drift towards the anode along the ionic debris left by earlier avalanches. Some recombination occurs but a more important process is attachment to electronegative molecules, such as O_2 or $O_2(H_2O)_n$, to form negative ions. The use of known attachment coefficients, of the order 100 to 500 per metre per electron [10], indicates an e-folding distance for the rate of loss of free electrons of the order of a few centimetres in air, a result that has been confirmed by recent experiment [33]. Thus, after drifting from the streamer head towards the anode, most of the free electrons are soon lost and the remainder of the streamer trail is made up of positive ions which are nearly matched, in density, by negative ions.

Since the mobility of positive and negative ions is low, of the order 10^{-4} $m^2V^{-1}s^{-1}$, the conductivity of the streamer trail is low, even though a small proportion of free electrons is able to reach the anode. Typically, the resistance of the trail is of the order of megohms. Thus, the existence of an assembly of streamers, as in a corona, does not constitute a breakdown, even though the corona may bridge the electrode gap.

1.2.8   The leader

When the electric field is sufficiently high, a new development can occur within the corona: the formation of one or two distinct highly conducting channels having properties different from those of the streamer channels. This is the leader, first photographed as a precursor to the main flash of the lightning discharge by Schonland and Collens [34] and later in laboratory studies by Allibone and Schonland [35]. In contrast to the streamer channel, the light emission is in the visible range of the spectrum and it has been observed to grow from both positive and negative electrodes, although more extensively studied in the former case, usually under impulse voltages.

1.2.8.1   Formation of the leader

In an extensive corona, having many branches over a length of several tens of centimetres, the number of electrons able to reach the stem of the streamer trails at the point of origin, close to the anode, becomes considerable. Extensive detachment of electrons from negative ions in that region is also believed to occur. As a result of energy exchange between the energetic electrons and neutral gas molecules, the rate of ohmic loss of energy increases and significant heating of the channel can occur. The temperature of the neutral gas thus rises, as a result of which it expands; the gas density falls. The quantity E/N therefore increases and ionisation becomes more efficient. The process is cumulative and a transition takes place to a highly ionised arc-like channel of high temperature and relatively high conductivity. The channel has now been transformed into a leader which proceeds to grow in the general direction of the electric field.

An idealised model of leader development is given in Figure 1.6, in which a rod–rod gap is imagined to be subjected to an impulse voltage with a time to peak of the order of a few hundred microseconds. The voltage at the positive electrode rises until the field at the tip exceeds 3×10^6 Vm^{-1} at time t_1 when a streamer corona forms. A burst of current is detected in the circuit. The corona injects a net positive charge into the region, so reducing the field at the tip and inhibiting further streamer formation for a 'dark' period until the voltage has increased. A second corona (often termed a secondary corona) then occurs at time t_2 which may be followed by others at short intervals of time thereafter (omitted for clarity). After another corona is formed, at t_3, sufficient heating has occurred in the streamer stem at the anode for a leader channel to form. Where the diameter of the electrode is relatively large, the leader may form immediately out of the secondary corona at time t_2. In either case, it extends in length across the gap towards the opposite electrode. Since the leader channel is highly conducting the potential of its tip remains high and a streamer corona forms ahead. Thus, the avalanches at the heads of the streamers provide the ionisation and, therefore, the electron current and consequent heating needed for further leader development.

The streamer coronas have formed more or less continuously as the leader has extended across the gap. Since the positive ions deposited remain immobile on the timescale involved, a roughly cylindrical volume of remanent positive charge surrounds the leader channel along the whole of its length.

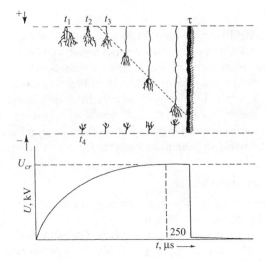

Figure 1.6   Simplified picture of streamer and leader development to breakdown in a rod–rod gap under an impulse voltage rising to peak in 250 μs

When the field at the negative electrode exceeds the order of $(5–6) \times 10^6$ Vm^{-1} at time t_4 negative streamers (section 1.2.9) can form and develop towards the anode, but, as these require an ambient field of about 10^6 Vm^{-1} to progress, they do not extend as far into the gap as do the positive streamers. Moreover, by contrast with the process at the positive electrode, the electrons are moving into a reducing electric field. At a later stage, transformation to a leader occurs but these processes occur at higher fields than is the case with the positive counterparts, and the distance traversed is smaller. When the two leader systems meet, a conducting channel bridges the gap and a low voltage arc can complete the breakdown of the gap.

A simple semiempirical argument can be used to relate the streamer and leader lengths and the respective average electric gradients needed for their propagation to the sparkover voltage V_s of the gap. For at the instant at which the systems meet, τ_4, we can write down the sum of the voltages across the gap:

$$V_s = E_s^+ L_s^+ + E_l^+ L_l^+ + E_s^- L_s^- + E_l^- L_l^- \tag{1.12}$$

where E_s^+, E_s^-, E_l^+ and E_l^- are the gradients for the positive and negative streamers and leaders, and L_s^+, L_s^-, L_l^+ and L_l^- are the corresponding lengths. Note also that the gap length d is:

$$d = L_s^+ + L_l^+ + L_s^- + L_l^- \tag{1.13}$$

Certain of these quantities, such as E_s^+, E_s^- are known from independent measurements and, in specific experiments, lengths of streamers and leaders estimated, so that other quantities in Equation (1.12) can be estimated from a sparkover measurement.

It may be expected from the foregoing descriptions that only a relatively large streamer corona is likely to develop into a leader, where fields near the positive

Figure 1.7   *Photograph of leaders developing in a 4 m rod–rod gap [38]. Note that the leader does not develop along the line of maximum field between the rods; also bifurcation at the tip of the rod*

electrode are high, detachment of electrons is rapid and large numbers of electrons are found in that region. Thus, leader initiation will occur most readily in large gaps, usually 0.5 m or more, where the sparkover voltage is high. Indeed, Reference 36 shows that the charge in the initial streamer corona, which subsequently produces a leader in a five metre gap, is several tens of microcoulombs.

The description of leader growth in Figure 1.6 is much simplified. For example, the leader rarely follows the axial path between electrodes. It tends to go off-axis in a path which may be much longer than the direct one, particularly if the electrode diameter is relatively large. Moreover, a bifurcation often occurs, originating at the electrode, although one of the two branches is usually dominant. An example of a leader in a long gap is shown in Figure 1.7. Waters [37] gives a more detailed description of leader development under impulse voltage in a practical case. Ross *et al.* [36], quoting results obtained by the Les Renardieres Group, give an interesting summary of quantities associated with leader development in a 10 m gap under the 50 per cent sparkover impulse voltage. For example, the total energy dissipated during the growth of the leader is about 25 J per metre of its length; the current is taken as 0.6 A, so that if the length of streamers in the leader corona, having a gradient of 500 kVm^{-1}, is 1 m, then the power input to the leader corona is 300 kW. Since the

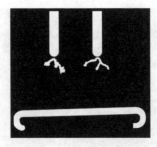

Figure 1.8   Two examples of leader development in a short (0.6 m) rod–plane gap showing on left, streamer formation (leader corona) at the leader tips. Impulse voltage ~330 kV (after Hassanzahraee, PhD thesis, University of Leeds, 1989)

system advances at the rate of 2 cmμs^{-1}, then the energy dissipated in the leader corona alone in 1 m of advance is about 15 J. The energy available to heat the leader channel itself is estimated at about 10 Jm^{-1}. As will be shown below, these are average values, indicating general features only, since the potential gradient in the leader varies along its length and the leader corona is also variable.

The simple heating process described above suggests also that leaders can be produced in short gaps provided a sufficiently high electric field can be created at the electrode. This can indeed be achieved by impulse overvolting of a gap; Figure 1.8 shows an example obtained in a gap of 0.4 m, to which an impulse of twice the threshold sparkover voltage was applied.

1.2.8.2   Leader properties

Much experimental data on leader properties has come from the work of the Les Renardieres Group [36, 38]. Although there is general agreement that the foregoing represents the main features of leader initiation and propagation, there are many aspects that are not fully understood by theory. Waters [37] gives a summary of the models of leader dynamics proposed by various authors up to 1978, and a refinement based on later experimental work is given in Reference 38. We now discuss those properties of the leader that are of the greatest significance to the breakdown process.

Several parameters have been measured. Of primary importance in breakdown is the velocity of propagation of the leader which, in laboratory tests, has been found to be of the order of 2×10^4 ms^{-1}. This value increases roughly in proportion to the current, but changes only slowly with the voltage across the gap. Indeed, the leader velocity is generally insensitive to changes in the electrical conditions or gap configuration, although it increases with increasing atmospheric humidity. By contrast, streamer velocities increase with electric field, usually exceeding 10^6 ms^{-1} at their point of creation, falling to 10^5 ms^{-1} at the ends of their trajectories. Thus, a streamer system, that is, the leader corona, ahead of the leader tip has time to form and progress along with the tip of the extending leader.

The charge injected by the leader includes not only the charge in the leader stem but also the charge in the streamer corona which everywhere precedes the leader tip. Thus, the charge injected per metre of length of the actual path is about 40 μC. This is independent of the diameter of the electrode, although the total charge injected increases with increasing rod diameter, for a given gap length [36]. This may be a reflection of a corresponding increase in the leader length. The charge injected in a long gap can thus be very appreciable and, when impulse testing, the inception of the leader can sometimes be recognised by a sudden drop in the voltage waveform measured by a capacitor potential divider.

The mean current in the leader is measured in a 10 m gap to be of the order of 1 A and estimates of the channel width of 5 to 10 mm lead to a value of current density of up to 10 A/cm. The mean electric field in the leader has been estimated by fluxmeter and probe measurements and depends upon its length [36]. At the instant of transition from streamer to leader, at the electrode tip, the gradient is nearly equal to the gradient of the streamer from which it develops, namely about 500 kV/m. After a few tens of microseconds, that is, after several tens of centimetres' growth, the effect of heating has greatly reduced the local resistivity and, hence, the gradient. Thus, the mean gradient is reduced as the leader progresses. The problem has been considered by Ross *et al.* [36] who, by relating ionisation and attachment rates and the energy input to the channel, showed that the gradient averaged over the whole time of growth of a leader decreases rapidly with time and, therefore, with length. The predicted result, which depended on an assumed channel radius, was verified by later experiments [38] and a comparison between the two is shown in Figure 1.9.

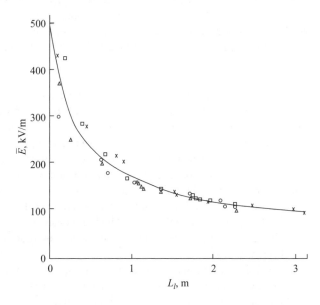

Figure 1.9   *Average electric field along the leader channel as a function of its length [38]*

The formation of the relatively highly conducting leader can be likened to an extension of the metal anode itself, for the potential drop along it has become comparatively small. The high field region at the electrode tip has now been transferred to the tip of the leader where conditions for the formation of further streamers now exist. Thus the leader corona is formed, which itself now provides the condition for further leader formation at its stem. In this way, the leader is able to grow outwards to a distance from the anode which is determined by the combination of applied and space charge electric fields.

The comparatively low velocity of advance of the leader is determined not only by the rate of heating of the channel; it is also checked by the amount of positive space charge produced in the leader corona. Where this is large, it may reduce the field around the leader tip to reduce heating effects or even to choke off leader development completely. On the other hand, high humidity tends to increase the rate of leader growth. In this case, streamer corona development is inhibited (section 1.3.4), resulting in a lower charge in the leader corona and a higher resultant field at the tip of the leader, so facilitating progressive development.

1.2.9   Negative discharges

Negative discharges are of generally lesser importance in the breakdown process than discharges in the diverging fields close to electrodes at positive polarity. The difference arises because, in the negative case, electrons are moving into an electric field of decreasing intensity, whereas the reverse is the case near a positive electrode. As a result, the drift velocity and ionisation efficiency decrease with increasing distance from the electrode. It is more difficult for a discharge to spread across the gap, unless a higher electric field is applied. Also, diffusive spread of an avalanche increases as it progresses down the field gradient so that the space charge field of the positive ions decreases.

As a consequence, a negative corona has a different, more complex, structure than a positive corona and tends to be more localised around the electrode.

A useful picture of the formation and propagation of negative streamers has been given by Hutzler *et al.* [20], shown in Figure 1.3(*ii*) where the contrast between positive and negative streamer mechanisms is drawn. Again, a single free electron *a* is postulated to start the first avalanche, which progresses into a decreasing electric field. The electron space charge so produced causes a local increase of electric field ahead, so that an electron liberated by ultra-violet radiation *c* can produce a further avalanche *d*. The electrons from the first avalanche move into the (almost stationary) positive ions left by the second avalanche, so producing a channel with conducting properties behind the leading avalanche *e*. This process can continue, provided the electric field is sufficiently high.

The dipolar channel illustrated in *e* has been termed a space stem, within which a reduced electric field component, due to the space charges, exists.

A moment's thought will show that, for a given magnitude of electric field, the diverging field at an anode will provide more favourable conditions for ionisation than the converging field at a cathode, since, in the former case, the electrons are

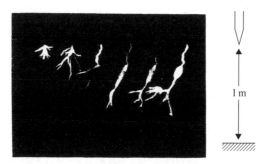

Figure 1.10   *Streak photograph of development of a negative leader system, showing negative streamers ahead of the leader tip and retrograde streamers towards the anode [20]*

always moving into an increasing field. Thus, a negative streamer requires a higher field for propagation than a positive streamer; the ratio is, in fact, about two to one.

Figure 1.3(*ii*) indicates that an augmented field also exists between the electrode and the positive space charge deposited by the first avalanche. This provides conditions entirely analogous to those of Figure 1.3(*i*) which are favourable to the formation of a positive streamer. Thus, in addition to the anode-directed negative streamer, a retrograde, cathode-directed streamer can also be set up. This has, in fact been observed, by means of image convertor streak photography, Figure 1.10 [20]. This shows a fairly diffuse initial streamer corona, less well defined than its positive counterpart, which quickly at its head appears concentrated and which then gives rise to filaments extending in both directions.

The space stem acts as a location for the onset of the negative leader; it appears as regions of high luminosity in Figure 1.10, gradually moving across the gap. The anode and cathode-directed streamers appear on either side, and the leader develops after some delay. It is assumed that the streamer–leader transition occurs by heating in the same way as in the positive case, and as in that case also, its rate of growth is determined by the corona at its head.

The velocity of negative streamers is not known. The negative leader grows towards the anode with a velocity less than 10^4 ms^{-1}, which is significantly less than that of the leader in a long positive rod–plane gap. However, the cathode-directed streamer has been shown to have the same velocity as in a positive discharge.

1.3   Applications

The applications discussed here are relevant mainly to high voltage power systems and associated apparatus. Most power systems employ alternating voltages, although direct voltage operation is becoming significant in special situations. However, the insulation against breakdown may be required to be at least twice that needed for the operating voltage, since transient overvoltages of higher levels may occur. These may be caused by lightning strikes, near to or upon an overhead line, by switching operations in the power system, or by temporary increases in the level of the operating voltage. Of these, the first two are the most frequent and dangerous to the system and the insulation is designed with this in view.

The lightning impulse approximates to the disturbance caused by a lightning strike: it is characterised by a relatively fast rising impulse reaching its peak in a few microseconds and decaying in a few tens of microseconds. The switching, or slow front impulse is considered to rise in a few hundred microseconds, decaying in a few thousand microseconds. In both cases, these impulses may be reflected from discontinuities in the transmission line, with consequent increase in voltage. For this reason, testing of high voltage apparatus is carried out under an impulse voltage related to, but much higher than, the proposed operating voltage.

1.3.1   Sparkover under lightning impulse voltage

The IEC definition of a standard lightning impulse voltage specifies a rise time of 1.2 μs and a decay to half the peak value in 50 μs [31]. A tolerance of +30 per cent is allowed on the rise time and of +20 per cent on the decay time (Figure 1.11).

The duration of voltage around the peak is thus short compared with the times required for a leader to advance a significant distance, but the time scales are long compared with that needed by streamers to propagate. The leader thus has a negligible role in the breakdown process. Taking as an example the positive rod–plane gap, the critical volume is expanding around the tip of the electrode during the rise time of the impulse voltage. Negative ions are attracted to the volume in this time, but their density is not sufficient to allow a significant number to enter the critical volume in the time available. Also, the number located in the critical volume for detachment of electrons to occur is subject to significant statistical variation. Thus, avalanches and

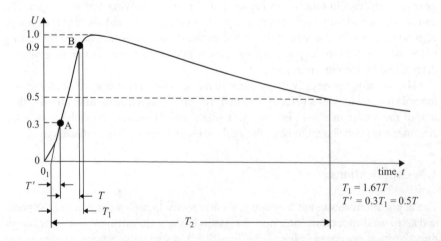

Figure 1.11   *Parameters of the lightning impulse voltage [31]. The front of 1.2 μs is defined as the time 1.67T; the tail of 50 μs is defined as the time T_2 to half the peak voltage*[1]

streamer corona will occur at statistically variable times during the rise of voltage to the peak. This first group of streamers may not be sufficiently extensive to cause immediate breakdown. Since appreciable leader growth cannot progress, breakdown can occur only if a second group of streamers crosses to the cathode, around or soon after peak voltage, producing sufficient secondary emission of electrons to cause increased ionisation and form a conducting channel across the gap.

It is evident, from this qualitative picture, that the breakdown stress in a gap with a single region of non-uniform field under a lightning impulse voltage depends directly upon the stress needed for streamers to cross the gap. This follows from Equation (1.12) where the following simplifications can be made:

- there is no leader development at either electrode
- there is no streamer development at the plane cathode.

Thus, Equation (1.12) reduces to:

$$V_s = E_s^+ d \qquad (1.14)$$

and the sparkover voltage increases linearly with the gap length.

This has been shown experimentally to be true with rod–plane gaps up to 8 m [39, 40]. For a gap with two non-uniform field regions, typified by the rod–rod gap, some negative streamer growth from the cathode is likely to occur. Here, the approaching positive streamer corona will generally set up a sufficient space charge field to enhance the field at the cathode to a higher value than that set up by the applied voltage alone. Thus, again noting the absence of leader formation, Equation (1.12) reduces to:

$$V_s = E_s^+ L_s^+ + E_s^- L_s^- \qquad (1.15)$$

so that the voltage at sparkover is now increased with respect to that in the rod–plane case.

Two important properties make the rod–plane gap, under positive impulse, a valuable reference in high voltage testing:

(i)   it has the lowest sparkover voltage of any gap configuration of the same length (from Equation (1.14))
(ii) it shows a linear increase of sparkover voltage with gap length.

Both of these properties arise from the lack of any negative discharge growth at the plane.

The absence of significant leader growth and consequent lack of ambiguities makes the positive lightning impulse an important test voltage, recommended in standards, to be applied to high voltage components and apparatus, such as bushings, insulators, transformers and so on.

Under negative impulse voltage, applied to a rod–plane gap, the sparkover voltage increases with gap length slightly less rapidly than linearly, but the magnitudes are a factor of nearly two greater than those for the positive case. These differences may be expected from the differences in the discharge propagation modes, discussed in section 1.2.9.

1.3.2 Sparkover under slow front impulse voltage

The IEC Standard 60060-1 (1989) [31] defines a standard slow front impulse having a nominal rise time of 250 μs and a decay time to half value of 2500 μs. In practice, a variety of rise and decay times is encountered and it is necessary to consider, in a general way, the nature of the processes involved in breakdown.

Consideration of Figure 1.6 and the associated qualitative arguments indicates at once that where the voltage rises relatively slowly, there is a relationship between the voltage rise time, the leader growth rate and the gap length which will determine the voltage at which breakdown occurs. A simple example will illustrate.

Consider a rod–plane gap of, say, 2 m, to which a positive slow front impulse voltage is applied. At a velocity of 2×10^4 ms^{-1} a leader would take of the order of 100 μs to cross the full gap (in fact somewhat less because its own leader corona would occupy a significant length). Taking a range of rise times from 1 μs to, say, 250 μs, it is clear that as they increase, the leader can traverse progressively larger proportions of the gap. The breakdown voltage V_s is correspondingly reduced on account of the fact that the gradient of the leader is lower than that of streamers. Experiment shows that for further increase in rise time, the trend reverses and the breakdown voltage increases. The reason is that successive streamer coronas develop during the relatively slow rise of voltage, none of which has injected sufficient charge at a sufficiently high stress to cause a leader to form. The resulting positive space charge reduces the local field at the anode so that a larger stress is ultimately needed to form the leader. Thus, a higher voltage is required and there is a minimum in the curve of V_s against time to peak. It follows from this argument that the time at which the minimum occurs depends also on the gap length.

This behaviour results in the so-called U-curve which has been established by testing over a wide range of gaps. It is illustrated in Figure 1.12, where U-curves obtained in rod–plane gaps in the range $1 < d < 25$ m are given. The time to peak impulse voltage at which the minimum occurs for a given gap is called the critical time to peak and it defines, therefore, a limiting minimum voltage at which that gap can be broken down under an impulse.

It is found that the critical time to peak shows a closely linear increase with the length of gap. Since the rate of leader growth has been shown to be approximately independent of its length [36], the U-curve minimum for a given gap is thus identified with optimum leader growth. For shorter times to peak, the reduced time for leader growth means that a higher voltage is needed for breakdown. For longer times to peak, the succession of streamer coronas also requires a higher stress for leader initiation, so that the breakdown voltage rises again. It follows from these facts that, where an impulse having a critical time to crest is applied to the gap, the breakdown occurs at or near the crest of the impulse.

Experimentally, it has been found [41] that the sparkover voltage of the rod–plane gap at the critical time to crest is given by the formula:

$$V = \frac{3400}{1 + 8/d} \tag{1.16}$$

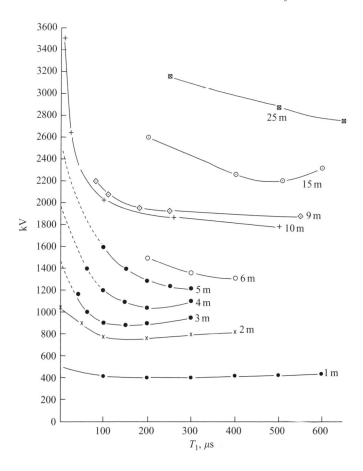

Figure 1.12 *A selection of U-curves for rod–plane gaps in the range 1 to 25 m. Note the trend for the minima to occur at longer times to peak as the front of the impulse increases*

where d is the gap length. This formula must be modified for other gap configurations (section 1.3.3) and also when variations in atmospheric conditions must be taken into account (section 1.3.4).

The existence of the U-curve was first realised about 1960 [41, 42] and it is now widely used in high voltage technology, for instance in determining minimum clearances in high voltage equipment such as overhead lines and in general questions of insulation coordination.

1.3.3 The influence of field configurations: the gap factor

Discussion so far has assumed the non-uniform field electrode to be of small radius, so that high fields exist in its vicinity, leading to ready formation of corona. When the radius is increased, so reducing the per unit field at the electrode boundary, higher

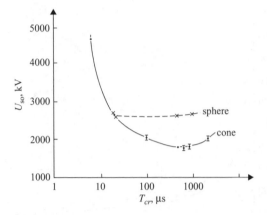

Figure 1.13   Effect of the change in profile of the energised electrode (plane earthed) on the U-curve; gap = 10 m [43]

- cone electrode of small tip radius
- × sphere electrode of radius 0.75 m

voltages are needed for the initial ionisation, resulting in a higher ultimate voltage for the final breakdown. The effect is shown in Figure 1.13 where the U-curves are shown for a 10 m air gap with two electrode geometries, namely a cone and a sphere. The former shows the U-curve minimum clearly, but with the sphere; production of a viable streamer corona requires a higher voltage in order that the transition to leader can occur. Therefore, the minimum becomes much flatter. For shorter times to peak, where leader formation is minimal, the two curves come together.

It is evident that, since the corona initiation depends on the field configuration around highly stressed electrodes, the variety of gap geometries that is inherent in practical situations will impose variations in the highly stressed regions and, therefore, in breakdown strength. This effect is particularly marked where there is a highly stressed region around the negative, as well as the positive, electrode. An argument is presented in the CIGRE guidelines [43] in which it is shown that the total voltage across the gap for maintenance of the predischarges in mid-gap prior to sparkover is increased by the insertion of a high field region around a pointed cathode, when compared with the case of a plane cathode. It follows also from Equation (1.12) that where significant negative streamer development occurs (which has gradient for propagation of the order of twice that for positive streamers), the voltage across the gap at the instant of breakdown is higher than it would have been in the absence of negative streamers.

Similar considerations apply for gaps of other geometries that may arise in practice. This has led to the concept of the gap factor k, which is defined as the ratio of the sparkover voltage of a particular gap to the positive rod–plane air gap sparkover voltage, for gaps of the same length and subjected to the same switching impulse. It is of value because the ratio holds good for nearly all lengths of gap that are of practical interest. Thus, adapting Equation (1.16) for the U_{50} sparkover voltage

configuration		k
rod–plane	$\big|\,d$	1
conductor–plane	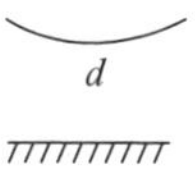 d	1.1 to 1.15
rod–rod	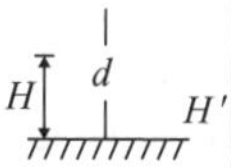 $H\;\;d\;\;H'$	$1 + 0.6\,\dfrac{H'}{H}$
conductor–rod	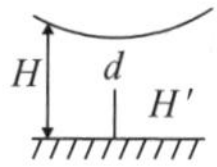 $H\;\;d\;\;H'$	$(1.1 \text{ to } 1.15)\exp\left(0.7\,\dfrac{H'}{H}\right)$
protrusions	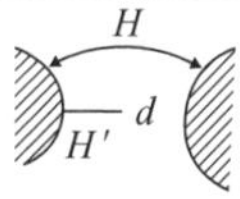 $H\;\;d\;\;H'$	$k_o \exp\left(\pm 0.7\,\dfrac{H'}{H}\right)^{*}$ $k>1$

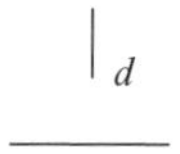

*sign + for protrusions from the negative electrode
sign – for protrusions from at the positive electrode
k_o : gap factor without protrusions

Figure 1.14   Some gap factors for a selection of simple electrode geometries [43]

at the critical time to peak, the sparkover voltage of any gap of gap factor k is:

$$V = \frac{3400\,k}{1 + 8/d} \tag{1.17}$$

Gap factors have been determined from experiment with a number of basic geometries and a summary, taken from Reference 43, is shown in Figure 1.14. More detailed empirical formulae for estimation of gap factors for a number of practical configurations are also given in the same reference.

Since gap factors are always >1, the value of k may be assumed to be an indicator of the increase in stress needed for leader formation, compared with that in the rod–plane gap. It would then be expected that the gap factor must influence the critical time to peak; the empirical formula below shows this to be the case:

$$T_p = [50 - 35(k - 1)]d \tag{1.18}$$

where T_p is in μs and d is in m [43].

The presence of an insulator in the gap also affects the gap factor, since the presence of its surface affects both streamer and leader properties (section 1.3.5). The following formula has been given [43]:

$$k_i = [0.85 + 0.15\exp -(k - 1)]k \tag{1.19}$$

where k_i and k are the gap factors with and without the insulator in place.

It is frequently possible to approximate a practical gap to a simple one for which the gap factor is already known and the concept is, therefore, of considerable value to power engineers concerned with design and insulation coordination.

1.3.4 Atmospheric effects

All insulation problems which involve air as a primary medium must take account of the variations in atmospheric conditions. These conditions are of pressure, temperature and humidity, but it is important to note that fog, coastal salt laden air and, in areas of high pollution, particulate matter in the atmosphere can dominate. Pressure and temperature changes are manifest in changes of air density; altitude can reduce the air density, in regions where power must be delivered, to as little as 0.65 of that at sea level [44]. Temperature changes occur, between ±40°, with resulting changes in air density to 0.9 and 1.1, respectively, of that at the standard atmospheric temperature of 20°C. These changes are expressed in terms of the relative air density (RAD) which is unity at 101.3 kPa pressure and 293 K. At all other conditions of pressure, b, and temperature, T, the relative air density δ is:

$$\delta = \frac{b}{101.3} \frac{293}{T} \tag{1.20}$$

Absolute humidity can vary between $1\,\mathrm{gm}^{-3}$ and $30\,\mathrm{gm}^{-3}$; the standard atmospheric humidity is taken as $11\,\mathrm{gm}^{-3}$. Changes in both air density and humidity result in changes in ionisation efficiency and, therefore, in sparkover voltage characteristics.

The prebreakdown processes have been described in sections 1.2.1 to 1.2.7. The effects of atmospheric changes are on avalanche initiation and development and the streamer trail. In the case of density variation, where caused by altitude change, the effect on the density of atmospheric negative ions, available for detachment in the critical volume (section 1.1.2.1), is not known. Indeed, the density of negative ions is more likely to be affected by local variations in geology and industrial activity, as well as by local weather, than by altitude. On the other hand, an increase of humidity increases the density of the $O_2(H_2O)$ ionic complexes which, therefore, increases the probability of detachment of electrons in a high electric field. The effect of air density change is considered first.

The most significant changes occur in avalanche development. As density decreases, ionisation efficiency can notionally be maintained at a constant level at constant E/N (section 1.1.2.6) so that the electric field can be reduced in proportion to N. However, electron diffusion effects, depending on N as well as E/N, become important, so affecting changes in avalanche radius and the conditions for streamer formation.

The foregoing remarks assume that the kinetic processes of ionisation, attachment and electron mobility remain unchanged at constant E/N as the density N changes. However, there is evidence that if N is reduced by increase of temperature at constant pressure, both the ionisation coefficient and the electron mobility are reduced at constant E/N [45]. This implies that, together with changes in diffusion already referred to, a further change will take place in the condition for achieving a critical avalanche.

The subject of avalanche formation as a function of density N requires solution of the continuity equations for electron flow in which experimentally determined parameters can be used.

Experiment has shown that the electric field required for stable propagation of a streamer varies with relative air density as $\delta^{1.3}$ [24, 28]. This appears to be true whether arising from pressure or temperature change within the range 0.7–1.0. Tests with long rod–plane air gaps (>1 m) have established, however, that under lightning impulse, where the sparkover voltage depends on the streamer gradient, this voltage varies linearly with δ [46]. This fact has been adopted as a reference in the IEC Standard 60060-1(1989) [31] which specifies how high voltage measurements of dielectric strength shall be adjusted to take account of density variations. Under slow front impulse, however, where there is significant leader growth, the dependence of sparkover voltage on density is less strong since the leader properties are less dependent than those of streamers upon relative air density. However, the conditions for leader inception and, therefore, development, do depend on the streamers from which it originates. The Standard has set out an empirical procedure for adjustment of sparkover voltages which implicitly takes account of the extent of leader growth; it is, however, beyond the scope of this chapter.

Where relative air density is linked to temperature change, the situation is less clear. Testing in long gaps, over an appreciable range of temperatures, is impracticable in the open atmosphere and this has precluded measurements in which significant leader development could be expected. In practice, the range of atmospheric RADs encountered is smaller than that arising from pressure variations. Therefore, the Standard adopts an *ad hoc* adjustment procedure which makes no distinction between the two. Researches in short gaps (<25 cm) over large temperature ranges have shown conflicting results despite the fact that the breakdown process involves streamers only. The results, again, conflict with those obtained with variable pressure [47, 48]. Further work is needed for a resolution of these problems.

Humidity change also affects avalanche and streamer development, since the attachment coefficient is increased relative to the ionisation coefficient [10]. Much more data is available in this area, from outdoor testing and from laboratory experiments, than is the case with density variation. Researches have shown that the electric field required for streamer propagation increases at the rate of about 1 per cent per gram of moisture content per cubic metre of air at the standard pressure and temperature of 101.3 kPa and 293 K [23, 50]. This is also the rate at which the sparkover of the rod–plane gap increases under lightning impulse [49] where, as noted earlier, the breakdown is determined solely by streamer growth in the gap.

Where significant leader growth occurs, the effect of humidity is smaller since humidity, while increasing the leader velocity, does not change the leader gradient. Thus, inspection of Equation (1.15) indicates that the relatively large effect on the sparkover voltage of the humidity effect on streamers is offset by the smaller effect on the leader. This results in a humidity coefficient of less than 1 per cent per gm per cubic metre when the leader occupies a significant part of the gap.

Again, the IEC procedure for adjustment of sparkover voltages for humidity change is given in IEC60060-1(1989) to which the reader is referred for details.

1.3.5   Corona at low air density

Some of the changes occurring to the prebreakdown corona at reduced air density have already been alluded to in the preceding section. However, it is likely that there will be future interest in the incidence of corona and breakdown under the low density conditions encountered in the aerospace arena, where electrical equipment may be called upon to operate in relative air densities of the order of <0.3. Much work has been carried out in the past on corona and breakdown at low pressures (see, for example, References 50 to 52), but a different focus is needed at high altitude. In this case, simultaneous large changes in pressure, temperature and humidity occur. Little data is available, but some relevant physical factors can be briefly presented.

The minimum voltage for corona inception is an important parameter. As pressure is reduced at room temperature, it has been shown that the inception voltage decreases approximately linearly with air density [47]. With increasing temperature, corona inception voltage decreases much more slowly, according to a relationship given in Reference 53. However, the effect of decreasing temperature in the presence of decreasing pressure is not known.

The inception voltage depends upon the E/N condition for ionisation as the density decreases and this, in turn, determines the volume and boundaries of the critical volume around an electrode, as discussed in section 1.2.3. However, the probability of formation of an avalanche also depends upon the probability of an initiating electron being found within the critical volume. This in turn depends on the local density of atmospheric negative ions. Discussion of the influence of atmospheric ion densities on corona initiation was given in Reference 54, and the effects of a change in the ambient atmospheric conditions were measured in Reference 18. The increase in ion density with altitude resulting from cosmic ray activity is well known and must be considered as a further variable in determining corona onset.

The change with air density of corona development after inception is also to be taken into account. Measurements made under both positive [50] and negative [52] polarity corona studies show that the charge injected in corona tends to increase with decreasing air density, even though the inception voltage decreases. This result is true whether the coronas are produced under impulse [50] or direct [56] voltage. The reasons for this trend have not yet been elucidated.

Finally, design of high altitude systems having relatively small dimensions of the order of centimetres must take into account the fact that Paschen's Law, relating sparkover voltage of a gap to the product of pressure and electrode separation (pd) [57], also applies to the onset of corona. Briefly, the Law indicates that where pd is small so that there is relatively little gas between electrodes, production of sufficient ionisation to initiate corona or breakdown requires that either the voltage needed to start corona must be raised or, if the electrode arrangement permits an alternative path of greater length, the corona can develop at a lower voltage over the longer path.

The use of power systems at very high altitudes, albeit on a relatively small scale, thus involves novel applications of the fundamental knowledge that is presently extant.

1.3.6 Sparkover over insulator surfaces

Conductors between which high potential differences are maintained are usually separated by solid insulators, frequently porcelain, glass or polymers. Insulators for outdoor installations such as overhead lines, while maintaining necessary clearances between conductors, must be designed primarily to cope with rain, fog and pollution conditions which dictate detailed design (see chapter 6 and Looms [58] for more details). However, under more specialised requirements for indoor insulation, choice of materials and their dielectric properties become important.

The development of avalanches and streamers in close proximity to insulating surfaces is likely to be affected in several ways. In addition to air processes, the electrons in an avalanche may have a probability of attaching to the insulator surface, with consequent removal from the ionisation process. This has been postulated to account for the fast rate of decay of current in a corona which is propagated over a surface [55]. Nevertheless, it has been established that streamers propagate more rapidly over a surface than in air, and this has been related to photoelectron emission from the surface, so effectively increasing the ionisation coefficient α and shortening the lengths of avalanches required to achieve a critical condition (section 1.2.3). Possibly overriding these influences is the effect of the increase in electric field around the head of a streamer due to the relative permittivity of the material. In the streamer trail, it is likely that negative or positive ions will attach to the surface, depending on the chemical nature and activation energies of the material.

Data is available on some of these effects. Thus, Verhaart *et al.* [59] have demonstrated that electrons are, indeed, photo-emitted at an insulating surface to an extent that depends on the quality of the ultra-violet light emitted by avalanches in the particular gas surrounding the insulator. In the experiments described, these photoelectrons had the effect of augmenting the avalanches that were initiated on the insulator surface. The UV light from SF_6 proved to be effective in causing photo-emission from PTFE, but no significant emission was detected when repeated with carbon dioxide. However, there was no attenuation of the avalanches in this case, indicating that unless counterbalanced by photo-emission, attachment was negligible. Generally, similar effects were obtained by Jakst and Cross [60].

The attachment of ions to the surface is, of course, very well studied by the Lichtenberg figure technique, which has been made the subject of quantitative measurement by several authors [5, 61, 62]. Very recently, however, scanning measurements by electrostatic probe have been integrated over the whole discharge region to show that, with positive impulse corona, the net charge detected on the surface is only a few per cent of the total charge injected from the circuit [55]. The result suggests that ions of both polarity settle on the surface after the passage of the discharge current, with a slight preponderance of the same polarity as the electrode producing the corona.

Further measurements have shown that the ambient electric field required to sustain streamer propagation over an insulating surface is of the order of 20 per cent higher than that needed in air. This appears at variance with the generally accepted reduction in the dielectric strength of an insulator surface, compared with that in air alone. However, the reduction is due to effects at a highly stressed electrode, where the combination of a (usually) imperfect contact between electrode and insulator in air (the so-called triple junction) produces very high localised fields in which low energy discharges trigger the larger discharge between electrodes. This subject is beyond the scope of this chapter.

1.4   Note

[1] Figure from British Standards reproduced with the permission of BSI under licence number 2003 SK/0157. British Standards can be obtained from BSI Customer Services, 389 Chiswick High Road, London W4 4AL (Tel +44 (0) 20 8996 9001).

1.5   References

1   FRANKLIN, B.: 'Experiments and observations on electricity made at Philadelphia' (Cave, London, 1751)
2   LICHTENBERG, G.C.: *Novi Comment*, 1777, **8**, pp. 168–177
3   TOWNSEND, J.S.: 'The conductivity produced in gases by the motion of negatively charged ions', *Phil. Mag. Servi.*, 1901, **1**, pp. 198–227
4   LOEB, L.B.: 'The problem of mechanism of spark discharge', *J. Franklin Inst.*, 1930, **210**, pp. 115–130
5   MERRIL, F.H., and von HIPPEL, A.: 'The atom physical interpretation of Lichtenberg Figures and their application to the study of gas discharge phenomena', *J. Appl. Phys.*, 1939, **10**, pp. 873–887
6   TEICH, T.H.: 'Emission gasioniseierender Strahlung aus Elektronenawinen', *Zeits. fur Phys.*, 1967, **199**, pp. 378–394
7   MEEK, J.M.: 'The electric spark in air', *JIEE*, 1942, **89**, pp. 335–356
8   RAETHER, H.: *Arch. Elektrotech.*, 1940, **34**, p. 49
9   BROWN, S.C.: 'Basic data of plasma physics' (Technology Press and Wiley, New York, 1959)
10 BADALONI, S., and GALLIMBERTI, I.: 'The inception mechanism of the first corona in non-uniform gaps', University of Padova, 1972, UPee 72/05, pp. 31–38
11 ROCHE, A.E., and GOODYEAR, C.C.: 'Electron detachment from negative oxygen ions at beam energies in the range 3 to 100 eV', *J. Phys. B, At. Mol. Phys.*, 1969, **2**, pp. 191–200
12 BOYLETT, F.D.A., and WILLIAMS, B.G.: 'The possibility of photodetachment in the impulse breakdown of positive point plane gaps in air', *Br. J. Appl. Phys.*, 1967, **18**, pp. 593–595
13 CROMPTON, R.W., HUXLEY, L.G.H., and SUTTON, D.J.: 'Experimental studies of the motion of slow electrons in air, with application to the ionosphere', *Proc. R. Soc. Lond. A, Math. Phys. Sci.*, 1953, **218**, pp. 507–519.
14 MORROW, R., and LOWKE, J.J.: 'Streamer propagation in air', *J. Phys. D, Appl. Phys.*, 1997, **32**, pp. 614–627

15   GALLIMBERTI, I., MARCHESI, G., and NIEMEYER, L.: 'Streamer corona at insulating surface'. Proceedings of 5th international symposium on *High voltage engineering*, Dresden, 1991, paper 41.10

16   SHOCKLEY, W.: 'Currents to conductors induced by a moving point charge', *J. Appl. Phys.*, 1938, **10**, pp. 635–636

17   SATO, N.: 'Discharge current induced by the motion of charged particles', *J. Phys. D, Appl. Phys.*, 1980, **13**, pp. L3–L6

18   ALLEN, N.L., BERGER, G., DRING, D., and HAHN, R.: 'Effects of humidity on corona inception in a diverging electric field', *IEE Proc. A Phys. Sci. Meas. Instrum. Manage. Educ.*, 1981, **128**, pp. 565–570

19   BURCH, D.S., SMITH, S.J., and BRANSCOMB, L.M.: 'Photodetachment of O', *Phys. Rev.*, 1959, **112**, pp. 171–175; correction, *Phys. Rev.*, 1961, **114**, p. 1952

20   HUTZLER, B., *et al.*: LES RENARDIERES GROUP: 'Negative discharges in long air gaps at Les Renardieres, 1978, *Electra*, 1981, (74), pp. 67–216, published by CIGRE, Paris

21   von HIPPEL, A., and FRANCK, J.: 'Der Elektrische durchschlag und Townsends Theorie', *Zeits. fur Phys.*, 1929, **57**, pp. 696–704

22   RAETHER, H.: 'Gasentladungen in der Nebelkammen', *Zeits. fur Phys.*, 1935, **94**, pp. 567–573

23   GALLIMBERTI, I.: 'A computer model for streamer propagation', *J. Phys. D, Appl. Phys.*, 1972, **5**, pp. 2179–2189

24   PHELPS, C.T., and GRIFFITHS, R.F.: 'Dependence of positive streamer propagation on air pressure and water vapour content', *J. Appl. Phys.*, 1976, **47**, pp. 2929–2934

25   ACKER, F.E., and PENNEY, G.W.: 'Some experimental observations of the propagation of streamers in low-field regions of an asymmetrical gap', *J. Appl. Phys.*, 1969, **40**, pp. 2397–2400

26   GELDENHUYS, H.J.: 'Positive streamer gradient and average breakdown voltage as functions of humidity'. Proceedings of 5th international symposium on *High voltage engineering*, Braunschweig, 1987, paper 14.02

27   ALLEN, N.L., and BOUTLENDJ, M.: 'Study of electric fields required for streamer propagation in humid air', *IEE Proc. A, Sci. Meas. Technol.*, 1991, **138**, pp. 37–43

28   ALLEN, N.L., and GHAFFAR, A.: 'The conditions required for the propagation of a streamer in air', *J. Phys. D, Appl. Phys.*, 1995, **28**, pp. 331–337

29   ALLEN, N.L., and MIKROPOULOS, P.N.: 'Dynamics of streamer propagation in air', *J. Phys. D, Appl. Phys.*, 1999, **32**, pp. 913–919

30   FESER, K., and HUGHES, R.C.: 'Measurement of direct voltage by rod–rod gap', *Electra*, 1988, (117), pp. 23–34

31   INTERNATIONAL ELECTROTECHNICS COMMISSION IEC 60060-1: 'High voltage testing', 1989

32   ALLEN, N.L., and DRING, D.: 'Effect of humidity on the properties of corona in a rod–plane gap under positive impulse voltages', *Proc. R. Soc. Lond. A, Math. Phys. Eng. Sci.*, 1985, **396**, pp. 281–295

33   FAIRCLOTH, D.C., and ALLEN, N.L.: 'High resolution measurements of charge densities on insulator surfaces', *IEEE Trans. Dielectr. Electr. Insul.*, 2003, **10**, pp. 286–290

34   SCHONLAND, B.F.J., and COLLENS, H.: 'Progressive lightning', *Nature*, 1933, **132**, p. 407

35   ALLIBONE, T.E., and SCHONLAND, B.F.J.: 'Development of the spark discharge', *Nature*, 1934, **134**, pp. 735–736

36   ROSS, J.N., *et al.*: LES RENARDIERES GROUP: 'Postive discharges in long air gaps at Les Renardieres', *Electra*, 1977, (53), pp. 31–153, Table 3-1

37   WATERS, R.T.: *in* MEEK, J.M., and CRAGGS, J.D. (Eds): 'Electrical breakdown of gases', (Wiley, Chichester, 1978), Chap. 5, pp. 385–532

38   WATERS, R.T. (Ed.): LES RENARDIERES GROUP: 'Double impulse tests of long air gaps', *Proc. IEE A, Phys. Sci. Meas. Instrum. Manage. Educ. Rev.*, 1986, **133**, pp. 393–483

39   PARIS, L., and CORTINA, R.: 'Switching and lightning impulse discharge characteristics of large air gaps and long insulator strings', *IEEE Trans.*, 1968, **PAS-87**, pp. 947–968

40   CARRARA, G.: 'Investigation on impulse sparkover characteristics of long rod/rod and rod/plane air gaps'. CIGRE report no. 328, 1964

41   BAZELYAN, E.M., BRAGO, E.N., and STEKOLNIKOV, I.S.: 'The large reduction in mean breakdown gradients in long discharge gaps with an oblique sloping voltage wave', *Dokl. Akad. Nauk SSSR*, 1960, **133**, pp. 550–553

42   HUGHES, R.C., and ROBERTS, W.J.: 'Application of flashover characteristics of air gaps to insulation coordination', *Proc. IEE*, Part A, 1965, **112**, pp. 198–202

43   CIGRE: 'Guidelines for the evaluation of the dielectric strength of external insulation'. Technical brochure no. 72, 1992

44   ZHANG, X., LHADIAN, and WANG, X.: 'Effect of humidity on the positive impulse strength at the high altitude region'. Proceedings of 8th international symposium on *High voltage engineering*, Yokohama, 1993, paper 40.01

45   FRIEDRICH, G.: 'Temperature dependent swarm parameters in N_2 and air'. Proceedings of 9th international conference on *Gas discharges and their applications*, Venezia, 1988, pp. 347–350

46   BOUTLENDJ, M., and ALLEN, N.L.: 'Assessment of air density correction for practical electrode systems', *Eur. Trans. Electr. Power*, 1996, **6**, pp. 267–274

47   DAVIES, A.J., MATALLAH, M., TURRI, R., and WATERS, R.T.: 'The effect of humidity and pressure on corona inception in a short air gap at breakdown voltage levels'. Proceedings of 9th international conference on *Gas discharges and their applications*, Venezia, 1988, pp. 185–188

48   ALLEN, N.L., LAM, D.S.K., and GREAVES, D.A.: 'Tests on the breakdown of air in non-uniform electric fields at elevated temperatures', *IEE Proc., Sci. Meas. Technol.*, 2000, **147**, pp. 291–295

49   FESER, K., and PIGINI, A.: 'Influence of atmospheric conditions on the dielectric strength of external insulation', *Electra*, 1987, (112), pp. 83–95

50   DAVIES, A.J., DUTTON, J., TURRI, R., and WATERS, R.T.: 'Predictive modelling of impulse corona in air at various pressures and humidities'.

Proceedings of 9th international conference on *Gas discharges and their applications*, Venice, 1988, pp. 189–192

51 LOEB, L.B.: 'Electrical coronas' (University of California Press, 1965)

52 SIGMOND, R.S.: 'Corona discharges' *in* MEEK, J.M. and CRAGGS, J.D. (Eds): 'Electrical breakdown of gases' (John Wiley, Chichester, 1978), Chap. 4

53 KONG, J., and ALLEN, N.L.: 'Temperature effect on air breakdown under switching impulse voltages'. Proceedings of 14th international symposium on *High voltage engineering*, Delft, 2003

54 WATERS, R.T., JONES, R.F., and BULCOCK, C.J.: 'Influence of atmospheric ions on impulse corona discharge', *Proc. IEE*, 1965, **112**, (7), pp. 1431–1438

55 ALLEN, N.L., and FAIRCLOTH, D.C.: 'Corona propagation and charge deposition on a PTFE surface', *IEEE Trans. Dielectr. Electr. Insul.*, 2003, **10**, pp. 295–304

56 ALLEN, N.L., and KONG, J.: 'Effect of temperature on corona onset characteristics'. Proceedings of 14th international conference on *Gas discharges and their applications*, Liverpool, 2002, **1**, pp. 271–274

57 DUTTON, J.: 'Spark breakdown in uniform fields', *in* MEEK, J.M., and CRAGGS, J.D. (Eds): 'Electrical breakdown of gases' (Wiley, 1978), Chap. 3

58 LOOMS, J.S.T.: 'Insulators for high voltages' (IEE Power Engineering Series, No. 7, Peter Peregrinus, London, 1988)

59 VERHAART, H.F.A., TOM, J., VERHAGE, A.J.L., and VOS, C.J.: 'Avalanches near solid insulators'. Proceedings of 5th international symposium on *High voltage engineering*, Braunschweig, 1987, paper 37.01

60 JAKST, A., and CROSS, J.: 'The influence of a solid dielectric spacer on electron avalanches in Nitrogen at atmospheric pressure', *Can. Electr. Eng. J.*, 1981, **6**, pp. 14–20

61 NASSER, E.: 'The nature of negative streamers in spark breakdown', *Dielectrics*, 1963, pp. 110–117

62 MUROOKA, Y., TAKADA, T., and HIDAKA, K.: 'Nanosecond surface discharge and charge density evaluation, Part 1, Review and experiments', *IEEE Electr. Insul. Mag.*, 2001, **17**, pp. 6–16

Chapter 2

SF$_6$ insulation systems and their monitoring

O. Farish, M.D. Judd, B.F. Hampton and J.S. Pearson

2.1 Introduction

Sulphur hexafluoride is used as an insulant in a wide range of power system applications, including switchgear, gas-insulated substation (GIS) components, transformers and gas-insulated cables. SF$_6$ is chemically stable, non-toxic and non-flammable, and has a high vapour pressure ($\sim$21 bar at room temperature). At pressures typical of GIS applications ($\sim$5 bar) it can be used down to $-35°$C without liquefaction occurring. Its wide use in power equipment is promoted by the fact that, in addition to high dielectric strength, SF$_6$ has good heat transfer characteristics and excellent arc-quenching properties.

Although its dielectric strength is three times that of air and, at a pressure of 6 bar is approximately the same as that of transformer oil, SF$_6$ is a 'brittle' medium, in that ionisation builds up very rapidly if a critical field strength is exceeded. In practical applications, this can happen in the vicinity of any small defects such as electrode surface protrusions or free conducting particles. Depending on the nature of the defect, the partial discharges (PD) which occur at these local regions of field enhancement may result in breakdown of the insulation system. There is, therefore, increasing interest in the use of techniques for monitoring the PD activity in SF$_6$-insulated equipment with a view to identifying critical defects before they lead to failure.

This chapter reviews the basic ionisation processes which occur in SF$_6$, the streamer mechanism which controls breakdown under relatively uniform field conditions, and the influence of electrode surface roughness on breakdown at high pressure. The characteristics of the partial discharges (corona discharges) which occur under the non-uniform field conditions associated with certain types of defect are then discussed. Following a discussion of the various PD diagnostic techniques that have been proposed for use in GIS, an account is given of the principles of the UHF technique for detection of PD in metal-clad equipment. Finally, the design and calibration of the sensors used in UHF monitoring are discussed and an explanation given of the interpretation of the PD patterns recorded in practical UHF monitoring systems.

2.2   Ionisation phenomena in SF_6

The high dielectric strength of SF_6 is due to its property of electron attachment. In this process, a free electron moving in the applied field, which collides with a neutral molecule, may be attached to form a negative ion:

$$SF_6 + e \longrightarrow (SF_6)^-$$

This process competes with that of collisional ionisation, by which an electron with sufficient energy can remove an electron from a neutral molecule to create an additional free electron:

$$SF_6 + e \longrightarrow (SF_6)^+ + 2e$$

Ionisation is a cumulative process and, provided that the field is high enough, successive collisions can produce ever increasing numbers of free electrons which can result in electrical breakdown of the gas. On the other hand, the heavy, slowly moving negative ions which are formed by attachment are unable to accumulate the energy required to cause ionisation and the attachment process therefore effectively removes electrons and inhibits the formation of 'avalanches' of electrons which might lead to breakdown.

Some gases, such as nitrogen, hydrogen and argon, do not form negative ions; others, such as oxygen and CO_2, are weakly attaching. SF_6, in common with a number of other gases containing fluorine or chlorine, is highly 'electronegative' (i.e. exhibits strong electron attachment). Some gases have a dielectric strength significantly greater than that of SF_6 (Table 2.1). However, most present problems of one kind or another, including toxicity, limited operating pressure range or production of solid carbon during arcing. SF_6 is therefore the only dielectric that is accepted as suitable for GIS applications, although some high strength gases have been considered for use in mixtures with SF_6.

In later sections of the chapter, the mechanisms governing the formation of partial discharges and breakdown in compressed SF_6 insulation are reviewed. It is therefore useful at this stage to consider how the (net) rate of electron production in SF_6 depends on the gas pressure and the applied field.

The average energy attained by charged particles moving in a gas varies linearly with both the applied field E and the average mean free path λ between collisions. Since λ is proportional to the reciprocal of the number density N, the energy varies directly with E/N or, if the temperature is constant, with E/p, where p is the gas pressure.

The frequency of collisions made by a charged particle of a given energy increases with number density (or pressure). Thus, for an electron moving in a gas, for example, the probability of ionisation for a single collision will be a function of the energy parameter E/p, and the rate of ionisation will depend both on pressure and on E/p.

Table 2.1   Dielectric strength of certain
gases relative to SF_6 [1]

Gas	Relative strength	
H_2	0.18	non-attaching
air	0.3	
CO_2	0.3	weakly attaching
CO	0.4	
C_2F_8	0.9	
CCl_2F_2	0.9	
SF_6	1.0	strongly or very
$C\text{-}C_4F_8$	1.3	strongly attaching
$C\text{-}C_4F_6$	~1.7	
C_4F_6	~2.3	

Considering a swarm of electrons moving in a gas under a constant field, the growth of ionisation rate is defined in terms of the number of ionising collisions per electron per cm travel in the gas parallel to the applied field. This parameter is known as the ionisation coefficient, α, where $\alpha = pf_1 (E/p)$, that is:

$$\frac{\alpha}{p} = f_1 \left(\frac{E}{p} \right)$$

Similarly, the removal of electrons from the swarm is determined by an attachment coefficient, η, defined as the number of attachments per electron per cm travel. Hence:

$$\frac{\eta}{p} = f_2 \left(\frac{E}{p} \right)$$

The net ionisation then depends on the balance between ionisation and attachment. If $\eta > \alpha$, attachment dominates over ionisation and no discharge growth is possible; if $\alpha > \eta$, cumulative ionisation can occur.

Figure 2.1 shows the net (pressure-reduced) ionisation coefficient $(\alpha - \eta)/p$ as a function of E/p for air and SF_6. It can be seen that the critical reduced field strength at which $(\alpha - \eta) = 0$ is about 89 kV/cm bar in SF_6, compared with only ~27 kV/cm bar in air. This explains the high dielectric strength of SF_6 relative to air as no build-up of ionisation can occur until the reduced field exceeds the critical value $(E/p)_{crit}$.

It is worth noting the steep slope of the curve of $(\alpha - \eta)/p$ versus E/p in SF_6. This means that SF_6 is a relatively brittle gas in that, once $(E/p)_{crit}$ is exceeded, the growth of ionisation is very strong. This is significant in situations where stress-raising defects are present in gas-insulated equipment as intense ionisation activity will occur in the regions where $E/p > (E/p)_{crit}$ and this may initiate complete breakdown of the insulation.

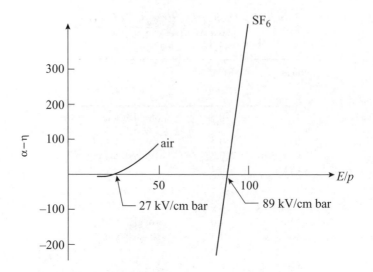

Figure 2.1   Effective ionisation coefficients in air and SF$_6$

Also worthy of note is the fact that the net ionisation coefficient in SF$_6$ can be represented by the linear relationship:

$$\frac{\alpha - \eta}{p} = A\left(\frac{E}{p}\right) - B$$

where $A = 27.7 \text{ kV}^{-1}$ and $B = 2460 \text{ bar}^{-1} \text{ cm}^{-1}$. The critical reduced field strength is therefore:

$$\left(\frac{E}{p}\right)_{\text{crit}} = \frac{Bp}{A} = 88.8 \text{ kV/cm bar}$$

This simple relationship is useful in estimating onset voltages in SF$_6$ insulation.

2.3   Breakdown mechanisms in low divergence fields

As discussed above, the build-up of ionisation in SF$_6$ is possible only under conditions where the (pressure-reduced) field exceeds a critical value $(E/p)_{\text{crit}}$ of ~ 89 kV/cm bar.

For highly divergent fields (as, e.g., for the case of a sharp protrusion on a high voltage conductor) ionisation will be confined to a critically-stressed volume around the tip of the protrusion. In this situation localised PD, or corona, will be the first phenomenon observed as the applied voltage is increased. Breakdown under these conditions is a complex process, because of the effects of the space charge injected by the prebreakdown corona.

As any stress-raising defect in gas-insulated equipment will result in PD activity, it is important to understand non-uniform field discharge mechanisms. However, GIS are designed for relatively low field divergence and it will be useful first to consider the simple case of breakdown in SF$_6$ under uniform field conditions, before reviewing the phenomena associated with particulate contamination or other defects.

2.3.1 Streamer breakdown

For a perfectly uniform field (plane–plane electrode geometry), no ionisation activity can occur for reduced fields less than the critical value. Above this level, ionisation builds up very rapidly and leads to complete breakdown of the insulation (formation of an arc channel).

The first stage of the breakdown involves the development of an avalanche of electrons. The growth of this avalanche from a single starter at the cathode can readily be found by computing the net electron multiplication. Considering a swarm that has grown to contain $n(x)$ electrons at position x in a gap of width d; then, in travelling a further incremental discharge dx, these will generate a net new charge:

$$dn(x) = (\alpha - \eta)n(x)\,dx = \bar{\alpha}n(x)\,dx$$

as a result of ionising and attaching collisions with neutral molecules, where α is the net ionisation coefficient.

Integration over the interval 0 to x gives the number of electrons in the avalanche tip at that stage in its growth:

$$n(x) = \exp\left(\int_0^x \bar{\alpha}\,dx\right) = \exp(\bar{\alpha}x)$$

In crossing the whole gap, an avalanche of $\exp(\bar{\alpha}d)$ electrons is created.

In itself, the occurrence of avalanches does not constitute breakdown. For example, if conditions were such that $\bar{\alpha} = 5$ then, in a 1 cm gap at 1 bar, the current gain would be $e^5 \sim 150$. The normal low background conduction current density (due to collection of free charges present in the gap) would be increased as a result of ionisation from $\sim 10^{-13}$ A/cm^2 to $\sim 10^{-11}$ A/cm^2, but the gap would still be a very good insulator. However, as illustrated in Figure 2.1, $\bar{\alpha}$ increases very quickly when the reduced field exceeds $(E/p)_{\mathrm{crit}}$ and the multiplication can rapidly reach values of 10^6 or greater, with most of the charge confined to a very small region at the head of the avalanche (approximately a sphere of typically ~ 10 μm radius).

The bipolar space charge generated by the ionisation process results in local distortion of the applied field such that ionisation activity ahead of, and behind, the avalanche tip is greatly enhanced. At a critical avalanche size ($\exp(\bar{\alpha}x) = N_c$), the space charge field is high enough to generate rapidly moving ionisation fronts (streamers) which propagate at $\sim 10^8$ cm/s towards the electrodes. When these bridge the gap, a highly conducting channel is formed within a few nanoseconds.

For pressures used in technical applications ($p > 1$ bar), the streamer process is the accepted breakdown mechanism in SF$_6$ under relatively uniform field conditions.

The critical avalanche size for streamer formation is found to be that for which the streamer constant $k = \ell n N_c$ is approximately 12. The breakdown voltage is then easily calculated using the linear relationship between $\bar{\alpha}/p$ and E/p:

$$\frac{\bar{\alpha}}{p} = A\left(\frac{E}{p}\right) - B$$

where $A = 27.7$ kV^{-1} and $B = 2460$ bar^{-1} cm^{-1}.

The minimum streamer inception or breakdown level will occur when the critical avalanche size is achieved at the anode. Thus:

$$\bar{\alpha}d = AEd - Bpd = k$$

The breakdown voltage V_s $(=Ed)$ is then:

$$V_s = \frac{B}{A}(pd) + \frac{k}{A} = 88.8\,(pd) + 0.43\ \text{(kV)}$$

where pd is in bar cm.

Note that the breakdown voltage is a function only of the product (pressure x spacing). This is an example of the similarity relationship (Paschen's Law) which allows gas-insulated equipment to be made more compact by increasing the pressure above atmospheric.

As indicated above, once the gap is bridged arc formation in SF$_6$ is extremely rapid. The voltage collapse time depends on the pressure (p), spacing (d) and geometry, and is typically $\sim 10\ d/p$ nanoseconds. In certain situations, this can present serious problems in GIS equipment. Sparking during closure of a disconnector switch, for example, can generate travelling waves in the GIS bus which have very fast rise times (up to 100 MV/μs). Doubling at open circuits elsewhere in the system can result in insulation being stressed with high amplitude (>2 p.u.) pulses with very short rise times (<10 ns). There are also problems with grounding and shielding, as very high fields (500 kV/m) can appear across parts of the grounded enclosure during the pulse transit.

2.3.2   Quasi-uniform fields (coaxial cylinders)

If the field is varying with position across the gap, the initial avalanche formation will occur within a critical volume for which $(\alpha - \eta) > 0$ (i.e. $E/p > 88.8$ kV/cm bar). Under these conditions, the streamer inception criterion is:

$$\exp\left(\int_0^{x_c} \bar{\alpha}(x)\,dx\right) = N$$

where x is the distance from the inner electrode along a field line and x_c is the position of the boundary of the ionisation region.

For coaxial electrode geometry (inner radius r_0, outer r_1) the field distribution is:

$$E(r) = \frac{V}{r}\ln\left(\frac{r_1}{r_0}\right)$$

Also, at onset, $\bar{\alpha} = 0$ at position r_c, so that:

$$E(r_c) = E_{crit} = \frac{Bp}{A}$$

Using these relationships, together with the streamer criterion, it can easily be shown that the surface field at onset is:

$$E(r_0) = \frac{Bp}{A}\left[1 + \left(\frac{k}{Bpr_0}\right)^{1/2}\right]$$

and that:

$$x_c = \left(\frac{kr_0}{Bp}\right)^{1/2}$$

With the above values of A and B, this yields:

$$\frac{E(r_0)}{p} = 89\left[1 + \left(0.07\sqrt{pr_0}\right)\right] \quad (pr_0 \text{ in bar cm})$$

and

$$x_c = \frac{0.07\sqrt{r_0}}{p} \quad (\text{cm})$$

For the large curvature electrodes and high pressures used in GIS, the field at the inner conductor at onset is therefore very close to the critical reduced field of ~89 kV/cm bar.

Note that the streamer forms when the primary avalanche has developed a relatively short distance. For $r_0 = 8$ cm, $p = 4$ bar, for example, x_c will be ~1 mm. The streamer will then propagate until the combination of the space charge field and the geometric field is unable to sustain further ionisation. In order for breakdown to occur, it will then be necessary to increase the surface field above the onset level. In the relatively low divergence field in a (clean) GIS system only a small increase above the onset voltage is necessary to initiate breakdown.

2.3.3 Effect of surface roughness

Although laboratory measurements using polished coaxial electrodes are in agreement with the theoretical criterion that the inner surface field at breakdown should be close to the critical field of ~89 kV/cm bar, this value cannot be sustained in large scale equipment with a practical (machined) surface finish.

One reason for this is the fact that increased ionisation occurs in the vicinity of microscopic surface protrusions (surface roughness). This results in reduction of the breakdown field strength by a factor ζ. Figure 2.2 shows calculated values of the factor ζ as a function of the product ph (pressure × protrusion height) for a range of spheroidal protrusions [2].

It can be seen (a) that the breakdown voltage can be reduced to a low level and (b) that there is a critical protrusion size for the onset of roughness effects.

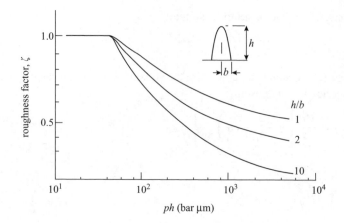

Figure 2.2   Roughness factor for uniform field breakdown in SF_6

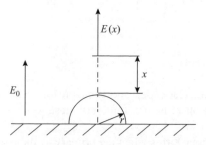

Figure 2.3   Hemispherical protrusion on a uniform field electrode

The existence of a threshold value of *ph* can readily be demonstrated [2] for the hemispherical protrusion shown in Figure 2.3. For this case, the field at an axial distance *x* above the protrusion is given as:

$$E(x) = E_0 \left[1 + \frac{2r^3}{(x+r)^3} \right]$$

For the protrusion to have no effect, E_0 (the macroscopic field) at onset must be equal to $(E/p)_{\text{crit}}$ $(=Bp/A)$.

Also:

$$\bar{\alpha}(x) = AE(x) - Bp = Bp \left[1 + \frac{2r^3}{(x+r)^3} \right] - Bp = \frac{2Bpr^3}{(x+r)^3}$$

Breakdown occurs when:

$$\int_0^{x_c} \bar{\alpha}(x)\,dx = k$$

that is

$$2Bpr^3\left[-\tfrac{1}{2}(x+r)^{-2}\right]_0^{x_c} = k$$

therefore

$$\frac{1r^2}{(x_c+r)^2} = \frac{k}{Bpr}$$

therefore

$$\frac{x_c}{r} = 1 - \left[\frac{1}{1-k/Bpr}\right]^{1/2}$$

With $k = 12$ and $B = 0.246$ bar^{-1} μm^{-1}:

$$x_c = r\left[1-(1-49/pr)^{-1/2}\right]$$

For x_c to be real, pr must be >50 bar μm. At a working pressure of 5 bar, surface roughness would therefore begin to affect the onset level for protrusion heights greater than ~ 10 μm.

Because of surface roughness effects (and other electrode phenomena such as micro discharges in charged oxide layers, etc.) practical SF$_6$-insulated equipment is designed such that the maximum field is everywhere less than ~ 40 per cent of the critical value. In a typical GIS, for example, the basic insulation level (BIL) will correspond to a peak reduced field of only ~ 35 kV/cm bar.

With a good technical surface finish, streamers should not form in a clean coaxial electrode system under these conditions. Further, if a local defect does cause streamer formation, the streamer should not be able to propagate into the low field region of the gap. The fact that breakdown can occur, even at the lower reduced field associated with the AC working stress (~ 15 kV/cm bar) indicates that an additional mechanism is operative. This is discussed in the following section on non-uniform field breakdown in SF$_6$.

2.4   Non-uniform field breakdown in SF$_6$

Highly divergent fields can exist in GIS under certain conditions as, for example, when a needle-like free metallic particle is attracted to the inner conductor or is deposited on the surface of an insulator. Such defects can result in very low breakdown levels and, with large defects (e.g. particles several mm long), failure can occur even at the working stress of the equipment. For this reason, there have been many laboratory studies of the breakdown characteristics of highly non-uniform field gaps in SF$_6$.

These studies have shown that there are two distinct types of breakdown, depending on the rate at which the voltage is applied to the gap. When the stress is applied relatively slowly, as with alternating voltage or long rise time switching surges, corona space charge plays an important part in controlling the field distribution by

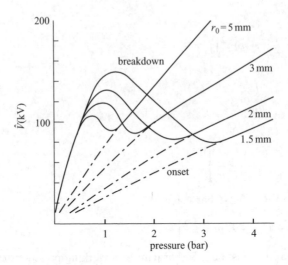

Figure 2.4   AC corona onset and breakdown characteristics for a 40 mm rod–plane gap in SF$_6$ [4]

the so-called corona stabilisation process [3]. With shorter rise time surges (lightning impulse or fast transients), breakdown occurs directly by a stepped leader mechanism [5]. For both cases, the breakdown voltage is lower when the high field electrode is positive and most attention has therefore been given to breakdown under positive point conditions.

2.4.1   Corona stabilised breakdown

Figure 2.4 shows AC voltage–pressure characteristics for point-plane gaps in SF$_6$ [4]. The shape of these curves is typical of all non-uniform field gaps with slowly varying voltage applied, in that there is (a) a broad pressure region over which the breakdown voltage is much higher than the (streamer corona) onset voltage and (b) a critical pressure at which breakdown occurs directly at onset (i.e. the first streamer leads directly to breakdown).

The peak in the mid-pressure range is due to the effects of space charge injected by streamer activity around the point.

For a positive point, for example, the electrons generated by the corona are quickly removed at the point while the positive ions diffuse relatively slowly into the low field region. This space charge tends to shield the point and stabilises the field there to a level close to the onset value.

As the voltage is raised, the space charge density (and the shielding effect) intensifies, and a voltage considerably above onset is required to cause breakdown. The breakdown usually occurs as a result of filamentary (leader) discharges developing around the shielding space charge, so that spark channels in the stabilisation region typically take a very irregular, curved path.

As the pressure is increased, the individual streamers become more intense, the corona region becomes confined to a smaller region at the tip of the point and the stabilisation becomes less effective, so that the breakdown voltage is reduced. Eventually, the shielding effect is lost, and the streamer which forms at onset is able to initiate a discharge which develops completely across the gap at the onset voltage. This discharge has been shown to be identical to the stepped leader discharge which is found to occur in non-uniform field gaps under fast pulse conditions.

2.4.2 Leader breakdown

With fast-fronted surges, where the voltage passes rapidly through the theoretical streamer onset level, the initial streamers can be very intense and may lead to the formation of a highly ionised leader channel before there is time for space charge stabilisation of the field at the tip of the protrusion [5].

In addition to the rate-of-rise of voltage, the statistics of appearance of initiatory electrons may play an important role. For negative-point conditions, electrons are produced by field emission; with the positive point, however, the trigger electrons result from detachment from negative ions in the vicinity of the point [6].

Before the stress is applied, the gas contains a negative-ion population of a few thousand ions per cm^3. (These are produced by the action of cosmic rays, which typically generate ~ 10 ion pairs per cm^3 per second in gases at atmospheric pressure.) For discharge initiation to occur, it is necessary to find one of these ions in the very small critical volume where ($\alpha > \eta$). This critical volume is vanishingly small at the theoretical onset level and increases with voltage.

For a fast-fronted wave, the field may therefore be well above the minimum onset level when inception occurs so that the streamer corona is more vigorous than would be the case for AC or DC stress.

If the streamer corona is large enough, a stepped leader discharge may be initiated. The mechanism of the stepped leader may be summarised, with reference to Figure 2.5, as follows.

During the dark period *a–b* which follows the initial corona, charge separation in the streamer filaments generates a succession of ionising waves which build-up their conductivity. Eventually, one of the streamer filaments is transformed into a highly conducting leader channel step; this behaves essentially as an extension to the point electrode and a new corona burst immediately occurs at its tip *b*. The range of this second corona determines the length of the second channel step *c*. During each streamer's dark period, there are regular reilluminations of the leader channel, probably associated with the relaxation processes which are occurring in the streamer filaments.

The leader propagates into the gap in steps typically of a few mm until the streamer activity is too weak for further channel steps to form. If the voltage is high enough, the leader can cross the gap, resulting in breakdown. As the interstep interval is ~ 100 ns for $p \approx 3$ bar, the breakdown formative time lag can be greater than 1 μs.

As the field along the leader channel is much lower than that in the streamer filament, breakdown can occur by the stepped leader process at much lower voltages

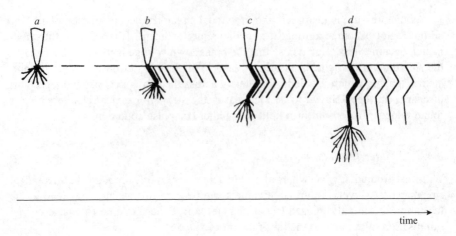

Figure 2.5   Schematic of leader development

than would be required for streamer breakdown. For point-plane gaps, the minimum leader breakdown voltage is found to be almost independent of pressure and average breakdown fields of $\sim$25 kV/cm are typical of short (20–50 mm) gaps [7].

In configurations where the background field in the low field region is falling less steeply (as, e.g., for the case of a particle fixed to the inner conductor of a GIS), leader breakdown can occur at average fields of only $\sim$15 kV/cm.

Figure 2.6 shows $V-p$ characteristics in SF$_6$ for corona-stabilised breakdown and for the minimum breakdown voltage under impulse conditions. As p_1 is typically only about 0.5 bar, the minimum breakdown voltage in non-uniform fields at pressures typical of GIS is determined by the conditions for leader propagation in the absence of preexisting corona space charge. Models have been developed [8, 9] which allow the conditions for direct leader inception and propagation to be predicted for a wide range of geometries.

It must be emphasised that, under surge conditions, the leader propagation field is the minimum level at which breakdown can occur. Depending on the statistics of initiation of the primary streamer, there is a probability of corona stabilisation occurring so that, even for lightning impulse, the 50 per cent probability voltage–pressure characteristic will exhibit a stabilisation peak. It is important therefore to determine the low probability breakdown level when carrying out surge breakdown tests in inhomogeneous fields.

2.5   Breakdown in GIS

2.5.1   Streamer-controlled breakdown

The design stresses used in GIS are low enough ($<$50 per cent E_{crit}) that streamer inception will not occur even at the full rated impulse level. However, scratches or other small defects on the inner electrode surface may result in streamer formation.

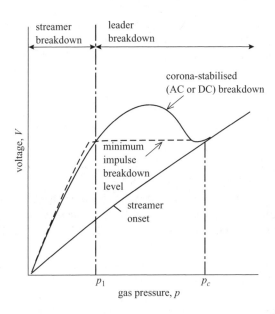

Figure 2.6   *Idealised V–p characteristics for minimum impulse (direct leader) break-down and corona-stabilised (AC or DC) breakdown in a point-plane gap in SF₆ [7]*

For a small scratch or protrusion ($h \approx 100\ \mu$m), the field perturbation is very localised and the geometric field has to be quite high in order to achieve streamer onset. Such defects would probably be detected only under impulse-voltage test conditions at levels close to the BIL.

At such high voltages, the conditions for breakdown (whether by the leader process or by a direct streamer channel mechanism) are automatically satisfied at onset. The breakdown voltage of the GIS may therefore be predicted on the basis of the streamer criterion:

$$\exp\left[\int_0^x (AE(x) - Bp)\,dx\right] = N_c$$

provided that $E(x)$, the spatial distribution of the perturbed field, is known.

As discussed earlier, the statistics of discharge initiation can play an important role in determining the probability of breakdown in perturbed quasi-homogeneous fields and this has significant implications for insulation coordination on SF₆-insulated equipment. Various models have been developed for calculating surge breakdown probability on the basis of the negative-ion density distribution and the evolution of the critically-stressed volume with time during the surge [10, 11].

2.5.2   Leader breakdown

For large defects, such as needle-like particles of several mm length attached to the high voltage conductor, the onset voltage for streamer corona will be low.

This means that the onset voltage is lower than the leader propagation voltage and breakdown is preceded by corona. This is the situation discussed in the section on non-uniform fields, where either corona-stabilised or direct leader breakdown may occur, depending on the voltage waveform.

2.5.3   Particle-initiated breakdown

Free conducting particles (FCPs) are the most common cause of failure in GIS, and long, thin particles are most dangerous because of the strong field enhancement associated with such defects.

If FCPs are present in a coaxial system they become charged by the applied field and, at a relatively low voltage, lift-off will occur. If the particle is rod shaped, it will stand up on the outer conductor and corona onset will occur. For DC stress, particles will cross the gap at the onset voltage. For AC conditions the particles initially make small hopping excursions at the outer electrode. As the stress is increased, the excursions become longer and, because of inertial effects and the bouncing action at the electrode, the bounce interval becomes longer. At each contact with the electrode, a PD occurs and the interval between contacts is an important parameter in the assessment of the severity of particle-induced PD activity in GIS.

As the voltage is increased, particles may cross the gap to the upper electrode where they will receive a new charge, which depends on the polarity and magnitude of the applied voltage at the instant of contact, and will then move back towards the lower electrode. The crossing does not necessarily lead to breakdown, and a voltage increase will usually be necessary for the conditions for leader breakdown to be achieved.

In coaxial electrode systems, breakdown is most likely to occur when a particle strikes the inner conductor just as the voltage reaches a positive maximum [12]. The particle then behaves like a needle fixed to the HV conductor.

Early studies of the mechanism of AC particle-initiated breakdown in GIS led to some confusion, as the particle-triggered breakdown voltage–pressure characteristics did not show the strong peak observed with AC stress applied to a coaxial electrode system having a particle fixed to the inner conductor [13]. This led to the consideration of mechanisms such as density reduction in the wake of the moving particle, and the possible triggering action of the microdischarge which occurs at the instant of contact.

The actual reason for the disparity between the fixed and free particle data may be inferred from the earlier discussion on corona-stabilised and direct leader breakdown. For the fixed particle, there is enough time under AC conditions to guarantee effective stabilisation at each voltage maximum, resulting in the typical peaked V–p characteristic. With the free particle, however, the sudden arrival of the particle, together with the step function increase in the field at the tip as a result of the small spark occurring on contact, means that the behaviour is more similar to that for a fixed particle subjected to a fast fronted impulse voltage. Under these conditions, the minimum breakdown voltage is that associated with direct leader breakdown, which is almost independent of pressure (see Figure 2.6).

In practice, the probability of breakdown in GIS at each contact will be a function of the particle charge, velocity and orientation, the phase and magnitude of the applied voltage, the statistics of breakdown of the particle–electrode microgap and the probability of formation of a stabilisation corona. It may therefore be necessary to wait for a relatively long period (20–30 s) to ensure that a test particle will not trigger breakdown at a given voltage.

Measurements made in a 125/250 mm coaxial system with fixed and free particles showed good agreement between the ($t = 20$ s) AC breakdown voltage with free particles and the minimum breakdown voltage with 1 µs rise time impulses applied to a particle fixed to the inner conductor [14]. As the minimum fixed point impulse breakdown level corresponds to breakdown by a leader mechanism, it is clear that models of leader inception and propagation can be used to predict the conditions of particulate contamination which will result in breakdown in GIS.

Present indications are that, for normal working stresses, particles of length less than ~4 mm should not be able to cause breakdown [9]. However, smaller particles may be scattered onto the surfaces of insulating barriers or spacers, where they may cause breakdown under subsequent impulse stresses. It is important, therefore, to ensure that free conducting particles of significant size (≥ 1 mm) are not present in GIS under working conditions.

2.6 Possible improvements in SF₆ insulation

The dielectric performance of present designs of GIS is probably close to the best that can be achieved with careful component design and good quality control, using existing materials and construction methods. Techniques which may offer further improvements in GIS insulation include.

2.6.1 Use of additives or gas mixtures

The dielectric properties of SF₆ can be significantly improved [15] by using leader-suppressing additives such as Freon 113, although the use of such additives in GIS would require confirmation that they have no effect on the solid insulation in the system. Gas mixtures containing buffer gases such as N_2 in concentrations of up to 80 per cent have dielectric strengths which are not much below that of SF₆ under clean conditions and may be less susceptible to particulate contamination. SF_6/N_2 mixtures are particularly attractive for use in gas-insulated transmission lines because of their lower cost and reduced environmental impact.

2.6.2 Improved spacer formulation and construction

There is continual progress in the development of resins and fillers with improved properties in terms of mechanical strength, resistance to tracking and compatibility with SF₆ discharge products. The use of semiconducting surface coatings to prevent the build-up of surface charge may also be beneficial, and the presence of ribs on the surface of the spacers may offer significant advantages, particularly in the presence of particulate contamination [16].

2.6.3   Particle control

Although every effort is made to ensure that FCPs are removed during preassembly cleaning of GIS, particles can be produced during operation, for example as a result of abrasion of sliding contacts. Simple slotted trays in the outer conductor make very effective particle traps and their use in the vicinity of solid spacers can offer useful protection. Other proposals for particle control have included techniques for covering FCPs with a sticky insulating coating by post-assembly polymerisation of an appropriate additive to the SF_6.

Even if improvements are made in one or more of the above areas, it is probable that it will never be possible to completely eliminate every defect that may result in local stress enhancement in GIS. It is important, therefore, to ensure that the procedures used in factory and/or site testing are able, where possible, to indicate the presence of particles or other defects and that the tests used do not in themselves cause damage to the insulation, or create conditions which may cause later failure (e.g., by scattering particles on to spacer surfaces).

These considerations have led to the development of diagnostic techniques which allow the presence of defects to be recognised as a result of their partial discharge activity so that action can be taken to remove them before a failure occurs. The remainder of this chapter is devoted to a discussion of these techniques.

2.7   Partial discharge diagnostic techniques for GIS

2.7.1   Introduction

Although the reliability of GIS is high [17], any internal breakdown that does occur invariably causes extensive damage and an outage of several days' duration is needed to effect the repair. During this time the associated circuit may be out of operation and the consequential losses can be high, especially if the GIS is operating at 420 kV or above. If in addition the GIS is connecting the output of a nuclear station to the transmission network and the breakdown leads to a reactor shutdown, the financial penalties could be most severe.

Modern designs of GIS, such as that shown in Figure 2.7, have benefited from the experience gained with earlier versions, and they have a high level of reliability. In the future it appears likely that further improved GIS designs will be supplemented by quite extensive diagnostic monitoring, which will have two main advantages; it will allow the user to adopt condition-based maintenance and reduce revenue expenditure, and will lead to the ideal situation of being able to detect a developing fault in time to prevent an unplanned outage.

Much progress in diagnostic techniques for GIS has been made in recent years, and they are used increasingly in factory testing, site commissioning and during the service life of the equipment [18–22]. In the UK, diagnostic couplers are specified for

Figure 2.7   A modern 400 kV GIS

all new GIS, and in some cases have been fitted retrospectively to existing substations. The insulation can then be monitored for signs of any incipient weakness, and action taken to prevent it developing into complete failure. With one technique, this is being done continuously and remotely, in line with the trend among utilities to operate GIS unattended.

So far the main objectives of GIS diagnostics have been to detect whether there is any defect in the GIS, to identify it as a particle, floating shield and so on, and to locate it so that it can be repaired. Various diagnostic techniques have been demonstrated in the laboratory, and with some of them several years' experience on site, both during commissioning tests and with the GIS in service, has been gained. The results of the on-site work have been very promising, and have shown the need for these further developments:

(i)   Complex and often intermittent discharges can be found in GIS, and a better understanding of the physical processes leading to breakdown is needed to allow the diagnostic data to be interpreted and its significance assessed.

(ii) Continuously monitoring one or more GIS can produce very large quantities of data, and it is important not to overburden the engineer with its interpretation. The discharge data needs to be analysed by an expert system, and the engineer informed only when some condition arises which needs attention.

(iii) A monitor installed in a GIS to detect defects in the insulation can in addition be used to record the condition of circuit breakers, transformers and other plant, and so provide a complete diagnostic system on which predictive maintenance of the GIS can be based [20].

It is these aspects which are becoming increasingly important, and where the main advances can be expected.

2.7.2   *The range of diagnostic techniques for PD detection*

2.7.2.1   Fundamental processes

The statistics of GIS reliability [23, 24] show that the most common cause of electrical failure is a free metallic particle, which can become attracted to the high voltage conductor and produce a microdischarge which triggers breakdown. Other causes of failure are discharges from any stress-raising protrusion, capacitive sparking from an electrode which is not properly bonded to either the high voltage conductor or earth, and so on; and the common feature of all these defects is that they generate PD activity in advance of complete breakdown. With the exception only of the mechanical noise from a bouncing particle, PD detection is the basis of all dielectric diagnostics in GIS.

A PD is the localised breakdown of gas over a distance of usually less than a millimetre. For surface defects such as small protrusions, the discharge takes the form of corona streamers which give rise to current pulses with very short rise times (<1 ns). Discharges in voids, and microsparks associated with poor contacts or with the transport of conducting particles, are also characterised by high rates of change of current. In all cases, the very short rise time of the PD pulse causes electromagnetic energy to be radiated into the GIS chamber, and the energy dissipated in the discharge is replaced through a pulse of current in the EHV supply circuit. In microsparks and intense coronas, the discharge is followed by rapid expansion of the ionised gas channel, and an acoustic pressure wave is generated. PD is also accompanied by the emission of light from excited molecules, and by the creation of chemical breakdown products. The PD therefore has many effects – physical, chemical and electrical – and in principle any of them could be used to reveal the presence of the discharge.

2.7.2.2   Light output

Detecting the light output from a discharge is probably the most sensitive of all diagnostic techniques, because a photo-multiplier can detect the emission of even a single photon. The radiation is primarily in the UV band, and since this is absorbed strongly both by glass and SF_6 it is necessary to use quartz lenses and a reasonably short path length. Although this is a powerful laboratory tool for finding the onset of activity from a known corona point, there are many difficulties in using it to detect a discharge which might be anywhere in a GIS.

2.7.2.3   Chemical byproducts

This approach initially appears attractive because chemical decomposition is immune to the electrical interference which is inevitably present in the GIS, and with any steady discharge the concentration of the diagnostic gas should rise in time to a level where it can be detected (this assumes, of course, that an absorbing reagent is not used in the chambers).

The main decomposition product of sulphur hexafluoride is sulphur tetrafluoride (SF_4), but this is a highly reactive gas. It reacts further, typically with traces of water vapour, to form the more stable compounds thionyl fluoride (SOF_2) and sulphuryl fluoride (SO_2F_2). These are the two most common diagnostic gases, and, by using a gas chromatograph and mass spectrometer, they may be detected with sensitivities down to 1 p.p.m. As a simpler but less sensitive alternative, chemical detection tubes can be used.

In small-volume laboratory tests, a reasonably small discharge of 10–15 pC can be detected, typically after some tens of hours. However, in a GIS the diagnostic gases would be greatly diluted by the large volume of SF_6 in which they occur, and much longer times would be needed. It therefore appears that the chemical approach is too insensitive to be considered for PD monitoring.

2.7.2.4 Acoustic emission

Acoustic signals arise both from the pressure waves caused by partial discharges and from free particles bouncing on the chamber floor. The latter is the only instance of a diagnostic signal not coming from a PD (although of course the particle generates a PD as well). The signals in GIS have a broad bandwidth, and travel from the source to the detector by multiple paths [21, 25]. Those originating at the chamber wall propagate as flexural waves, at velocities which increase with the square of the signal frequency to a maximum of approximately 3000 m/s. Propagation through the gas is at the much lower velocity of 150 m/s, and the higher frequencies in the signal are absorbed quite strongly. The alumina- or silica-filled epoxide barriers which are used to support the inner conductor also attenuate the signal markedly.

The different propagation velocities of the wave as it passes through various materials, and the reflections occurring at boundaries between them, give rise to a complex acoustic waveform. This signal can be picked up by accelerometers or acoustic emission sensors attached to the outside of the chamber. The acoustic signal from a particle bouncing on the chamber floor is characterised by a signal not correlated with the power frequency cycle. It also has other features, such as the crest factor (ratio of the peak/r.m.s. value), the impact rate and the ratio of the lift-off/fall-down voltages, from which the particle shape and its movement pattern can be inferred. Other sources of discharge may be identified in a similar way from their own characteristics [26].

One advantage of acoustic measurements is that they are made non-intrusively, using external sensors which may be moved from place to place on the GIS. Because of the rather high attenuation of the signals, the sensors should preferably be on the chamber containing the source. This in itself gives the approximate location of the defect, but a more accurate position can be found to within a few cm using a second sensor and a time of flight method.

The acoustic technique is not suited to a permanently installed monitor, because too many sensors would be needed.

2.7.2.5   Electrical methods

There are two approaches to detecting the electrical charge in a partial discharge; in the external circuit by a conventional PD measurement system, and internally by detecting the resonances set up in the GIS chambers.

(i)   *Conventional method* The test circuit is that given in IEC Publication 270, and the charge flowing through a coupling capacitor fitted in parallel with the GIS is measured using a quadrupole and detector. The PD current pulse at the defect has a duration of less than 1 ns, and propagates as a travelling wave in each direction along the chambers. The pulses are attenuated and undergo multiple reflections, but do not appear immediately in the external circuit. After about a microsecond or so, the pulses die away and the GIS appears to the external circuit as a lumped capacitor with a depleted charge. From then, a replacement charge flows into the GIS, and is measured by the detector.

To obtain the maximum sensitivity of measurement, a completely shielded test arrangement is required [27], which is possible for a test assembly but may be inconvenient when testing a complete GIS. Also, the total capacitance of a GIS is high, and it must be divided into sections for test. In addition, there is no means of locating the discharge, and since a coupling capacitor is not normally provided in a GIS the technique cannot be used for in-service measurements.

(ii)   *UHF method* The current at the PD site rises in less than a nanosecond, and can radiate EM waves with energy spectra extending to frequencies of 2000 MHz or more. This excites the GIS chambers into various modes of electrical resonance, which because of the low losses in the chambers can persist for up to a microsecond. The resonances are indicative of PD activity, and if they are picked up by couplers installed in the GIS may be displayed on a spectrum analyser.

The resonant technique was developed in the UK [22], where over the past 15 years or so much experience has been gained using UHF for PD detection at frequencies from 300 to 1500 MHz [28].

The UHF technique will later be described in detail, when it will be seen that its advantages are its high sensitivity; the ability to locate discharges accurately by time of flight measurements; and that it can readily be used in a continuous and remotely operated monitoring system.

2.7.3   Comparison of the techniques

It is difficult to compare the various diagnostic techniques from results gained under different experimental conditions, so a CIGRE Working Group (15-03) arranged for them to be used simultaneously to detect a range of artificial defects [29]. The tests were made in a 6 m long section of 420 kV GIS chambers into which was placed one of the following defects:

- a free metallic particle
- a particle attached to the surface of a barrier

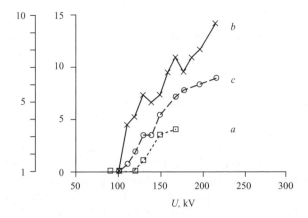

Figure 2.8   Signal to noise ratio of PD from a needle on the busbar (taken from [29])

    a   IEC 270 (pC/pC)
    b   UHF (dB)
    c   acoustic (mV/mV)

- a corona point at the HV conductor
- a similar point on the chamber wall.

The vessel was energised by a 0–510 kV metalclad test transformer, the voltage being increased slowly until breakdown occurred. During this time, diagnostic measurements were made using the following techniques:

a   conventional electrical detection according to IEC Publication 270 with either a standard detector at 1 MHz, or the phase-resolved partial discharge (PRPD) evaluation system at 200 kHz [30]
b   UHF, using an internal coupler at frequencies up to 1500 MHz
c   acoustic, using an external acoustic emission sensor at 34 kHz
d   chemical, using detector tubes; this technique proved too insensitive to give a result over the limited test period.

To illustrate the results reported in Reference 29, those for a needle attached to the busbar are reproduced as Figure 2.8, in which the data from the various techniques has been expressed as signal/noise ratios, so that the results can be compared.

The general conclusions of this investigation were that:

- the acoustic, IEC 270, and UHF techniques all show good sensitivity
- acoustic measurements are non-intrusive and can be made on any GIS, but the attenuation of the signal across barriers and along the chambers is rather high
- conventional PD measurements need an external coupling capacitor, and cannot be used on GIS in service
- the UHF technique is suitable for in-service monitoring.

2.7.4   Overview of UHF technology

A GIS installation consists of a network of coaxial transmission lines which acts as a waveguiding structure for UHF signals, with an inherently low loss. In the absence of barriers and discontinuities, the attenuation at 1 GHz in a waveguide of this size (typically 0.5 m diameter) would be only 3–5 dB/km. In practice, reflections at discontinuities within the GIS chamber cause a reduction in signal strength which has been observed to be in the region of 2 dB/m [28]. These reflections can cause resonances to appear, such as those set up between dielectric barriers [31].

When coaxial lines are used for signal transmission, the usual mode of signal propagation is the transverse electromagnetic (TEM) mode, in which the electric and magnetic field components are transverse to the direction of propagation. The frequency of operation is kept below the cut-off frequencies at which higher-order transverse electric (TE) and transverse magnetic (TM) modes begin to be excited, thus ensuring non-dispersive propagation, a normal requirement for the maintenance of signal fidelity.

In the case of a GIS, the coaxial structure is a consequence of the need to contain the gaseous insulation, and its dimensions are accordingly defined by high voltage requirements. At UHF (300–3000 MHz), the GIS dimensions are such that the TE and TM modes of propagation cannot be neglected [32]. Excitation of a purely TEM mode signal would require symmetrical excitation of the waveguide, whereas the location of a PD current pulse is always asymmetrical with respect to the coaxial cross-section and therefore couples strongly with higher-order modes. These modes are closely related to those of the hollow cylindrical waveguide, and are therefore capable of propagating across gaps in the HV busbar, which would block TEM signals. For these reasons, it is necessary to account for all modes of propagation within the measurement bandwidth to describe adequately the UHF signal resulting from a PD.

The design of internal couplers for detection of UHF signals in GIS involves a compromise between the conflicting requirements of minimising the field enhancement while maximising the UHF sensitivity. The coupler must not create an additional risk of breakdown, and is normally mounted in a region of relatively weak HV field, at an inspection hatch for example, where it is shielded in a recess in the outer conductor. A disadvantage is that the UHF fields also tend to be weaker in these regions, since they are subject to the same boundary conditions as the HV field. UHF antennas of a form which would be desirable for good sensitivity, such as a radial monopole, are unacceptable as they would invite breakdown. However, other forms of broadband planar couplers, such as the spiral, have been investigated [33] and shown to have good sensitivity. Circular plate couplers have proved useful, and are more readily accepted in GIS because they are similar to capacitive dividers, and can be seen not to cause stress enhancement. Circular couplers are themselves resonant structures at UHF frequencies, and the effect of design parameters has been investigated [33], showing how their sensitivity can be enhanced within the design constraints.

A more detailed description of UHF theory and the generation and transmission of UHF signals will be found in the next section.

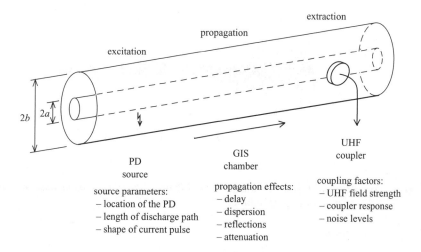

source parameters:
– location of the PD
– length of discharge path
– shape of current pulse

propagation effects:
– delay
– dispersion
– reflections
– attenuation

coupling factors:
– UHF field strength
– coupler response
– noise levels

Figure 2.9   Transfer functions involved in the UHF detection of PD in GIS

2.8   The generation and transmission of UHF signals in GIS

2.8.1   Introduction to UHF theory

Detection of PD by the UHF method involves the stages of energy transfer that are shown in Figure 2.9.

To take full advantage of the UHF technique, an understanding of the basic processes involved is important. In the following notes, a represents the radius of the inner conductor of the GIS, and b represents the radius of the outer conductor. A system of cylindrical coordinates (r, ϕ, z) will be used to describe the electromagnetic field components.

2.8.2   Excitation

The shape of the streamer current pulse $i(t)$ at the PD source is most important in determining the characteristics of the UHF signal. The energy radiated in the UHF range is highly dependent on the rate of change of PD current. However, for a given pulse shape, the UHF signal amplitude scales linearly with the current flowing at the defect.

For small defects, the UHF signal amplitude is proportional to the product ql when the pulse shape is constant [34]. Here, q represents the charge contained in the PD current and l is the length over which it flows. Because the length of the streamer itself rarely exceeds 1 mm, l is predominantly a function of the defect size (e.g., particle or protrusion length).

The UHF signal excited by a PD source depends on the position of the defect in the transverse plane. This is because the coupling coefficients to each of the waveguide modes vary across the coaxial cross-section of the GIS.

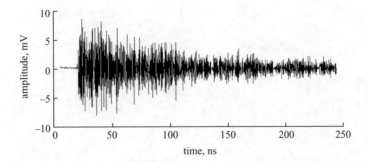

Figure 2.10   Typical UHF signal excited by PD in a 400 kV GIS, as measured at the output of a UHF coupler

2.8.3   Propagation

The electromagnetic waves radiated from the defect region begin to propagate in the GIS chamber. Different frequency components of the PD pulse propagate at different velocities, causing dispersion of the pulse. The overall effect of dispersion is to cause the signal to appear as a long, oscillating waveform with a somewhat random appearance (Figure 2.10). The UHF signals obtained from a GIS coupler typically have a duration of 100–1000 ns. Some of the signal is rapidly attenuated because it is below the cut-off frequency of the mode in which it is propagating. The highest frequency components travel along the coaxial lines with a velocity approaching c. Propagation through barriers takes place at a lower velocity, $c/\sqrt{\varepsilon_r}$, where ε_r is the relative permittivity of the insulating material (typically 5–6). The relative arrival times of the wavefronts at couplers on either side of the PD source can often be used to locate the defect.

Any non-uniformities in the GIS will cause partial reflections of the UHF signals. Most discontinuities inside the GIS have a complicated reflection pattern, because they do not reflect the signal at a plane, but over a distributed volume (e.g., a conical gas barrier). The attenuation of the UHF signal along the GIS duct (between one coupler and the next) is mainly due to the confining of signal energy within the chambers by the partially reflecting discontinuities. These effects cannot be analysed theoretically, except in a greatly simplified form [31]. However, numerical techniques and experimental measurements have resulted in guideline figures for the attenuation caused by reflecting obstacles in the GIS.

2.8.4   Extraction

Internal UHF couplers are normally mounted at a recess in the outer conductor. Because of the boundary conditions in this region, the radial component of electric field is predominant. The intensity of this electric field is therefore the primary factor affecting the signal level that can be obtained from the coupler. Externally mounted couplers (e.g., at an inspection window) will be affected by the field patterns in the structure on which they are mounted. In this case, the mounting arrangement should be considered as part of the coupler, so that the reference plane is still the inner surface of the outer conductor. The coupler's function is to maximise the output voltage for a given radial component of UHF electric field [35].

The frequency response of the coupler should be suitable for the frequency range of the UHF signal. In a 400 kV GIS, the UHF energy is normally concentrated between 500 and 1500 MHz. UHF signals from low level PD can only be detected if they are of sufficient amplitude to be distinguished from electrical background noise. Internal couplers are best from this point of view, as the noise levels are low, especially if the GIS is cable fed. Where external couplers must be used, they should be screened from interfering signals [36].

2.8.5   Waveguide modes and UHF propagation

This section introduces the basic field patterns that can exist in a coaxial waveguide, and describes their relevance to the UHF detection of PD.

2.8.5.1   TEM and higher-order modes

The transverse electromagnetic (TEM) mode is the familiar mode of propagation in a coaxial line, in which the electric and magnetic fields are directed wholly in the plane transverse to the direction of propagation. Higher-order modes are classified as transverse electric (TE) or transverse magnetic (TM) types. TE modes have only E_r and E_ϕ components to their electric fields, and TM modes have E_r, E_ϕ and E_z components. An infinite number of these higher-order modes exists, designated using subscripts such as TE_{nm} and TM_{nm}. Each of these modes has a unique cut-off frequency, below which it does not propagate.

2.8.5.2   Calculating the cut-off frequencies of TE and TM modes

Cut-off frequencies of the higher-order modes are related to the mode eigenvalues u_{nm} and v_{nm} by the following equations:

$$f_{TE_{nm}} = \frac{c}{2\pi} u_{nm} \qquad (2.1)$$

$$f_{TM_{nm}} = \frac{c}{2\pi} v_{nm} \qquad (2.2)$$

The mode eigenvalues are dependant on the radii a and b of the GIS conductors [35] and are roots of the following equations, which involve the Bessel functions $J_n(x)$, $Y_n(x)$ and their first derivatives $J'_n(x)$, $Y'_n(x)$:

$$J'_n(ua)Y'_n(ub) - Y'_n(ua)J'_n(ub) = 0 \qquad (2.3)$$

$$J_n(va)Y_n(vb) - Y_n(va)J_n(vb) = 0 \qquad (2.4)$$

For a given value of the integer n, these equations possess an infinite series of roots, which are numbered $m = 1, 2, 3, \ldots$. Thus, u_{nm} is the mth root of Equation (2.3) and v_{nm} is the mth root of Equation (2.4). These values can be determined numerically using standard mathematical software packages.

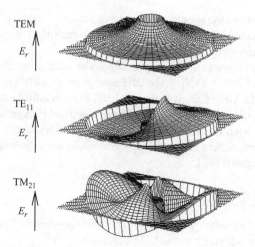

Figure 2.11   Patterns of the radial electric field over the GIS cross-section for some coaxial waveguide modes. The rectangular plane represents the position of zero electric field

2.8.5.3   Electric fields of propagating modes in GIS

In cylindrical coordinates, the electric field of the TEM mode can be expressed in terms of the voltage V on the inner conductor as:

$$E_r = \frac{V}{r \ln(b/a)} \tag{2.5}$$

The radial electric field components of TE and TM modes vary in a more complex manner, as defined by the following equations:

$$\text{TE modes:} \quad E_r = K \frac{n}{r} \cos(n\phi)[J_n(ur)Y'_n(ua) - Y_n(ur)J'_n(ua)] \tag{2.6}$$

$$\text{TM modes:} \quad E_r = Kv \cos(n\phi)[J'_n(vr)Y_n(va) - Y'_n(vr)J_n(va)] \tag{2.7}$$

In these equations, K is an arbitrary constant defining the amplitude of the excitation. Variation of the radial field components defined by Equations (2.5)–(2.7) is illustrated in Figure 2.11 for a mode of each type.

2.8.5.4   Importance of higher-order modes for PD detection

The total field radiated by a PD current can be represented by the sum of the field patterns of the waveguide modes that it excites. This is analogous to the representation of a non-sinusoidal signal by a Fourier series in the time domain. In the case of PD, at any instant in time the field pattern is represented in the three-dimensional space by the sum of three-dimensional fields. By its nature, a PD source is usually located asymmetrically within the coaxial line. During the early stages of propagation, any excitation of the TEM mode must be associated with considerable contributions from other modes to produce a field that is only non-zero in the region close to the PD source. Hence, the higher-order modes are excited in greater proportion.

As the duration of the PD pulse decreases, the spatial variation of the radiated fields must increase because the propagation velocity c is fixed. The number of modes required to adequately represent the field increases as a result. Conversely, if the PD pulse is relatively slowly varying compared with the time taken for the field to traverse the waveguide cross-section, the higher-order modes become less significant in the representation of the field pattern and the TEM mode is predominant.

A PD source such as a protrusion excites more higher-order modes when it is at the outer conductor than at the inner conductor [34]. This is because the spatial variation of the modes is inherently greater at the inner conductor, so fewer are required to represent the rapidly changing field in this region. In certain circumstances, this property can be used to discriminate between different PD locations [37].

2.8.5.5   Relative contributions of TEM and higher-order modes

The VHF content of PD signals in the frequency range up to 300 MHz is usually below the cut-off frequencies of any higher-order modes in 400 kV class (or smaller) GIS. PD detection that is carried out at VHF therefore depends upon the level of the TEM mode signal for its sensitivity. The loss of sensitivity that results from this restriction can be illustrated by comparing the higher-order and TEM mode contributions to the electric field at a coupler for two excitation pulses having equal amplitudes but different pulse widths. The configuration used to make the comparison is that of a 10 mm PD path (e.g., a protrusion) located at the outer conductor in the centre of a 3.6 m long 400 kV GIS chamber ($a = 0.05$ m, $b = 0.25$ m). The resulting radial electric field at a distance of 1.2 m from the PD source can be determined by computer simulation. A Gaussian current pulse $i(t)$ at the PD source was defined as:

$$i(t) = I_{max}\, e^{-(t-t_0)^2/2\sigma^2} \tag{2.8}$$

For the first example, a 1 pC pulse was defined by $I_{max} = 2.0$ mA, $t_0 = 2$ ns and $\sigma = 200$ ps. The contributions to the total field are shown in Figure 2.12a. The higher-order modes make a much larger contribution to the total field in response to this pulse, with a peak-to-peak amplitude of 75 mV m^{-1} compared with 12 mV m^{-1} for the TEM mode. The power available from each signal is proportional to the square of the field strength, which increases the significance of this amplitude difference. Also, the higher-order modes do not have the dead time that occurs between reflections of the TEM mode pulse, during which it does not deliver any energy to the coupler.

The second example, shown in Figure 2.12b, was generated by changing σ to 500 ps while leaving t_0 and I_{max} unchanged in Equation (2.8). This increases the pulse width at half-amplitude from 0.50 ns to 1.25 ns. The peak amplitude of the TEM mode field remains unchanged but the width of the reflected pulses has increased accordingly. In contrast, the higher-order mode field has altered significantly. The amplitude is now comparable to that of the TEM mode, and the high frequency content has decreased.

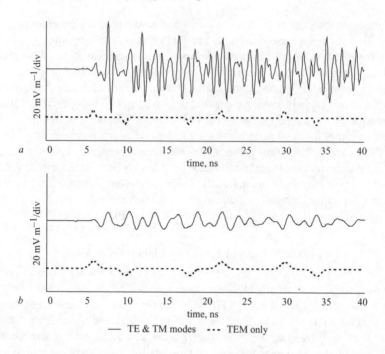

Figure 2.12   Simulation results comparing the mode contributions to the radial electric field E_r at 1.2 m from a PD source. The PD current has a path length of 10 mm and the peak current is 2 mA.

a   Gaussian PD pulse of half-amplitude width 500 ps
b   Gaussian PD pulse of half-amplitude width 1.25 ns

E_r has been shown in Figure 2.12 rather than the coupler output voltage, to illustrate the typical electric field strengths to which UHF couplers are subjected. The couplers are normally capacitive, and their sensitivity decreases at lower frequencies [33]. Although the amplitude of the TEM mode field is unchanged in Figure 2.12b, the resulting coupler output would be significantly reduced because the rate of change of the electric field is lower. Making the coupler larger to counteract this effect would have the undesirable consequence of increasing the amount of low frequency noise coupled from the GIS. Measurements of the PD current pulses generated by small defects such as particles and protrusions in SF_6 have shown typical values of less than 500 ps for the half-amplitude width. Theoretical studies have indicated that for the short PD pulses typical of small defects, limiting PD detection to VHF frequencies (TEM mode only) results in the majority of the available signal energy being neglected.

2.8.6   Attenuation of UHF signals

The GIS can be considered as a series of loosely coupled chambers in which the UHF resonances occur. Over the shorter timescale (10–100 ns), energy is transferred between adjacent chambers, with those nearest the PD source maintaining the higher energy levels. Dissipation losses (skin effect) become significant in the longer timescale (100–1000 ns), and are the reason for the ultimate decay of the signal.

2.8.6.1   Losses due to the skin effect

The skin effect causes signal attenuation through dissipation in the surface resistance of the conductors. The losses tend to increase with increasing frequency. However, attenuation at UHF in GIS is theoretically quite low (typically 3–5 dB per km), due to the high ratio of the cross-sectional area to the surface area in the waveguide.

2.8.6.2   Attenuation caused by barriers and discontinuities

In long gas-insulated lines (GIL), which have relatively few discontinuities (insulating posts at intervals to support the inner conductor and the occasional gas barrier), attenuation of UHF signals can be quite low, allowing PD sources to be detected at distances of >100 m [38]. However, in the more compactly constructed GIS, the number of bends, junctions, gas barriers and circuit breakers is such that the empirical figure of 1–2 dB m^{-1} is more appropriate.

Various research groups have attempted to assign specific attenuation values to individual components, but it is now generally accepted that this cannot be done. The reason is that the overall signal level at a given position in the GIS arises from interactions between signals reflected from discontinuities. These interactions are dependent on the distances between the discontinuities, and the signal levels also depend on the position of the coupler relative to barriers, junctions, etc. [37]. This can lead to situations where a UHF coupler at a greater distance from a PD source detects a larger signal than does one closer to it.

As a consequence of these effects, it is not possible to calculate with any accuracy the attenuation that will be experienced by a PD signal propagating between two UHF coupling points on a GIS. In order to ensure that a UHF monitoring system has sufficient sensitivity to detect a 5 pC discharge located midway between a pair of couplers, a practical approach has been proposed by CIGRE [39]. The two-stage procedure involves first determining an artificial pulse, which, if injected into one of the couplers, will radiate a UHF signal equivalent to a real 5 pC PD. An on-site test is then carried out, injecting the calibrated pulse into each coupler and ensuring that it can be detected at adjacent couplers by the UHF monitoring system.

2.9   Application of UHF technique to PD detection in GIS

2.9.1   *Design and testing of UHF couplers*

For radially-directed PD currents, the majority of the energy available to an electric field sensor (coupler) is in the radial component of the electric field. In the region close to the outer conductor where couplers are normally mounted, the other components of the electric field are very small because of the waveguide boundary conditions. Consequently, the purpose of a coupler is to provide the maximum transfer of energy from the incident radial electric field to the 50 Ω input of the monitoring system. Couplers can be classified according to whether they are mounted internally or externally:

(i)   *Internal couplers* must be fitted to the GIS during construction or retrofitted during a planned outage, because degassing of the GIS chambers is necessary. These couplers often take the form of a metal disc insulated from the GIS enclosure by a dielectric sheet. The measurement connection is made through a hermetically sealed coaxial connector that is usually connected to the centre of the disc.

(ii) *External couplers* are usually portable sensors that are fitted to an aperture in the metal cladding such as an inspection window or exposed barrier edge. These couplers are suitable for periodic insulation testing of GIS for which a permanently installed monitor is not economically viable or for older GIS that cannot be retrofitted with internal couplers. External couplers are sometimes less sensitive than their internal counterparts because the UHF signal is attenuated by impedance discontinuities at the surfaces of the barrier and window materials. They may also be more prone to electrical interference signals when they are not shielded as well as internal couplers. For these reasons, it is preferable to fit internal couplers during the construction phase, and this is common practice in the UK.

2.9.1.1 Coupler calibration system

The UHF excitation model developed in Reference 32 allows the electric field at the outer conductor of the GIS to be determined. This is the radial electric field at the point where the coupler is to be mounted. In practice, the presence of the coupler and the mounting arrangement will distort the field pattern. For this reason, the mounting of the coupler for calibration purposes should duplicate the GIS mounting structure. Couplers are calibrated by measuring their transfer function in terms of their output voltage in response to a defined incident electric field. To allow for the many types of coupler and mounting arrangement that exist, a standard measurement scheme has been developed which is independent of the GIS in which the coupler is to be used.

The calibration system (see Figure 2.13) measures the frequency response of the coupler and its mounting arrangement when subjected to a known electric field normal to the ground plane in which the coupler is mounted. The incident field is first calibrated using a monopole probe having a known frequency response. The probe is then replaced by a mounting plate suitable for holding the coupler to be tested. A digitiser records the signal from the coupler under test, and a continuously updated display of the coupler gain is provided by the signal processing unit. Figure 2.14 gives the response of an external window coupler used on a 500 kV GIS. Note that the coupler sensitivity has units of length and is represented by an effective height h. This arises because the transfer function relates the output voltage (V) to the incident field (Vm^{-1}). This particular coupler is shown, mounted on a window, in Figure 2.15.

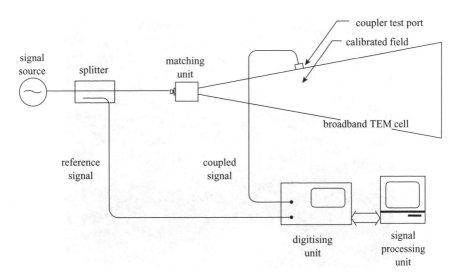

Figure 2.13   Diagram of the UHF coupler calibration system

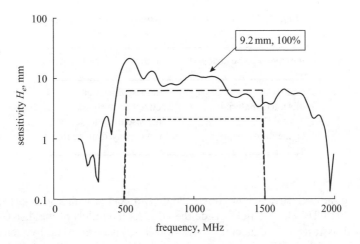

Figure 2.14   Frequency response plot for an external, window coupler as measured by the calibration system

2.9.1.2   Disc couplers – design guidelines

Studies of resonance in disc couplers [33] have shown that the UHF signal level can be increased by making the connection to the disc close to its edge rather than in the centre. This also improves the bandwidth of the coupler by increasing sensitivity at lower frequencies. For a given disc size, the UHF signal level increases when the dielectric height is increased or a material of lower dielectric constant is used. Both of these changes lead to a reduction in the Q factor of the disc resonances, and

Figure 2.15   An external window coupler on a 500 kV GIS

Table 2.2   Optimising sensitivity of disc coupler

Parameter to be changed	To increase sensitivity
size	increase diameter
dielectric material	reduce permittivity
	increase thickness
connection	move away from centre

increasing the dielectric height allows the disc to couple with a greater proportion of the radial UHF electric field inside the GIS. As the radius of the disc coupler is decreased its resonant frequencies move upwards, away from the frequency range where most of the UHF energy is concentrated. Generally, the disc should be made as large as is practical for the size of the GIS in which it is to be installed. This information is summarised in Table 2.2.

2.9.1.3   External couplers

External couplers, such as the one shown in Figure 2.15, are usually based on antenna designs and fabricated using normal PCB techniques. At UHF, the losses in glass fibre substrates are insignificant and the cost of low loss microwave substrates cannot be justified. In some cases, external couplers can be calibrated using the system described above. However, in some circumstances the coupler may be so far removed from the GIS chamber (e.g., in a CT housing) that the mounting arrangement cannot be

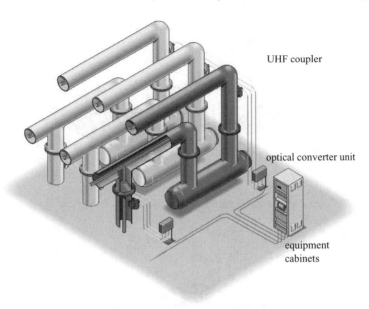

UHF coupler

optical converter unit

equipment
cabinets

Figure 2.16   General arrangement of the PDM system

reproduced reliably. In these cases, the coupler must be optimised in a GIS test rig, using the pulse injection technique described below.

When the coupler is not in the HV chamber of the GIS, more complicated designs can be used because there is no concern about field enhancement and the risk of flashover. Some designs are described in Reference 40, which also describes the test procedure. Optimisation should take place in the time domain, because it is difficult to make comparisons based on a spectrum analyser plot of the coupler output in response to an injected pulse.

2.9.2   Design of a PDM system for GIS

The general arrangement of one particular partial discharge monitor (PDM) is shown in Figure 2.16.

The monitoring system consists of the following basic parts:

- UHF couplers – to take the UHF signals from the GIS.
- Optical converter units (OCUs) – each OCU contains circuitry to detect and process the UHF signals from one three-phase set of couplers. The UHF data is then transmitted via an optical fibre link back to the equipment cabinets, which are located in a central place.
- The central data handling, processing, storage and display are carried out within equipment cabinets located in the relay room. The racks contain the electronics to receive and handle the streamed data from the OCUs, the PC and control unit for data storage and display.

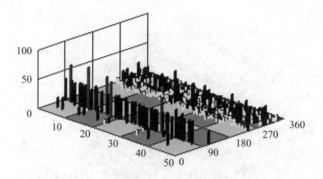

Figure 2.17   Busbar corona, streamers and leaders

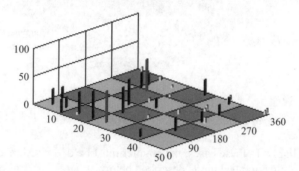

Figure 2.18   Free metallic particle

2.9.3   Display and interpretation of PD data

The features of the UHF discharge pulses that are most useful for interpretation purposes are their amplitude, point on wave and the interval between pulses. These parameters enable typical defects such as fixed point corona, free metallic particles and floating electrodes to be identified. Other defects occur less commonly, but have their own distinctive features.

The UHF data may be displayed in any way which reveals the characteristic patterns typical of the defects causing them, as, for example, in the three-dimensional patterns shown in Figures 2.17 and 2.18. Here, the pulses detected in 50 (60) consecutive cycles are shown in their correct phase relationships over the cycle. In the three-dimensional displays, 0° corresponds to the positive-going zero of the power frequency wave, 90° to the positive peak, and so on.

2.9.3.1   Corona

Three distinct phases in which corona develops from a protrusion are seen as the voltage is raised:

(i) *Inception*, where discharges occur first on the half-cycle that makes the protrusion negative with respect to the other electrode. Inception is therefore on the negative half-cycle when the protrusion is on the busbar, and on the positive half-cycle when it is on the chamber wall. The initial discharges are of very low magnitude (less than 1 pC), and are centred on one of the voltage peaks.

(ii)   *Streamers* start at a slightly higher voltage, and appear as a regular stream of pulses on the peak of the positive half-cycle. At the same time, the negative discharges become larger and more erratic. This difference between the positive and negative discharges reveals whether the protrusion is on the busbar or chamber wall. Streamer discharges will not lead to breakdown.

(iii) *Leaders* follow further increases of the applied voltage, and appear every few cycles as large discharges on the positive half-cycle. They propagate in steps until either they become extinguished, or reach the other electrode and cause complete breakdown. Figure 2.17 shows a typical display of PD from a protrusion on the busbar at this stage of the discharge process. Leaders are the precursors of breakdown, and there is always a risk of failure when they are present.

2.9.3.2   Free metallic particle

A particle lying on the chamber floor becomes charged by the electric field, and if the upward force on it exceeds that due to gravity it will stand up and dance along the floor. This generates a discharge pulse each time contact is made with the floor, since the particle then assumes a new value of charge. The pulses occur randomly over the complete power frequency cycle, but their peak amplitudes follow the phase of the voltage. A typical PD pattern for a free metallic particle is shown in Figure 2.18. At higher voltages, the particle will start to jump towards the busbar, and after several cycles may reach it. Two factors combine to make this an especially serious condition which often leads to breakdown: (i) as the particle approaches the busbar it discharges and generates a voltage transient which increases still further the stress at its tip; (ii) breakdown can occur by leader propagation well before any space charge has had time to develop and shield the tip.

2.9.3.3   Floating electrode

This arises if the contact to, for example, a stress shield deteriorates and sparks repetitively during the voltage cycle. The sparking is energetic because the floating component usually has a high capacitance, and this degrades the contact further. Metallic particles are produced, and may lead to complete breakdown. The discharges are concentrated on the leading quadrants of the positive and negative half-cycles, and their amplitude does not vary with the applied voltage.

Often the gap is asymmetrical, and sparks over at different voltages on the two half-cycles. Then, a different charge is trapped on the floating component, and this gives rise to the characteristic wing-shaped patterns shown in the plan view of Figure 2.19. These provide a positive confirmation that the defect is a floating component.

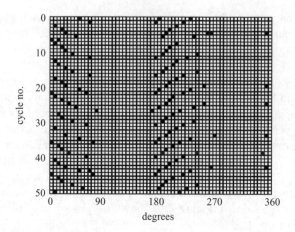

Figure 2.19   Floating electrode, plan view

2.9.4   AI diagnostic techniques

PDM systems can generate large amounts of data, which needs to be interpreted auto-matically so that the results are immediately available to system operation engineers. Various artificial intelligence (AI) techniques may be used to classify the type of defect present, which gives a guide to any action that might need to be taken.

The real need, however, is for an expert system that will give the probability and time to failure of the GIS. The expert system will need to include additional information on the severity of the defect, its location in the GIS chamber and the electric field at that point. This full risk analysis is not possible at present, but much effort is being devoted to its development.

2.9.4.1   Classification algorithms

PD signals may be classified using analytical techniques such as distance classi-fiers, artificial neural networks (ANNs) and fuzzy algorithms. In general, features are extracted from the PD signatures and the techniques trained to recognise them. Since no single technique appears adequate to recognise all types of defect, the combined outputs from a number of techniques may be used to give a higher prob-ability of success in the final classification. All the analytical techniques need to be trained using extensive databases of PD signals, usually recorded in labora-tory tests, and with sufficient care the confidence level in the results may exceed 90 per cent.

A classification technique that is used with much success is the ANN. This can be applied to identify the single-cycle patterns recorded by the monitoring system out-lined above; for example, the particle and floating components shown in Figures 2.18 and 2.19. As shown in Figure 2.20, groups of multiple ANNs are used first to separate the signal from noise, and then to identify the source of each with a known confidence level. This technique may itself be used in conjunction with others.

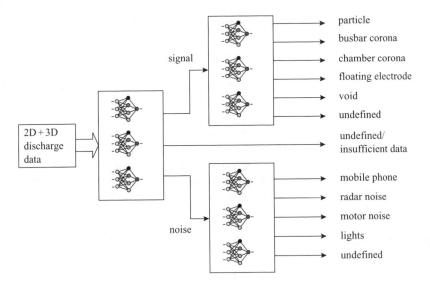

signal

particle

busbar corona

chamber corona

floating electrode

void

undefined

2D + 3D
discharge
data

undefined/
insufficient data

mobile phone

radar noise

motor noise

noise

lights

undefined

Figure 2.20   Classification of PD signals by ANNs

2.9.4.2   Trend analysis and alarms

Often, the PD signals recorded on site are intermittent in nature, and it is necessary to build-up a history of the activity over a period of time. The historical records are used to determine the trends in the activity of a PD source, such as its amplitude and repetition rate. Averaging is applied to eliminate spurious results and reveal any underlying change in the PD activity, and this can be a further basis for sending an alarm signal to the operator.

2.9.5   Service experience

The first online UHF system for PD monitoring in GIS was commissioned in Scotland in 1993 and the technology has since been adopted by other utilities in the UK, South-East Asia, the Middle East, South America and the USA. In the case of new substations, PDM systems are often specified for key installations, as, for example, where the GIS is supplying a major population centre, or is connected to a critical generation source such as a nuclear power plant.

In many cases, the use of UHF monitoring is also seen as being cost effective for GIS which have been in service for a number of years, where the benefits are improved availability and increased plant life. The availability of suitable external couplers which allow a PDM system to be retrofitted without the need for an outage is an important factor in this situation, although planned maintenance or refurbishment operations provide the opportunity for internal couplers to be installed where there are no suitable dielectric apertures at which external sensors can be mounted.

Service experience with UHF monitoring has been excellent. Several utilities have reported the identification of defects that would have resulted in major failures, any one of which would have incurred costs considerably in excess of the cost of the PDM system. Typical examples have included sparking at internal shields, tracking on solid insulation and, in one case, failure of a joint between two sections of busbar.

For the future, it seems inevitable that the use of monitoring in all types of high voltage equipment will become increasingly common. PD monitoring of the insulation in GIS will merely be one component in integrated systems which will monitor a range of mechanical, thermal and electrical parameters in all power plant, including transformers, switchgear, CTs, bushings and cables. There will be a growing need for the development of sophisticated information management techniques to coordinate and prioritise the output from such systems in order to provide a basis for effective remedial action.

2.10   References

1   CHRISTOPHOROU, L.G., and DALE, S.J.: 'Dielectric gases'. 'Encyclopedia of Physical Science and Technology', 1987, **4**, pp. 246–262

2   SOMERVILLE, I.C., TEDFORD, D.J., and CRICHTON, B.H.: 'Electrode roughness in SF_6 – a generalised approach'. Proceedings of XIII international conference on *Phenomena in ionised gases*, Berlin, 1977, pp. 429–430

3   FARISH, O.: 'Corona-controlled breakdown in SF_6 and SF_6 mixtures'. Proceedings of XVI international conference on *Phenomena in ionised gases*, Dusseldorf, 1983, pp. 187–196

4   SANGKASAAD, S.: 'AC breakdown of point-plane gaps in compressed SF_6'. Proceedings of 2nd international symposium on *HV technology*, Zurich, 1975, pp. 379–384

5   CHALMERS, I.D., GALLIMBERTI, I., GIBERT, A., and FARISH, O.: 'The development of electrical leader discharges in a point-plane gap in SF_6', *Proc. R. Soc. Lond. A Math. Phys. Sci.*, 1987, **412**, pp. 285–308

6   SOMERVILLE, I.C., FARISH, O., and TEDFORD, D.J.: 'The influence of atmospheric negative ions on the statistical time lag to spark breakdown', *in* CHRISTOPHOROU, L.G. (Ed.): 'Gaseous dielectrics IV' (Pergamon Press, New York, 1984), pp. 137–145

7   CHALMERS, I. D., FARISH, O., GIBERT, A., and DUPUY J.: 'Leader development in short point-plane gaps in compressed SF_6', *IEE Proc. A, Phys. Sci. Meas. Instrum. Manage. Educ. Rev.*, 1984, **131**, pp. 159–163

8   PINNEKAMP, F., and NIEMEYER, L.: 'Qualitative model of breakdown in SF_6 in inhomogeneous gaps', *J. Phys. D: Appl. Phys.*, 1983, **16**, 1293–1302

9   WIEGART, N. *et al.*: 'Inhomogeneous field breakdown in GIS – the prediction of breakdown probabilities and voltages', *IEEE Trans. Power Deliv.*, 1988, **3**, pp. 923–946

10   KNORR, W., MOLLER, K., and DIEDERICH, K.J.: 'Voltage–time characteristics of slightly nonuniform arrangements in SF_6 using linearly rising and oscillating lightning impulse voltages'. CIGRE paper 15.05, Paris, 1980

11 BOECK, W.: 'SF₆-insulation breakdown behaviour under impulse stress', in 'Surges and high voltage networks' (Plenum Press, New York, 1980)

12 WOOTTON, R.E., COOKSON, A.H., EMERY, F.T., and FARISH, O.: 'Investigation of high voltage particle-initiated breakdown in gas insulated systems'. EPRI reports RP7835,1977; EL-1007,1979

13 COOKSON, A.H., and FARISH, O.: 'Particle-initiated breakdown between coaxial electrodes in compressed SF₆', *IEEE Trans.*, 1973, **PAS-92**, pp. 871–876

14 FARISH, CHALMERS, I.D., and FENG, X.P.: 'Particle-initiated breakdown in a coaxial system in SF₆/air mixtures'. Proceedings of IEEE symposium on *Electrical insulation*, Washington DC, 1986, pp. 206–209

15 CHALMERS, I.D. *et al.*: 'Extended corona stabilisation in SF₆/Freon mixtures', *J. Phys. D: Appl. Phys.*, 1985, **18**, pp. L107–112

16 VANAGIWA, T., ISHIKAWA, T., and ENDO, F.: 'Particle-initiated breakdown characteristics on a ribbed spacer surface for SF₆ gas-insulated switchgear', *IEEE Trans. Power Deliv.*, 1988, **3**, pp. 954–960

17 CIGRE WG 23.10: 'GIS in service – experience and recommendations'. CIGRE paper 23-104, Paris, 1994

18 IEEE Substations Committee, Working Group K4 (GIS diagnostic methods): 'Partial-discharge testing of gas-insulated substations', *IEEE Trans. Power Deliv.*, 1992, **7**, (2), pp. 499–506

19 BARGIGIA, A., KOLTUNOWICZ, W., and PIGINI, A.: 'Detection of partial discharge in gas-insulated substations', *IEEE Trans. Power Deliv.*, 1992, **7**, (3), pp. 1239–1249

20 JONES, C.J., IRWIN, T., HEADLEY, A., and DAKERS, B.: 'The use of diagnostic based predictive maintenance to minimise switchgear life cycle costs'. Proceedings of 9th conference on *Electric power supply industry*, Hong Kong, 1992, **5**, pp. 225–234

21 LUNGAARD, L.E., TANGEN, G., SKYBERG, B., and FAUGSTAD, K.: 'Acoustic diagnosis of GIS: field experience and development of expert systems', *IEEE Trans. Power Deliv.*, 1992, **7**, (1), pp. 287–294

22 HAMPTON, B.F., and MEATS, R.J.: 'Diagnostic measurements at UHF in gas insulated substations', *IEE Proc. C, Genes. Trans. Distrib.*, 1988, **135**, (2), pp. 137–144

23 FUJIMOTO, N., and FORD, G.L.: 'Results of recent GIS fault survey'. IERE workshop on *Gas insulated substations*, Toronto, September 1990

24 KOPEJTKOVA, D., MOLONY, T., KOBAYASHI, S., and WELCH, I.M.: 'A twenty-five year review of experience with gas-insulated substations'. CIGRE paper 23-101, Paris, 1992

25 LUNGAARD, L.E., and SKYBERG, B.: 'Acoustic diagnosis of SF₆ gas insulated substations', *IEEE Trans. Power Deliv.*, 1990, **5**, (4), pp. 1751–1759

26 SCHLEMPER, H.D.: 'TE-Messung und TE-Ortung bei der Vor-Ort-Prüfung von SF₆-Anlagen'. Haefely Trench symposium, Stuttgart, March 1995

27 SCHLEMPER, H.D., KURRER, R., and FESER, K.: 'Sensitivity of on-site partial discharge detection in GIS'. Proceedings of 8th international symposium on *HV engineering*, Yokohama, 1993, **3**, pp. 157–160

28   HAMPTON, B.F., PEARSON, J.S., JONES, C.J., IRWIN, T., WELCH, I.M., and PRYOR, B.M.: 'Experience and progress with UHF diagnostics in GIS'. CIGRE paper 15/23-03, Paris, 1992

29   CIGRE WG 15.03: 'Diagnostic methods for GIS insulating systems'. CIGRE paper 15/23-01, Paris, 1992

30   FRUTH, B., and NIEMEYER, L.: 'The importance of statistical characteristics of partial discharge data', *IEEE Trans. Electr. Insul.*, 1992, **27**, (1), pp. 60–69

31   KURRER, R., FESER, K., and HERBST, I.: 'Calculation of resonant frequencies in GIS for UHF partial discharge detection', *in* CHRISTOPHORU, L.G., and JAMES, D.R. (Eds): 'Gaseous dielectrics' (Plenum Press, New York, 1994)

32   JUDD, M.D., FARISH, O., and HAMPTON, B.F.: 'Excitation of UHF signals by partial discharges in GIS', *IEEE Trans. Dielectr. Electr. Insul.*, 1996, **3**, (2), pp. 213–228

33   JUDD, M.D., FARISH, O., and HAMPTON, B.F.: 'Broadband couplers for UHF detection of partial discharge in gas insulated substations', *IEE Proc., Sci. Meas. Technol.*, 1995, **142**, (3), pp. 237–243

34   JUDD, M.D., and FARISH, O.: 'Transfer functions for UHF partial discharge signals in GIS'. Proceedings of 11th international symposium on *High voltage engineering*, London, 1999, **5**, pp. 74–77

35   JUDD, M.D., and FARISH, O.: 'A pulsed GTEM system for UHF sensor calibration', *IEEE Trans. Instrum. Meas.*, 1998, **47**, (4), pp. 875–880

36   JUDD, M.D., FARISH, O., PEARSON, J.S., and HAMPTON, B.F.: 'Dielectric windows for UHF partial discharge detection', *IEEE Trans. Dielectr. Electr. Insul.*, 2001, **8**, pp. 953–958

37   MEIJER, S.: 'Partial discharge diagnosis of high-voltage gas-insulated systems' (Optima Grafische Communicatie, Rotterdam, 2001), ISBN: 90-77017-23-2

38   OKUBO, H., YOSHIDA, M., TAKAHASHI, T., and HOSHINO, T.: 'Partial discharge measurements in a long distance SF_6 gas insulated transmission line (GIL)', *IEEE Trans. Power Deliv.*, 1998, **13**, (3), pp. 683–690

39   CIGRE Task Force 15/33.03.05: 'Partial discharge detection system for GIS: sensitivity verification for the UHF method and the acoustic method', *Electra*, (CIGRE), 1999, (183), pp. 74–87

40   WANNINGER, G.: 'Antennas as coupling devices for UHF diagnostics in GIS'. Proceedings of 9th international symposium on *High voltage engineering*, Graz, 1995, **5**

Chapter 3

Lightning phenomena and protection systems

R.T. Waters

3.1 From Franklin to Schonland

Benjamin Franklin was born in 1706, and was fascinated in his youth by electrical phenomena. In 1732 he founded with his collaborators the Library Company of Philadelphia with the support of the Penn family, and achieved a new insight into electrical science with his definition of a 'single electric fire'. This is equivalent to the free-electron concept, and implies the principle of the conservation of charge.

In 1750 he showed that a needle brought near to a charged conductor caused a spark, but when further away discharged the conductor silently (a mechanism today known as a glow corona). Franklin wrote that: 'this power of points may be of some use to mankind'. Foremost in his mind was to achieve protection against lightning, and in the famous Philadelphia Experiment (performed under Franklin's direction by d'Alibard in France in 1752) point discharges were used to prove the electrification of the thundercloud and to identify correctly the direction of the field formed by the cloud charges [1]. The public effect of these successes was great, and they were regarded as the most important work since Newton.

Franklin, in fact, had envisaged from his laboratory tests two complementary concepts for the function of a lightning rod:

- the possible harmless discharge of the cloud
- the attraction of the flash and the conduction of the charge safely to ground.

Although the first process is not feasible, the second is very efficient, at least for negative lightning. Franklin [2] clearly showed the need to earth the down conductor. During the well known controversy about the optimum design of a lightning rod (or air termination in present terminology), experiments by Benjamin Wilson in 1777 indeed suggested that sharp points might be too attractive to the flash, and he favoured spheres. Consequently, George III had points replaced on government buildings – with some enthusiasm as these had become associated with the American colonists. In a letter dated October 14, 1777, to a friend, Franklin wrote that: 'The King changing his rods to blunt conductors is a matter of small importance to me. It is only since he thought himself and his family safe from thunder of Heaven that he dared to use his own thunder in destroying his innocent Subjects'. As with so much concerning the lightning flash, the design of air terminations continues to be a subject of research [3].

It is noteworthy that the term 'striking distance', still in current use to calculate the range of attraction of the rod termination to the lightning flash, was coined by Franklin. In addition, from his experiments with point discharges, he correctly deduced that the cloud appeared from the ground to be negatively charged [4].

It was over a century later that Preece [5] proposed a quantification of the zone of protection that is afforded by a vertical mast; he anticipated the modern rolling-sphere concept, and set the sphere radius equal to the striking distance, which was itself equated to the mast height. This leads to a protection angle of 45°.

It was a further century and a half later that the electrical structure of the thunder-cloud, as a dipole with an upper positive charge [6], and the manner of growth of the lightning flash [7] were clarified.

The pioneering observations of the lightning flash structure achieved by Basil Schonland in the 1920s and 1930s constituted the biggest advance in the field since Benjamin Franklin's work in the late 18th century [8, 9]. Schonland read mathematics at Cambridge and his research on lightning from 1925 at Capetown University capitalised on South Africa's active thunderstorms. He made electric field measurements and, having collected data from some 23 storms, showed the cloud base to be negatively charged as Franklin had proposed. This resolved more modern disputes, and also confirmed the dipole nature of the cloud charge, with a high altitude positive charge in the cloud. He also recorded field changes due to lightning strikes to the ionosphere; such flashes are now commonly seen from orbiting vehicles [10].

He carried out high speed photography of lightning using, at first, a streak camera designed for the purpose by Charles Boys, who had not succeeded in obtaining records. It displayed the multiplicity and duration of the lightning strikes, and Allibone and Schonland [11] soon demonstrated the similarity of the lightning growth and streak photographs of sparks later obtained in the laboratory. The spatial growth of the lightning discharge, particularly as controlled by the physics of the stepped first leader and the upward discharge from ground, determines the probability of strikes to grounded structures. The statistics of the lightning current and its rate of rise, on the other hand, determine the severity of the resultant strike. Our knowledge of these statistics was largely established by the formidable work from 1946 to 1974 by Berger at Mt San Salvatore in Italy. These spatial and electrical properties, which are not independent, together comprise the risk of lightning damage by direct current injection to power lines and industrial plant, or by coupled surges in electronic systems.

Recent years have seen an acceleration in all aspects of lightning research, both experimental and theoretical, and in the techniques of protection, which will be described here.

3.2 Phenomenology of lightning

3.2.1 Characterisation of the flash

Data from Anderson and Eriksson, Berger, Gary, Uman and many others have quantified lightning parameter characteristics in respect of:

1   incidence of ground flashes
2   flash polarity
3   structure height
4   multiple strokes
5   flash duration
6   peak currents in first and subsequent strokes
7   current shapes.

3.2.2 Incidence

Substantial advances have been made since 1989 in the location and tracking of lightning, using radio direction finding technology. Before these developments, data on ground flash density N_g (flashes per square kilometre per annum) was primarily obtained from lightning flash counters triggered by lightning radiation fields, particularly with the CIGRE 10 kHz counter which is mostly responsive to cloud-to-ground flashes and which has an effective range of 20 km. For example, Anderson and coworkers [12, 13] aimed to relate the measured N_g to the commonly available data on T_d, the mean number of thunderstorm days in a given locality as evidenced by audible thunder. Some 400 counters over five years in South Africa suggested an approximate relationship

$$N_g = 0.04T_d^{1.25} \tag{3.1}$$

where T_d varied regionally between 4 and 80, corresponding to a range of ground flash densities between 0.2 and 10 flashes/km²/annum. In tropical areas such as Indonesia, the average T_d values are about 270. Lightning activity is infrequent for latitudes greater than 50°N or S, and recent orbital observations suggest that 90 per cent of flashes occur over land.

Surveys in the UK based upon observer reports over several decades have indicated N_g values in the range of 0.1–0.7 flashes/km²/annum with large seasonal and geographical variations within that range. Gary *et al.* [14] reported that the keraunic level T_d over France averages 20, but can be as high as 30 in the Alps, and CIGRE counter studies found N_g to range geographically between 2 and 4 flashes/km²/annum.

The number of ground flashes in a given region can be greatly increased by the frequency of strikes to tall structures. Eriksson [15] found that the incidence of upward-initiated flashes increased exponentially with structure height h. Structures of $h > 400$ m involved 95 per cent upward flashes (where the lightning flash was initiated by a leader from the structure). For $h < 100$ m, upward leaders caused initiation of only ten per cent of strikes. For 3000 flashes studied, Eriksson [16] quoted for the number of strikes per annum to a structure of height h:

$$N = N_g 2.4 \times 10^{-5} h^{2.05} \tag{3.2}$$

He determined that for structures of height greater than 100 m a significant number N_u of the strikes represented by Equation (3.2) were upward flashes:

$$N_u = N_g 3 \times 10^{-9} h^{3.53} \tag{3.3}$$

Many tall structures furthermore are located on elevated terrain, and these equations then much underestimate the risk. Diendorfer and Schulz [17] found the 100 m tower at Peissenberg, Austria (elevation 1287 m), to be struck 83 times in two years. This is likely to be the result of large thunderstorm fields at high locations rather than the effect of altitude on the flash (section 3.4.4.5), and all flashes were of the upward type [18].

Dependence on observer reports and lightning counter measurements has become almost obsolete in many countries since ground flash density data have become more widely available from real-time lightning location detection using networks of sensors. The online display of ground flash activity, usually with only seconds of delay, and with a capacity to record up to 100 flashes per second, provides economic advantages for the scheduling of alternative power routes and protecting vital installations in, for example, the telecommunications, petrochemical, defence and aviation industries. The UK Lightning Flash Location System (EA Technology) [19] has monitored cloud-to-ground flashes since 1989. The four-loop detection system installed at six recording stations in the UK and Ireland can discriminate and locate ground flashes down to 3 kA and to within a claimed 100 m. The important capability of discriminating cloud-to-ground strikes from intercloud flashes is achieved by utilising the low attenuation of the mainly vertically-polarised radiation associated with the ground strikes. At 1.1 kHz, radiated waves in the earth-to-ionosphere waveguide are effectively recorded.

Although the data from such stations generally confirm overall levels (Figure 3.1), they give much higher localised values of N_g than had been expected from earlier observation and detection methods, with up to eight ground flashes/km^2/annum in some areas of the UK. One flash in three is found to strike ground. The shape and amplitude of the signals also give important indications of peak current magnitude and flash polarity.

The National Lightning Detection Network (Global Atmospherics Inc) [20, 21] employs 107 stations over the 10^7 km^2 of the USA with a satellite link to calculate the time, location, polarity and current magnitude of each flash. Resolution of component return strokes is also possible. This system uses IMPACT sensors which combine magnetic field direction finding with time-of-arrival (TOA) sensors (these latter comprising the LPATS Lightning Position and Tracking System). TOA measurement is particularly useful at long range. VHF TOA systems also allow three-dimensional reconstruction of lightning paths. In Japan a SAFIR VHF-DF network detects total lightning (cloud–cloud and cloud–ground).

Since 1989 an average of 20 million ground strikes per annum, corresponding to a mean value of $N_g \approx 2$, have been recorded in the USA with a location best accuracy

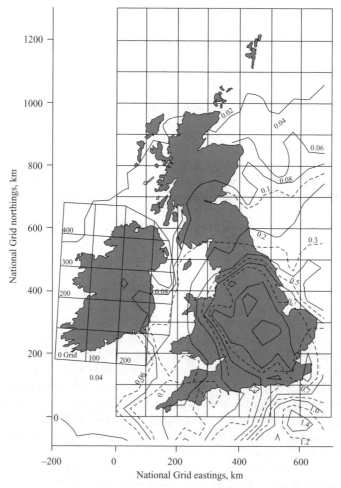

Note 1 This lightning density map was compiled by E.A. Technology Ltd.
from data accumulated over 10 years

Note 2 A linear interpolation should be used to determine the value of the
lightning flash density, N_g, for a location between two contour lines

Figure 3.1   Ground flash density in the UK (BS 6651:1999)[1]

of about 500 m. In a recent development, new sensors enabling simultaneous detection
of both cloud–cloud and cloud–ground flashes provide an earlier warning of active
storms. These new radio detection techniques provide not only vital contemporaneous
information, but also the considerable benefits of a growing archive of lightning
statistics for risk analysis.

A recent addition to global lightning observations derives from the development
of the NASA optical transient detector (OTD) [22]. This detects lightning events
from low orbit, sweeping a given surface location three times every two days. The

detection efficiency is about 60 per cent, with no discrimination between cloud flashes and cloud-to-ground strikes. A study using the OTD system in combination with the US ground sensor network [23] has enabled estimates to be made for the first time of the ratio Z of cloud flashes and ground strikes. The results give $Z = 2.94 \pm 1.28$ (standard deviation), but local values can be as high as 9. Low Z values of unity or less are often observed in mountainous regions, where this high probability of ground strikes is perhaps to be expected. The orbiting OTD records of total lightning flash density $N_c + N_g$ reached 30 flashes/km^2/annum or higher.

3.2.3   Polarity

Flash polarity is of relevance for two reasons. In the first place, field studies suggest that the highest peak currents are associated with positive flashes. Second, analogies with the long laboratory spark, and some generic models of the mechanism of strikes to grounded structures (section 3.4.4), both indicate that the protection of structures from positive direct strikes with standard air terminations may sometimes be much less effective than for the more common negative flash. For the same reason, the probability of a direct strike to a structure may be lower because of a smaller attractive area for positive flashes (section 3.4.5). This may also have led in the past to an underestimate of the ratio of positive to negative flashes.

Before recent lightning tracking networks were developed, the flash polarity could be deduced, even in the absence of lightning current recording, by means of magnetic links on transmission lines and structures and from the polarity of transient local electric field changes. Most of these results suggested that a minority (12 per cent) of downward flashes is of positive polarity. Over 29 years of observation, Berger [24] recorded 1466 negative flashes and 222 positive flashes. At Peissenberg, Fuchs [18] found that 95 per cent were of negative polarity. There is evidence that winter storms in some regions (e.g. Japan and northeast USA [25, 26]) or lightning to high altitude locations may more commonly result in positive flashes. In the last decade, lightning location networks have improved the statistics, with in addition peak current estimates of 50 per cent accuracy. UK data shows that a surprising 30 per cent of flashes are positive, with peak currents up to 250 kA. Geographical variations are also apparent. In the northern and western UK, positive flashes are more frequent than negative; the opposite is true for southern England [27]. The OTD/ground network studies in USA [23] show a correlation between the high Z values (indicating a low ground strike ratio) and a high positive-to-negative flash ratio. The authors suggest that a high altitude cloud dipole, together with an enhanced lower positive charge centre, could account for this. The positive strikes were associated with high ($\approx$275 kA) peak currents.

It is notable that bipolar lightning, where successive return stroke currents are of opposite polarity, is not uncommon in both natural and triggered flashes. This probably indicates the involvement of different cloud charge regions in a multistroke flash.

3.2.4   Flash components

The total flash duration incorporating all the individual strokes of a flash is particularly important in protection design for power systems, since multiple strokes to an overhead line can prejudice the overvoltage protection based upon autoreclose switchgear or surge arresters. Restrikes can be caused in switchgear and excessive power dissipation in arresters by long duration overvoltages.

Schonland and Collens [7] showed from electric field change measurements that the flash often comprised multiple successive strokes. The mean strokes/flash value was 2.3, and the points of strike to ground of individual strokes may be significantly displaced (section 3.2.8). They found probability values of:

$$P(\text{single stroke}) = 45\%$$
$$P(> 10 \text{ strokes}) = 5\%$$

(3.4a)

More recent studies by Rakov *et al.* [28] find that 80 per cent of negative flashes contain two or more strokes, but that few positive flashes do so.

In Schonland's work the mean flash duration was estimated to be 200 ms, and:

$$P(< 64 \text{ ms}) = 5\%$$
$$P(> 620 \text{ ms}) = 9\%$$

(3.4b)

3.2.5 Peak current

The basic knowledge of peak lightning currents i_0 from direct measurements is owed mainly to Berger [24], Anderson *et al.* [13, 29], Garbagnati and Piparo [30], Diendorfer *et al.* [17, 31] and Janischewskyj *et al.* [32]. Again, the function of modern lightning tracking systems allows new, more extensive semiquantitative indications of current magnitudes.

Berger's measurements of the peak current probability distributions for return strokes preceded by negative stepped and dart leaders, and for positive lightning, were of log-normal form and ranged from a few kA to above 100 kA with a median value of about 30 kA. As pointed out by Berger himself, his 70 m towers at 914 m altitude often initiated negative lightning by upward positive leaders of at least 1 km in length; positive lightning almost always began with an upward stepped (negative) leader, so that the peak current values for his recorded strikes might differ from normal downward flashes. In Diendorfer's measurements, the median peak currents to the tower were 12.3 kA, compared with 9.8 kA to the surrounding terrain (both values obtained from radiated signal analysis). Direct measurement of lightning currents is also feasible from rocket-triggered flashes. However, such lightning strikes are again initiated by upward leaders and are thus atypical of natural downward flashes [33, 34].

When structures of height 60 m or less are struck, this is usually the result of the approach of a downward leader to within the striking distance before the upward leader is launched. The number of analysable flashes is consequently not large. Anderson and Eriksson [12] have examined a worldwide sample of 338 flashes, where over a third of these were obtained by Berger at Mt San Salvatore. The overall median peak current in the first stroke of a flash was 34 kA, and for high current flashes:

$$P(> 100 \text{ kA}) = 2.5\%$$

(3.4c)

The current in subsequent strokes of a multiple flash was about 60 per cent lower than this.

Anderson and Eriksson point out that the statistical distribution of peak currents below 20 kA can be well represented by a log-normal distribution, with a second, narrower log-normal distribution for higher currents. The practical merit of this approach in respect of overhead line protection is that the lower current flashes can be identified with the risk of shielding failure, whereas high current flashes may cause backflashovers. In the UK Standard BS 6651, the cumulative probability range is taken to extend from 1 per cent for currents to exceed 200 kA to 99 per cent for currents to exceed 3 kA.

Data from the EA Lightning Location System in the UK [35] estimate current magnitude and polarity. A signal strength scale of 1 to 5 is used, corresponding to a range of current from less than 10 kA to more than 80 kA. There is a possibility that the bimodal log-normal distribution derives from separate log-normal distributions for each polarity. The observed positive flash majority in Scotland has a median peak current of 80 kA, whereas the more common negative flashes in southern England have peak currents of 12–15 kA [27].

The electric field component of the radiated signal will have a magnitude proportional to the peak current and the return stroke velocity, and inversely proportional to the distance of the sensor from the flash. Errors in estimated currents will arise from propagation losses and uncertainties in relation to the return stroke process. The precision is judged to be about ±25 per cent.

Current measurements in USA during 1995–1997 gave median values of peak current of between 19 kA and 24 kA, depending upon the geographical region [36]. The network has a low detection efficiency for flashes with a peak current below 5 kA. This has been confirmed by Diendorfer *et al.* [31], from direct comparisons of strike currents to their tower and simultaneous records from the ALDIS location system in Austria.

On the basis of recent data from such systems, Darveniza [37] has suggested that a median current of about 20 kA is more appropriate than the long accepted values of 31 kA (CIGRE) or 33 kA (IEC).

3.2.6 Current shape

Data on the temporal characteristics of the return stroke current are largely owed to the work of Berger, with additional estimates of current front duration from the USA [38]. Berger's risetimes were of median value 5 μs for the first stroke current; for subsequent strokes steeper fronts of up to 100 kA/μs were measured. More recent values from the literature are 9–65 kA/μs (median 24 kA/μs) for the first stroke, and 10–162 kA/μs (median 40 kA/μs) for subsequent strokes, and it is in the nature of extreme value statistics that as new research becomes more intensive, higher maximum values are recorded. BS6651 [39] gives (di/dt) max $= 200$ kA/μs.

The rate of change of current measured by Berger in the first and later strokes of the flash may be represented by:

$$\frac{di}{dt} = 3.9 i_0^{0.55} \quad [\text{kA}/\mu\text{s}] \qquad \text{(first stroke)} \qquad (3.5)$$

$$\frac{di}{dt} = 3.8i_0^{0.93} \quad [kA/\mu s] \qquad \text{(subsequent strokes)} \tag{3.6}$$

Berger's measurements also provide mean values for the current tail duration and its statistical range:

first stroke: mean 75 μs, range 30–200 μs

later strokes: mean 32 μs, range 6.5–140 μs.

Up to half of all ground strikes also show a follow-through continuous current component of about 100 A which can persist for some hundreds of milliseconds. The long continuous current in the positive leader initiating an upward flash from a tall mast can carry several kiloamperes for tens of milliseconds [18].

3.2.7 Electric fields

3.2.7.1 Field below the thundercloud

The electric field at ground level is observed to change from the fair weather value of about $+130$ V/m (created by the -0.6 MC negative charge on the earth) to a maximum -15 or -20 kV/m below the bipolar charge of a thundercloud, the negative charge centre being at a lower altitude (about 5 km) than the main positive charge (10 km). These are large charges of about ±40 C, which create a cloud to ground potential of the order of 100 MV. There is frequently a smaller low altitude positive charge at about 2 km that probably resides on precipitation from the upper cloud and is sometimes responsible for positive downward leader initiation.

The negative thundercloud field has been shown from balloon measurements to increase linearly from ground level at a rate of about -0.1 kV/m^2 to reach, for example, a value of about -65 kV/m at 600 m altitude [40, 41]. This increase arises because positive corona with a current density of up to 10 nA/m^2 from grounded objects creates a screening effect from positive space charge and an intensification with increasing height. An extensive region of charge density of about $+1$ nC/m^3 is responsible. The screening effect is thought also to be the cause of very long (10 km) horizontal leaders before a final leader-to-ground jump. Horizontal leader channels are commonly observed in Japanese winter storms; the effect is clearly relevant to the side flash risk.

There has also been interest in the localised strike inhibition effects possibly associated with multipoint air terminations. This arises from the generation in time t, from a termination of radius R_0, of an expanding, spherical, positive ion cloud of charge Q, mobility k and an increasing radius:

$$R(t) = \left(1 + \frac{3kQt}{4\pi\varepsilon_0 R_0^3}\right)^{1/3} R_0 \tag{3.7}$$

The inhibition effect of a positive space charge layer would be subject to wind dissipation, and the retardation of a positive upward leader from a ground structure is not yet quantified [42, 43].

The increase of field with altitude, together with the concentration of field at tall structures, accounts for the increased frequency of lightning strikes at such structures, especially in mountain environments, and to rockets and aircraft. At the St Privat d'Allier research centre of EDF, upward positive lightning leaders were initiated from a 100 m rocket wire when the ground field value was −10 kV/m. There is some doubt of course whether measured characteristics well represent natural lightning. In 1988 tests at Kennedy Space Center showed upward positive leader growth from the rocket wire for several milliseconds before a downward negative stepped leader was triggered. Nevertheless, despite these reservations, simultaneous current and electric and electromagnetic field data can be accumulated which are impossible to obtain from natural lightning. In Florida electric field changes have been recorded within tens of metres of the flash [28].

3.2.7.2   Electrostatic and electromagnetic field changes during the flash

Although measurement of radiated electric field is useful for peak current estimates (section 3.2.2), close-in (within 20 km) electric field transient changes are more informative for the study of the development of the flash. This is especially so when combined with time-resolved photography of the flash, as in the work of Schonland and Berger. These field changes have been measured by rotary field mills, with a resolution time of 0.1–1.0 ms, or fast response capacitive probes with buffer ampli-fiers. The main characteristics of the field changes observed for negative lightning are shown in Table 3.1.

These field change studies, with synchronised photography, have provided much detail on spatial growth. The triggered lightning data of Rakov *et al.* [28] on R and M changes have allowed physical modelling of conduction in the return stroke.

At the present time, the vulnerability of electronic equipment to LEMP (lightning electromagnetic pulses) produced close to lightning strikes is of much concern (section 3.5). Fuchs [18] found the maximum values of fields and their rates of change at a distance of 200 m from the Peissenberg tower to be 30 kV/m, 84 kV/m/μs,

Table 3.1   Electrostatic field changes for negative lightning

Nomenclature	Source	Character
L	downward or upward leader of first or subsequent strokes	close to leader: gradual negative-going field change at distance: gradual positive-going change charge on leader 3–20 C
R	first and subsequent return strokes	positive step field changes
C	continuous current component	slow positive change
M	continuous current component	positive pulse transients
J	interstroke intervals	slow positive change
K	interstroke intervals	positive pulse transients

−35 A/m and −237 A/m/μs. Concerning close-in magnetic fields, Schnetzer *et al.* [44] found induction by the displacement current density $\varepsilon_0 dE/dt$ to dominate for distances beyond 50 m from the flash. Close-in electric fields are much amplified by the radiation from the return stroke. Calculations by Schwab and Cvetic [45] of the magnetic potential derivative $-dA/dt$ in the region of the insulators on a 110 kV tower show these fields to be 50 to 120 kV/m, larger than the backflashover stresses arising from return stroke current in the tower surge impedance (section 3.5.5).

3.2.8 Spatial development

The spatial growth of the lightning flash was described by Schonland and his colleagues [7, 46, 47] (Figure 3.2). The overall behaviour of a negative strike to ground, as summarised by Berger [24] is that in most cases a down-coming first negative leader is initiated at an altitude of about 5 to 10 km, advancing apparently discontinuously in steps of 10–200 m length (mean value 25 m) at intervals of 10–100 μs (mean 50 μs), and establishes a path for the ensuing upwardly-directed first return stroke. The mean downward negative leader velocity is 0.1–0.8 m/μs, but Schonland could not measure the velocity of the stepped advance, which exceeded 50 m/μs.

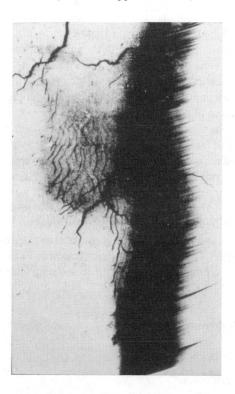

Figure 3.2   *Spatial development of the negative downward flash (courtesy of 'Progressive lighting: A comparison of photographic and electrical studies of the discharge', by B.F.J. Schonland and H. Collens, Proceedings of the Royal Society. Series A, vol. 143, 1934, pp. 654–674)*

These apparent (two-dimensional) velocities may be about 30 per cent lower than the real velocities. Because the transient local electric field changes recorded during the negative leader growth showed no marked stepping, this suggested to Schonland that the majority of the negative charge lowered by the leader was transported during the intervals between the visible leader steps. He postulated that the stepped first leader was itself preceded by a non-visible pilot leader which progressed continuously rather than in steps; laboratory evidence for this is discussed in section 3.3.1.6. The downward negative leader visible below the cloud base is undoubtedly accompanied by an upward-directed leader, from the same negative charge centre of the cloud, that is not observable from ground level but whose development can be modelled [48].

If the downward leader approaches within striking distance of the position of a prospective earth termination, an ascending positive leader of a velocity in the range 20–60 mm/μs is launched from that position. Such low velocities are also found in triggered upward leaders [49] and in cloud flashes [50], which suggest propagation at low electric field. After the formation of the bright return stroke by the junction of the descending and ascending leaders, the velocity of the upwardly-directed luminous front along the leader track increases from 15 to 150 m/μs (0.05*c* to 0.5*c*, where *c* is the light velocity) with increasing peak current. After a delay of about 50 ms, further component strokes of the flash may occur along the same path for most of its length, and for these strokes a continuous downward dart leader with a velocity of 1–20 m/μs precedes its return stroke. Multiple stroke flashes can be of one second duration.

From the point of view of estimating the hazard associated with direct strikes to earth, the risk factor (section 3.5) is significantly increased in the light of observations that more than one ground termination may occur in multiple stroke flashes. This results from the creation of a new leader path from some point along the downward dart leaders that precede the second and later strokes of a flash. Ishii *et al*. [50] used a fast antenna network to locate ground termination points to within 500 m, and found that for negative flashes the average distance between stroke terminations was 2 km (interstroke time intervals 15–130 ms) and for positive flashes 13 km (30–190 ms).

For mountain environments [18], tall structures [51] or airborne objects [52], the negative flash is frequently initiated by an upward positive leader. In these situations, the less common positive polarity flash is also observed to begin with an upward negative leader, which shows all of the stepping characteristics of the negative downward leader [24]. Berger's rare records of a positive downward leader to a low altitude termination showed a continuous growth with no stepping characteristic.

3.3   Physics of lightning

3.3.1   *Long sparks in the laboratory*

3.3.1.1   Leader mechanisms

The physical similarities between the small-scale electric spark and natural lightning inspired Franklin's exploration of these phenomena and led to his innovative safety measures. Modern research continues to use controlled laboratory tests for both physical studies and improvements in the design and protection of electronic and power plant against both natural and internally generated voltage and current surges [53, 54]. These advances have indicated answers to three basic questions about the lightning flash:

(i)   How can a lightning leader develop in electric fields as low as 50 kV/m?
(ii) What is the mechanism of the stepped negative leader?
(iii) Why does an upward leader initiate from a grounded structure when a given striking distance is achieved?

3.3.1.2   Spark leaders

The process that establishes a long discharge path in the atmosphere by the longitudinal growth of a leader channel was revealed first in the lightning flash. The work of Allibone and Meek [55] made a detailed study of the leader channel that also preceded the laboratory spark, using a moving film camera in order to obtain time-resolved streak photographs analogous to the rotating lens Boys camera lightning records. By the use of resistive control of the impulse generator, a step-like progression of the spark leader could be created. This current limitation of the bright full sparkover path also enabled the fast, low luminosity leader phase to be discerned on the streak picture. Waters and Jones [56, 57] avoided the difficulty of the masking effect of the bright phase by employing a high aperture (f/1.0) reflector lens, equipped with a rotating mirror synchronised with the impulse generator, to photograph arrested leader strokes in which the impulse voltage was chopped before full sparkover. With no circuit limitation of current, the leader progressed continuously until sparkover or its prior arrest.

These laboratory tests were usually made with positive impulses, since the positive leader is more easily initiated than the negative. The minimum breakdown voltage for a 2 m gap with a positive impulse is about 800 kV. However, very long gaps of up to 32 m show a rate of increase of breakdown voltage with gap length of only 55 kV per additional metre [58]. For gaps of some metres the negative polarity breakdown voltage is significantly greater than the positive value, but recent tests by Mrazek [59] for gaps of up to 50 m showed that the polarity difference disappears in long gap sparkover.

It is the ratio of electric field to air density within the leader channel that is the important parameter to maintain ionisation, so that the field of 500 kV/m in the positive streamers ahead of the leader can fall to the 55 kV/m in the long leader if the gaseous heating of the leader channel can result in such a density reduction. The leader channel is distinguished from the cold plasma in the streamer filaments that precede it by its temperature elevation to 1000–3000 K. Leader growth in long gaps can then resemble closely the continuous growth in lightning. In 50 m gaps Mrazek observed the negative leader to progress in 3.6 m steps at intervals of 24 μs (mean velocity 0.15 m/μs).

3.3.1.3   Leader gradient

The electrical gradient within the leader channel is influential on the stability of leader propagation. Successful growth of the spark leader requires that its tip potential remains high enough to support corona ionisation ahead of itself, and that the conductivity of the established leader channel is maintained. Waters and Jones [56] showed that the positive leader (without external current limitation) is accompanied by an extensive corona discharge at the leader tip. A deceleration of the leader to about 10 mm/μs occurs that might result in its extinction if the applied voltage is insufficient. Otherwise an acceleration to 100 mm/μs or more bridges the gap and leads to the arc phase that is analogous to the lightning return stroke.

The work of the Les Renardieres Group [60–62] in 10 m gaps using electric field probes has enabled the mean gradient g over the length of the leader to be deduced. It is clear that as the length of the leader grows, the mean gradient within the channel decreases. For example, in a 10 m gap at critical sparkover (1.8 MV), the mean gradient along the leader is 0.5 MV/m when the leader is 2 m long, and this falls to 0.18 MV/m when it has grown to 5 m. This decreasing gradient characteristic means that the potential at the tip of the leader varies little during its development, which is consistent with the uniform velocity and current (15–20 mm/μs, 0.6–1 A) that is observed in this case. It is also, of course, the underlying cause of the non-linear breakdown voltage and the falling breakdown-gradient/gap-length characteristic of long sparks.

All the recent theories of the leader channel attempt to account for this negative gradient characteristic, which is also the key to understanding how the lightning leader of tens of kilometres length is able to maintain its conductivity. These theories need a quantitative knowledge of the leader diameter as a function of current and time. As in lightning, measurement of this diameter by direct photography is difficult because of the limited resolving power of this method. This problem in long sparks has largely been solved by the use of strioscopy (Schlieren photography), combined with high speed image recording. Ross [63] found that the leader channel thermal boundary was so clearly defined by this technique that the diameter could be measured with a precision of ±0.1 mm (Figure 3.3). The diameter of a given section of the channel was found to increase with time. No strong shock wave was observed under these conditions, and the rate of radial expansion was less than 100 m/s in all cases. Later work [61] shows that a pressure wave is generated from the leader tip, but confirms that the subsequent radial expansion of the channel is subsonic. Figure 3.4 shows, via a 0.66 mm slit, part of a 2.2 m gap. A sweep-mode image converter time-resolves the image in the axial direction, showing the birth of the leader with an associated pressure wave, the channel diameter growth at 18 m/s and the instant of sparkover with a strong shock wave resulting from the arc channel expansion. This indicates that the gas pressure within the leader channel remains constant at the ambient value throughout its temperature rise. The leader diameter typically increased in 10 m gaps from an initial value of about 1 mm or less to about 5 mm for a spark leader of 6 m length just before sparkover.

The expansion of the leader channel and the decreasing axial gradient characteristic are related parameters. The absence of radial shock waves and the nature of the strioscopic records of the channel show that the neutral-particle density within the channel is a decreasing function of time, since a constant mass of gas is involved. This reduced density is the basic explanation of how the leader channel conductivity can be

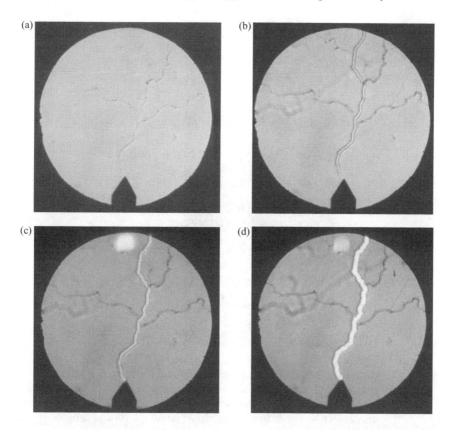

Figure 3.3   *Leader channel expansion in the long spark [46]. Schlieren photographs of exposure time 1 μs per frame. Field of view 100 mm diameter in a 1.5 m rod–plane spark gap. Successive frames at*

   a   5.2 μs
   b   8.1 μs
   c   15.2 μs
   d   36.5 μs

sustained at low overall electric fields. At atmospheric density a field of about 3 MV/m is necessary to maintain direct-impact ionisation by electrons, but an expansion of the leader radius by a factor of four will reduce the required field to below 0.19 MV/m. This approaches the average field available in long gap sparkover (the critical impulse sparkover voltage for a 10 m rod–plane gap is 1.8 MV).

3.3.1.4   Energy dissipation and storage in the leader growth

The observed current shows that at critical breakdown the mean power input to the leader growth is about 1 MW. Some of the energy supplied, amounting under these

130/3000 µs

$T_B = 60.2\,\mu s$

Figure 3.4   Shock wave generation at the leader head (Gibert). Schlieren record of section of vertical spark channel viewed through a horizontal slit. Image swept downwards by an electronic streak camera. Conical image shows leader expansion to time of sparkover at 60 µs. Radial shock waves are visible

conditions to about 40 J/m extension of the leader, is stored in the redistributed electrostatic field, and most of the remainder is utilised in discharge processes. This division of energy was studied by examination of the space charge in the gap using a rotating fluxmeter [61], which indicated an approximate equipartition of energy. Much of the expended energy causes heating and expansion of the gas; calibrated photomultiplier measurements show that very little energy is lost by radiation.

3.3.1.5   Physical theories of the leader channel

Kline and Denes [64] considered the influence of the gas density on the electric gradient at which ionisation and electron attachment rates become equal. Extrapolation of their estimates to a hot gas column at 5000 K gives a value for the product ET of this critical electric gradient and temperature of 5×10^8 VK/m. Two separate theories suggest that the value of ET at each point in the expanding leader remains approximately constant. In the approach of Ross *et al.* [63], it is noted that the constant current condition that is observed in the leader channel at critical breakdown implies:

$$\left(\frac{n_e}{n}\right) v_e = \text{constant} \tag{3.8}$$

where n_e, n are the electron and neutral particle volume densities at any instant during leader growth, and v_e is the electron drift velocity. This equation supposes that the mass of gas within the expanding leader channel remains constant. Now the equilibrium electron density in the leader, with a positive ion density of n_i, is given by the condition:

$$k_1 n n_e - k_2 n^2 n_e - k_3 n_i n_e = 0 \qquad (3.9)$$

where k_1, k_2 and k_3 are, respectively, the rate coefficients for ionisation, the three-body collisions that control the electron loss process of attachment to form negative ions and electron–ion recombination. Since the leader channel can certainly be regarded as a quasi-neutral plasma with $n_e \approx n_i$, we have:

$$\frac{n_e}{n} = \frac{k_1 - k_2 n}{k_3} \qquad (3.10)$$

If the ratio of electric gradient to neutral gas density in the leader E/n were to increase as n decreases because of channel expansion, then the ratio n_e/n would also increase strongly. This is because with the channel expansion k_1 is much increased, and k_2 and k_3 are decreased. Furthermore, the electron drift velocity v_e increases approximately linearly with E/n in the range of interest. An increasing value of $n_e v_e/n$ would imply an increasing leader current with time, which is contrary to Equation (3.8). The constant current condition therefore indicates a constancy of E/n and ET. If we define a leader constant K of value:

$$ET = K = 5 \times 10^8 \text{ VK/m} \qquad (3.11)$$

and assume a constant specific heat, then the rate of rise of temperature in any element of the leader is:

$$\frac{dT}{dt} = BEi_L \qquad (3.12)$$

where B is also constant and i_L is the leader current. Combining these equations:

$$i_L = -\frac{K/BE^3}{dE/dt} \qquad (3.13)$$

Under critical breakdown conditions, where the leader current and velocity $v_L = dL/dt$ are constant, the velocity is:

$$v_L = -\frac{K/BqE^3}{dE/dt} \qquad (3.14)$$

where q is the injected charge for unit leader length.

   If the initial gradient is taken as 0.5 MV/m (corresponding to the initial condition in the positive streamer corona) and an initial leader radius of 1 mm, then by integration the electric gradient E can be calculated at the position $x = 1$ m from the leader origin. Table 3.2 shows the variation in both the gradient and radius a during the leader growth from a length $L = 1$ to 7 m.

Table 3.2   Leader gradient E with increasing leader length L at a position 1 m along channel

L (m)	1	2	3	4	5	6	7
E (MV/m)	0.39	0.28	0.23	0.10	0.18	0.17	0.15
a (mm)	1.0	3.8	4.6	5.0	5.4	5.7	6.0

Gallimberti [61] has described a hydrodynamic/thermodynamic analysis of the expansion of the leader channel, again based upon the expansion of a constant mass of gas within the channel. It quantifies such physical considerations as the conservation of neutral particles, heat transport at the boundary, energy transfer from charge carriers to neutrals, ambipolar diffusion and a radial distribution of gas temperature. It is possible to obtain, as a function of time, the rate of the channel expansion, the increase in its temperature and the weak shock generation at the leader boundary. The value of ET within the channel again remains constant, and the same value of 5×10^8 VK/m emerges.

This relationship between electric gradient and channel temperature is a useful concept for lightning modelling, but gas temperature estimates for the lightning leader channel are obviously tentative. Orville [65] quotes the work of Maecker, who derived from spectrosopic analysis a temperature of about 1500 K for this phase and concluded also that the leader was at atmospheric pressure. This would be consistent with an electric gradient $E = 0.3$ MV/m which is comparable with long spark leaders (section 3.3.1.2).

3.3.1.6   Space leaders in the negative leader channel

The negative leader channel is easily observed in air gaps such as rod–rod electrodes, where the positive leader from the anode preionises the gap with an active high current corona and promotes the formation and growth of a negative leader from the cathode. The negative leader can be initiated without the assistance of an associated positive discharge only for gaps over 3 m in length, and has been studied for gaps up to 7 m at Les Renardieres [62]. The negative leader, like its lightning counterpart, progresses in steps. The detailed structure of this progress in the negative polarity rod–plane gap, which is almost certainly present but is inaccessible in lightning photographs because of its low luminosity, throws much light on the probable stepping mechanism of the negative lightning leader:

(i)   A negative long spark leader grows continuously with a velocity of 10 mm/μs and a current of about 1 A, accompanied by regular fast bright elongations, current pulses of up to 100 A and large negative corona bursts. These are seen in time-resolved streak mode (Figure 3.5a) and framing mode (Figure 3.5b) photographs, interpreted in Figure 3.5c.

(ii) Between these elongation events, one or more space leaders form well ahead of the negative leader tip, connected to the main leader only by corona streamers (as in the static photograph of Figure 3.5d). As represented schematically in Figure 3.5e, the extension of each successive space leader takes place at

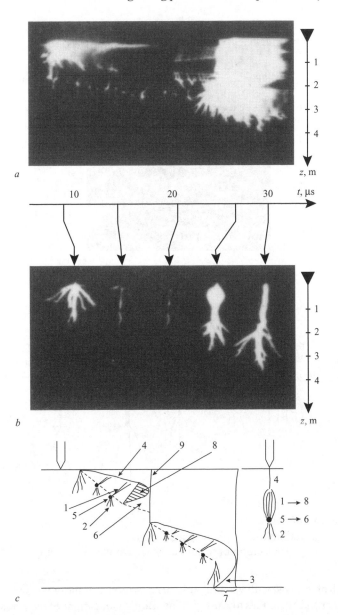

Figure 3.5   *Negative space leaders in the 7 metre spark (Les Renardieres)*

a Image-converter streak display; b Image-converter frame display
(Courtesy of IEE Proceedings, Vol. 125, 1978, p. 1160);
c Streak display interpretation (Courtesy of IEE Proceedings,
Vol. 125, 1978, p. 1160): 1 retrograde positive corona; 2 negative
corona; 3 upward positive leader; 4 downward negative leader;
5 space stem; 6 space leader elongation; 7 final jump duration;
8 leader corona region

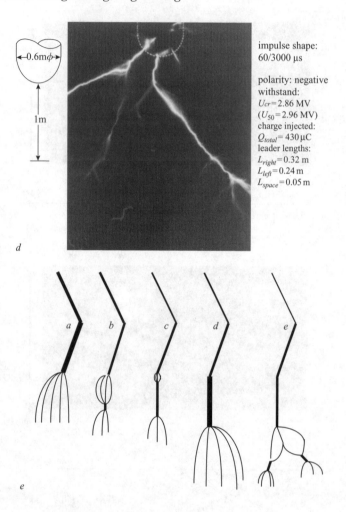

Figure 3.5   Continued.
    d   still photograph of space leaders
    e   visualisation of negative leader growth in lightning

both of its ends, with velocities of 30 mm/μs towards the downcoming negative leader and 10 mm/μs towards ground. Bright elongations of the whole length of the negative leader occur whenever it connects with the space leader(s) ahead of its tip.

(iii)   New space leaders then form in the new corona burst from the tip of this elongated channel. The leader current does not show any significant discontinuities during this stepped leader growth.

This unanticipated space leader mechanism was first reported by Stekolnikov and colleagues [66, 67], and is easily overlooked experimentally, since the sudden elongation (i) is a much brighter phenomenon than the space leaders (ii). The local gas heating which triggers the formation of space leaders within the negative corona is a stochastic process, and their existence is clearly relevant to any model of the negative leader in both the long spark and lightning. The transformation of the corona discharge ahead of the down-coming leader into a space leader will determine channel tortuosity, branching and the path taken in the later stages of the lightning strike to earth. It is clearly possible to draw a parallel between the pilot leader, postulated by Schonland to precede the lightning stepped leader in order to account for the absence of large step field changes, and the space leaders ahead of the laboratory negative leader.

3.3.2 Lightning leader propagation

3.3.2.1 Correlation of leader channel charge, current and velocity

In the same way that the electric gradient and temperature in the long spark leader informs physical studies of the lightning leader channel, so the mode of propagation can be expected to develop from similar ionisation processes ahead of the leader in both cases. The current in the lightning leader is several orders of magnitude smaller than the tens of kiloamperes in the return stroke that neutralises the leader charge. Petrov and Waters [68] showed that a comparison of the charge q per unit length of the leader and the leader current is possible by considering the ionisation at the head of the high temperature leader channel itself. By representing the head of the leader to have a simple circular face, then within its radius r_0 a charge density ρ is maintained by the leader current i_L, so that:

$$i_L = \pi r_0^2 \rho v_L \tag{3.15}$$

where v_L is the leader velocity, and for current continuity:

$$i_L = q v_L \tag{3.16}$$

The velocity v_L depends upon ionisation processes at the surface r_0, and for propagation along the x axis:

$$v_L = \left[\frac{dx}{dt}\right]_{r_0} = \frac{[dn_e/dt]_{r_0}}{[dn_e/dx]_{r_0}} \tag{3.17}$$

If n_0 and v_0 are the values of electron density n_e and drift velocity v_e at the leader tip surface r_0, the current density at that surface is:

$$j_0 = e n_0 v_0 \tag{3.18}$$

The continuity condition at r_0, taking account of the ionisation coefficient α (photoionisation neglected), is:

$$\left[\frac{dn_e}{dt}\right]_{r_0} = \left(\frac{1}{e}\right)\left[\frac{dj_0}{dx}\right] + \alpha n_0 v_0 \approx \frac{n_0 v_0}{r_0} + \alpha n_0 v_0 \tag{3.19}$$

for the approximation $dj_0/dx \approx j_0/r_0$.

If the electron density behind the leader head is considerably smaller than n_0, then we can also approximate:

$$\left[\frac{dn_e}{dx}\right]_{r_0} = \frac{n_0}{r_0} \tag{3.20}$$

Then the leader velocity is:

$$v_L = v_0 + \alpha v_0 r_0 \tag{3.21}$$

In the intense field at the leader head, α can become very large and the leader velocity v_L can greatly exceed the electron drift velocity. This allows the further simplification:

$$v_L \approx \alpha v_0 r_0 = v_i r_0 \tag{3.22}$$

where v_i is the ionisation rate coefficient. These approximations lead to unique basic relationships between current i_L, velocity v_L and linear charge density q for the lightning leader:

$$i_L = \frac{\pi \rho v_L^3}{v_i^2} \tag{3.23}$$

and

$$q = \left[\frac{\pi \rho}{v_i^2}\right]^{1/3} i_L^{2/3} \tag{3.24}$$

Here ρ is the charge density in the leader head that is necessary to maintain its propagation [68]. This will be approximately constant during the leader propagation, so that it is possible to deduce the important proportionalities:

$$i_L \propto q^{3/2} \propto v_L^3 \tag{3.25}$$

Plausible values can be obtained from the model on the basis of laboratory measurements. The ionisation frequency is typically $v_i = 5.6 \times 10^6$ per second and the leader head charge density has been estimated as $\rho \approx 1$ C/m^3. These values give:

$$q \approx 46 \times 10^{-6} i_L^{2/3} \approx 10^{-13} v_L^2 \quad \text{[C/m, A, m/s]} \tag{3.26}$$

Table 3.3 gives numerical examples of this leader propagation model, together with return stroke calculations (section 3.3.2.2).

A downward leader length of say 10 km would lower a total charge of 3.9 C during a stroke at the median current of 31 kA. This agrees well with lightning observations. The leader velocity of 0.062 m/µs is low compared with the slowest observed velocity of 0.1 m/µs. This would be consistent with the formation of the leader step by the simultaneous growth of two or more space leaders in series ahead of it (so doubling or more the velocity). The median leader current of 25 A in Table 3.3 compares with the 40 to 50 A range estimated by Mazur and Ruhnke [48].

Equation (3.26) also scales down effectively to long laboratory sparks. During the stage of stable leader development in a 10 m gap, the measured leader current is 0.8 A, the propagation velocity 18 mm/µs and the charge flow 45 µC/m. Equation (3.26)

Table 3.3   *Leader current, velocity and charge with associated return stroke current and velocity*

i_0 (kA)	5	31	100	300
v_r	0.14c	0.26c	0.39c	0.55c
q (mC/m)	0.12	0.39	0.86	1.80
i_L (A)	4.1	25	82	245
v_L (m/μs)	0.035	0.062	0.092	0.13

for $i_L = 0.8$ A gives 20 mm/μs and 40 μC/m. This accord is important to counter the frequently expressed doubt that the laboratory discharge can represent adequately the lightning event.

3.3.2.2   Return stroke current

Although the properties of the leader channel can be used to model the strike process to ground structures (section 3.4), the consequential damage caused by lightning is strongly current dependent via the so-called action-integral $\int i_0^2 dt$ of the return stroke. For this reason, the probability density of the lightning peak current distribution has been extensively studied in the field. The connection between the properties of the leader development and the prospective peak current i_0 in the lightning stroke is not known; however, it is possible to relate the much smaller current in the leader phase i_L to the return stroke current i_0 by noting that the return stroke results in the neutralisation of the leader channel charge q. Much of this charge certainly resides in the ionised region surrounding the channel, as a result of the corona discharge ahead of the leader tip propagation [69, 70]. Using the simplest of assumptions that both charge density and velocity are constant along the channel, then:

$$q = \frac{i_0}{v_r} \tag{3.27}$$

where i_0 is the return stroke current and v_r is the effective return stroke velocity. Lightning field studies of the return stroke [71] indicate $i_0 \propto v_r^3$, similar to the relationship in Equation (3.25) for leader current and velocity, which suggests that $i_0 \propto i_L$. A precise cubic relationship between i_0 and v_r can also be deduced from recent return stroke models [72].

It is recalled in section 3.2.8 that lightning observations show that return stroke velocities lie within a range from $0.05c$ to $0.5c$, and return stroke currents mainly from 5 to 300 kA. If we correlate these values, then the proportionality $i_0 \propto v_r^3$ becomes:

$$i_0 = 1.75 \times 10^6 c^{-3} v_r^3 \tag{3.28}$$

As a consequence of Equations (3.26)–(3.28) we can deduce $i_0 = 1220 i_L$. Table 3.3 includes numerical examples of return stroke current and velocity using these relationships.

It is also noteworthy that the charge density variation with peak current is now obtained from Equations (3.26) as:

$$q = 40i_0^{2/3} \quad [\mu C/m, kA] \tag{3.29a}$$

This may be compared with the empirical equation derived from Berger's field data by Dellera and Garbagnati [73]:

$$q = 38i_0^{0.68} \tag{3.29b}$$

3.4   Lightning termination at ground

3.4.1   Striking distance

The probability of a downward lightning flash terminating on a structure can be calculated by empirical or by physical methods. At a critical point in the development of a downward leader in the vicinity, an upward leader may be launched from the grounded structure, and this will determine the location of the strike – unless another competing upward leader precedes it to make the first connection to an adjacent structure. The distance at this moment of launch between the tip of the downward leader and the origin of the successful upward leader is defined as the striking distance.

Calculation of the striking distance enables both a risk assessment to be made of the probability of a flash to a structure and also of the efficacy of the protection afforded by a grounded air termination, overhead ground wire or Faraday cage.

It is useful to classify into three types the striking distance calculations that are presently used:

(i)   Geometric models, in which the striking distance is assumed to be independent of both the prospective peak current i_0 in the return stroke and the geometrical contours of the ground structures. This simplest of models is nevertheless the basis of international standards for lightning protection because of its convenience and utility.

(ii)   Electrogeometric models, where the striking distance is represented by a function only of i_0, and is again supposed independent of the local geometry of the ground structures. Because of the prime influence of i_0 in the prospective magnitude of injected lightning overvoltages, this method was developed for and is widely used in insulation coordination of overhead line power systems.

(iii)   Generic models, which take account more realistically that both i_0 and the structure geometry will together determine the striking distance. With these more refined concepts, such factors as flash polarity and ground elevation can be incorporated, and calculations can employ electric field and statistical packages. These are recent advances but, except in specialised risk assessment requirements, generic models have so far remained in the scientific rather than the engineering domain. There is a case for their increased use as data on lightning parameters improve, and an extensive review is included here.

3.4.2   Geometric models and lightning standards

As envisaged in Preece's early work, the height h of a structure can be used to estimate its exposure risk (or the efficiency of an air termination). Suppose that a mast (or a long horizontal conductor) of height h is approached by a downward leader, whose tip has a horizontal distance component r from the mast axis (or conductor position). If the leader tip reaches its striking distance r_s at the instant when it is equidistant from the mast top (or the conductor) and the ground surface, then this horizontal component will define the attraction radius $r = r_a$:

$$r_a = h \left(\frac{2r_s}{h} - 1 \right)^{1/2} \tag{3.30}$$

In present international standards this concept is simplified to the proportionality $r_a = kh$, where the choice of $k = 1$ or 3 may be used (section 3.5.2), with the unstated implication from Equation (3.30) that $r_s = h$ or $5h$.

In the BS 6651:1999 *Code of practice for the protection of structures against lightning* [39] are three fundamental recommendations each linked to the phenomenology of the flash:

(i)   The conventional 30° or 45° cones of attraction are retained only for structures less than 20 m in height. The risk of side flash is estimated by the rolling sphere method (Figures 3.6a and b) [74].

(ii)   The increased vulnerability of electronic equipment (for telecommunications, telemetry, computing, control and instrumentation, and power electronic installations) to transient fields is recognised by a recommended increase in the number of downconductors (which may be structural steel members).

(iii) The risk factor for direct strikes is quantified.

These practical measures are discussed in section 3.5, but from the point of view of the physics of the lightning flash, developments (i) and (iii) are particularly relevant. The adoption of the rolling sphere criterion has arisen from the pragmatic desire for a straightforward alternative geometrical construction to the cone that will better predict the side flash risk. It also offers a generally useful application to other problems such as attachment points to aircraft. The discovery of space leaders in the recent work on long gap negative breakdown suggests that the rolling sphere is indeed a better physical analogy than the cone. As stated in section 3.2.8, the original concept of Schonland that a non-visible pilot leader develops from the previous tip to establish the next step can be invoked. It can be postulated that, as occurs in the laboratory long spark, the negative step is preceded by a space leader developing in both forward and retrograde directions from an origin that lies ahead of the leader tip. This bidirectional growth will thus be well simulated by a spherical region; in this interpretation, the centre of the rolling sphere can be regarded as the origin of the space leader ahead of the previous step rather than the tip of the step itself, as envisaged in the Standard. Since step lengths of 10 to 200 m are observed, a radius for the rolling sphere of 5 to 100 m seems supportable. This compares with the recommended radii of 20 or 60 m in the Standard. The smaller choice of rolling sphere radius is the more onerous for

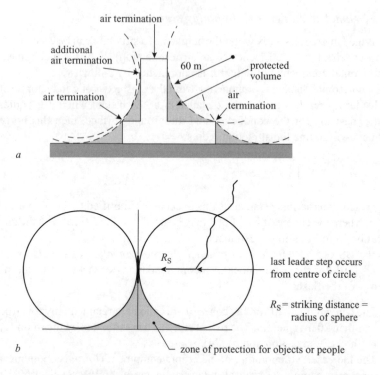

Figure 3.6   Rolling-sphere representations of the ground strike (BS 6651)

the protection system design. The US NFPA Standard 780 [75] gives the values 46 m and 30 m, and the IEC Standard 61662 [76] specifies the four radii 60, 45, 30 and 20 m as options.

This flexibility in the choice of rolling sphere radius is intended both to recognise the expected effects of structure height on the strike process and to allow a choice of security rating for the structure. In addition, it implicitly admits that a purely geometric model does not incorporate the reality that the striking distance will be smaller, and protection more difficult, for smaller prospective peak currents [77].

In the long laboratory negative spark, several space leaders can develop simulta-neously in series ahead of the leader tip, all subsequently causing bright leader channel elongations and stepping. So it is possible to envisage, as a geometrical representation of this process, not merely a single rolling sphere, but a string of small radius spher-ical regions ahead of a descending leader. This method would imply a higher rate of predicted shielding failures (Figure 3.7).

3.4.3   Electrogeometric models

3.4.3.1   Effect of peak current

A simplistic geometrical model has obvious limitations in not quantifying the initia-tion of the upward leader in terms of the field enhancement associated with the electric

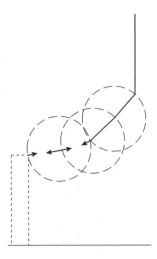

Figure 3.7   Spherical-string representation of the ground strike

charge on the downward leader. Quantitative estimates must rely on modelling, and the first authors to develop the now generally used relationship:

$$r_s = ki_0^n \tag{3.31}$$

for a given structure were Armstrong and Whitehead [78, 79]. They based this relationship upon physical concepts due to Wagner [71], and

$$r_s = 9.4i_0^{2/3} \tag{3.32}$$

is commonly used [80, 81].

The striking distance as a function of current was calculated by these authors on the basis of the assumptions that a 1 C leader charge was equivalent to a 20 kA peak current. Lightning observations by Berger, Anderson and Kroninger [82] suggest a leader charge of 4.5 C for a median lightning current of 30 kA. Armstrong and Whitehead [58] also needed to specify the mean electric field to complete the lightning channel over the striking distance. They used values of 5 kV/cm (for negative flashes) and 3 kV/cm (for positive flashes).

The electrogeometric model developed by Armstrong and Whitehead [58] provides a more realistic representation of the protection principle than a simple shielding angle. It is in extensive use for the design of lightning protection not only for power lines and substations, but other important structures such as chemical plant and aviation and rocket installations.

3.4.3.2   Electrogeometric boundary for strikes to a vertical mast

The electrogeometric model makes the same important simplifying assumption as before that the striking distance is independent of the geometry of the grounded

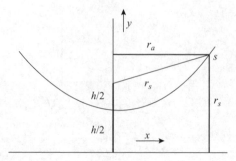

Figure 3.8   Electrogeometric model for a vertical mast

structure. The value of r_s at a given peak current is assumed to be the same for a strike to the mast as for one to the surrounding terrain, so the calculations are as for Equation (3.30). For example, for a vertical mast of height h, and for large striking distances where $r_s > h/2$, the limiting boundary for strikes to the mast top (Figure 3.8) is defined by the parabola:

$$y = \frac{x^2 + h^2}{2h} \tag{3.33}$$

This parabola is the locus of the point S which defines, for each value of prospective peak current i_0, the limiting condition for a strike to the top of the mast. Here $y = r_s = f(i_0)$, since r_s is assumed to be the striking distance to both mast and ground. From the above parabolic equation,we can define the attraction radius of the mast for a flash with this current:

$$r_a(i_0, h) = x(r_s) = h \left(\frac{2r_s}{h} - 1 \right)^{1/2} \tag{3.34}$$

and the attraction area of the mast, i.e. the protected area at ground level:

$$a(i_0, h) = \pi x(r_s)^2 = \pi h^2 \left(\frac{2r_s}{h} - 1 \right) \tag{3.35}$$

The attraction distance and attraction area are termed collection distance and area in lightning standards (section 3.5.2). If the mast is a lightning rod (or air termination) then the terms 'protection distance' and 'area' are appropriate. It should be noted that the attraction distance is smaller than the striking distance except for $r_a = h = r_s$.
    The shielding angle concept can be retained where:

$$\theta_s(i_0, h) = \tan^{-1} \left(\frac{x(r_s)}{h} \right) = \tan^{-1} \sqrt{\frac{2r_s}{h} - 1} \tag{3.36}$$

Calculation of the attraction area, together with a probability distribution for the lightning currents, enables the risk factor for the mast to be found (section 3.5.1, Equation (3.58)).

Table 3.4 *Electrogeometric model: striking distance, attraction area and shielding angle for a vertical mast*

i_0 (kA)	5		31		100		300	
			(median value)					
h (m)	20	60	20	60	20	60	20	60
r_s (m)	27.5		93		203		422	
a (m² × 10³)	2.2	2.4	10	24	24	64	52	150
θ_s	53°	25°	71°	55°	77°	67°	81°	75°

For $r_s < h/2$, the protected area is simply:

$$a(\text{min}) = \pi x (r_s)^2 = \pi r_s^2 \tag{3.37}$$

Examples of these calculations are shown in Table 3.4. These also illustrate the inherent implication of the electrogeometric approach that elevated structures are more selectively struck by higher current flashes. It is based on representation of a downward flash and takes no account that for structures over 100 m tall or at high elevation the local geometrical enhancement of the thundercloud field by the structure may be sufficient for the lightning flash to be triggered by an upward leader even before any downward leader is observed.

3.4.3.3 Shielding calculations for an overhead line

The same simplifying assumption of striking distance to be a function only of peak lightning current is the basis of a well known approach to the calculation of the shielding of the phase conductors that is provided by an overhead ground wire (or wires).

The probability of a shielding failure, defined by a direct strike to a phase conductor, is in the case of a single ground wire a function of the volume of the prospective strike zone bounded by (Figure 3.9a):

(i) the parabola defining equidistance between the phase conductor and the ground plane
(ii) the linear locus PR equidistant between the ground wire and the phase conductor
(iii) the circular locus centred on the phase conductor, defining the maximum safe striking distance $r_s(\text{maximum}) = r_m$.

This last value is determined by the basic insulation level (BIL) of the line, since this defines a maximum allowable peak current for such a direct strike to a phase conductor. From the unprotected volume, the risk factor for a direct strike to a phase conductor (section 3.5) can be calculated using the N_g value for the region.

Complete shielding is rarely justified economically. For the line of Figure 3.9, complete shielding is achieved, with a critical shielding angle θ_c, for QR = PR. The procedure is then to specify, according to the system insulation coordination requirements, the maximum allowable overvoltage, the anticipated peak lightning

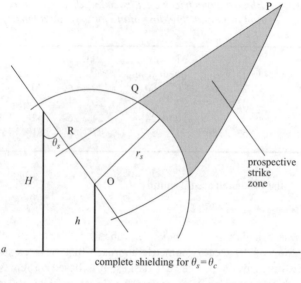

a

complete shielding for $\theta_s = \theta_c$

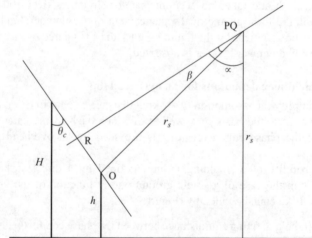

b

Figure 3.9   Electrogeometric modelling for overhead line shielding
  a   Strike zone for line with shield wire
  b   Evaluation of critical shielding angle

current and corresponding striking distance. Then, using Figure 3.9*b*, the associated critical shielding angle is calculated from:

$$\theta_c = \left(\frac{\pi}{2} - \alpha\right) - \beta = \sin^{-1}\left(1 - \frac{h}{r_s}\right) - \sin^{-1}\left[\frac{H - h}{2r_s \cos\theta_c}\right] \tag{3.38}$$

where h and H are the mean heights of the phase conductor and the ground wire.

3.4.4 Generic models

Since the late 1980s, models of the striking distance, based to a large extent upon the improved knowledge of the physics of long sparks, have taken account of both prospective peak lightning current and the geometry of the grounded structure. Some examples follow.

3.4.4.1 Eriksson model

Following a review of field data, Eriksson [16] developed a quantitative model to calculate the attraction radius for a vertical mast on flat terrain. The criterion for launch of the upward leader is based upon the critical radius concept developed from long spark studies [83]. Effectively, this represents the apex of the mast as having a radius of curvature of 0.35 m, since positive leader inception in the laboratory varies little for radii smaller than this. An induced field of 30 kV/cm at the surface of this notional anode is required for upward leader inception. The magnitude of the induced field depends upon the charge on the downward leader and the prospective peak current, and for a downward leader directly above the mast a strike is then inevitable. For a vertical downward leader displaced horizontally from the mast, the possibility of completion of a strike to the mast is calculated from the downward (negative) and upward (positive) leader velocities. The model does not directly utilise the physics of leader and leader corona growth, but is a powerful and simple approach for which a regression analysis yields the relationship between attraction radius r_a (m), lightning return current i_0 (kA) and mast height h (m):

$$r_a = 0.84 i_0^{0.74} h^{0.6} \tag{3.39}$$

The resulting attraction radius is calculated for different lightning return stroke currents i_0 and mast heights h, as shown in Table 3.5, where the values deduced for the electrogeometric model are also shown.

3.4.4.2 Dellera and Garbagnati model

This also uses the critical radius approach to determine the upward leader inception [73, 84], and can be applied to both downward and upward lightning. The model computes the spatial–temporal development of both downward and upward leaders numerically, using a charge simulation program to calculate the electric field distribution, and simulates incrementally their directional and velocity development in the direction of maximum field. The inclusion of the streamer zone between the leader tips and the use of field computation techniques facilitate predictive engineering applications such as a realistic representation of the local ground topology at the mast or transmission line. Examples of the output of this model are shown in Table 3.5. They generally indicate a smaller attraction radius than the Eriksson model and a stronger variation with mast height. The charge simulation software also allows the cloud charges to be included, and the probability of upward-directed flashes is computed as a function of structure height. Equal probability of upward or downward flashes is estimated for a height of 230 m.

Table 3.5   Generic models of attraction radius for a single mast
(downward negative flash)

i_0 (kA)	5		31		100		300	
h (m)	20	60	20	60	20	60	20	60
r_a (m) electrogeo. model	26	28	56	87	87	143	129	218
r_a (m) Eriksson	17	32	64	125	147	296	345	668
r_a (m) Dellera	20	30	35	95	90	240	170	535
r_a (m) Rizk	22	45	95	150	180	280	315	560
r_a (m) Petrov	17	28	57	95	124	206	258	429

3.4.4.3   Rizk model

Here, a critical potential criterion determines inception of the upward leader, and then the completion of the lightning strike is determined from two aspects of long spark studies [61]. The first aspect sets a critical streamer gradient between the downward and upward leaders of 5 kV/cm at normal air density. Second, the voltage gradient along the leaders, which is known from the work of the Les Renardieres Group [43] to decrease temporally from 5 kV cm 1 to 0.5 kV/cm or less, is also represented, together with the effect of reduced air density at high altitudes. An analytical evaluation of the electric field allows simple ground topologies (based upon a semiellipsoidal terrain) to be simulated. The predictions for a mast on flat terrain are shown in Table 3.5, where the attraction radii are similar to those deduced from the Eriksson model. On elevated terrain, the effect of the thundercloud field is predicted to increase the attractive radius of tall masts by over 100 per cent for low peak currents. A 50 per cent probability of an upward flash is expected for $h = 160$ m.

3.4.4.4   Petrov and Waters model

This aims to be a flexible model, based upon the physics of the leader channel described in section 3.3.2, which is adaptable for stroke polarity and terrain altitude. The procedures which will be described here lead to the values of attraction radius for negative flashes at sea level that are shown for comparison in Table 3.5. A mast height of 190 m is calculated for 50 per cent probability of upward flashes (Table 3.8, section 3.5.4). This model agrees with values deduced from the electrogeometric approach for currents up to the median, but generic models predict significantly greater attraction radii for the larger peak currents.

(i) Upward leader criterion

A physical condition for the launch of a successful upward leader that will complete the junction with the down-coming negative leader [68] is not in this case based on a critical radius for leader inception, but upon a critical interaction between a putative upward discharge and the downward leader. In Figure 3.10, the leader channel is represented by a vertical linear charge of length L with a charge per unit length q and leader tip charge Q. The mast is shown as an ellipsoid of height h and half width b.

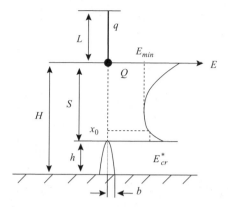

Figure 3.10 Analytical modelling of a downward leader above a ground mast [53]

At large distances S between the lightning channel and the mast, the range of the electric field intensification above the top of the mast may be insufficiently extensive to support a successful upward leader, although there may well be corona streamer activity and weak leader growth as the downward leader approaches. The streamer corona from the top of the mast propagates to a distance where the electric field falls to the minimum streamer gradient E_s, at a distance x_0 from the top of the mast. For standard sea level conditions, the electric field E is equal to about 5 kV/cm for a streamer of positive polarity and about 10 kV/cm for negative polarity. The criterion for the lightning strike to the mast is thus a critical upward streamer length so that an upward leader can be successfully developed.

Evaluation of the critical streamer length is made from long spark studies: if an upward positive leader of charge q_u per unit length grows simultaneously with a hemispherical upward leader corona region of radius L_u, the charge per unit radius within the leader corona will be equal to that in the leader. Thus the charge within the upward leader corona zone is given by $q_u L_u$. This charge produces at the hemispherical surface of the upward leader corona a field:

$$E_S = \frac{q_u}{2\pi\varepsilon_0 L_u} \tag{3.40}$$

The critical streamer length L_u is associated with a critical value of q_u. In the case of a positive upward leader, long spark studies show that the minimum charge for positive leader inception is about 20–40 μC/m (section 3.3.1.1), so enabling the minimum L_u to be found. It is known also from optical and electronic measurements that the minimum length of the streamer zone of the positive leader in long air gaps is about 0.7 m [53]. This corresponds to a critical streamer charge q_u of 20 μC/m.

(ii) *Termination to a plane ground*

For specific cases of gaps with simple geometry, analytical expressions for the potential and electric field may be obtained. In particular, the axial electric field distribution created by the vertical downward leader channel represented in Figure 3.10 at a height H above an earth plane surface has the form:

$$E(x, H, L) = \left(\frac{Q}{4\pi \varepsilon_0}\right)\left[\frac{1}{x^2} + \frac{1}{(2H - x)^2}\right]$$

$$+ \left(\frac{q}{4\pi \varepsilon_0}\right)\left[\frac{1}{x} - \frac{1}{x + L} + \frac{1}{2H - x} - \frac{1}{2H - x + L}\right] \quad (3.41)$$

In the absence of a structure the field at ground level $(x = H)$ is, for $H \ll L$:

$$E_g = \left(\frac{1}{2\pi \varepsilon_0}\right)\left[\frac{Q}{H^2} + \frac{q}{H}\right] \quad (3.42)$$

If the leader is sufficiently close to ground so that this field is equal to the critical streamer propagation field E_s, then the leader will complete the strike to ground without any upward discharge growth.

Although the streamer is rarely visible in lightning photographs, it can be inferred from long gap studies to be essential for the propagation of the leader. For analytical purposes, we may represent the net charge in the streamer zone as the charge Q (Figure 3.10). Petrov and Waters [53] showed in the following way the correspondence between the streamer charge Q and the linear charge density q consequently established upon the leader channel. If the streamer zone of the downward leader is represented as a hemisphere, with a radius equal to the streamer zone length L_S, then we can write:

$$Q = qL_S \quad (3.43)$$

But the field at the head of the streamer zone is:

$$E_S = \frac{Q}{2\pi \varepsilon_0 L_S} \quad (3.44)$$

The important relationship between the charges Q and q is therefore from Equations (3.43) and (3.44):

$$Q = \frac{q^2}{2\pi \varepsilon_0 E_S} \quad (3.45)$$

As a result, the charge Q of the leader head and the leader channel charge density q are both determined by the streamer zone length. For example, the charge q on the channel is 0.39 mC/m for a median current stroke (Table 3.3). Equation (3.45) gives the charge Q on the negative streamer system as 2.7 mC and the streamer zone length $L_S = 7$ m. So the charge Q can be expressed in terms of q, and in the case of $E_g = E_S$ the height H represents the striking distance r_s. From Equation (3.42):

$$E_S = \left(\frac{1}{2\pi \varepsilon_0}\right)\left[\frac{q^2}{2\pi \varepsilon_0 E_s r_s^2} + \frac{q}{r_s}\right] \quad (3.46)$$

which yields a value of striking distance:

$$r_s = \left(\frac{q}{4\pi\varepsilon_0 E_S}\right)[1 + \sqrt{5}] \tag{3.47}$$

Using now the relationship between q and i_0 obtained from Equations (3.26) and (3.29a), and a value of $E_S = 10$ kV/cm for the propagation field of the downward negative leaders, we get:

$$r_s = 1.16 i_0^{2/3} \quad [\text{m, kA}] \tag{3.48}$$

These striking distances shown in Figure 3.11a must be regarded as a lower limit for the striking distance to a plane ground, since no upward positive discharge has been assumed. In practice, such upward leaders are observed from local asperities on level terrain, and r_s may consequently be larger than in Equation (3.48).

(iii) *Strikes to a mast*

When a vertical mast is approximated by a semiellipsoid of half width b as in Figure 3.10, the electric field between it and a vertical coaxial downward leader can also be represented analytically. The striking distance is then calculated by employing the criterion for a critical upward streamer length from (i) earlier. Petrov and Waters [68] showed that for a negative flash the striking distance is:

$$r_s = 0.8[(h + 15)i_0]^{2/3} \quad [\text{m, kA}] \tag{3.49}$$

the factor $0.8 (h + 15)^{2/3}$ giving a convenient numerical representation (for $b = 1$ m) of the analytically calculated field enhancement in the leader-to-mast space for the attainment of the strike condition. Striking distances calculated from Equation (3.49) are shown in Figure 3.11b, together with those for the electrogeometric model. These give comparable values for a 20 m mast but the Petrov–Waters model suggests that an electrogeometric approach significantly underestimates the striking distance for a 60 m mast. Most importantly, both the electrogeometric and generic modelling show that the minimum rolling sphere radius of 20 m recommended in international standards is optimistically larger (for reliable protection) than the striking distance for low current strikes of 3 kA and below.

3.4.4.5   Effect of altitude on striking distance

Phelps and Griffiths [86] studied the effect of air density and humidity on positive streamer growth. From this work a relationship between the streamer gradient E_S and the relative air density δ and absolute humidity γ was obtained by Eriksson *et al.* [87]:

$$E_S = 425\delta^{1.5} + (4 + 5\delta)\gamma \quad [\text{kV/m}] \tag{3.50}$$

where the relative air density δ with respect to standard sea level values of 760 torr and 293 K is:

$$\delta = \frac{293p}{760T} \tag{3.51}$$

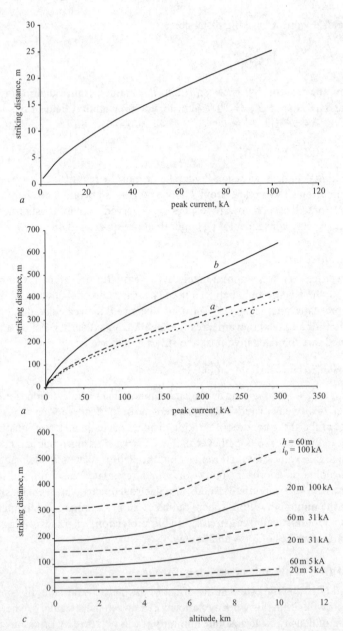

Figure 3.11   Petrov–Waters model

    a Striking distance calculations for a negative flash to plane ground

    b Mast at sea level. Curve a: electrogeometric model; curve b: generic model (60 m mast); curve c: generic model (20 m mast) [53]

    c Mast at altitude [53]

Pressure, temperature and humidity change with increasing altitude z and, for the range $0 < z < 10$ km, Petrov and Waters [68, 88] represented the meteorological data by the equations

$$p(z) = p(0) \exp\left(\frac{-z}{z_0}\right) \tag{3.52}$$

where $p(0)$ is the sea level pressure and $z_0 = 8$ km:

$$T(z) = T(0) - kz \tag{3.53}$$

where $T(0)$ is the sea level temperature and $k = 6$ K/km:

$$\gamma(z) = \gamma(0) \exp\left(\frac{-z}{z_H}\right) \tag{3.54}$$

where $\gamma(0)$ is the sea level absolute air humidity and $z_H = 3$ km. The standard value for $\gamma(0)$ in high voltage testing is 11 g/m^3.

At an elevation of z km, the reduction of the critical field E_S implies a significant increase of striking distance. At the San Salvatore measuring station altitude of 914 m, for example, the critical positive streamer field is 4.4 kV/cm, which partly contributes to Berger's observations of flashes of either polarity resulting from upward first leaders.

Incorporation of altitude effects into Equation (3.49) for a vertical mast gives for an altitude of z km:

$$r_s = 0.8[(h + 15)i_0]^{2/3} \left[\frac{1 + z^2}{h + 80}\right] \tag{3.55}$$

This greater striking distance (Figure 3.11c) represents a significant increase of risk factor in high mountainous regions and in aviation. At $z = 5$ km, the critical streamer propagation field is predicted to be 2 kV/cm compared with 5 kV/cm at sea level.

3.4.4.6 Air termination geometry

In recent years, there has been considerable interest in early streamer emission air terminals for lightning protection [89, 90]. These generate a locally triggered streamer discharge from the terminal with a shorter delay than a standard device. However, the efficiency of such systems is not proved experimentally [43, 91–95] or by field observations. On the contrary, there are physical grounds to believe that early streamer initiation (also claimed to be encouraged by air terminations incorporating radioactive sources) would make more difficult a successful upward leader, that is to say one which would propagate over the distance required to intercept the downward negative leader [96]. The influence of the geometry of the air terminal on the striking probability has been investigated experimentally by several authors [68, 85, 97–100]. These investigations showed a higher striking probability to a grounded rod with a blunt top which probably results from a delayed streamer onset. The determination of a possible optimum radius of curvature of the top of a lightning conductor is clearly of practical interest. Petrov and Waters [101, 102] have calculated the striking distance

*Table 3.6   Optimum half width b(opt) and radius of curvature ρ for
semiellipsoid mast: negative flash striking distances for mast
(b = 1 m), sphere and horizontal cylinder*

i_0 (kA)	h (m)	b(opt) (m)	ρ (m)	Striking distance r_s (m)			
				$b = b$(opt)	$b = 1$ m	sphere $r = h$	cylinder $r = h$
10	20	3.5	0.6	42	40	17	13
	60	4.8	0.4	75	66	16	12
31	20	3.5	0.6	100	85	41	29
	60	4.8	0.4	174	141	37	26
100	20	3.5	0.6	210	185	99	74
	60	4.8	0.4	384	307	87	62

for a semiellipsoid mast with a half width b. In Table 3.6, for various lightning currents and mast heights, numerical examples are given of optimum half widths b(opt) and radii of curvature $\rho = b$(opt)$^2/h$ at the ellipsoidal mast top that result in the maximum striking distances. Additional calculations for a rod lightning conductor with a sphere on the top showed also that the optimum radius of the sphere for which the striking distance is a maximum is of the order of 0.7 m. This approach has a similar outcome to the equivalent radius concept [83].

Additional results are included for grounded hemispheres and long semicylinders, with their bases lying on a plane earth, as a guide to the behaviour of basic building shapes. These calculations show sensitivity to the structure geometry for negative lightning. There is little influence on the striking distance r_s in the case of positive lightning (section 3.4.5).

3.4.5   Positive lightning

3.4.5.1   Striking mechanisms for positive flashes

The lightning detection techniques of the 1990s indicate that positive polarity flashes may occur more frequently than had been supposed previously. Although undoubtedly less common than the negative flash, positive lightning is significant because such strikes often carry particularly damaging peak currents exceeding 100 kA. Observations of positive lightning by Berger [20] (section 3.2.8) showed that, at an elevation of 914 m, the initiation was by a stepped upward negative leader that could be of great length. However, positive downward leaders initiated from the cloud have been recorded by Berger and others, in which case the determination of the striking distance based solely upon a criterion for upward leader inception cannot always be used. Laboratory experiments with high positive impulse voltages show that the negative upward leader may not propagate until a very late stage, in which case the striking distance is determined instead by the streamer zone length of a positive downward leader at the final jump phase preceding sparkover. The efficiency of lightning rods against positive polarity downward leaders can thus be expected in some cases to be substantially lower than against negative polarity leaders. The necessity of further clarification of the mechanism of the positive lightning flash to determine striking distances was emphasized also by Golde [103].

The following calculations by Petrov and Waters [101, 102] indicate that a significantly lower sensitivity of striking distance $r_s(+)$ to the structure height can be expected for positive polarity downward lightning. At the same time, the variation of striking distance with prospective peak current can be expected to be the same ($r_s \propto i_0^{2/3}$) as shown for negative lightning in sections 3.3.2 and 3.4.4.4, since the charge q on the channel and the peak return current will be similarly related as in the negative flash.

Two mechanisms of lightning strike for downward flashes are envisaged for this polarity, depending upon the criterion that succeeds to establish the striking distance.

(i) *Positive streamer criterion*

For positive lightning, the downward leader may not need to initiate a negative upward leader before the striking distance has already been achieved by the successful bridging of the gap to the grounded structure by the positive streamers. In this case, the striking distance $r_s(+)$ is determined by the streamer zone length of the downward leader in the transition to what in long spark observations is known as the final jump phase. To calculate $r_s(+)$, it is necessary again to calculate the electric field distribution between the lightning channel and the grounded structure. Streamers of the downward positive leader will propagate to a distance where the electric field falls to the minimum streamer gradient $E_S \approx 5$ kV/cm at sea level. At the approach of the lightning channel to the mast, the electric field intensity and the range of its intensification are increased, and this will extend the streamer zone length of the positive leader. When the minimum value of the electric field between the downward leader and the grounded structure becomes equal to E_S, then positive streamers can successfully propagate to the structure. The instant at which this condition is reached will define the striking distance $r_s(+)$. Of course, the ongoing development of the leader towards the structure will almost inevitably initiate a late upward negative leader before the ultimate establishment of the return stroke.

(ii) *Negative streamer criterion*

For tall structures or high lightning currents, the conditions for negative upward leader initiation may well be fulfilled before the minimum electric field between a downward positive leader and the grounded structure becomes equal to $E_S(+)$. Here, the striking distance $r_s(+)$ will instead be defined by that upward leader initiation condition, in the same way as for the negative flash. As already discussed, negative leaders in very long spark discharges are also known to propagate by steps, which are themselves initiated by space leaders ahead of the main negative channel. The length of the steps in long air gaps is practically independent of both the gap length and the shape of the high voltage impulse, and takes values of between 2.5 and 3.6 m [59, 62, 66, 104, 105]. The critical range of field intensification in the vicinity of the grounded structure is then important, and the intensified field must, at an adequate distance x_0 from the structure, reach a critical value $E_S(-)$. Measured values for this electric field in the negative streamer zone have been variously reported as 11 kV/cm by Volkova and Koriavin [67] and as 10–16 kV/cm by the Les Renardieres Group [62]. For a minimum value of 10 kV/cm, and a corresponding value of critical charge per unit length of $q = 160$ µC/m, the value of x_0 for a vertical mast is 2.9 m. This represents physically the minimum length of the upward negative streamer zone and the negative leader step length.

The mast height and prospective lightning current will determine which of these criteria (i) or (ii) is achieved the earlier. The discrimination boundary calculated for mechanisms (i) and (ii) is shown in Figure 3.12*a*.

3.4.5.2 Striking distance for positive lightning as a function of peak current and mast height

Figure 3.12*b* shows the striking distance dependence on the lightning peak current and mast height resulting from these calculations. The solid lines correspond to the application of the positive streamer criterion (i), and the broken lines are obtained from the negative streamer criterion (ii).

The striking distances indicated by criterion (i) show a weak dependence on mast height h. The calculations can be approximated by the numerical relationship:

$$r_s(+) = 1.08 i_0^{2/3} \ln \left(\frac{h}{15} + 10 \right) \quad \text{[m, kA]} \tag{3.56}$$

As already noted the increase of striking distance with return stroke current is again of the form $r_s \propto i_0^{2/3}$ since it is proportional to the charge on the leader for either polarity.

It is seen from Figure 3.12*b* that for lightning currents less than a median value of 31 kA, the first criterion (i) is fulfilled earlier (at a larger striking distance) than criterion (ii). However, for tall structures and high current flashes, large striking distances will frequently follow the prior achievement of criterion (ii). The striking distances indicated by criterion (ii) may be approximated by the relationship:

$$r_s(+) = 0.103[(h + 30)i_0]^{2/3} \quad \text{[m, kA]} \tag{3.57}$$

These calculations for downward flashes from positive cloud charges show that the striking distance is merely 10–20 per cent of that associated with negative lightning. However, as Berger [20] has shown, for tall masts at altitude the upward leader is usually the initiation mechanism for flashes of either polarity. This is discussed in section 3.5.4.4.

3.5 Risk factors and protection

3.5.1 Risk assessment

As in the case of striking distance calculations, the assessment of lightning risk can be made at three levels of refinement. The probability of a strike is usually expressed

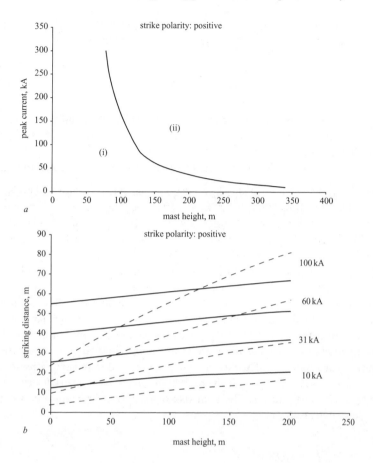

Figure 3.12   Positive flash modelling

 a criterion boundary for upward negative leader from a mast. Region (i): striking distance achieved before upward negative leader inception; region (ii): striking distance achieved by upward negative leader inception

 b striking distance to a mast [53]. Solid curves: striking distance from criterion (i); broken curves: striking distance from criterion (ii)

as the risk factor, which is quantified as the estimated number of strikes per annum to the structure. This can be calculated:

(i) independently of the prospective lightning current magnitude, and taking into account only the collection area of the structure; this is the procedure used in international standards

(ii) noting that the risk factor will increase with increasing prospective lightning current; in this case, an attraction area $a(i_0)$ is calculated from $r_s(i_0)$ by geometrical relationships, and the annual number of lightning strokes to the structure for all peak currents then gives the risk factor:

$$R = N_g A_c = N_g \int_0^\infty a(i_0)p(i_0)\,di_0 \tag{3.58}$$

where $p(i_0)$ is the normalised probability density of the lightning current distribution, the integration gives the collection area A_c, and N_g is the local density of lightning strikes to earth

(iii)   using a generic model of a criterion for a successful upward leader to find $a(i_0)$.

A more refined assessment of risk goes further than the risk factor R. This approach estimates the probability that the strike is sufficiently severe to cause an electrical, thermal, electromagnetic or mechanical shock that is prejudicial to the structure. An example of these risk statistics is the calculation of the risk of flashover R_F of an insulated electrical system as a result of a lightning overvoltage [106]. The per-unit probability density $p(V)$ defines the range and magnitudes of such overvoltages. The shape of the $p(V)$ curve depends both on the statistics of natural lightning and on the structural and other characteristics of the system, including the effect of protective measures. The cumulative probability function $P(V)$, on the other hand, is system dependent and defines the increasing risk of flashover of the system insulation with increasing voltage V for a voltage shape that is representative of lightning overvoltages (usually a 1.2/50 impulse but sometimes a chopped impulse or a non-standard shape). The product $p(V)P(V)$ is the probability of an overvoltage V arising which results in a system flashover. The total risk of flashover per annum in a region of ground flash density N_g is:

$$R_F = N_g \int_0^\infty p(V)P(V)\,dV \tag{3.59}$$

3.5.2   Standard procedure for the calculation of risk factor

Using the simple geometric method, the risk-of-strike assessment is determined in the Standard BS 6651:1999 [39] by the calculation:

$$R = (ABCDE)A_c N_g 10^{-6} \tag{3.60}$$

where the important weighting factors A to E (in the range 0.3–2.0) concern the structure and its contents and location. The collection area A_c (m^2) is calculated in BS 6651 by adding an attraction radius $r_a = h$ (structure height) to the plan dimensions. IEC 61662 [76] adds an attraction radius $r_a = 3h$, so increasing the risk factor by almost an order of magnitude.

As far as collection area is concerned, however, a much greater impact on BS 6651 arises from its advice for the protection of electronic equipment. This reasonably suggests that a lightning hazard to vulnerable equipment is caused by strikes to surrounding ground, associated structures and incoming or outgoing mains services and signal lines. On the basis of a collection distance $d(m)$ equal numerically to the earth resistivity (Ωm), say typically 100 m ($\equiv 100$ Ωm), the values of A_c would be increased by factors of 10 to 1000 above those calculated by use of the attraction radius r_a. Consequently, high risk factors in the range 0.05–0.1 for the UK are then found. The risk of lightning damage often justifies expenditure on the protection of power supplies, signal lines and telephone cables. The Standard BS 6651 classifies four types of structure, with a consequential loss rating that ranges from one for domestic dwellings to four for major industrial infrastructure. The acceptability of a given risk factor will depend on the individual structure. A value of 10^{-5} is sometimes quoted as a guide, which corresponds to a very cautious single annual failure in 100 millennia.

3.5.3 Electrogeometric calculation of risk factor

Armstrong and Whitehead [78] used their electrogeometric model to take account of the structure of an overhead transmission line to design its lightning shielding. They deduce for the flash rate to an overhead line of effective width w and height h:

$$R = 0.1 N_g (2h + w) \tag{3.61}$$

which implies an attraction distance $r_a = h$ and a 10^3 weighting factor. Whitehead [107] was able to test the utility of the electrogeometric model by an eight-year study involving 51 cases of shielding failure and 52 cases of backflashover. Table 3.7 shows results from this study giving specific tripout rates (lightning outages per 100 km years) at thunderstorm day values of $T_d = 40$ (hence Whitehead's STR-40 nomenclature).

Whitehead [108] importantly noted that application of the electrogeometric model should take account of the variability of both:

a   striking distance: statistically low values for a given peak current will increase the risk of shielding failure; one standard deviation allowance for this purpose indicates that r_s (statistical) $= 0.9 r_s$ (i_0) should be used
b   terrain: the mean height of the phase conductor can be much higher in exceptional terrain such as deep valleys.

Induced overvoltages occur on phase conductors from nearby ground strikes. Whitehead concluded that such overvoltages are generally harmless in respect of

Table 3.7   Line outage statistics [86]

Line voltage (kV)	Impulse flashover voltage U_{50} (kV)	Shielding angle, θ_S	STR-40
345	1600	31°	5.7
		22°	3.44
		−15°	0.19
500	1800	20°	0.23

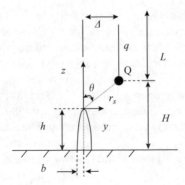

Figure 3.13   Modelling of a laterally displaced leader [53]

outages on UHV lines, but are significant with respect to a BIL of less than 60 kV at distribution voltages.

3.5.4   Generic modelling of risk factor for a negative flash

3.5.4.1   Strikes to the top of a mast

From the attraction radii calculated in Table 3.5, risk factors are found from generic models using Equation (3.58). For the determination of an attraction area $a(i_0)$ that takes account of both lightning current and structure geometry, it is first necessary to calculate the striking distance as a function of lateral displacement of the downward lightning channel (Figure 3.13) [109].

As before for the Petrov–Waters model, the criterion for determination of the striking distance uses analytical field formulae to find the range of field intensification by the grounded structure that is equal to a critical upward streamer length. Figure 3.14 shows, for a mast height of 60 m, the striking distance calculated for various angles θ between the vertical axis and the lightning channel. For the range $0 < \theta < 90°$, the striking distance dependence on the mast height and lightning current can be approximated for slender structures, in extension of Equation (3.49), by the relationship:

$$r_s(i_0, h, \theta) = [(h + 15)i_0]^{2/3}[0.8\cos\theta + 0.24\sin\theta] \qquad \text{[m, kA]} \qquad (3.62)$$

There is a maximum collection angle θ_m, above which lightning will not strike the mast. The maximum lateral displacement Δ_m is the attraction radius r_a and the collection area of the mast is:

$$a(i_0) = \pi \Delta_m^2 \qquad (3.63)$$

Since

$$\Delta = r_s \sin\theta \qquad (3.64)$$

the differentiation with respect to θ of Equations (3.62) and (3.64) gives:

$$r_a = \Delta_m = 0.54[(h + 15)i_0]^{2/3} \qquad (3.65)$$

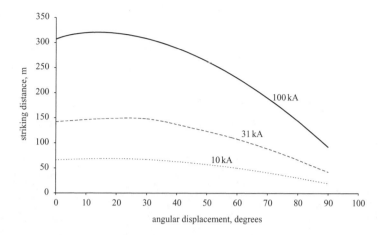

Figure 3.14 *Striking distance dependence on angular displacement (negative flash). Mast height 60 m [53]*

and

$$\theta_m = 53° \tag{3.66}$$

The values of attraction radius are shown in Table 3.5 for comparison with other generic models. Although some differences are seen between these models, they nevertheless allow significant conclusions to be drawn on the estimation of risk factor. Present day standards give acceptably conservative estimates of risk factor for low current flashes, but greatly underestimate strike frequency for high current flashes. Even the electrogeometric model, which recognises the increased risk for high current flashes, is shown to underestimate the risk factor for a mast structure by about four-fold.

For a Franklin rod air terminal, a large striking distance gives efficient protection and a large sheilding angle. The value Δ_m determines also the conventionally used shielding angle of the mast:

$$\theta_s = \arctan \frac{\Delta_m}{h}$$
$$= \arctan \left\{ \frac{0.54[(h+15)i_0]^{2/3}}{h} \right\} \tag{3.67}$$

Figure 3.15 shows the shielding angle dependence on the lightning current and mast height. Comparison with the angle deduced from an electrogeometric model (from Table 3.4) suggests that the electrogeometric approach may overestimate shielding efficiency for low current flashes and underestimate it for high current flashes.

By incorporating the downward leader velocity into this model, Petrov and D'Alessandro [100] have calculated the time interval between the upward leader inception from the mast top and its stable propagation phase. For $h = 60$ m and $i_0 = 100$ kA, this interval is almost 200 μs. They conclude that, contrary to the early

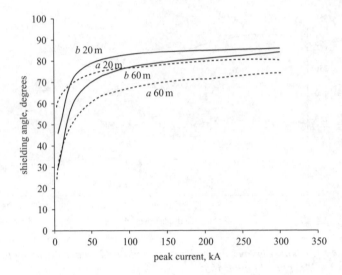

Figure 3.15   *Shielding angle variation with peak current for mast height 20 and 60 m (negative flash)*

curves a: electrogeometric model
curves b: generic model [53]

streamer emission concept (section 3.4.4.6), a delayed upward leader inception may be more effective for lightning protection.

3.5.4.2   Side flashes to a mast

The above calculations are for strikes to the top of a mast. However, a side flash below the top of tall structures can arise, especially as a result of negative downward leaders that have a horizontal path in the vicinity of a high tower [110]. Petrov and Waters [109] have shown that the striking distance for such side flashes is:

$$r_{ss} \approx 2.4 i_0^{2/3} \quad \text{[m, kA]} \tag{3.68}$$

The striking distance for side flashes is significantly less than that for downward lightning to the mast top because of the low field intensification near the side of the mast. The value of $r_s = 50$ m, corresponding to the rolling sphere concept of the standards, is suitable only for a peak current $i_0 = 95$ kA according to this model.

3.5.4.3   Inverted cone strike zones

Because the maximum collection angle of $53°$ (Equation 3.66) does not depend on the mast height or the lightning current, the collection region of the mast will be an inverted cone of this angle whose apex is at the mast top (Figure 3.16). As Table 3.5 shows few major numerical differences between generic models, the maximum collection angle concept would appear to have good validity. A recent Fractal model of leader tortuosity [111] has shown that 95 per cent of strikes will occur within a half-angle of $60°$.

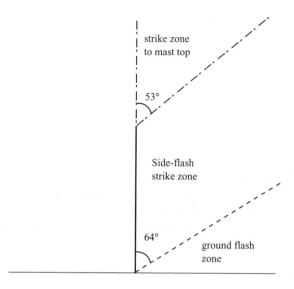

Figure 3.16 Collection cones for apex strikes and side flashes in negative lightning
    a   attraction zone boundary for vertical downward flashes
    b   side flash boundary for horizontally oriented flashes

Noting also that the side flash striking distance is larger than that to the plane earth surface, which is (section 3.4.4.4):

$$r_s = 1.16i_0^{2/3} \quad [\text{m, kA}] \tag{3.69}$$

we can by means of Equation (3.68) define the side flash zone boundary, for all values of peak current, as an inverted cone of half angle 64° whose apex is at the mast base. This together with the collection angle derived in the treatment of strikes to the mast top gives an easily applied current-dependent concept that is simpler to apply to both side flash and mast top risk calculations than a fixed radius current-independent rolling sphere locus (Figure 3.16).

3.5.4.4   Lightning to tall structures

Only generic models can provide an estimation of the risk of lightning initiation from tall structures. The critical ambient thundercloud field necessary for upward leader inception decreases with the increase of structure height h because of the intensification of the local field. For structures with height $h > 200$ m, the critical ambient field becomes lower than 20 kV/m. The electric field E_{cl} created by a storm cloud at the earth's surface before a lightning flash can approach this critical value (section 3.2.7). Because of space charge effects, the electric field increases with height above ground (section 3.2.7.1) and the field intensity before a lightning flash can be simultaneously $E_{cl} = 5$ kV/m at the earth's surface and $E_{cr} = 65$ kV/m at 600 m. The electric field created by the storm cloud may be substantially less than the critical electric field needed for upward leader inception in the case of structures with height $h < 100$ m, but for higher structures an upward leader may form and initiate a strike without a preceding downward negative leader. To incorporate this effect in the Petrov–Waters model for a downward flash requires the addition of the electric field E_{cl} created by the thunderstorm cloud to the electric field created by the downward negative leader (if present). Such a consideration allows an estimate to be made of the proportion of downward and upward flashes occurring to very high structures.

Application of the Petrov–Waters model then gives a maximum lateral displacement for a strike to the mast top that is approximated by the relationship:

$$\Delta_{max} \approx 0.54[(h + 15)i_0]^{2/3} \exp(\alpha h) \tag{3.70}$$

where, for example, for a storm cloud field of $E_{cl} = 4.2$ kV/m the mast height enhancement coefficient $\alpha = 0.0021$. The coefficient α will of course increase with the intensity of the storm cloud field E_{cl}.

Since the risk factor of lightning strikes to a structure calculated from Equation (3.58) is:

$$R = \pi \int_0^\infty \Delta_{max}^2(i_0, h) p(i_0) \, di \tag{3.71}$$

where $p(i_0)$ is the probability density function for the current amplitude distribution, then in the case when the lightning current amplitude is, for example, log-normally distributed this integral may be calculated analytically. Substituting the values for lateral displacement and calculating the integral we have for the collection area:

$$A_c = 153(h + 15)^{4/3} \exp(0.0042h) \quad [\text{m}^2, \text{m}] \tag{3.72}$$

Here, the values $i_0 = 31$ kA and $\sigma = 0.7368$ for the mean and standard deviation of the peak current statistical distribution have been used [109]. The risk factor for a structure with height h can then be expressed as:

$$R = 153 \times 10^{-6} N_g (h + 15)^{4/3} \exp(0.0042h) \tag{3.73}$$

In Table 3.8, the risk factor is presented for a ground flash density $N_g = 1$ km^{-2}year^{-1}. It is seen from the Table that the influence of the storm cloud field becomes important for structures with heights $h > 100$ m, indicating that the contribution of upward, structure-initiated flashes increases with the structure height. The results describe well the observed data of Petrov and D'Alessandro [100, 112], including the probability of strikes to very high structures. It is known [66], that the average number of strikes per year to the 540 m television tower in Moscow

Table 3.8   *Risk factor for a negative flash to a tall mast (for $N_g = 1$)*

h (m)	20	50	100	200	300	400	500
R (downward flashes)	0.02	0.04	0.09	0.2	0.3	0.5	0.6
R (all flashes)	0.02	0.05	0.13	0.45	1.2	2.5	5.1

is equal to $R \approx 30$, using a value of ground flash density for the Moscow area of $N_g \cong$ 3–4/km^2/annum.

3.5.5   Protection of overhead power lines

The vulnerability of power systems to lightning is illustrated by considering that a direct lightning strike of median current (31 kA) to an 11 kV line (surge impedance 260 Ω) could produce a peak voltage of 4 MV. Modern lightning location techniques show that even in Northern Europe there were 540 000 ground strikes in 1992. On the UK 132 kV system, the lightning fault rate is between 0.3–1.5 faults/l00 km/year, with damage limited to 5–10 per cent of cases because of a high BIL and fault current interruption. 11 kV fault rates are lower, but the much larger system and lower BIL results in many more incidents and a higher consequential damage in 40 per cent of cases on lines and 100 per cent on cables and equipment.

Power supply plant such as pole-mounted transformers or substation equipment have to be protected against high voltage surges resulting from shielding failures and backflashovers. Lightning overvoltages can be predicted using electromagnetic transient programs for which the main parameters are the peak current, earth resistivity, line geometry, arrester characteristics and cable type. The return stroke current source is often modelled as a transmission line. Even SF$_6$-pressurised GIS are vulnerable to very fast transients. The last two decades have seen advances in applications of metal-oxide surge arresters, solid-state overcurrent and distance protection, line sectionalisers and automatic reclose circuit breaker technology. ZnO gapless arresters in polymeric housings are effective provided that they are properly specified and are positioned close to the protected equipment. This will limit the overvoltage at the equipment terminals to $V_a + 2ST$, where V_a is the residual voltage at the arrester, S is the surge steepness and T the transit time from arrester to the equipment terminals (see chapter 5, section (i)). The IEC Standard [113] gives important recommendations on the specification of arrester rated voltage, maximum continuous operating voltage, discharge current, energy rating and residual voltage.

The availability of online lightning tracking data offers transmission system engineers new possibilities for improved plant asset management and security of supply [21]. Early warning of severe storms is being improved, and the high precision of ground strike location within 500 m is often useful in fault location after a lightning event. Archival data of lightning activity also allows realistic estimates of risk factor and seasonal and geographical variations.

Where a direct strike is prevented successfully by shielding, overvoltages from backflashover to the phase conductor will be proportional to the peak current and are determined by the surge impedances of the tower and shield wire and the footing resistances [114]. The classical lattice diagram method remains useful to take account of factors causing variability in backflashover voltages, such as the steepness of the lightning current front, the point-on-wave of the supply voltage and the line shielding by the upward leader. A useful and substantial reduction of steepness and amplitude results from corona attenuation during the surge propagation along the line. Until recently, this effect has often been misinterpreted as either a reduction in surge propagation velocity or a change in the surge impedance of the line. In fact, the corona discharge affects neither of these, but absorbs energy during the travelling voltage front because of the ionisation of the air around the conductor. Al-Tai *et al.* [115] have shown how the corona attenuation factor can be predicted in terms of the line conductor geometry.

When transient analysis of the network provides an estimate of the overvoltage probability density $p(V)$, the level of protection required can be chosen on the basis of the risk factor R (section 3.5.3) and the risk of flashover R_F (section 3.5.1). Close-in or direct strikes to substations must be prevented because the magnitude and rate of rise of the unattenuated overvoltage would disable protection by gaps or autoreclosure. Gary *et al.* [14] recommend a rolling sphere design of the shielding. Complete shielding will be achieved, even for the minimum observed lightning peak current of 2 kA, with a choice of sphere radius $R = 15$ m. Alternatively, a design current of 5 kA or greater would include 97 per cent of flashes, and is less onerous with its value $R = 27$ m. For 0.2 flashes/year to the substation, this would be equivalent to one failure in 150 years.

The dependence of flashover voltages on the complex overvoltage shapes arising on power systems [116], and the consequent critical values of lightning peak current have been considered by Darveniza *et al.* [117] who deduced empirical equations for time to flashover for partly chopped impulses, and Hutzler and Gibert [118] who both modelled and tested backflashover impulse shapes. In more recent work, Haddad *et al.* [119] have quantified the flashover voltage for air gaps in parallel with ZnO surge arresters. Another factor is that following fault protection operation, an air gap can suffer significant arc erosion of the electrodes which will alter its volt–time breakdown curve.

3.5.6   Protection of electronic equipment

3.5.6.1   Strategy

Lightning electromagnetic pulses (LEMP) are a particular hazard for electronic systems. The indirect surges arising from inductive coupling and resistive voltage drops, and to a lesser degree capacitive and radiative coupling, are sufficient to cause severe damage; a pulse energy of less than 1 microjoule can easily destroy an integrated circuit. As in power equipment the requirements for the protection are speed and reliability in overvoltage control, survivability and system compatibility in normal operation. The strategy of the protection is the same in both cases: where power engineers speak of main and backup protection, electronic engineers specify primary and secondary protection. A recent comprehensive review of the standards for the protection of low voltage systems against lightning and other surges has been published by Hasse [120].

Three stages are generally necessary to achieve immunity to lightning problems:

(i)   transient amplitudes must be controlled by means of the building design and equipment layout
(ii)   the equipment must meet electromagnetic compatibility standards
(iii) surge protector devices should be used to minimise let-through voltages.

3.5.6.2 Mitigation of surges

IEEE C62.41 [121] reviews field experience for transient overvoltages and overcurrents recorded on internal power supplies. This and its companion Standard IEEE C62.45 [122] are presently being updated [123]. Transient magnitudes were found to be as high as 6 kV and 10 kA. Resistive coupling is an important cause of such transients when multiple earth points are present. Single-point grounding is ideal, but because a number of down conductors are usually involved, bonding of all conducting paths that may share lightning current is recommended.

Annexe C of BS 6651 (*General advice on protection against lightning of electronic equipment within a structure*) [39] gives practical advice on transient control. As specified in BS 6651, the effective earth resistance should if possible not exceed 10 Ω. However, with good bonding between N down-conductors, a resistance to earth of $10N$ Ω per conductor should give a satisfactory earth. Removable links are advisable to enable testing of individual conductors. For structures taller than 20 m, down conductor spacing around the periphery of the structure should be no greater than 10 m because of the risk of side flashes. Horizontal bonding conductors should be present every 20 m of elevation, and on all ridges, eaves and parapets. Large flat roofs on high risk structures justify a conductor mesh of 5 m $\times$ 10 m. Other standards differ in detail from these specifications, and are sometimes more stringent. All such conductors can be structural members of good conductivity, reinforcement rods and stanchions, or lightning conductor strip or rod. Sharp bends should be avoided if possible, and low resistance joints and connections are vital.

Whether internal installations such as heating systems should be bonded to the lightning earth will depend on the risk of sparkover between the installations and earth. BS 6651 defines the minimum clearance distances between down conductors and internal metalwork above which bonding is optional rather than mandatory. It is also important to limit to a safe level the step voltage to which persons who are in the vicinity during a direct strike are exposed. If necessary, electrical insulation should be added to the grounding system.

Inductive coupling to electronic systems, arising from the steep current fronts in down conductors and by induction from nearby flashes, can be minimised by suitable routing of cabling and the avoidance of loops. Incoming and outgoing power feeds, LAN connections and signal and telephone lines can be subject to both resistive and inductive transients, and common entry points and single point earthing are important. For some applications optical isolation is worthwhile.

3.5.6.3 Electromagnetic compatibility

Much engineering effort has aimed to identify the types of disturbance that can be experienced from both near and distant storms and to specify the performance tests that must be carried out to meet these satisfactorily. In the last decade, EMC Standards 61000 and 61312 [124, 125] have defined these hazards and test techniques. In particular, multiple screening of structures, interior bonding, grounding requirements, deployment of surge protection devices (SPD) and comprehensive testing of systems and devices are justified.

3.5.6.4   Protector devices

Lightning transients can be effectively limited by protection so as to avoid data and software corruption, partial discharges and irreversible damage to hardware. Such protection is arranged so that the transient control level (TCL) limits overvoltages to below the equipment transient design level (ETDL).

The type of protector device will relate to its location in the electronic system. Locations are defined in the standards as falling into three categories, from a low exposure category A (e.g. plug-in equipment that is expected to experience only low transient levels) to category C which is appropriate for major exposure to transients, such as an incoming mains supply. High vulnerability data, signal or telephone lines are also included in category C because transient attenuation is weak in such networks.

Mains power protectors are tested during manufacture with 1.2/50 impulses of up to 20 kV peak and 8/20 impulse currents of up to 10 kA. Data system protectors are tested up to 5 kV and with 10/700 125 A impulses.

In order to achieve the required TCL, the let-through voltage of the protector must be specified [126]. For mains supplies, the primary series fuse, residual current and miniature circuit breaker boards have secondary shunt varistor protectors together with L and C filters. Filter design, whether high or low pass, is not straightforward for fast transients because of the reduction of inductance by magnetic saturation effects and the effect of series and shunt capacitance.

For electronic devices, two-tier protectors using gas discharge tubes (GDT) followed by metal oxide varistors (MOV) provide transient control. For steep front transients, the let-through voltage of the GDT can be as high as 1 kV before the tube fires and reduces the follow-through arc voltage to about 20 V.

Fortunately, the solid-state technology which has increased the vulnerability of electronic equipment has also offered new protection techniques. The metal oxide varistor with its almost ideal non-ohmic current–voltage characteristic:

$$I = AV^n \tag{3.74}$$

where n is 25 to 30 has found widespread application after its development in the 1970s. Its high capacitance and limited speed remain a limitation for high bandwidths and bit rates, and ageing and condition monitoring are also problems. Recent developments in thyristor technology offer crowbar protection with a fast response as an alternative to the GDT for robust primary protection. These devices can now be manufactured with a drain current of less than 10 μA, a response time below 50 ns and a peak current of 750 A with good resealing.

The efficiency of fortress screening, combined with conventional hybrid GDT/diode and GDT/MOV protection of electronic components even against full triggered lightning currents of up to 52 kA was proven in a test programme of triggered lightning in Japan at a 930 m altitude site [127].

3.5.7   Strikes to aircraft and space vehicles

Lightning strikes to airborne structures are essentially triggered events, and are thus common, especially since altitude effects are also present. On average, civilian airliners receive one stroke per annum [128], 90 per cent of which arise from positive leaders triggered from the aircraft. The standards for the protection of aircraft against lightning strikes define zone 1 as the initial lightning attachment region. This zone is usually taken as being restricted to aircraft extremities where electric field enhancement will tend to favour attachments to such sites, which are defined as the extremity plus a region 0.5 m aft or inboard of it. Severe lightning strikes with currents and action integrals greater than 100 kA and 0.25×10^6 A^2s, respectively, should be avoided outside zone 1 regions. Reported flight experience, however, indicates that occasionally very severe strikes do occur outside this zone. A known hazard to aircraft arises from the nose to tail sweeping of the lightning attachment points due to the aircraft motion. This has led to new schemes for determining the initial attachment zone such as the rolling sphere method and the swept leader method. The definition of aircraft attachment zones has been improved by the work of the FULMAN program [129]. Methods for determining the initial attachment zone are electrical field analysis or test methods such as arc attachment to scale models.

Aircraft flight safety is sometimes discussed in terms of the probability of a catastrophic incident per flying hour. As an example, to illustrate aircraft zoning, it is essential [130] to assume a vanishingly small probability of a hazardous lightning strike, for example less than one in 10^9 flying hours, assuming:

- one strike every 1000 flying hours
- that the most severely damaging lightning strikes are associated with cloud to ground strikes and there is only one strike to ground involved in every ten aircraft strikes
- only one in ten of ground strikes has severe parameters exceeding the protection levels appropriate to a swept stroke zone (zone 2).

These assumptions give a lightning strike to aircraft exceeding zone 2 protection requirements every 10^5 flying hours. Hence, in order to achieve the required probability of a potentially hazardous strike to a zone 2 region, zone 1 has to contain 99.99 per cent of all cloud to ground strikes.

To model the electric field of a real aircraft, the boundary element method derives a solution to the Laplace equation over the surface. The method has the advantage that only the surface of the aircraft has to be meshed. The far-field boundary must also be included as a second two-dimensional surface, but this outer mesh can be remote and simple.

Radomes to protect aircraft radar systems from the elements do not prevent occasional but costly damage from lightning strikes, even after being tested to industry-agreed standards [130]. Radomes usually form the nose cones of aircraft, a zone 1 location which makes them susceptible to damaging strikes. To conduct these lightning currents and prevent damage to the insulating surface of the radome, while maintaining effective transparency to radar signals, thin metal diverter strips consisting of a large number of gaps in series may be fitted.

The new lightweight composite materials which are now being used for radomes, and the introduction of a new generation of airborne weather radar with forward-looking wind shear detection, require increased radar transparency from the radome. Since 1993 all aircraft carrying more than 30 passengers are required to have windshear detection capabilities. Other non-conductive aircraft parts include a new generation of satellite communication and antenna fairings installed on the exterior of aircraft.

Strikes that prejudice air worthiness are fortunately rare, but a Boeing 707 was lost to a lightning strike in the USA in 1963. A survey of USAAF aircraft loss and damage between 1977 and 1981 revealed a financial cost of M\$10. Boulay [52] has described inflight data obtained from strikes, both triggered and intercepted, to suitably instrumented aircraft. These strikes confirmed that at an altitude above 3 km these usually involved cloud discharges. This suggests that both low altitude military sorties and helicopter flights are most at risk from high current cloud-to-ground lightning. Uhlig *et al.* [131] express the need for an agreed test waveform for aeronautical equipment, since the mechanism of the flow of return stroke current in mid channel is unclear.

Lightning research directly associated with the NASA space shuttle activity, both at the launch pad and in space using the optical transient detector (section 3.2.2), followed serious lightning incidents in Apollo 12 and the total destruction of an Atlas-Centaur rocket in 1987. Thayer and colleagues [132] analysed the considerable distortion of the ambient electric field by a vertical space vehicle at launch, and the consequent risk of triggered lightning. An interesting example of the use of the generic Rizk model to the design of lightning protection for a satellite launch pad is given by Joseph and Kumar [133].

3.6 Note

[1] Figure from British Standards reproduced with the permission of BSI under licence number 2003SK/0157. British Standards can be obtained from BSI Customer Services, 389 Chiswick High Road, London W4 4AL (Tel +44(0)20 8996 9001).

3.7 References

1   COHEN, I.B.: 'Benjamin Franklin's experiments' (Harvard University Press, 1941)
2   FRANKLIN, B.: Letter to the Royal Society, London, December 18th, 1751
3   MOORE, C.B., RISON, W., MATHIS, J., and AULICH, G.: 'Lightning rod improvement studies', *J. Appl. Meteorol.*, 2000, **39**, pp. 593–609
4   AUSTIN, B.: 'Schonland: scientist and soldier' (Institute of Physics Publishing, London, 2001)
5   PREECE, W.H.: 'On the space protected by a lightning conductor', *Phil. Mag.*, 1880, **9**, pp. 427–430
6   WILSON, C.T.R.: 'On some determinations of the sign and magnitude of electric discharges in lightning flashes', *Proc. R. Soc. A*, 1916, **92**, pp. 555–574
7   SCHONLAND, B.F.J., and COLLENS, H.: 'Progressive lightning', *Nature*, 1933, **132**, p. 407

8   SCHONLAND, B.F.J.: 'Benjamin Franklin: natural philosopher', *Proc. R. Soc. A*, 1956, **235**, pp. 433–444

9   SCHONLAND, B.F.J.: 'The lightning discharge', *Handbuch der Physik*, 1956, **22**, pp. 576–628

10   BOECK, W.L.: 'Lightning to the upper atmosphere as seen by the space shuttle'. International conference on *Lightning and stat elec*, Florida, 1991, paper 98

11   ALLIBONE, T.E., and SCHONLAND, B.: 'Development of the spark discharge', *Nature*, 1934, **134**, pp. 735–736

12   ANDERSON, R.B., and ERIKSSON, A.J.: 'Lightning parameters for engineering applications', *Electra*, 1980, **69**, pp. 65–102

13   ANDERSON, R.B., ERIKSSON, A.J., KRONINGER, H., MEAL, D.V., and SMITH, M.A.: 'Lightning and thunderstorm parameters', IEE Conf. Publ. 236, 1984, pp. 57–61

14   GARY, C., LE ROY, G., HUTZLER, B., LALOT, J., and DUBANTON, C.: 'Les proprietes dielectriques de l'air et les tres hautes tensions' (Editions Eyrolles, Paris, 1984)

15   ERIKSSON, A.J.: 'Lightning and tall structures', *Trans. South Afr. Inst. Electr. Eng.*, 1978, **69**, pp. 238–252

16   ERIKSSON, A.J.: 'The incidence of lightning strikes to power lines', *IEEE Trans.*, 1987, **PWRD-2**, (3), pp. 859–870

17   DIENDORFER, G., and SCHULZ, W.: 'Lightning incidence to elevated objects on mountains'. International conference on *Lightning protection*, Birmingham, 1998, pp. 173–175

18   FUCHS, F.: 'Lightning current and LEMP properties of upward discharges measured at the Peissenberg tower'. International conference on *Lightning protection*, Birmingham, 1998, pp. 17–22

19   LEES, M.I.: 'Measurement of lightning ground strikes in the UK', *in* 'Lightning protection of buildings, structures and electronic equipment', (ERA, London, 1992), pp. 2.1.1–2.1.10

20   KRIDER, E.P.: 'Wideband lightning detection and mapping system', *in* 'Lightning protection of buildings, structures and electronic equipment' (ERA, London, 1992), pp. 2.2.1–2.2.11

21   CUMMINS, K.L., KRIDER, E.P., and MALONE, M.D.: 'The US National Lightning Detection Network and applications of cloud-to-ground lightning data by electric power utilities', *IEEE Trans. Electromagn. Compat.*, 1998, **40**, (4)

22   CHRISTIAN, H.J., DRISCOLL, K.T., GOODMAN, S.J., BLAKESLEE, R.J., MACH, D.A., and BULCHLER, D.E.: 'The optical transient detector'. 10th international conference on *Atmos elec*, Osaka, 1996, pp. 368–371

23   BOCCIPIO, D.J., CUMMINS, K.L., CHRISTIAN, H.J., and GOODMAN, S.J.: 'Combined satellite and surface based estimation of the intra-cloud-cloud-to-ground lightning ratio over the continental United States', *Mon. Weather Rev.*, 2001, **129**, pp. 108–122

24   BERGER, K.: *in* GOLDE, R.H. (Ed.): 'Lightning' (Academic Press, 1977), chap. 5

25   ISHII, M., SHINDO, T., HONMA, N., and MIYAKE, Y.: 'Lightning location systems in Japan'. International conference on *Lightning protection*, Rhodes, 2000, pp. 161–165

26   RAKOV, V.A.: 'A review of positive and bipolar lightning discharges,' *Bull. Amer. Met. Soc.*, June 2003, pp. 767–776

27   WAREING, B.: 'The effects of lightning on overhead lines'. IEE seminar on *Lightning protection for overhead line systems*, London, 2000, paper 1

28   RAKOV, V.A., UMAN, M.A., WANG, D., RAMBO, K.J., CRAWFORD, D.E., and SCHNETZER, G.H.: 'Lightning properties from triggered-lightning experiments at Camp Blanding, Florida (1997–1999)'. International conference on *Lightning protection*, Rhodes, 2000, pp. 54–59

29   ERIKSSON, A.J., and MEAL, D.V.: 'The incidence of direct lightning strikes to structures and overhead lines', IEE Conf. Publ. 236, 1984, pp. 67–71

30   GARBAGNATI, E., and PIPARO, G.B.L.: 'Parameter von Blitzstromen', *Electrotech. Z*, 1982, **a-103**, pp. 61–65

31   DIENDORFER, G., MAIR, M., SCHULZ, W., and HADRIAN, W.: 'Lightning current measurements in Austria'. International conference on *Lightning protection*, Rhodes, 2000, pp. 44–47

32   JANISCHEWSKYJ, W., HUSSEIN, A.M., SHOSTAK, V., RUSAN, I., LI, J.X., and CHANG, J-S.: 'Statistics of lightning strikes to the Toronto Canadian National Tower', *IEEE Trans.*, 1997, **PD-12**, pp. 1210–1221

33   FIEUX, R.P., GARY, C.H., and HUTZLER, B.P., *et al.*: 'Research on artificially triggered lightning in France', *IEEE Trans.*, 1978, **PAS-97**, pp. 725–733

34   LAROCHE, P.: 'Lightning flashes triggered at altitude by the rocket and wire technique'. International conference on *Lightning and stat elec*, Bath, 1989, paper 2A.3

35   EA TECHNOLOGY PLC: www.eatechnology.co.uk

36   GLOBAL ATMOSPHERICS INC: www.LightningStorm.com

37   DARVENIZA, M.: 'Some lightning parameters revisited'. International conference on *Lightning protection*, Rhodes, 2000, pp. 881–886

38   UMAN, M.A., and KRIDER, E.P.: 'Natural and artificially initiated lightning', *Science*, 1989, **246**, pp. 457–464

39   BRITISH STANDARD 6651: 'Code of practice for protection of structures against lightning', 1999

40   CHAUZY, S., MEDALE, J.C., PRIEUR, S., and SOULA, S.: 'Multilevel measurement of the electric field underneath a thundercloud: 1. A new system and the associated data processing', *J. Geophys. Res.*, 1991, **96**, (D12), pp. 22319–22236

41   SOULA, S.: 'Transfer of electrical space charge from corona between ground and thundercloud: measurements and modelling', *J. Geophys. Res.*, 1994, **99**, pp. 10759–10765

42   HAMELIN, K.J., HUBERT, P., STARK, W.B., and WATERS, R.T.: 'Panneau a pointes multiples pour la protection contre la foudre'. CNET research report 81/230, 1981

43 UMAN, M.A., and RAKOV, V.A.: 'A critical review of nonconventional approaches to lightning protection,' *Bull. Amer. Met. Soc.*, December 2002, pp. 1809–1820

44 SCHNETZER, G.H., FISHER, R.J., RAKOV, V.A., and UMAN, M.: 'The magnetic field environment of nearby lightning'. International conference on *Lightning protection*, Birmingham, 1998, pp. 346–349

45 SCHWAB, A.J., and CVETIC, J.M.: 'Radiated field stresses in the environment of lightning current'. International conference on *Lightning protection*, Birmingham, 1998, pp. 341–345

46 SCHONLAND, B.F.J., and COLLENS, H.: 'Progressive lightning', *Proc. R. Soc.*, 1934, **143**, pp. 654–674

47 SCHONLAND, B.F.J., MALAN, D.J., and COLLENS, H.: 'Progressive lightning II', *Proc. R. Soc.*, 1935, **152**, pp. 595–625

48 MAZUR, V., and RUHNKE, L.H.: 'Model of electric charges in thunderstorms and associated lightning', *J. Geophys. Res.*, 1998, **103**, pp. 23200–23308

49 HUBERT, P.: 'Triggered lightning in France and New Mexico'. *Endeavour, J. Geophys. Res.*, 1984, **8**, pp. 85–89

50 ISHII, M., SHIMIZU, K., HOJO, J., and SHINJO, K.: 'Termination of multiple-stroke flashes observed by electromagnetic field'. International conference on *Lightning protection*, Birmingham, 1998, pp. 11–16

51 McEACHRON, K.B.: 'Lightning to the Empire State Building', *J. Franklin Inst.*, 1939, **227**, pp. 149–217

52 BOULAY, J-L.: 'Triggered and intercepted lightning arcs on aircraft'. International conference on *Lightning and stat elec*, Williamsburg, 1995, pp. 22.1–22.8

53 WATERS, R.T.: 'Breakdown in non-uniform fields', *IEE Proc.*, 1981, **128**, pp. 319–325

54 WATERS, R.T.: 'Lightning phenomena and protection systems: developments in the last decade', *in* 'Lightning protection of buildings, structures and electronic equipment' (ERA, London, 1992), pp. 1.2.1–1.2.8

55 ALLIBONE, T.E., and MEEK, J.M.: 'The development of the spark discharge', *Proc. R. Soc. A*, 1938, **166**, pp. 97–126

56 WATERS, R.T., and JONES, R.E.: 'The impulse breakdown voltage and time-lag characteristics of long gaps in air I: the positive discharge', *Philos. Trans. R. Soc. Lond. A, Math. Phys. Sci.*, 1964, **256**, pp. 185–212

57 WATERS, R.T., and JONES, R.E.: 'The impulse breakdown voltage and time-lag characteristics of long gaps in air II: the negative discharge', *Philos. Trans. R. Soc. Lond. A, Math. Phys. Sci.*, 1964, **256**, pp. 213–234

58 PIGINI, A., RIZZI, G., BRAMBILLA, R., and GARBAGNETI, E.: 'Switching impulse strength of very large air gaps'. International symposium on *High voltage eng.*, Milan, 1979, paper 52.15

59 MRAZEK, J.: 'The thermalization of the positive and negative lightning channel'. International conference on *Lightning protection*, Birmingham, 1998, pp. 5–10

60 LES RENARDIERES GROUP: 'Long air gaps at Les Renardieres: 1973 results', *Electra*, 1974, **35**, pp. 49–156

61   LES RENARDIERES GROUP: 'Positive discharges in long air gaps', *Electra*, 1977, **53**, pp. 31–132

62   LES RENARDIERES GROUP: 'Negative discharges in long air gaps', *Electra*, 1981, **74**, pp. 67–216

63   ROSS, J.N.: 'The diameter of the leader channel using Schlieren photography', *Electra*, 1977, **53**, pp. 71–73

64   KLINE, L.E., and DENES, L.J.: 'Prediction of the limiting breakdown strength in air from basic data'. International symposium on *High voltage eng.*, 1979, paper 51.02

65   ORVILLE, R.E.: *in* GOLDE, R.H. (Ed.): 'Lightning' (Academic Press, 1977), chap. 8

66   GORIN, B.N., and SHKILEV, A.V.: 'Electrical discharge development in long rod–plane gaps in the presence of negative impulse voltage', *Elektrichestvo*, 1976, **6**, pp. 31–39

67   BAZELYAN, E.M., and RAIZOV, Y.P.: 'Lightning physics and lightning protection' (Institute of Physics Publishing, London, 2000)

68   PETROV, N.I., and WATERS, R.T.: 'Determination of the striking distance of lightning to earthed structures', *Proc. R. Soc. Lond. A, Math. Phys. Sci.*, 1995, **450**, pp. 589–601

69   DIENDORFER, G., and UMAN, M.A.: 'An improved return stroke model with specified channel base current', *J. Geophys. Res.*, 1990, **95**, pp. 13621–13664

70   COORAY, V.: 'A model for subsequent return strokes', *J. Electrost.*, 1993, **30**, pp. 343–354

71   WAGNER, C.F.: 'The relationship between stroke current and the velocity of the return stroke', *IEEE Trans.*, 1963, **PAS-68**, pp. 609–617

72   COORAY, V.: 'The Lightning Flash' (The IEE, London, 2003)

73   DELLERA, L., and GARBAGNATI, E.: 'Lightning stroke simulation by means of the leader progression model', *IEEE Trans. Power Deliv.*, 1990, **PWRD-5**, pp. 2009–2029

74   JONES, C.C.R.: 'The rolling sphere as a maximum stress predictor for lightning attachment zones'. International conference on *Lightning and stat elec*, Bath, 1989, paper 6B.1

75   NFPA STANDARD 780: 'Installation of lightning protection systems', 1997

76   IEC STANDARD 61662: 'Assessment of the risk of damage due to lightning', 1995

77   GRZYBOWSKI, S., and GAO, G.: 'Protection zone of Franklin rod'. International symposium on *High voltage engineering*, Bangalore, 2001

78   ARMSTRONG, H.R., and WHITEHEAD E.R.: 'Field and analytical studies of transmission line shielding', *IEEE Trans.*, 1968, **PAS-87**, pp. 270–281

79   WHITEHEAD, E.R.: *in* GOLDE, R.H. (Ed.): 'Lightning' (Academic Press, 1977), chap. 22

80   GOLDE, R.H.: 'Lightning and tall structures', *IEE Proc.*, 1978, **25**, (4), pp. 347–351

81   GILMAN, D.W., and WHITEHEAD, E.R.: 'The mechanism of lightning flashover on high-voltage and extra-high-voltage transmission lines,' *Electra*, 1973, **27**, pp. 69–89

82 BERGER, K., ANDERSON, R.B., and KRONINGER, H.: 'Parameters of lightning flashes', *Electra*, 1975, **41**, pp. 23–27

83 CARRARA, G., and THIONE, L.: 'Switching surge strength of large air gaps: a physical approach', *IEEE Trans.*, 1976, **PAS-95**, pp. 512–520

84 DELLERA, L., and GARBAGNATI, E.: 'Shielding failure evaluation: application of the leader progression model', IEE Conf. Publ. 236, 1984, pp. 31–36

85 RIZK, F.A.M.: 'Modelling of lightning incidence to tall structures', *IEEE Trans. Power Deliv.*, 1994, **PWRD-9**, pp. 162–178

86 PHELPS, C.T., and GRIFFITHS, R.F.: 'Dependence of positive corona streamer propagation on air pressure and water vapor content', *J. Appl. Phys.*, 1976, **47**, pp. 1929–1934

87 ERIKSSON, A.J., ROUX, B.C., GELDENHUYS, H.J., and MEAL, V.: 'Study of air gap breakdown characteristics under ambient conditions of reduced air density', *IEE Proc. A, Phys. Sci. Meas. Instrum. Manage. Educ. Rev.*, 1986, **133**, pp. 485–492

88 PETROV, N.I., and WATERS, R.T.: 'Conductor height and altitude: effect on striking distance'. International conference on *Lightning and mountains* (SEE), Chamonix Mont-Blanc, 1994, pp. 52–57

89 ALEKSANDROV, G.N., BERGER, G., and GARY, C.: CIGRE paper no. 23/13-14, 1994

90 VAN BRUNT, R.J., NELSON, T.L., and STRICKLE, H.K.L.: 'Early streamer emission lightning protection systems: an overview', *IEEE Electr. Insul. Mag.*, 2000, **16**, (1), pp. 5–24

91 ALLEN, N., HUANG, C.F., CORNICK, K.J., and GREAVES, D.A.: Sparkover in the rod–plane gap under combined direct and impulse voltages', *IEE Proc. Sci. Meas. Technol.*, 1998, **145**, pp. 207–214

92 ALLEN, N.L., CORNICK, K.J., FAIRCLOTH, D.C., and KOUZIS, C.M.: 'Tests of the "early streamer emission" principle for protection against lightning', *IEE Proc. Sci. Meas. Technol.*, 1998, **145**, pp. 200–206

93 ALLEN, N.L., and EVANS, J.C.: 'New investigations of the "early streamer emission" principle', *IEE Proc. Sci. Meas. Technol.*, 2000, **147**, pp. 243–248

94 CHALMERS, I.D., EVANS, J.C., and SIEW, W.H.: 'Considerations for the assessment of early streamer emission lightning protection', *IEE Proc. Sci. Meas. Technol.*, 1999, **146**, pp. 57–63

95 HARTONO, Z.A., and ROBIAH, I.: 'A study of non-conventional air terminals'. International conference on *Lightning protection*, Rhodes, 2000, pp. 357–360

96 MACKERRAS, D., DARVENIZA, M., and LIEW, A.C.: 'Review of claimed enhanced protection of buildings by early streamer emission air terminals', *IEE Proc. Sci. Meas. Technol.*, 1997, **144**, pp. 1–10

97 ALEKSANDROV, G.N., and KADZOV, G.D.: 'On increasing of efficiency of lightning protection', *Elektrichestvo*, 1987, (2), pp. 57–60

98 RISON, W., MOORE, C.B., MATHIS, J., and AULICH, G.D.: 'Comparative tests of sharp and blunt lightning rods'. International conference on *Lightning protection*, Birmingham, 1998, pp. 436–441

99   D'ALESSANDRO, F., and BERGER, G.: 'Laboratory studies of corona current emissions from blunt, sharp and multipointed air terminals'. International conference on *Lightning protection*, Birmingham, 1998, pp. 418–423

100 PETROV, N.I., and D'ALESSANDRO, F.: 'Theoretical analysis of the processes involved in lightning attachment to earthed structures', *J. Phys. D, Appl. Phys.*, 2002, **35**, pp. 1–8

101 PETROV, N.I., and WATERS, R.T.: 'Striking distance of lightning to earthed structures: effect of stroke polarity'. International symposium on *High voltage eng.*, London, 1999, pp. 220–223

102 PETROV, N.I., and WATERS, R.T.: 'Striking distance of lightning to earthed structures: effect of structure geometry'. International symposium on *High voltage eng.*, London, 1999, pp. 393–396

103 GOLDE, R.H.: 'Lightning' (Academic Press, 1977), chap. 17

104 LUPEIKO, A.V., and SYSSOEV, V.S.: 'Dependence of probability of strikes to aircraft models on the parameters of spark discharges', *Proceedings MEI* (in Russian), 1990, (231)

105 GAYVORONSKY, A.S., and OVSYANNIKOV, A.G.: 'New possibilities of physical modelling of orientation process and the influence of lightning leader on the protected object'. International conference on *Lightning protection*, Florence, 1996, pp. 440–443

106 JONES, B., and WATERS, R.T.: 'Air insulation at large spacings', *IEE Proc.*, **125**, pp. 1152–1176

107 WHITEHEAD, E.R.: 'Edison Electric Institute Report Pathfinder Project, 1972

108 WHITEHEAD, E.R.: 'CIGRE survey of the lightning performance of EHV transmission lines', *Electra*, 1974, **33**, pp. 63–89

109 PETROV, N.I., PETROVA, G.N., and WATERS, R.T.: 'Determination of attractive area and collection volume of earthed structures'. International conference on *Lightning protection*, Rhodes, 2000, pp. 374–379

110 YUMOTO, M.: 'Technology of electrical discharges ranging from nanometer scale to megameter scale,' *IEEJ Trans. Fundamentals and Materials*, 2004, **124**, pp. 13–14

111 PETROV, N.I., PETROVA, G.N., and D'ALESSANRO, F.: 'Quantification of the possibility of lightning strikes to structures using a fractal approach,' *IEEE Trans. Diel. Elec. Insul.*, 2003, **10**, pp. 641–654

112 D'ALESSANDRO, F., and PETROV, N.I.: 'Assessment of protection system positioning and models using observations of lightning strikes to structures', *Proc. R. Soc. Lond. A*, 2002, **458**, pp. 723–742

113 IEC STANDARD 60099-5: 'Surge arresters part 5 – selection and application recommendations', 1996

114 IEEE WORKING GROUP: 'A simplified method for estimating lightning performance of transmission lines', *IEEE Trans. Power Appar. Syst.*, 1985, **104**, pp. 919–932

115 AL-TAI, M., GERMAN, D.M., and WATERS, R.T.: 'The simulation of surge corona on transmission lines', *IEEE Trans.*, 1989, **PD-4**, pp. 1360–1368

116 LES RENARDIERES GROUP: 'Double impulse tests of long air gaps', *IEE Proc. A, Phys. Sci. Meas. Instrum. Manage. Educ.*, 1986, **133**, pp. 395–479

117 DARVENIZA, L.D., POPOLANSKY, F., and WHITEHEAD, E.R.: 'Lightning protection of UHV lines', *Electra*, 1975, **41**, pp. 39–69

118 HUTZLER, B., and GIBERT, J.: 'Breakdown characteristics of air insulations exposed to short-tailed lightning impulses'. IEE Conf. Publ. 236, 1984, pp. 158–162

119 HADDAD, A., GERMAN, D.M., WATERS, R.T., and ABDUL-MALEK, Z.: 'Coordination of spark gap protection with zinc oxide surge arresters', *IEE Proc. Gener. Transm. Distrib.*, 2001, **148**, pp. 21–28

120 HASSE, P.: 'Overvoltage protection of low voltage systems' (IEE Publishing London, 2000)

121 IEEE STANDARD C62.41: 'Recommended practice on surge voltages in low-voltage AC power circuits', 1991

122 IEEE STANDARD C62.45: 'Guide on surge testing for equipment connected to low-voltage AC power circuits', 1992

123 MARTZLOFF, F.: 'The trilogy update of IEEE Standard C62.41'. International conference on *Lightning protection*, Rhodes, 2000, pp. 887–892

124 IEC STANDARD 61000: 'Electromagnetic compatibility (EMC)', 1995

125 IEC STANDARD 61312-1: 'Protection against lightning electromagnetic pulses', 1995 (parts 2&3 in draft)

126 HASBROUCK, R.T.: 'Performance of transient limiters under laboratory simulated and rocket-triggered lightning conditions'. International conference on *Lightning and stat elec*, Bath, 1989, paper 12B1

127 NAKAMURA, K., WADA, A., and HORII, K.: 'Long gap discharge to an EHV transmission tower by a rocket triggered lightning experiment'. International conference on *Lightning and stat elec*, Florida, 1991, paper 62-1

128 LARSSON, A., LALANDE, P., BONDIOU-CLERGERIE, and DELANNOY, A.: 'The lightning swept stroke along an aircraft in flight – phenomenology and numerical simulations'. International conference on *Lightning protection*, Rhodes, 2000, pp. 819–824

129 ZAGLAUER, H., WULBRAND, W., and DOUAY, A.: 'Definition of lightning strike zones in aircraft and helicopters'. International conference on *Lightning stat elec*, Toulouse, 1999

130 HARDWICK, C.J.: 'Review of the 1999 Joint Radome Programme'. International conference on *Lightning and stat elec*, Toulouse, 1999, paper 01-2322

131 UHLIG, F., GONDOT, P., LAROCHE, P., LALANDE, P., and HARDWICK, J.: 'A basis for a new improved definition of the lightning external environment in the aeronautical field'. International conference on *Lightning protection*, Rhodes, 2000, pp. 909–914

132 THAYER, J.S., NANAVICZ, J.E., and GIORI, K.L.: 'Triggering of lightning by launch vehicles'. International conference on *Lightning stat elec*, Bath, 1989

133 JOSEPH, N.T., and KUMAR, U.: 'Evaluation of the protective action of LPS to Indian satellite launch pad'. International symposium on *High voltage eng.*, Bangalore, 2001

Chapter 4

Partial discharges and their measurement

I.J. Kemp

4.1 Introduction

Partial discharges are localised gaseous breakdowns which can occur within any plant system provided the electric stress conditions are appropriate. Because the breakdown is only local, failing to result in a following current flow, it is described as partial.

Why are partial discharges (PD) of importance to high voltage engineers? Partial discharge activity is both a symptom of degradation in the insulating systems of power plant – irrespective of the causative stress – and a stress mechanism in itself. Wherever degradation occurs in an insulating system, be it due to electrical, mechanical, thermal or chemical/environmental conditions, it is generally accompanied by the generation of partial discharges. Once present, these then tend to dominate as the stress degradation mechanism. Irrespective of whether the insulating system is gaseous, liquid, solid or a combination of these, partial discharge activity will cause degradation. It can therefore be appreciated why understanding the processes by which partial discharges cause degradation is so important to the development of new insulating systems capable of withstanding this stress mechanism. In addition, it can also be appreciated why understanding the correlations among the measurable parameters of discharge activity and the nature, form and extent of degradation present is so important to the engineer responsible for the maintenance and asset management of existing plant systems.

4.2 Partial discharge degradation mechanisms

The electrons, ions, atoms, radicals and excited molecular species produced in a partial discharge move under the influence of the following forces variously:

- thermal excitation
- the electric field

- electrostatic forces
- the electric wind, generated by the collision of the ionic species, moving under the influence of the electric field, with the molecules of the surrounding gas.

The distribution of the reactive species within the gas discharge, and their resulting impact at the discharge surfaces, will be complex. The following sections discuss this complete interaction by considering the different stress mechanisms likely to be prevalent.

4.2.1   Particle impact stress

As has been explained earlier in this book, a gas discharge consists generally of electrons, positive and negative ions and photons. In relation to partial discharges, when these particles impact on a surface at the ends of the discharging channel, they may cause degradation at that surface. Any of these particle types may contain sufficient energy to cause bond scission, often with an associated electron release.

An impacting ion at an insulating surface may result in local molecular changes as a result of either an electronic interaction between the incoming charged particle and the shell electrons of the molecules of the insulating material or through a tight interaction between the ionising ion and one (or more) ions of the surface lattice.

As detailed by Hepburn [1], the interaction which occurs when an electron collides with a molecular surface will depend upon the structure and energy state of the impacted species and upon the electron energy.

An energetic electron impacting upon an uncharged molecular species can interact in four ways:

(i)   an electron, with velocity v_1, colliding with a molecule, M, can lose part of its kinetic energy to the molecule without becoming attached:

$$e^-(v_1) + M \rightarrow e^-(v_2) + M^*$$

the electron continues on with a lower velocity, v_2, and the excited molecule M^* either emits the extra energy as a photon:

$$M^* \rightarrow h\nu + M$$

or loses the energy by collision with a second molecule:

$$M^* + M_1 \rightarrow M + M_1^*$$

(ii)   an electron can collide with a molecule and become attached:

$$e^- + M \rightarrow M^-$$

forming a negative ion

(iii)   an electron can be energetic enough to detach an electron from a molecule:

$$e^- + M \rightarrow e^- + e^- + M^+$$

increasing the number of free electrons in the system and creating a positive ion

(iv) an electron can become attached to a molecule and cause a division into charged and neutral subspecies:

$$e^- + M \rightarrow M_1 + M_2^-$$

In a similar manner to the possible reactions between electrons and molecules, there are a variety of interactions possible between an electron and an ion:

(i) an electron colliding with a negative molecular ion can cause a reaction similar to that given above [3] but the resultant molecule is neutral:

$$e^- + M^- \rightarrow e^- + e^- + M$$

(ii) for a number of molecular species, it is possible for the molecule to become doubly negatively charged:

$$e^- + M^- \rightarrow M^{2-}$$

(iii) when an electron interacts with a positive ion the electron can become attached to the molecule and any excess energy may be released as a photon:

$$e^- + M^+ \rightarrow M + h\nu$$

As with electron impact, the transfer of energy from a photon to a molecule will cause changes in a number of ways:

(i) the energy transfer can cause photoionisation: where the ionisation energy is less than that of the photon, the excess can be released as a less energetic photon or as increased kinetic energy in the electron:

$$h\nu_1 + M \rightarrow M^+ + e^-(\nu_1) + h\nu_2$$

$$h\nu_1 + M \rightarrow M^+ + e^-(\nu_2)$$

(ii) a molecule can be split into ionic subspecies:

$$h\nu + M \rightarrow M_1^+ + M_2^-$$

(iii) a molecule can be split into ionic and neutral species and an electron:

$$h\nu + M \rightarrow M_1 + M_2^+ + e^-$$

(iv) a molecule can be divided into free radical species, which are highly reactive

$$h\nu + M \rightarrow M_1^. + M_2^.$$

In relation to photon impact with an ion, the energy transferred from a photon to an ionic molecular species can cause changes to occur in the following manner:

(i) release of an electron from a negative ion:

$$h\nu + M^- \rightarrow M + e^-$$

(ii)   release of an electron from a positive ion:

$$h\nu + M^+ \rightarrow M^{2+} + e^-$$

(iii)   splitting a negative ion into neutral and charged species:

$$h\nu + M^- \rightarrow M_1^+ + M_2 + 2e^-$$

(iv)   splitting a positive ion into neutral and charged species:

$$h\nu + M^+ \rightarrow M_1 + M_2^+$$

The photon energy and the energy state of the molecular or ionic species involved in the reactions given above will determine which of the reactions will occur.

Particle impact has been attributed variously as the mechanism of partial discharge degradation [2, 3]. Interestingly, in polyethylene, Mayoux [4] has shown that degradation due to ion bombardment occurs at a significant rate only if the charge density of ions exceeds $\approx 1.5 \times 10^2$ cm^{-2}. Further, he has demonstrated that although, in theory, electrons may produce degradation, electron energies in excess of 500 eV are required to cause substantial damage. However, once again, the synergetic nature of discharge degradation mechanisms cannot be overstressed; that one form of stressing does not in itself result in degradation does not mean that it can be dismissed as a contributory mechanism to degradation.

In respect of the potential for particle impact damage at a given insulating surface, it is useful to consult one of the many excellent texts on gas breakdown, e.g. Reference 5, to determine, for a given gaseous situation, the nature of the particles involved and for a given set of conditions (partial pressure etc.) the statistical distribution of energies associated with those particles. Having obtained this information, consultation of materials texts will provide details on molecular structure and bond energies. Comparison of the two sets of information should provide some guidance of the likelihood of bond scission etc. and where it might occur. An example of the latter form of data is provided in Table 4.1.

Table 4.1   Chemical bond energies associated with epoxy resin

Bond type	Group structure	Energy (kJ/mol)
C–H	aromatic	435
C–H	methyl	410
C–H	methylene	400
C–O	ether	331
C=O	ketone	729
C–C	aliphatic–aliphatic	335
C–C	aromatic–aliphatic	347

4.2.2 Thermal stress

The energy injected into the gaseous environment by the discharge will increase the temperature of the gas in the local vicinity. In turn, this thermal differential will cause the gas molecules in the hotter region to migrate to the cooler regions. Although attempts have been made to ascertain the gas temperature under partial discharging [6], in general these results must be treated with caution. In addition, the combinations of particle input energy transfer, chemical bond restructuring and other potentially exothermic reactions will result in temperature increases at a discharging surface. The author's personal experience has suggested that the thermal stress created by a partial discharge may be sufficient to cause damage to polymeric materials but not to other forms of solid insulating material. Even in the case of polymers, the degradation sustained due to thermal stressing alone is likely to be insignificant compared with other stresses present. However, since the degradation is due to a synergetic interaction of a number of stresses, it must still be considered where partial discharges are present.

The reaction of polymeric materials to a purely thermal stress is dependent on the structure of the material [1].

For example, simple structures, such as polyethylenes, degrade by random chain scission:

$$-CH_2-CH_2-CH_2-CH_2-CH_2-CH_2- \rightarrow -CH_2-CH_2-CH_2-CH_2-CH_2^\cdot + \,^\cdot CH_2-$$

the left-hand side of the reaction continues:

$$\rightarrow -CH^\cdot-CH_2-CH_2-CH_2-CH_3 \rightarrow -CH^\cdot + CH_2=CH_2-CH_2-CH_3$$

thus producing subgroups of length determined by the fold back length of the chain transferring the radical element.

Other simple polymers, e.g. polyvinyl chloride, degrade not by chain scission but by loss of side constituents:

$$-CH_2-CHCl-CH_2-CHCl-CH_2-CHCl- \rightarrow -CH_2-CH=CH-CH_2-CHCl-$$
$$+ HCl$$

thus although the polymer chain remains the same length it loses stability by developing unsaturated sites in the chain.

Depolymerisation will occur in situations where the polymer molecule contains no easily abstracted atoms or groups. Polymethylmethacrylate is an example of this type of reaction:

$$-[CH_2-C(CH_3)(COOCH_3)]-[CH_2-C(CH_3)(COOCH_3)]$$
$$-[CH_2-C(CH_3)(COOCH_3)]$$
$$\rightarrow -CH_2-C(CH_3)(COOCH_3)-CH_2-C^\cdot(CH_3)(COOCH_3)$$
$$\rightarrow -CH_2-C(CH_3)(COOCH_3) + CH_2=C(CH_3)(COOCH_3)$$

here, it can be seen that the polymer chain unzips, i.e. each segment of the polymer will return to the prepolymer state.

The presence of aromatic rings, i.e. benzene-type structures, in a polymer stiffens the chain and raises the glass transition temperature, i.e. the temperature at which the structure changes from a glassy to a plastic state. This can be seen in the work by Black [7] who has determined that the glass transition temperature of polymer (1) below, which has aliphatic rings in the chain, is 80°C whereas polymer (2), which has aromatic rings, has a glass transition temperature of 380°C:

$$[-NH-(CH_2)_6-NH(C=O)-(CH_2)_6-(C=O)-]_n \tag{1}$$

$$[-NH-(CH)_6-NH(C=O)-(CH)_6-(C=O)-]_n \tag{2}$$

The reaction of a polymer to thermal stress in air will also result in thermal oxidation of the material. Main chain scission or side group removal will create a radical species which reacts readily with oxygen to form peroxy radical species:

$$-CH_2-CH_2-CH_2- \rightarrow -CH_2-CH_2 + O_2$$
$$-CH_2-CH_2O_2$$

The peroxy radical can abstract hydrogen from a polymer group in the vicinity to form a hydroperoxide and a second radical species:

$$-CH_2-CH_2O_2 + R-H \rightarrow -CH_2-CH_2O_2H + R$$

The hydroperoxide species can also decompose due to the application of heat to form radical species:

$$ROOH \rightarrow RO + OH$$

From the preceding discussion, it can be seen that the application of heat to a polymeric insulating system will produce a number of reactive sites and species given the application of sufficient heat.

These are just some of the reactions which can occur due to the thermal effects. However, as indicated earlier, all effects must be considered to occur in a synergetic manner.

4.2.3 Mechanical stress

A vibrational mechanical stress will be set up in a solid insulating material subject to partial discharge stressing under normal AC operating conditions due to the interaction of trapped charge in the solid matrix from the discharge interacting with the applied AC electric stress field. In addition, trapped charged particles of similar polarity will be repelled from each other and dissimilar charges attracted to each other, again resulting in a local mechanical stress within the solid matrix. There will also be a mechanical stress resulting from the impact of larger particles at the discharging surface depending on the mass number and collision velocities of these particles. It is unlikely that this will have sufficient energy to cause fracture, as the shock wave is likely only to have energies of the order of 10^{-12} J [8]. However, once again the synergetic effects of such a process cannot be dismissed.

Particle impact from a partial discharge can also result in bond breakage and the production of ionic and radical species, as indicated earlier. These species may react with the gas but may also react with the solid to produce, for example, in the case of polymers, extra crosslinks. These extra crosslinks may produce a stiffer section in the polymer making it less resistant to shear, tensile and compressive forces induced in the polymer by the electric stress/trapped charge effects. Arbab, Auckland and Varlow [9] have been long proponents of mechanical stress damage to materials via partial discharge/AC electric field stressing, and their various papers on this subject are extremely illuminating.

4.2.4 Chemical stress

As indicated earlier, particle impact, thermal stressing and mechanical stressing can all result in changes to the chemical structure of a solid insulating system subject to partial discharge stressing. In addition, the species generated in the gaseous environment of the discharge may also interact chemically with the solid material when they impinge at its surface. Given the range of potential interactions (on the basis of the range of gases, liquids and solids plus contaminants) involved, it is impossible within the confines of this chapter, to detail all possible effects. However, to give the reader some sense of the issues involved, some examples are presented.

Air is the most common atmospheric medium through which partial discharges propagate and, as such, it is worthy of consideration from a chemical viewpoint. Air is a complex mixture of gases, of which the major components are nitrogen and oxygen with minor concentrations of argon, water vapour and oxides of carbon. The molecular species, ions, etc., generated by discharges in air [10–14] are, therefore, most likely to be combinations of nitrogen, oxygen, carbon oxides and hydrogen (from breakdown in atmospheric moisture). The gaseous species produced will be, for an AC stress situation, of both positive and negative polarity e.g. O^- and O_2^-, N^+ and O_2^+. In addition, in a discharge atmosphere the polar nature of water molecules causes them to be attracted to charged species in the discharge.

The ionic species formed in positive and negative DC corona in air are shown in Table 4.2 to illustrate these differences.

Column A shows the principal species formed during discharges where the high voltage point electrode is negatively charged. The species listed are all hydrated, i.e. had attached $(H_2O)_n$ groups. The principal species generated when the point is positively charged are found in columns B and C; the species listed in column B are not hydrated, those in column C are hydrated.

The difference in the ionic species generated in the discharges is significant in that the character of the chemical reactions which will occur on the material surface, due to the impact of the species produced in discharges from positive and negative points, will be different. In the case of an AC field stress condition, in which both potential discharge surfaces may act to form discharges, all species may be present.

The production of reactive oxygen species and oxides of nitrogen in the discharge atmosphere is particularly important when considering degradation processes. The triplet form of oxygen, ozone (O_3), is a strong oxidising agent and oxides of

Table 4.2   *Ionic species formed in DC coronas*

Column A	Column B	Column C
CO_2^-	O^+	H^+
CO_3^-	O_2^+	N^+
O_2^-	N^+	NO^+
O_3^-	NO^+	N_2^+NO
NO_3^-		NO_2^+
		NO^+NO

nitrogen are known to react in air to form nitric acid, also strongly degrading. This phenomenon and, indeed, the importance of chemical degradation in partial discharge stressing in general is exemplified by the work of Shields and the present author [15, 16] in comparing discharge degradation of mica in an air and a nitrogen environment. Under similar experimental conditions, including discharge repetition rate and magnitude, it was found that degradation was much more severe over a given time period in an air environment. Given the physical similarities between the two gases, it would be expected that the only differences between degradation in the two atmospheres would be attributable to their chemical differences. In this respect, the formation of nitric acid (HNO_3) at the mica surface in air appears to offer the most likely explanation of the variation in degradation. Although surface reactions are possible with active nitrogen, no nitric acid will be produced in a nitrogen atmosphere. The active nitrogen will transfer energy to the mica structure in order to return to the ground state. In air, however, where oxygen and water are also present the following reactions are likely to occur:

$$3O_{2(g)} + h\nu \Rightarrow 2O_{3(g)}$$

and

$$N_{2(g)} + O_{2(g)} + h\nu \Rightarrow 2NO_{(g)} \qquad NO_{(g)} + \tfrac{1}{2}O_{2(g)} \Rightarrow NO_2$$

and finally

$$2NO_{2(g)} + H_2O_{(g,l)} \Rightarrow HNO_{3(g,l)} + HNO_2$$

It is therefore suggested that the increased erosion of mica in air can best be explained by nitric acid, formed in the discharge environment, causing surface erosion by an acid reaction mechanism on the mica. This mechanism would also account for the presence of metallic elements from the electrode at the degraded surface, found during the experimental programme, as the electrode too would suffer similar erosion.

As damage was observed in both gaseous environments, however, a second degradation mechanism had to be postulated to account for degradation in nitrogen. Given the stress conditions prevalent, chemical, bulk thermal and surface/bulk field effects could be rejected, and, on this basis, the most likely source of degradation was considered to be energetic particle bombardment, as discussed earlier. Bearing in mind that mica consists of a lattice of SiO_4 units, this mechanism would involve the energetic particles within the gas transferring their energy to the mica surface causing either direct bond scission:

$$M^* + -Si-O-Si- \Rightarrow -Si-O + Si- + M$$

where M^* is the energetic particle, or bond scission by cumulative localised heating:

$$N M^* \Rightarrow \Delta T + NM$$

where N is the number of energetic particles and ΔT is the increase in temperature, then:

$$-Si-O-Si- + \Delta \Rightarrow -Si-O + Si-$$

where Δ is the heat applied.

As air contains a large proportion of nitrogen, and the presence of other molecules does not preclude this mechanism, it was assumed that a similar reaction was occurring in air, concomitantly with, but secondary to, the acid degradation mechanism.

Another example of the importance of chemical stressing under partial discharge stressing is exemplified by the work of Hepburn *et al.* [17] but this time in an organic polymeric material, i.e. epoxy resin, as opposed to the inorganic, crystalline mica structure.

Examination of the epoxy resin degraded surfaces following partial discharge stressing in air indicated the presence of various nitrogen compounds, carbonaceous anhydrides, acids and peracids and led to the following reactions being proposed for the resin degradation.

Nitric acid is known to react with organic compounds as follows:

nitric acid breaks down into a nitrous oxide and a hydroxyl radical

$$HNO_3 \rightarrow OH + NO_2$$

an organic radical is formed when hydrogen is extracted by the hydroxyl radical

$$R-H + OH \rightarrow R + H_2O$$

the organic radical then reacts with nitrogen oxide to form nitrate

$$R + NO_2 \rightarrow R-NO_2$$

nitrite rather than nitrate may be formed

$$R + NO_2 \rightarrow R-O-N-O$$

The reactions described would account for the presence of nitrogen compounds but do not explain the other reactions taking place.

It is known [10, 11, 18] that reactive carbonaceous compounds are present in air discharges. Formation of reactive carbon species and possible routes to production of anhydrides are thought to rely on either:

a activated oxygen attack on the methyl group:

$$R\text{–}CH_3 + O_2 \rightarrow R\text{–}CH_2 + OOH$$

or
b hydroxyl radical attack on the methyl group:

$$R\text{–}CH_3 + OH \rightarrow R\text{–}CH_2 + H_2O$$

both of these initiating reactions produce a methylene radical on the resin chain.

The radical reacting with oxygen produces an aldehyde:

$$R\text{–}CH_2 + O_2 \rightarrow R\text{–}CH_2O_2 \rightarrow R\text{–}(C{=}O)\text{–}H + OH$$

the aldehyde reacts with oxygen as follows:

$$R\text{–}(C{=}O)\text{–}H + O_2 \rightarrow R\text{–}C{=}O + OOH$$

Carbonyl and hydroxyl radicals interact as follows:

$$R\text{–}C{=}O + OH \rightarrow R\text{–}(C{=}O)\text{–}OH$$

A second interaction with a hydroxyl radical produces another radical as shown:

$$R\text{–}(C{=}O)\text{–}OH + OH \rightarrow R\text{–}(C{=}O)\text{–}O + H_2O$$

Interaction of the two radicals highlighted will produce a linear anhydride, as detected on the epoxy resin surface after electrical stressing:

$$R\text{–}(C{=}O)\text{–}O + R\text{–}C{=}O \rightarrow R\text{–}(C{=}O)\text{–}O\text{–}(C{=}O)\text{–}R$$

Given that activated oxygen species are less prevalent in a moist atmosphere [11] and that anhydrides are widespread following stress in a moist atmosphere but less so in a dry atmosphere, reaction b was considered the more probable initiating step. The production of nitrated species following normal air discharges is explained by the higher levels of nitrogen oxides in a moist atmosphere [11].

Interactions between the radical species involved in development of anhydrides can also be used to postulate reactions for the production of acids and peracids detected on the stressed resin surface.

The production of radical species, R, on the bisphenol chain (by removal of the methyl group) was postulated earlier. Interaction with oxygen and carbon species allows the following reaction:

$$R + O_2 + R\text{–}H \rightarrow RO_2 + R\text{–}H \rightarrow RO_2H + R$$

Peroxides can thus be formed on the resin surface.

Where the radical species, RO_2 and R, are formed in close proximity, additional crosslinks can be formed in the resin matrix by the oxygen molecule:

$$R + RO_2 \rightarrow R\text{–}O\text{–}O\text{–}R + h\nu$$

Once again the importance of chemical stressing is exemplified.

A careful trawl through the literature (e.g. Goldman, Mayoux, Bartnikas, Wertheimer and the present author plus appropriate general chemical texts, variously) will provide much information and data from which potential reactions can be hypothesised based on the specific materials and gaseous environment in use.

4.2.5   Electrical stress

The superposition of an electric field due to charge deposition from a partial discharge at a solid insulating surface will result in both local microscopic and macroscopic effects which may cause degradation. Electric fields can be responsible for dissociation and transport of ionised and ionisable byproducts resulting in increased losses and local stress enhancements. Charge trap filling and other forms of charge capture will result in local field effects. In turn, these may result in local electronic breakdowns around the stress enhanced site. For further details of this form of degradation/breakdown, the reader is referred to the excellent book by Dissado and Fothergill [19].

4.2.6   Synergetic interaction of stresses

As has been emphasised throughout this section, although, for the sake of convenience, the various forms of stress which apply during partial discharge stressing can be compartmentalised, it is the synergetic interaction of these stresses which results in degradation. Different materials, gaseous atmospheres, contamination levels, discharge magnitudes and orientation – all will result in a unique combination of stress effects at a discharging surface.

This synergy is epitomised by the work variously by Mayoux [4, 20, 21] for polyethylene, as noted earlier. Using an electron gun, an ion gun and an ultra-violet radiation source, he attempted to separate the different particle effects using infrared and ultra-violet spectroscopy and scanning electron microscopy. Unfortunately, although on a quantitative basis the damage observed using each of the above techniques individually yielded damage similar to that observed under discharge action, the energies required to produce similar damage, under similar conditions, were well in excess of those found in a gas discharge. For example, under electron bombardment, polyethylene was found to exhibit negligible degradation below 500 eV and with ion bombardment, under 100 eV. Mayoux was obliged, somewhat inevitably, to conclude that the structural transformations found when polyethylene is subjected to partial discharge stressing cannot be considered as the superposition of the effects due to the individual components of the discharge acting independently.

The reader interested in finding out more about the mechanisms and phenomena associated with partial discharge degradation would do worse than read variously the early papers of Mason and Garton and, more recently, the papers involving Densley, Bartnikas, Mayoux, Wertheimer, Kemp and, for underpinning mechanisms associated with the materials (in particular polymers), the book by Dissado and Fothergill [19].

4.3   Partial discharge measurement

The various techniques that can be applied, either directly or indirectly, to determine the presence of, and characterise, partial discharge activity are described in this section.

4.3.1   Electrical detection

The electrical detection of partial discharge activity falls within three distinct approaches:

(i)   measurement of each individual discharge pulse
(ii) measurement of the total, integrated loss in the insulating system due to discharge activity
(iii) measurement of electromagnetic field effects associated with discharge activity using antennae and capacitor probes.

4.3.1.1   Individual discharge pulse measurement

There are two broad approaches to making this kind of measurement, i.e. connecting a clampon current transformer (CT) to the neutral strap of the plant item and taking the output to an oscilloscope or similar recording instrument or connecting a transducer (typically a capacitor divider type assembly) to the high voltage terminals of the plant item and measuring the output in a similar way to the CT approach (see Figure 4.1). Each has its own advantages and disadvantages.

The CT approach is extremely cheap, simple and safe to use, utilising, as it does, the neutral strap on the plant item (see Figure 4.1*a*). No disconnections need to be made since the CT is simply clamped around the neutral and is supplied with a suitable output connector compatible for coaxial cable. It suffers, unfortunately, three major disadvantages:

- it cannot be effectively calibrated to determine the magnitude of any discharges present
- it is prone to interference from external sources such as pulses from power electronics circuitry and, indeed, corona discharging from elsewhere in the system
- it does not provide effective phase information on the location of discharges on the AC voltage power cycle.

In addition, there may not be a neutral available. However, despite these disadvantages, it is used, particularly for motors and, to a lesser extent, for transformers, as a first pass technique by some engineers. In the USA, the technique that is used employs a frequency spectrum analyser rather than an oscilloscope and is described as a radio

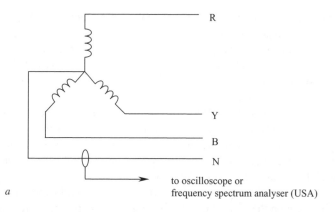

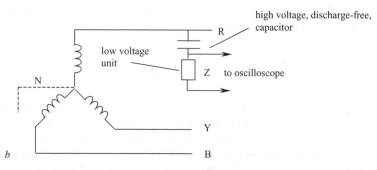

Figure 4.1 *PD pulse measurement on motors (Red/Yellow/Blue phase notation)*
 a CT connected to motor neutral
 b Capacitor coupler connected to high voltage phase terminal (each
     measured in turn)

interference (RI) measurement [22]. In this form, it is used to assess spectra on a comparative basis among spectra measured at different time intervals e.g. annually, on the presumption that changes in the spectrum may be indicative of discharge activity.

   In use, care should be taken to make a reference measurement with the CT disconnected from the neutral prior to making the actual measurement to ensure as far as is practicable that interference is not compromising the measurement. However, it must be remembered that the neutral will often act to pick up extraneous signals especially in a noisy environment, and this nullifies the validity of this approach. Only with the CT connected to the neutral, with the plant deenergised, can a true comparison of this type be made.

   The second approach is to connect the discharge transducers to the high voltage terminals of the plant item, e.g. the individual phase terminals of a motor, in turn (see Figure 4.1*b*). Typically, this discharge transducer consists of a discharge-free high voltage capacitor connected to a low voltage impedance circuit (RC or RLC) which in turn is connected to an oscilloscope or similar instrument. By careful choice of component values, the high voltage is reduced to a safe level at the low voltage impedance (typically 1000 : 1 ratio) and individual pulses from discharges can be displayed superimposed on the AC power cycle voltage. This system can be calibrated by injecting a discharge-simulating pulse, of known magnitude, into the detector circuit. All commercial instrumentation carry such a calibrator on board. It should be noted that for transformers with bushings containing tapping points, the bushing can act as the high voltage capacitor.

The second method utilising the high voltage phase terminals is the Rogowski coil. Essentially a form of CT, but not to be confused with the earlier type used at the neutral, its design concentrates the magnetic flux more effectively than in the standard CT. Its principle of operation is based on Ampere's Law. An air-cored coil is connected around the conductor in a toroidal fashion. The current flowing through the conductor produces an alternating magnetic field around the conductor resulting in a voltage being induced in the coil. The rate of change of this voltage is proportional to the rate of change of current. To complete the transducer, this voltage is integrated electronically to provide an output which reproduces the current waveform.

It is light, flexible and easy to connect to terminals since the coil can come in a form which can be opened and closed. In general, it is less sensitive than the capacitive coupler approach, but, broadly, there is little to choose between them.

Individual discharge pulse measurement, either by capacitive coupler or Rogowski coil, can be applied to most items of plant. Generally, it is applied as an online technique (although it can be used offline) with the exception of cables and, in this mode, care must be taken to ensure that any discharges detected are coming from the item of plant under investigation and not from some other item further away with the discharges coupling electrically through the conductors to the terminals where the transducer system is located. In this context, Rogowski coils have the advantage over capacitor couplers, providing an indication of pulse direction in reaching them.

The discharge pattern produced by individual discharge events is generally recognised as comprising the amplitude of individual discharge events, the number of discharge pulses per power cycle and the distribution of these pulses within the power cycle, i.e., their phase relationship. In addition to the discharge pattern produced by these parameters at a given time and under a given applied electrical stress, one would normally also be interested in any changes that occur in the pattern as a function of the magnitude of applied electrical stress and time of application. Armed with this information, it may be possible to evaluate the nature of the degradation sites and thereby provide an assessment of insulation integrity. However, although the preceding paragraphs might imply that this process can be relatively straightforward, in practice a variety of factors conspire to make interpretations based on such patterns a highly complex affair.

Of these factors, the major culprits are external interference and the complexity of the discharging insulating system. External interference, in general, can take the following forms [23]:

- PD and corona from the power system which can be coupled directly to the apparatus under test (in an online test) or radiatively coupled (in online or offline tests)

- arcing between adjacent metallic components in an electric field where some of the components are poorly bonded to ground or high voltage
- arcing from poor metallic contacts which are carrying high currents
- arcing from slip ring and shaft grounding brushes in rotating machinery
- arc welding
- power line carrier communication systems
- thyristor switching
- radio transmissions.

It can generally be minimised through the use of inline filters and some form of discrimination circuit if the problem is in the high voltage line. (An example of this is the PDA system developed for turbine generators [23], utilising two couplers in each phase.) In addition, as indicated earlier, Rogowski coils should discriminate, on the basis of polarity, the direction of a pulse reaching them. If the problem is airborne interference, e.g. from rectifiers, it can be more difficult to eliminate but an assessment of noise activity and its characterisation can be made prior to the high voltage insulation test. In relation to the complexity of the discharging insulating system, this is generally beyond one's control. The problem, in this case, lies in the potentially vast number of discharging sites and their variety. This combination tends to swamp out the characteristic patterns associated with specific discharge site conditions. Having indicated these caveats, however, there is no doubt that the information contained in a discharge pattern can be extremely useful in assessing insulation integrity.

Another aspect of pattern interpretation that should be noted at this stage is the importance of regular measurements on a given insulating system. Ideally, patterns should be obtained at regular intervals throughout the life of the insulating system since there is no doubt that the trend in the discharge pattern of a given insulating system with time provides far more useful information on insulation integrity than any measurement at only one point in time.

Although it is always possible to cite examples where a single measurement can be extremely beneficial, there are many situations where it provides relatively little information. As a diagnostician (be it of plant or human beings), once one has established that the absolute levels of the vital characteristic parameters do not indicate the imminent death of the patient, one is primarily interested in the rate at which these characteristic parameters are changing through life in comparison with similar systems (or humans) of comparable design and stress history.

That said, in relation to the interpretation of the discharge pattern, the starting point would be the magnitudes of the discharges being detected and their repetition rate. Broadly, the larger the discharge magnitude, the larger the site of degradation with which it is associated and the greater the likely rate of degradation. Similarly, the greater the repetition rate or number of discharges per cycle or unit time, the greater the number of discharging sites. Once these issues have been considered, the next aspect for interpretation would be the location of the discharges on the power cycle waveform. Cavity-type discharge sites generally yield a discharge pattern within which most pulses are in advance of the voltage peaks, i.e., $0°-90°$ and $180°-270°$

of the AC power cycle. In contrast, discharge sites containing a sharp metal surface generally produce a pattern with pulses symmetrically spaced on both sides of the voltage peak(s).

The next consideration would be the relative magnitudes of discharges on the positive and negative half cycles of applied voltage. Broadly, similar magnitudes on both half cycles imply that both ends of the discharge are in contact with similar physical surfaces. This contrasts with the situation which is prevalent if the pattern indicates (say) different magnitudes of pulses on the two half cycles, i.e. quadrants one and three. In this instance, it would be likely to assume that the discharge was active between two insulating surfaces but perhaps of different surface topology resulting in varying stress enhancements, thus changing the supply of initiatory electrons when the necessary cross-gap stress is reached to produce a spark and the time which an insulator takes to dissipate surface charge (potential) build-up.

Having decided whether the discharge site(s) are insulating-bound, insulating-metallic or metallic (corona), one would next consider the variation in discharge magnitude with test voltage and then with the time of application of voltage should an offline measurement be a possibility. If the discharge magnitudes remain constant with increasing test voltage, this tends to imply that one is observing discharge activity within fixed cavity dimensions. This suggests either a most unusual condition where all the cavities are similarly dimensioned or, far more probably, one is observing a single cavity situation – possibly of quite large surface dimensions. This contrasts with the condition in which the discharge magnitudes of the pattern rise with test voltage. In this instance, one is generally observing cavities, or gaps between insulating surfaces, of differing size. As the test voltage is increased, in addition to those cavities of low inception voltage discharging more often per cycle, cavities of higher inception voltage are also starting to discharge. In this case, one would be thinking of internal discharges in a number of insulation-bound cavities of different size, external discharge across a varying length gas gap (between, for example, two touching insulated radial conductors) or perhaps even surface discharges in an area of high tangential stress.

Having decided that one is dealing with (say) a cavity-type degradation condition and that the cavity surfaces are either insulation-bound or metallic-insulation bound, and having determined something of the cavity size/number distribution, observing the pattern at a fixed voltage over a period of time may yield still more information. For example, if the activity tends to lessen with time, then, depending on the insulation system under investigation, one might be observing a pressure increase in the cavities (caused by gaseous byproduct formation and resulting in a higher breakdown voltage for given cavity dimensions – see Paschen's Law), a build-up of surface charge within the cavity structure (thus inhibiting the realisation of sufficient potential drop across the cavity to produce further discharges following discharge activity and potential equalisation) or perhaps a build-up of water or acids within the cavity structure (increasing surface conductivity and allowing charge to leak away).

The preceding discussion has largely been related to cavity-type degradation sites – identified by the location of discharges primarily in advance of the voltage peaks – however, the same approach holds true for gaps with a sharp metallic boundary. For example, depending on the numbers of discharges, their spacing and magnitude on one half cycle compared with the other, one should be able to say something about the relative sharpness of any points. This relates to the much higher electric stress realised around a sharp point for a given applied voltage, rather than that at a plane surface, and the related spark activity when the point and then the plane are alternately the cathode within a cycle of applied voltage.

There are many more conditions that could be discussed within this section relating discharge magnitude, phase distribution, test voltage and time of application to specific conditions. However, it is not the purpose of this chapter to provide a comprehensive guide of this type. Other papers within the literature are available for this purpose [24]. Rather, it is to indicate to those unfamiliar with the interpretation of partial discharge patterns an effective and efficient way forward in their use. Although it may be sufficient to either memorise or have available a look-up table of specific discharge pattern characteristics from which one can make an interpretation, one will have a much greater possibility of making an accurate interpretation if one understands the physical and chemical processes that can produce a given discharge pattern (or result in a change in discharge pattern) and can relate these to the specific form and design of the insulation system under investigation.

Again, the importance of establishing a trend in the discharge pattern for a given insulating system over its lifetime cannot be stressed too strongly. In general, the particular degradation characteristics are often developed relatively early in the life of an insulating system. Thereafter, the parameters of primary importance are the absolute discharge magnitude and, more importantly, its rate of rise over given times – both relative to the values obtained from other insulating systems of similar design and history.

Modern instruments acquire, store and process this information digitally and provide, in general, a three-dimensional plot of this data on screen for analysis purposes. The three axes are phase (0°–360°), discharge magnitude (usually in pC) and number of discharges (or frequency) resulting in the so-called ϕ–q–n plot.

The x axis and y axis, which represent the phase angle of the PD pulse occurrence and the magnitude of the PD pulse, respectively, form the floor of the histogram, see Figure 4.2. The phase notation is conventionally that of a positive sine wave – as opposed to positive or negative cosine – the 0° is the zero crossing point of the rising edge (of a signal with no DC bias). The z axis, n, represents the number of pulses per unit time that have occurred within the specific phase-magnitude window.

Figure 4.3 demonstrates how a ϕ–q–n pattern is constructed. The x axis of the PD time domain signal is divided into a number of time windows, each corresponding to a number of degrees, or a phase window, of the power cycle. Likewise, the amplitude or y axis is divided into discrete windows. The phase and amplitude windows correspond to the row and column indices of a two-dimensional matrix. The value at each index in the matrix corresponds to the number of PD occurring in that phase and amplitude window, per unit time. In Figure 4.3, it can be seen how the PD a, b, c, d and e are fitted into their particular phase amplitude window and the $H_n(\phi, q)$ graph modified accordingly.

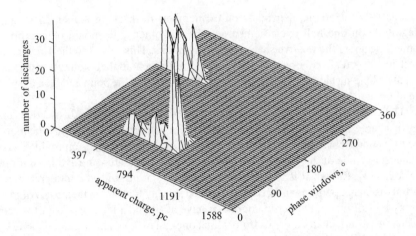

Figure 4.2   *φ–q–n plot of discharge activity*

The distribution is then the accumulation of every PD, in its respective phase/charge window. If the PD pulses occur statistically, then the distribution is a probability distribution. The value at each index can be correlated to the probability of a PD occurring within that particular phase/apparent charge window. The more cycles making up the distribution the more (statistically) accurate the distribution will be, with the one caveat that the PD pattern must be relatively stable over the sampling period.

Clearly, the ability of an instrument of this type to accurately portray the statistical pattern of PD activity depends on its sampling rate and memory size.

It should be noted, however, that this type of pattern constitutes only the primary data of PD activity, i.e. magnitude, phase, and number. Changes in the shape of the pattern with time are now also being recognised as potential indicators of the nature, form and extent of PD activity. Changes in the statistical moments of the distribution – mean, variance, skewness and kurtosis, when considered across both the x and the y axes – are beginning to show correlations with specific forms of degradation associated with PD activity.

It should also be noted that this is not the only form of PD data under consideration. The power of digital systems to acquire, store and process the primary data is being utilised to generate new, and potentially more useful, information. Parameters generated in this way include [25]:

- the discharge energy (p) where $p = q \cdot V$ for any discharge, V being the instantaneous voltage at which the discharge (q) occurs
- the discharge phase inception voltage, V_i, where the discharge pulse sequence starts
- the discharge phase extinction voltage, V_e, where the discharge pulse sequence ends

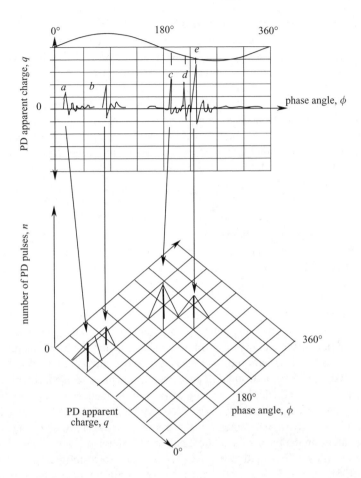

Figure 4.3   Build-up of φ–q–n distribution

- the discharge current, $I = 1/T \times \Sigma |q_i|$, where T is the duration of the power frequency half cycle and i is the number of discharges observed during T
- the discharge power, $P = 1/T \times \Sigma |q_i V_i|$
- the discharge intensity N, the total number of discharges as observed during time T
- the quadratic rate $D = 1/T \times \Sigma |q_i^2|$.

All these quantities can be analysed either as a function of time or phase angle. However, there has, as yet, been insufficient empirical data gathered to enable correlation to be made unambiguously between any of these parameters and the nature and extent of PD activity.

Those readers interested in learning more about this form of partial discharge measurement and its interpretation might read the following papers and their citations [26–30].

It should be noted at this stage that the standard (IEC60270) is not immune to inherent errors. For example, Zaengl [31–33] has analysed the effects of detection circuit integration error and sensitivity for various integration filter bandwidths and rise times of pulses. Zaengl also introduced the concept of a parasitic inductance, distinct from the integration circuit, capable of producing additional variations in the detector response circuit. The standard makes no reference to such a parasitic inductance. This phenomenon has been carried forward through simulation studies [34], which show that the calibration of any measurement PD system depends on the parasitic inductance, that it influences the measurement sensitivity according to rise time and filter bandwidth and that, if not considered, it may therefore result in erroneous measurements.

4.3.1.2   Pulse sequence analysis

In addition to providing statistical patterns of activity from acquired PD data, e.g. ϕ–q–n patterns, there has been increasing interest in recent years in the nature of the sequence of pulse events. This approach explores the deterministic nature of PD events. Generally, a given PD event terminates when the local voltage drops below some critical value. Although this can be purely a function of the external macroscopic electric field, it is far more likely to be due to a combination of this macroscopic field and the local field in the vicinity of the PD event due to space or surface charge effects. Factors such as the distribution of (say) surface charge and its rate of decay will, in such circumstances, have a strong influence over the occurrence of a further PD event. On this basis, it is not difficult to see why a given PD sequence may be deterministic and not random in nature. Equally, over relatively long time periods, changes in the nature of PD activity may be evident if this approach is taken. Such changes would not be apparent on the basis of statistical patterns of activity. For example, bursts of PD activity followed by relatively quiet periods were noted in treeing experiments on resin [35]. The underlying mechanisms responsible for degradation under PD stressing may be elucidated more readily using the pulse sequence approach, it can be argued, since, from the deterministic nature of the results produced, it provides a much greater insight into the underlying mechanisms than does the statistical, random, approach.

Rainer Patsch of the University of Seigen, often in collaboration with Martin Hoof when he too worked there, has been a key advocate of this approach since the early 1990s. Their work [36–39] has demonstrated that the sequence of discharges and, in particular, the change of the external voltage between consecutive discharges, provides important information from which the nature and form of the degradation can be inferred. In addition to aiding in the classification of PD defects, they have also demonstrated that the technique can be utilised to separate several sites discharging contemporaneously.

The increasing speed and memory capacity of digital signal acquisition, storage and processing systems is making the pulse-sequence-analysis approach more readily available than previously.

4.3.1.3 Calibration – a word of caution

It might strike anyone investigating, for the first time, the range of instruments available commercially for the electrical detection of individual partial discharge events, as somewhat incongruous that there are almost as many different bandwidths, centre frequencies etc. as there are detectors. Being essentially an impulse characteristic, a partial discharge pulse will contain frequencies across the spectrum from almost DC through to GHz. On this basis, any bandwidth of detector is likely to detect some energies contained within the discharge pulse. However, as can be appreciated, a broad bandwidth detector will detect far more frequencies than a narrowband detector, i.e. it will be more sensitive. How are these differences resolved to ensure that, irrespective of detector characteristics, an accurate measurement of PD activity is achieved? This is the role of the calibration process.

Essentially, to calibrate an individual event partial discharge detector, a discharge-simulating pulse of known magnitude (the calibration pulse), is injected into the detector at the point where the real discharge pulse enters the detector. By utilising this strategy, irrespective of the detector's individual characteristics, however those characteristics modify the PD pulse they will also modify the calibration pulse in exactly the same way – and comparability will be maintained. For example, if a given detector modifies a PD pulse to be of only half the magnitude it would have been if detected by a very broadband detector, the detector circuitry will modify the calibration pulse to also be of only half the magnitude, and the correct magnitude of discharge will be inferred.

Unfortunately, as has been demonstrated, this calibration strategy does not necessarily work for generators [40], motors [41] or transformers [42]. In the case of these items of plant, by injecting a discharge-simulating pulse of known magnitude at various points through their windings, and detecting the response at the phase terminals with detectors of different bandwidth, it has been demonstrated that the detected magnitude is a function of both the location of the injected pulse and of the bandwidth of the detector utilised. To make matters worse, it has been shown that no direct relationship exists between the magnitude that one detector will suggest and that of others. Indeed, there is not even an indirect relationship with (say) one detector consistently suggesting a larger discharge magnitude than another for the same injected discharge-simulating pulse. It is all a function of location.

What is going wrong and, specifically, why is the calibration strategy failing to ensure that detectors with different characteristics produce the same result? Quite simply, there is a fallacy within the calibration strategy. The strategy is predicated on the notion that the injected discharge-simulating pulse emulates the real PD pulse as it enters the detector. This is indeed true when, in general terms, the discharge site is close to the detector terminals. However, in the case of generators, motors and transformers, this may not be true. In these circumstances, the partial discharge pulse may have to propagate tens of metres to reach the detector. This propagation path may be complex – both electrically and literally! It is composed of inductive, capacitive and resistive components distributed, in all probability, in a non-linear fashion. Being complex, the impedance paths will be dependent on frequency and will vary as a function of the individual frequencies which make up the discharge pulse. As the PD pulse propagates, frequencies will be lost (especially high frequencies), there will be resonance effects etc. In short, having propagated any significant distance either electrically or physically to the detector, the PD pulse will, in all likelihood, no longer resemble the pulse that set out from the discharge site. Therein lies the fallacy in the calibration process. The calibration pulse does indeed emulate the PD pulse – as it occurred at its site of origin – but after it has propagated any significant distance this equality may be lost and the detected discharge magnitude will become a function of location of discharge site (controlling the frequencies of the PD pulse which reach the detector) and, of course, the detector bandwidth [41, 43].

Figure 4.4 illustrates the issue. The calibration pulse contains essentially similar frequencies across the measured spectrum. Note that the discharge pulse, at its site of origin, would also appear similarly. However, as can be seen, following propagation, the frequency characteristics of the PD pulse may look quite different, as illustrated. Although it is simplifying the argument greatly, by inspection of Figure 4.4, it can be appreciated that in this case a narrowband detector detecting around (say) 800 kHz would, relative to the detected calibration pulse, suggest an extremely small PD pulse whereas a detector utilising a narrowband detection system around 260 kHz would suggest that the PD pulse is large!

For those unfamiliar with PD measurements of this type, it might appear that the obvious solution is to use only wideband detectors. Unfortunately, although this would decrease errors introduced due to this phenomenon and result in an increased detector sensitivity, it would prove impractical in field measurements. The greater the bandwidth of detector, the greater the likelihood of detecting pulses which do not originate from discharges. Given the broad range of frequencies available from PD pulses, a balance must be sought between detector sensitivity and erroneous information.

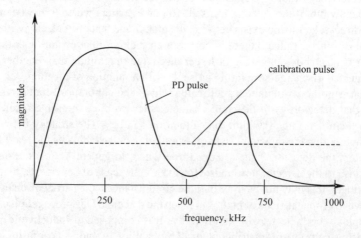

Figure 4.4   Calibration pulse characteristic compared with PD pulse characteristic following propagation

The field of partial discharge measurements has had a number of salutary experiences where noise has been interpreted as PD activity and plant has been removed from service unnecessarily with serious economic consequences.

Now, if the situation as described in the preceding paragraphs were very prevalent, no discharge measurements made on generators, motors or transformers could be trusted – and, fortunately, that is not the case! The reasons for this lie, in part, in the fact that much of the degradation suffered in such plant (and hence associated PD activity) occurs near the high voltage terminals. The PD pulses thus have only a relatively short distance to propagate to the detector, resulting in only minor errors. Also, most diagnosticians would rely on trend analysis at least as much as any absolute measure of PD magnitude and, with a range of caveats, this is probably unaffected provided the same detector is used consistently.

This is merely a word of caution for those instances when, in comparing two items of plant, possibly of different design and manufacture, one is tempted to suggest that one item is in poorer condition than the other because it is displaying higher discharge levels. The odds are in favour of this being true – but it may not be!

4.3.1.4 Noise and wavelet analysis

As indicated earlier, the electrical detection of PD can be affected – sometimes seriously – by various forms of noise [23], with measurements in the field rendered ineffectual or severely compromised.

Over the years, methods employing discrimination circuits [23], traditional filtering techniques [44], neural networks [45] etc. have been designed to suppress noise with limited success. However, a new, more powerful tool is now being applied to this problem – wavelet analysis. It is likely that this technique will bring radical improvement to noise reduction and suppression.

Since its introduction in practical applications in the mid-1980s, as a powerful tool for signal analysis and processing, wavelet analysis has been increasingly applied to solve engineering problems [46–48]. Essentially, the technique allows the user to obtain two-dimensional information on PD pulses in both the time and frequency domain, and to extract features of PD pulses in measurement data. By careful choice of wavelet transform, ideally coupled to a clear understanding of the PD detector characteristics, sophisticated feature extraction is possible for PD pulses [49, 50]. The required properties for a wavelet in this application include compactness, limited duration, orthogonality and asymmetry for analysis of fast transient irregular pulses, e.g. the Daubechies wavelet family. This, in turn, makes it possible to extract PD pulses from extraneous noise in a way which would be quite impossible with traditional filtering techniques [51, 52]. Indeed, it is possible to extract PD pulses from measurement data where the PD pulse is embedded within the noise, i.e. below the noise plane. This is illustrated in Figure 4.5, which shows raw measurement data and processed data following the application of the wavelet transform. As can be seen, there are three PD pulses within the measurement timescale, with one of these below the noise plane. Time taken for analysis is short, making the technique attractive for field analysis.

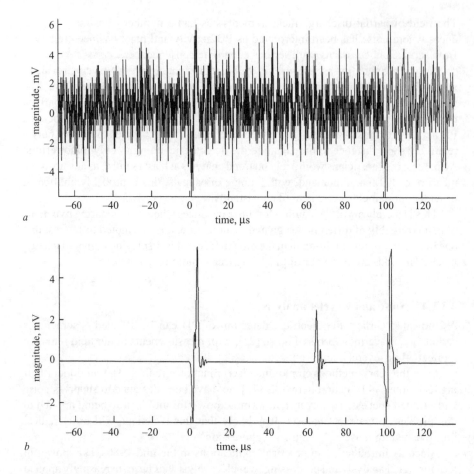

Figure 4.5   PD measurement in presence of noise
a   Raw data, PD activity and noise superimposed
b   Processed data to show PD activity

Another benefit of the technique is its ability to compress PD measurement data [53]. Because only limited coefficient data related to PD events need to be used, following wavelet analysis, to reconstruct precisely the actual PD signal extracted, the amount of data storage space can be greatly reduced, i.e. ≈ five per cent of the original data stream. The technique may, when fully developed to this application, revolutionise PD measurements in the field.

Electrical PD detection as discussed in the preceding section has clear advantages in terms of sophistication over most other approaches, but is generally viewed as difficult in measurement and, in particular, in interpretation. It is also relatively expensive and is normally only applied as a front line approach for motors and generators where most other techniques, with the exception of tan δ (see section 4.3.1.5), cannot be applied. For motors/generators, many users have chosen to install permanent high voltage capacitors and low voltage impedance units at the phase terminals of their plant, thus removing the need to deenergise the plant to make the high voltage connections. All that is required is to connect the measuring instrument to the low voltage connections. Capacitors for this application are relatively inexpensive.

4.3.1.5   Loss measurements associated with discharge activity

This technique operates on the principle that the current loss in an insulating system will increase markedly in the presence of discharges. By monitoring the tan δ of the system (δ being the angular quantity indicative of the relative values of resistive to capacitive current), relative discharge activity can be inferred.

The insulation system should, ideally, behave as a perfect capacitor, i.e., there should only be capacitive current through the system. However, inevitable losses in the system mean that there is also a small component of resistive current. As losses increase, the resistive current becomes larger and the angle created between the capacitive current and the resultant vector-summated current (δ) increases, as does the tan δ. In the presence of partial discharge activity, the losses increase enormously and dominate as the loss mechanism.

Traditionally, tan δ measurements would be made offline, allowing the insulation system to be energised progressively to full rated voltage (generally in 0.2 voltage steps to full working voltage). This has enabled a plot to be made of tan δ versus applied voltage similar to the ones shown in Figure 4.6.

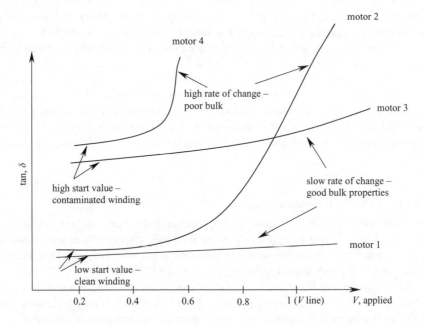

Figure 4.6   Tan δ versus applied voltage

Any sudden change in tan δ (at the so-called knee point or tip-up point on the tan δ plot) would be considered indicative of PD inception and the rate of change with increasing voltage would be indicative of the relative severity of the PD. In addition, with experience, the starting value of tan δ informs on the level of contamination present on the system – usually on stator end windings since it is in motors and generators that the technique has found greatest favour although it has also been applied to transformers and bushings [54]. On this basis, by considering the plots shown in Figure 4.6, motor 1 plot would be indicative of a clean winding (low start tan δ, and minimal rate of change of tan δ with voltage), motor 2 a clean winding but with serious discharge problems (low start tan δ, low inception voltage (knee point) and high rate of change of tan δ with voltage), motor 3 a contaminated winding (high start tan δ) but one which has low PD activity (low rate of change of tan δ with voltage) and motor 4 a contaminated winding which also suffers from serious PD activity (even more serious than motor 1).

As can be seen, tan δ informs on PD activity and thereby on the appropriate action to be taken. Traditionally, the measurement employed a Schering Bridge type circuit being connected at the high voltage terminals of the plant but, latterly, instruments have made direct measurements of the different current components.

The technique has the advantages of being simple in measurement and clear in interpretation. Its major disadvantage lies in it being an integrated measure of degradation in the insulating system. In motors and generators, quite large levels of discharge activity can be sustained both in terms of magnitude and number/cycle without fear for the insulation integrity. A few specific sites of degradation which are much worse than anywhere else will provide much larger magnitudes of discharge but these may be lost in the integration process with all other activity. That said, most instruments incorporate a peak magnitude detector to alleviate this problem.

In recent years, there has been an increased pressure to only apply online techniques for economic reasons. In relation to tan δ, this means that only a single value can be obtained. As can be seen from Figure 4.7, this would make it impossible, by this measurement alone, to distinguish between a contaminated but good winding with low PD activity and a clean but poor winding with high PD activity (see point X on Figure 4.7). However, trending the value with time and comparing results among different machines, backed-up by visual inspection, should make the situation much clearer.

4.3.1.6   Antenna techniques

The radio interference noise generated by partial discharges has been recognised since the 1920s. Using a range of different antennae, attempts have been made to quantify and characterise the noise produced by partial discharges – in particular corona associated with overhead lines and insulators [55–60].

Although the results from these studies would suggest that antenna measurements of partial discharges should enable characterisation of the nature, form and extent of discharge activity, no such rigorous study has yet been undertaken. This is disappointing given the attractiveness of a technique which is both non-intrusive

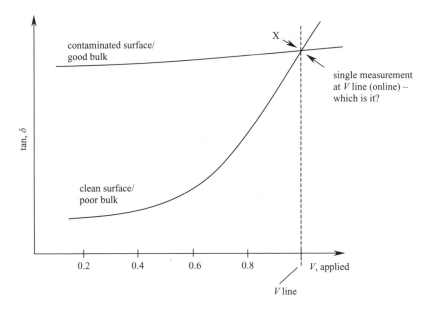

Figure 4.7 *Tan δ measured only at V line – problem of online measurement*

and requires no connections to plant under test. Recent measurements by the present author, as yet unpublished, on rotating machines and on air-cooled transformers have proved extremely successful in this respect. Using several specifically designed antennae coupled to a high bandwidth oscilloscope, not only could the presence of discharges be detected but pulse shapes obtained could be correlated with specific types of discharge. In addition, an estimate of discharge magnitudes in terms of picocoulombs could be obtained via an indirect laboratory calibration of the antennae. This technique will be much more widely used in the future.

4.3.1.7 Capacitive probe techniques

When a discharge occurs, an electromagnetic wave is produced which propagates away from the PD site. Where the plant is metal clad, the wave will propagate towards the earthed metal enclosure. Provided there is a gap somewhere in the enclosure, e.g. a gap in the gasket or gap at the busbar chamber cover in the case of metal clad switchgear, the electromagnetic wave is free to travel to the atmosphere outside the switchboard. The action of the wave connecting with the earthed metalwork produces a transient earth voltage (TEV) which can be detected by a capacitive probe if positioned at the gap. The amplitude of the detected signal is normally in the millivolts to volts range, and this is generally translated into dB for measurement. The principle is shown in Figure 4.8.

Although this method of detection will give some indication of severity, there is no detail of exact location. When more than one capacitive probe is used, however, it is possible to gain some knowledge of location using the time of flight principle, as

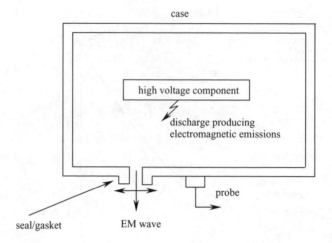

Figure 4.8   Transient earth voltage (courtesy of EA Technology Limited)

the probe nearest the source should detect the discharge first. This may indicate the panel from which the source is emanating.

An advantage of this type of testing is that the non-intrusive nature of the measurements allows for no disruption or outage to the plant under test. All items of plant within a substation can be monitored, such as circuit breakers, busbars, current and voltage transformers and cable end joints. Installing the test equipment while the substation is online has obvious benefits, and savings are also made due to the removal of outage costs and the relevant manpower reduction. The advantages and disadvantages of this technique are detailed in the work of Brown [61].

It is possible that external electromagnetic noise will interfere with the readings taken, and it is therefore essential that a background survey be completed prior to commencement. As with all types of PD measurement, if the interference is too severe, it may not be possible to complete the test, as results may not be sufficiently analysed. Problems are encountered when these noise sources produce voltages on other metallic surfaces within the substation using the TEV principle. The background reading should be taken from a metallic surface that is not attached to the switchboard such as a battery charger or doorframe. If a reading above a certain level is detected in this circumstance, the probe is unsuitable for monitoring the plant as it is not possible to differentiate between discharges from the plant or from external sources.

This type of probe can be purchased as a light, portable handheld unit or as a system with a number of probes connected to (typically) an event counter.

Clearly, the handheld unit is convenient, easy to use and relatively cheap. However, it will only detect discharges at the moment when the test is conducted. Although this is useful, it does not include any discharges that may present themselves at some other time due to load changes, humidity or temperature changes etc. Partial discharges are often intermittent and therefore a more thorough test regime may be required. In this event, the multiprobe system can be utilised.

Typically, such a system might have eight to ten capacitive probes and these would be connected to the plant using magnetic clamps. The probes are normally threaded and screw into the clamps until flush with the earthed enclosure. In turn, the probes would be connected by separate channels to (typically) an event recorder. Additional channels should be kept free for antennae. The purpose of the antennae is to detect any external electromagnetic noise that may filter into the plant environment, be detected by one or more of the probes and interpreted as partial discharge activity. By subtracting the events detected by these antennae from those detected by the probes, interference effects are reduced. These antennae should be positioned in the corners of the substation and extended vertically at the same height as the plant under test.

4.3.2 Acoustic detection

Partial discharges produce acoustic noise, as anyone who has listened to the crackling noise in electrical substations will confirm. Although directional microphone systems have been used to detect partial discharges from their airborne acoustic emissions, their application has been largely linked to external busbars, connectors and insulator assemblies. Acoustic detection has found much greater success and wider application through the use of piezoelectric sensors.

Piezoelectric polymers, such as PVDF (polyvinylidene fluoride), when compressed, result in the production of an external voltage proportional to the force applied to the polymer. As such, when built into the head of a handheld probe or when fixed, with a suitable paste or clamping mechanism, to an appropriate enclosure, they offer a simple means of detecting acoustic signals. Typically, handheld acoustic probes are coupled to an analogue voltmeter. Unfortunately, although it has been shown that for simple geometries the resultant voltage is proportional to the size of the discharge, due to the complex acoustic impedances associated with the propagation of an acoustic pulse to the probe, no effective calibration is possible for high voltage plant. The intensity of the emitted acoustic waves is proportional to the energy released in the discharge. On this basis, the amplitude of the wave is proportional to the square root of the energy of the discharge and, since energy may be taken as proportional to the charge squared, there should be a linear relationship between discharge magnitude and acoustic signal. However, acoustic measurements are more about detecting the presence of discharges, irrespective of magnitude, and locating these within the plant item. Typical applications for a handheld probe would be distribution circuit breaker boxes and small transformers. As the probe is moved around the enclosure, the larger the voltage detected, the closer the probe is to the source of discharge activity. On a much larger scale, and using more sophisticated acquisition instrumentation but the same sensor technology, with a minimum of three probes fixed to the earthed tank of a large transformer, and a reference signal, it is possible to determine both the presence and the accurate location of any discharges present. By measuring the relative times of arrival of the pulse(s) from the discharging site at the three probes, and by assuming a constant velocity of acoustic propagation through the transformer structure, the relative distance from each of the probes to the discharging site can be computed and triangulated in three dimensions. Commercial software is available to do this and can be readily utilised with a laptop computer for portability.

Knowing the times of arrival of t_1, t_2 and t_3 from the different sensors, and assuming a given velocity of propagation, since distance = velocity/time, three distances can be computed. Knowing these distances, and the location of the probes, a three-dimensional plot can be made; where these spheres intersect is the discharge source. The approach is illustrated in Figures 4.9*a* and 4.9*b*.

Lundgaard [62, 63] has produced a useful review of acoustic detection of partial discharges and, as he points out, changes in both signal amplitude and shape occur as the acoustic signal propagates to any sensors. He cites reduced signal amplitude as a result of:

- geometric spreading of the wave
- division of the wave down multiple pathways
- transmission losses in propagating from one medium to another and at discontinuities within a given medium
- absorption in materials.

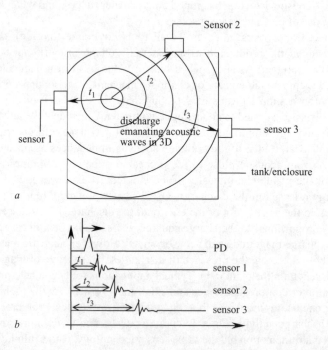

Figure 4.9

    a Partial discharge acoustic emissions arriving at different times at each sensor according to distance from source

    b Arrival of pulses at detector at different times according to distance travelled

In relation to changes in signal shape, he cites:

- frequency-dependent velocity effects resulting in different frequency components of a given signal arriving at the transducers at different times
- frequency-dependent propagation paths, again resulting in different wave components arriving at the transducer at different times
- absorption in materials removing high frequency components preferentially.

In addition, there are significant differences in the acoustic velocities in typical types of media encountered in (say) a transformer, i.e.:

- transformer oil at 25°C, 1415 ms^{-1}
- core steel, >5000 ms^{-1}
- impregnated pressboard, 1950 ms^{-1}

When one considers the complexity in both structure and materials in (say) a large oil-filled paper insulated transformer, the factors which can affect the signal propagation, and the variation in signal velocities with medium, it is impressive that this technique can be applied to complex structure, large plant.

Typically, in a quiet factory-type environment, discharges in a large power transformer can be located to within a volume the size of a football within a couple of hours and to within the size of a fist within a working day. That said, there is some variation in the estimated success rate using this technique according to manufacturer. Some claim total success whereas others are more circumspect admitting that they would be unlikely to detect a source, using this technique, embedded deep in a winding.

In the field, utilities also report a useful success rate (typically 50 per cent) using this technique with large power transformers. Indeed, a good example of this is presented in the work by Jones [64]. He reports that faults, using this approach, can be categorised as, for example, coming from:

- bushing connection stress shields
- windings
- winding jacking screws
- core bolts
- winding lead clamps
- tapchanger components.

The key issue here is that, although the absolute magnitude of discharge events cannot be determined from this technique, knowing the location of the source may be just as, if not more, important. For example, a source identified as coming from a winding might give serious cause for concern whereas, if the source is corona from a core bolt, it might not. A source within the winding will erode the paper and could lead to a catastrophic failure in time. A core bolt suffering corona will result in change to the dissolved gas levels but will not age the transformer in any significant way.

The single probe approach can also be used for capacitors and bushings but great care must be taken to ensure that placement of the probe on the capacitor or bushing surface does not distort the electric field resulting in a flashover to the probe.

Finally, it is worth noting that, in the case of distribution circuit breaker boxes, discharges are often intermittent and a single measurement over a short period of time may not be sufficient to ensure an accurate assessment. In these situations, an alternative to the hand held approach is to fix a sensor to the box and leave it on site for a number of days coupled to an event counter (in the same fashion as per capacitive probes). In a substation, a number of probes can be used simultaneously in this way, one to each box, with some form of reference probe also in place, to ensure that any detected events are coming from discharges within one or more of the distribution boxes and not from some external source.

In summary, acoustic techniques are relatively cheap and simple to apply, are utilised online and can detect the presence and location of discharges in the various items of plant discussed. Their disadvantages include their inability to be applied to intrinsically noisy plant, e.g. motors/generators, the need for the sensor to be relatively close acoustically to the discharge source (so the technology cannot, for example, be applied to cables) and their inability to be calibrated in terms of voltage output versus size of discharge.

4.3.3   Thermography and other camera techniques

Given that partial discharges are generally hotter than their surrounding media (see earlier), it might reasonably be thought that the thermal imaging camera could be applied in their detection and measurement. Unfortunately, most partial discharges are enclosed in some way, e.g. within solid insulation or within metal clad enclosures. Given the relatively low temperatures of partial discharges and the high thermal impedances likely to be present between such discharges and the imaging camera, its use is very limited in this application.

Certainly where the discharges are external to associated plant, e.g. on overhead lines/busbars or post/string-type insulators, such techniques can be used but not in any quantitative way. However, the desire to detect PD from these structures is somewhat limited compared with other items of plant.

Interestingly, in a similar vein, a daylight corona camera has recently been developed [65]. This incorporates independent UV video and visible cameras to capture separate video images of discharges and of associated high voltage plant. The system detects corona in the 240–280 nm region. Corona discharges emit in air mainly in the 230–405 nm range of the UV spectrum. Although the corona emission lines between 240 and 280 nm are not as strong as in the 290–400 nm range, this region is also called the UV solar blind band, i.e. there is no background radiation in this region. Despite the weaker intensity, the UV solar blind imager is able to provide high contrast images due to the complete absence of background radiation.

4.3.4   Chemical detection

Chemical techniques rely on the measurement of byproducts associated with PD activity and thereby from which PD activity can be inferred. This necessarily requires that these byproducts can be detected in some way. Clearly, PD activity associated

Table 4.3   Typical gases absorbed in the oil
under the action of PD

H_2	hydrogen	(H–H)
CH_4	methane	(CH$_3$–H)
C_2H_6	ethane	(CH$_3$–CH$_3$)
C_2H_4	ethylene	(CH$_2$=CH$_2$)
C_2H_2	acetylene	(CH≡CH)

with a closed, unvented void within a section of solid insulation would not lend itself to detection by this method.

Although chemical detection has been applied to various items of plant involving gas circulation over the years, e.g. hydrogen-cooled generators and, with relative success, gas insulated substations, it is primarily in oil-filled equipment that chemical detection has found favour.

Under the action of partial discharges (and, indeed, other fault conditions), oil will degrade through bond scission to form characteristic gases absorbed in the oil. Typical gases produced are given in Table 4.3.

The quantity and mix of gases produced depends on the nature of the fault, its severity and the associated temperature. The weakest C–H bond can be broken with relatively little energy, i.e. ionisation reactions, with hydrogen being the main recombination gas. As the strength of the molecular bond increases, more energy and/or higher temperature is required to create scission of the C–C bonds and the resulting recombination into gases which have either a single C–C, double C=C or triple C≡C bond.

Being essentially a low energy type fault, partial discharge activity tends to favour the breaking of the weakest C–H bond with the production of hydrogen. Carbon monoxide and carbon dioxide will also be present if the discharge occurs in the presence of cellulose, i.e. paper insulation (as is generally the case in large power transformers).

As little as 50 ml of oil suffices for analyses to be performed. This is important since, although dissolved gas analysis (DGA) has primarily been used for screening of large, oil-filled transformers (due to the capital involved in such assets) and where the loss of sampled oil would be insignificant, the increasing use of the technique with small oil volume plant such as bushings, CVTs etc., has made the volume of oil to remove critical.

4.3.4.1   Dissolved gas extraction and measurement

Once the sample has been obtained, it can be sent to one of the commercial laboratories which performs such analyses. It should be stressed at this stage that this is a cheap method of screening for faults and this too makes it extremely attractive to end users of oil-filled plant.

The gas can be extracted by a range of methods including the use of Toepler pump (vacuum extraction), partial degassing, stripping using argon, direct injection and head space analysis.

Measurements are made via the use of a gas chromatograph, infra-red spectrometer or, indeed, semiconductor sensors or miniature fuel cells.

For field measurements, a range of commercially available fault gas detectors is available, designed for use with large oil-filled transformers. These include the use of selectively gas permeable membranes with a miniature fuel cell or Fourier transform infra-red spectrometer as the gas detector, with portable gas chromatographs. A fault gas probe is also available, designed for instrument transformers where it is difficult to obtain an oil sample due to their low oil volume and location. The probe is best factory fitted and can be used with other forms of low oil volume plant.

4.3.4.2   Interpretation strategies

Several interpretation techniques have been developed and are used in the interpretation of dissolved gas in oil. These tend to be based on a combination of the quantity of individual gases present (in parts per million by volume, p.p.m.v.) and the ratios of these characteristic gases.

Although the presence of partial discharges in oil-filled plant can be inferred from the absolute levels of different dissolved gases measured, it is primarily through the ratios of these gases that PD is indicated.

Gas ratios have been in use since 1970 when Dornenburg utilised them to differentiate between fault types. The use of ratios had the advantage that oil volume did not affect the ratio and hence the diagnosis. Dornenburg first used pairs of gases to form ratios, to differentiate between electrical and thermal faults. In his first ratios, an electrical fault was indicated if the ratio of ethylene to acetylene exceeded unity, and the ratio of methane to hydrogen indicated a thermal fault if greater than 0.1 or a corona discharge if less than 0.1.

The ratios were developed further and significant levels, for each gas, known as L1 limits, were introduced. The technique was only to be applied if one of the gas levels exceeded its L1 limit (see Figure 4.10). As can be seen in Figure 4.10, the various ratios are indicative of, and capable of differentiating between, low intensity and high intensity partial discharges.

The development of a thermodynamic model [66] indicated that different temperatures favoured certain fault gases. The order of gas evolution with increasing temperature was found to be hydrogen, methane, ethane, ethylene and acetylene, respectively.

Rogers [67], used the order of gas evolution to form the gas ratios methane/hydrogen, ethane/methane, ethylene/ethane and acetylene/ethylene. A diagnosis table was created based on nearly ten thousand DGA results, together with examination of units with suspected faults and failed units. The table went through several evolutions and was produced in two formats. In the first, Figure 4.11, diagnosis was based on codes generated by ratios, and in the second, Figure 4.12, diagnosis was based on the value of the ratio. As can be seen, different types of partial discharge can be identified from the gas ratios.

L1 limit (significant value)

Gas	H_2	CH_4	CO	C_2H_2	C_2H_4	C_2H_6
L1 limit	100	120	350	35	50	65

Ratio				Diagnosis
$\dfrac{CH_4}{H_2}$	$\dfrac{C_2H_2}{C_2H_4}$	$\dfrac{C_2H_2}{CH_4}$	$\dfrac{C_2H_6}{C_2H_2}$	
>1.0	<0.75	<0.3	>0.4	thermal decomposition
<1.0	not significant	<0.3	>0.4	corona (low intensity partial discharge)
>0.1 and <1.0	>0.75	<0.3	>0.4	arcing (high intensity partial discharge)

Figure 4.10 Dornenburg's gas ratios (based on Joseph B. DiGiorgio's 'Dissolved gas analysis of mineral oil insulating fluids', published by NTT 1996–2002)

In 1975, CIGRE working group 15.01 assessed over 100 sets of DGA results from faulty transformers using interpretation schemes that were in use in Europe. As a result of this assessment, the IEC issued Code 599 [68] in 1978. This code was based on the Rogers' ratios, however, the ethane/methane ratio was omitted as it only covered a limited temperature range of decomposition but did not assist in further identification of the fault. The diagnosis table is code based with a rating from 0 to 2 being given depending on the value of the ratio (Figure 4.13).

In use, the code generated by the ratios of dissolved gas data being collected in the field at times did not match any given in the Standard's diagnosis table. This led various bodies such as utilities, transformer manufacturers, testing laboratories and consultants to develop their own interpretation techniques based on the IEC Standard; they were widely used to complement the IEC Code rather than replace it. A shortcoming of these techniques was that although they were developed to give more individual diagnosis for equipment, the diagnosis obtained was often not comparable between techniques.

Following a review of IEC 599, IEC Code 60599 [69] was issued in 1999. The code is based on a diagnosis table, Figure 4.14, which retains the original gas ratios although the diagnosis is no longer code based. As can be seen, once again the presence of partial discharges of various types can be inferred from the fault gas ratios.

4.3.4.3 Graphical techniques

As a consequence of IEC Code 599 being inconsistent in producing reliable diagnoses, additional schemes were introduced to complement the Code.

CH_4/H_2		C_2H_6/CH_4		C_2H_4/C_2H_6		C_2H_2/C_2H_4	
Range	Code	Range	Code	Range	Code	Range	Code
≤ 0.1	5	<1	0	<1	0	<0.5	0
$> 0.1 < 1$	0	≥ 1	1	$\geq 1 < 3$	1	$\geq 0.5 < 3$	1
$\geq 1 < 3$	1			≥ 3	2	≥ 3	2
≥ 3	2						

$\dfrac{CH_4}{H_2}$	$\dfrac{C_2H_6}{CH_4}$	$\dfrac{C_2H_4}{C_2H_6}$	$\dfrac{C_2H_2}{C_2H_4}$	Diagnosis
0	0	0	0	normal
5	0	0	0	partial discharge
1, 2	0	0	0	slight overheating $<150°C$
1, 2	1	0	0	slight overheating 150–200°C
0	1	0	0	slight overheating 200–300°C
0	0	1	0	general conductor overheating
1	0	1	0	winding circulating currents
1	0	2	0	core and tank circulating currents, overheated joints
0	0	0	1	flashover, no power follow through
0	0	1, 2	1, 2	arc, with power follow through
0	0	2	2	continuous sparking to floating potential
5	0	0	1, 2	partial discharge with tracking (note CO)
$CO_2/CO > 11$				higher than normal temperature in insulation

Figure 4.11   Rogers fault gas ratios (code based on Joseph B. DiGiorgio's 'Dissolved gas analysis of mineral oil insulating fluids', published by NTT 1996–2002)

Duval [70] developed a triangle, Figure 4.15, based on the relative percentage of methane, ethylene and acetylene gas. The triangle is divided into six regions representing high energy arcing, low energy arcing, corona discharge and hot spots. The triangle has the advantage that a diagnosis is always given – but, of course, this will always imply a fault! Hence it must only be used in conjunction with individual levels of gases which imply the possibility of a fault.

Other graphical techniques include the Church Nomograph Method, based on data published by Dornenburg and Strittmatter. The data are plotted on sliding logarithmic scales, with each scale representing a different gas. Data points are then joined together and the slope of the line between adjacent scales is indicative of the type of fault.

$\dfrac{CH_4}{H_2}$	$\dfrac{C_2H_6}{CH_4}$	$\dfrac{C_2H_4}{C_2H_6}$	$\dfrac{C_2H_2}{C_2H_4}$	Diagnosis
$\dfrac{>0.1}{<1.0}$	<1.0	<1.0	<0.5	normal
≤ 0.1	<1.0	<1.0	<0.5	partial discharge – corona
≤ 0.1	<1.0	<1.0	$\dfrac{\geq 0.5}{<3.0}$ or >3.0	partial discharge – corona with tracking
$\dfrac{>0.1}{<0.1}$	<1.0	≥ 3.0	≥ 3.0	continuous discharge
$\dfrac{>1.0}{<0.1}$	<1.0	$\dfrac{\geq 1.0}{<3.0}$ or >3.0	$\dfrac{\geq 0.5}{<3.0}$ or >3.0	arc – with power follow through
$\dfrac{>1.0}{<0.1}$	<1.0	<1.0	$\dfrac{\geq 0.5}{<3.0}$	arc – no power follow through
$\dfrac{\geq 1.0}{<3.0}$ or >3.0	<1.0	<1.0	<0.5	slight overheating – to 150°C
$\dfrac{\geq 1.0}{<3.0}$ or >3.0	≥ 1.0	<1.0	<0.5	overheating 150–200°C
$\dfrac{>0.1}{<1.0}$	≥ 1.0	<1.0	<0.5	overheating 200–300°C
$\dfrac{>0.1}{<1.0}$	<1.0	$\dfrac{\geq 1.0}{<3.0}$	<0.5	general conductor overheating
$\dfrac{>1.0}{<3.0}$	<1.0	$\dfrac{\geq 1.0}{<3.0}$	<0.5	circulating currents in windings
$\dfrac{>1.0}{<3.0}$	<1.0	≥ 3.0	<0.5	circulating currents core and tank; overload joints

Note: several simultaneously occurring faults can cause ambiguity in analysis

Figure 4.12   *Rogers fault gas ratios (ratio value based) (based on Joseph B. DiGiorgio's 'Dissolved gas analysis of mineral oil insulating fluids', published by NTT, 1996–2002)*

As can be imagined, artificial intelligence in the form of expert systems, artificial neural networks and fuzzy-logic-based systems are finding increasing application in this area.

Before leaving chemical detection of partial discharges, it is appropriate to note that for oil/paper insulated plant, the detection of degradation byproducts of the paper

Range	C_2H_2/C_2H_4	CH_4/H_2 Code	C_2H_4/C_2H_6
<0.1	0	1	0
0.1–1	1	0	0
1–3	1	2	1
>3	2	2	2

$\dfrac{C_2H_2}{C_2H_4}$	$\dfrac{CH_4}{H_2}$	$\dfrac{C_2H_4}{C_2H_6}$	Diagnosis	Typical examples
0	0	0	no fault	normal ageing
0	1	0	partial discharges of low energy density	discharges in gas filled cavities resulting from incomplete impregnation, or supersaturation or high humidity
1	1	0	partial discharges of high energy density	as above, but leading to tracking or perforation of solid insulation
1 → 2	0	1 → 2	discharges of low energy	continuous sparking in oil between bad connections of different potential or to floating potential; breakdown of oil between solid materials
1	0	2	discharges of high energy	discharges with power follow through; arcing – breakdown of oil between windings or coils or between coils to earth; selector breaking current
0	0	1	hot spots $T < 150°C$	general insulated conductor overheating
0	2	0	hot spots $150°C < T < 300°C$	local overheating of the core due to concentrations of flux;
0	2	1	hot spots $300°C < T < 700°C$	increasing hot spot temperatures; varying from small hot spots in
0	2	2	hot spots $T > 700°C$	core, shorting links in core, overheating of copper due to eddy currents, bad contacts/joints (pyrolitic carbon formation) upto core and tank circulating currents

Figure 4.13   IEC 599 1978 fault gas ratios[1]

Case	Characteristic fault	$\dfrac{C_2H_2}{C_2H_4}$	$\dfrac{CH_4}{H_2}$	$\dfrac{C_2H_4}{C_2H_6}$
PD	partial discharges (see notes 3 and 4)	NS [1]	<0.1	<0.2
D1	discharges of low energy	>1	0.1–0.5	>1
D2	discharges of high energy	0.6–2.5	0.1–1	>2
T1	thermal fault $T < 300°C$	NS[1]	>1 but NS [1]	<1
T2	thermal fault $300°C < T < 700°C$	<0.1	>1	1–4
T3	thermal fault $T > 700°C$	<0.2[2]	>1	> 4

note 1 – in some countries, the ratio C_2H_2/C_2H_6 is used, rather than the ratio CH_4/H_2, also in some countries, slightly different ratio limits are used

note 2 – the above ratios are significant and should be calculated only if at least one of the gases is at a concentration and a rate of gas increase above typical values

note 3 – $CH_4/H_2 < 0.2$ for partial discharges in instrument transformers; $CH_4/H_2 < 0.07$ for partial discharges in bushings

note 4 – gas decomposition patterns similar to partial discharges have been reported as a result of the decomposition of thin oil film between overheated core laminates at temperatures of 140°C and above

[1] NS = non-significant whatever the value

[2] an increasing value of the amount of C_2H_2/C_2H_6 may indicate that the hot spot temperature is higher than 1000°C

Figure 4.14   IEC 60599 1999 edition fault gas ratios interpretation[1]

in the oil may also imply the presence of partial discharges. Paper degrades to form several furans and these can be detected using various techniques of oil analysis. Although absolute levels of these byproducts are important, as with DGA, as the ratios of the different byproducts may prove more important in the longer term. It has been argued that, from these ratios, the temperature resulting in the paper degrading can be inferred and, in turn from this, the integrity of the paper [71–74]. Given that the temperature of the fault can be inferred, as with DGA, it should be possible to infer the presence (and, indeed, type) of partial discharges. However, this form of monitoring is not nearly so well established as DGA; the relationship to the detection of partial discharges has yet to be made. However, it is a potential technique for the future and most people would now analyse an oil sample for both gas and furan content.

4.3.5 Comparison among different PD measurement techniques relative to type of plant under investigation

From the foregoing it can be seen that, in general terms, most types of measurement of PD can be made on most types of plant – with some obvious exceptions as mentioned

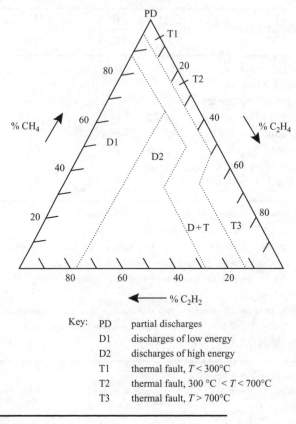

Key: PD   partial discharges
        D1   discharges of low energy
        D2   discharges of high energy
        T1   thermal fault, $T < 300°C$
        T2   thermal fault, $300 °C < T < 700°C$
        T3   thermal fault, $T > 700°C$

Limits of zones

PD   98% CH$_4$
D1   23% C$_2$H$_4$ 13% C$_2$H$_2$
D2   23% C$_2$H$_4$ 13% C$_2$H$_2$ 38% C$_2$H$_4$ 29% C$_2$H$_2$
T1   4% C$_2$H$_2$ 10% C$_2$H$_4$
T2   4% C$_2$H$_2$ 10% C$_2$H$_4$ 50% C$_2$H$_4$
T3   15% C$_2$H$_2$ 50% C$_2$H$_4$

triangle coordinates:

$$\%C_2H_4 = \frac{100x}{x+y+z} \qquad \%C_2H_4 = \frac{100y}{x+y+z} \qquad \%CH_4 = \frac{100z}{x+y+z}$$

where $x = C_2H_2$ $y = C_2H_4$ $z = CH_4$ in p.p.m.

Figure 4.15   IEC 60599 1999 edition Duval's triangle[1]

Type of plant	PD measurement technique							
	IEPD	tan δ	C.T.	capacitive probes	antenna	acoustic	chemical	thermography, etc.
Generators	✓✓✓	✓✓	✓		✓✓*			
Circuit breaker boxes	✓			✓✓		✓✓		
Transformers	✓✓		✓		✓✓*	✓✓	✓✓✓**	
Motors	✓✓✓	✓✓	✓		✓✓			
Cable end boxes				✓✓		✓✓		
Bushings		✓✓✓			✓✓		✓✓	
Capacitors		✓✓			✓	✓✓	✓✓**	
CVTs	✓✓	✓✓			✓✓	✓✓	✓✓**	
Overhead busbars, insulators, etc.								✓✓

* provided the plant item is air-cooled (i.e. has access vents)
** provided the plant item is oil filled

Figure 4.16   A comparison of different PD measurement techniques in relation to various items of plant

in the text. The decision on which to choose in a given situation becomes a balancing act among the costs of plant failure (including health and safety considerations, replacement/repair costs and outage losses), the cost in making a given measurement (including the frequency of measurement dictated over given time periods and the cost of potential outages dictated by the need to make solid connections with some measurement systems) and the quality of information provided by the technique. This type of decision is unique in each situation. However, for those embarking on the measurement of PD for their plant, the table shown in Figure 4.16 may prove helpful.

4.3.6   Other items of plant

In addition to the various types of plant highlighted in the foregoing sections, there are two other plant systems which, because of their specialist nature, have had evolved for them special techniques to measure PD. This is not to say that they cannot utilise the techniques described earlier, merely that better techniques are available specifically related to them. The two items of plant are cables and gas-insulated substations (GIS).

4.3.6.1   Cables

Although other techniques can be applied to cables with limited success in PD measurements, the technique described as cable mapping [75] has evolved specifically for use with cables where not only is the magnitude and nature of PD required but also their location. Without location, refurbishment of a plant item hidden underground and extending potentially over many kilometres would be extremely daunting.

The approach is as follows. First, the cable must be disconnected at both ends. One end is then used for measurement purposes and the other is open-circuited. The choice is arbitrary. The measurement end also acts as the input for the high voltage AC supply since, having been disconnected from its inherent supply, an external supply must be connected. This supply must be sufficiently portable to be carried in a van to the test site. Due to the impedance of a cable at 50 Hz, this would not be possible and so measurement on cables is made at between 0.01 Hz and 0.1 Hz.

$$I = \frac{V}{Z}$$

where

$$Z = \tfrac{1}{2}\pi f C$$

therefore

$$I = V 2\pi f C$$

The detector is similar to that described previously for electrical detection of discharge activity at the high voltage terminals of plant, i.e. a discharge-free capacitor coupled to a low voltage detector impedance but, in this case, the low voltage impedance is connected to instrumentation which can record both the magnitude and time of arrival of discharge pulses. When a degradation site discharges, part of the discharge pulse propagates towards the detector end and part to the open circuit end. When the latter encounters the open circuit, it is reflected and propagates back towards the detector end. In consequence, at the detector, a pulse arrives representing that part of the discharge which originally propagated from the discharge site followed by a second pulse representing that part of the pulse which propagated in the opposite direction, was reflected at the open circuit, and then propagated back along the cable to the detector (see Figure 4.17).

Knowing the relative times of arrival for the cable as a whole (obtained by injecting a calibration pulse at the detector end and detecting its time of arrival back to that point from the open-circuited end), the percentage location of the degradation site along the cable can be ascertained as can its length, knowing the velocity of propagation (Figures 4.17a and b) i.e. percentage location of PD along cable $= (1 - (\Delta t_1 - \Delta t_2)/\Delta t$ cable$) \times 100$, where Δt_1 and Δt_2 are the times of arrival of the two parts of the pulse and Δt cable is the total transit time (obtained by the calibration pulse injection) for the cable. Note, in this equation, Δt_1 and Δt_2 are taken as negative values.

What is produced is a map of the cable with the cable length on the horizontal axis and any discharge magnitudes on the vertical axis. By inspection, the size of the degradation site, its presence in only one or in more phases and its location in the cable or in a connector, can be obtained (Figure 4.18).

In addition, for each degradation site, a ϕ–q–n plot can be obtained from which the nature of the degradation can be inferred as indicated previously.

This approach can be used to measure PD activity on cables from around 4 m to 3 km in length (the latter figure arising from the attenuation of discharge pulses in

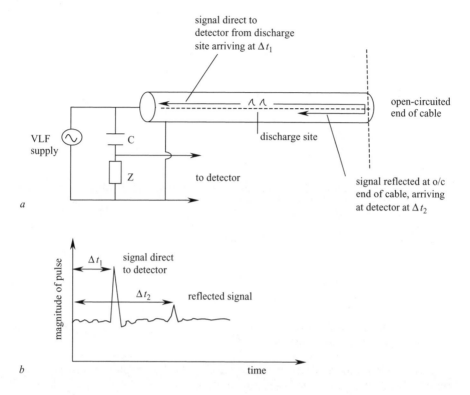

Figure 4.17

    a Partial discharge cable mapping
    b Signal's time of arrival used for location

paper-insulated cables). For cables with attenuation constants substantially different from paper cables, different figures may apply. It should be noted that a double-ended measurement is also possible which effectively doubles the cable length which can be tested. Errors associated with this approach have been addressed by Kreuger *et al.* [76].

4.3.6.2   Gas-insulated substations (GIS)

The criticality of GIS to electrical power generation/transmission has made it imperative that PD activity within their enclosures is identified (see Chapter 2). Chemical byproducts of SF_6 (as used in GIS) are particularly active chemically and will readily result in degradation within the system.

    Due to their nature, and in particular the very high electric fields within their assemblies, GIS suffer specific and relatively unique problems. These issues are neatly summarised in a two-part series of papers [77, 78]. As indicated there, the

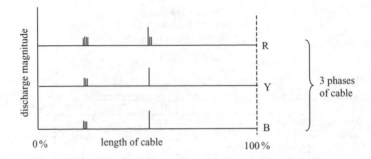

Figure 4.18   Cable map: PD as a function of location and magnitude for three phases of cable

range of possible PD inducing defects include:

- moving particles
- electrode protrusions
- fixed particles on insulating surfaces
- floating electrodes
- loose, non-floating electrodes
- voids in solid insulation of spacers.

To detect PD associated with these various defects, three approaches are possible, namely chemical, acoustic and electrical detection.

In relation to chemical detection, as indicated above, under PD stressing SF_6 will degrade to produce characteristic byproducts [79]. By gas sampling, it is possible to detect these and thereby infer the presence and, indeed, the nature of PD activity. Unfortunately, the technique is not quantitative and its effectiveness can be compromised by desiccants in the GIS chambers which absorb water – but also these byproducts.

Acoustic detection can be effective [80–82] but, due to the high attenuation of acoustic signals at flanges etc., each section must be tested in turn.

In relation to electrical detection, within the factory environment, a standard IEC60270 type approach can be used as described earlier in this chapter. Alternatively, rather than using large coupling capacitors, electric field sensors can be incorporated at a standard flange cover/inspection hatch within the GIS. Essentially, the gas gap capacitance between the busbar in the GIS and the chamber wall acts as the coupling/high voltage capacitor. The low voltage impedance coupled to the sensor is essentially as described previously.

These methods have all been described previously, although it is interesting to see how they are applied in GIS. However, a novel method of electrical detection has been developed for field measurements on GIS and it is this which is of particular interest in this section.

The principle of this UHF technique (further details can be found in Chapter 2) is that the current pulse which forms the partial discharge has a very fast risetime and this excites the GIS chambers into multiple resonances at frequencies up to 1.5 GHz or so (for more details see Chapter 3). These resonances can last for several microseconds (compared with the nanosecond(ish) duration of the originating current pulse). Once again, suitable UHF couplers can be fitted at flange covers etc. (thus permitting retrofitting). The PD signals can then be amplified and processed to produce (typically) the ϕ–q–n type patterns described earlier. The patterns produced are indicative of the nature of the PD-inducing defect. The work of Pearson *et al.* [83] and Judd *et al.* [84] is particularly illuminating in this context [85].

4.4   Concluding remarks

The study of partial discharges is not an academic pursuit. It is driven by a very practical desire, a desire to understand a phenomenon which can be utilised to infer the level of integrity within the insulation systems of high voltage plant and which, in itself, constitutes a serious stress degradation mechanism. The role of partial discharge studies remains twofold: to enable the development of new insulating systems which are resistant to partial discharge stressing and to predict remnant life for existing plant systems. To this end, studies in partial discharges involve a balance – a balance between understanding the phenomenon and being able to measure it. On one side lies the understanding of how partial discharges cause degradation, and which are the key parameters of activity to be measured. On the other side lies the ability to make the measurement, often under very adverse conditions.

Partial discharge study is a minor, but important, field of research. The continuing desire to find new, cheaper materials or subject existing materials to ever higher stress levels in a quest to cut costs will keep it so. The 1950s and 1960s saw the emergence and development of PD measurement techniques culminating in correlations being identified among PD activity and specific forms of degradation. These decades also witnessed a developing understanding, based on strong physical principles, of the mechanisms of partial discharge degradation and breakdown. In many senses this was the golden age of partial discharge study. The 1970s saw steady progress in electrical and acoustic measuring techniques and progress in chemical techniques for oil analysis. However, in degradation studies, the optimism of the 1950s and 1960s was waning. With a few conspicuous exceptions, many researchers were moving to empirical rate studies to breakdown. Accelerated frequency ageing of materials to breakdown under partial discharge stressing was replacing the more fundamental studies designed to elucidate the physical and chemical mechanisms of degradation. There appeared to be a recognition that an understanding of partial discharge degradation mechanisms was more complex – and more unattainable – than had been thought in earlier years. A more pragmatic, if limited, approach had emerged. The 1980s and 1990s saw another major shift in activity. Fuelled by the advent of digital microprocessor technologies, the ability to capture, store and process data on partial discharges – to measure partial discharge activity – made an enormous leap forward. The excitement of these new technologies and how they could be applied to partial discharge measurement far outpaced the research on degradation studies – again with a few, conspicuous, exceptions relating principally to the understanding of chemical stress reactions at a discharging surface.

Now, as we move into the new millennium, our ability to measure, store, process etc. data related to partial discharge activity far outstrips our understanding of what to do with these data – irrespective of whether the data was acquired by electrical, acoustic or chemical means.

Despite our progress, the synergetic nature of interactions present under partial discharge stressing continues to ensure that a clear understanding of discharge degradation remains elusive. Without that understanding, the quest for new materials capable of suppressing partial discharge activity will remain unfulfilled. Equally, without that understanding, the ability of database techniques and associated analytical/statistical tools to permit an accurate and objective assessment of insulation integrity and remnant life for existing plant insulating systems will remain flawed. It is perhaps time to recognise that a conscious shift to more fundamental degradation studies, to improve our understanding of the mechanisms involved, is the most efficient and effective way forward.

And finally – to those readers who are perhaps embarking on a study of partial discharges for the first time – a thought. As with most fields of endeavour, partial discharge research is complex. Understanding the mechanisms of partial discharge degradation, parameters of partial discharge activity and thereby inferring the rate, form and extent of degradation to predict remnant life is extraordinarily difficult. In a career's research of partial discharges, most of us can only hope to contribute one small piece to the puzzle of this phenomenon. However, once in every few generations, a researcher of true stature emerges. This person is capable of taking the many and varied pieces of the puzzle and bringing them together to form an intelligible picture. That has yet to happen for partial discharges. The most that each of us can hope to do, until then, is to ensure that the small piece which we contribute to the study of partial discharges has clarity, objectivity and accuracy.

The major breakthrough in partial discharge research – the solution to the puzzle – must not be hindered by errors in the pieces of research that have gone before. Contribute your piece but be open to the thought that maybe, just maybe, you are the one to solve the puzzle!

4.5   Note

[1] The author thanks the International Electrotechnical Commission (IEC) for permission to reproduce: the table 'Fault Gas Ratios Interpretation' and the illustration of 'Duval's Triangle' from its International Standard IEC 60599 2nd edition 1999–03 Mineral oil-impregnated electrical equipment in service – Guide to the interpretation of dissolved and free gases analysis. All such extracts are copyright of IEC, Geneva, Switzerland. All rights reserved. Further information on the IEC is available from www.iec.ch. IEC has no responsibility for the placement and context in which the extracts and contents are reproduced by the author; nor is IEC in any way responsible for the other content or accuracy therein.

4.6   References

1   HEPBURN, D.M.: 'The chemical degradation of epoxy resin by partial discharges'. PhD thesis, Glasgow Caledonian University, October 1994

2   TANAKA, T.: 'Internal partial discharge and material degradation', *IEEE Trans. Electr. Insul.*, 1986, **EI-21**, pp. 899–905

3   GAMEZ-GARCIA, M., BARTNIKAS, R., and WERTHEIMER, M.R.: 'Synthesis reactions involving XLPE subjected to partial discharges', *IEEE Trans. Electr. Insul.*, 1987, **EI-22**, pp. 199–205

4   MAYOUX, C.J.: 'Partial discharge phenomena and the effect of their constituents on PE', *IEEE Trans. Electr. Insul.*, 1976, **EI-11**, pp. 139–148

5   MEEK, J.M., and CRAGGS, J.D. (Eds.): 'Electrical breakdown of gases' (Wiley, Chichester, New York, 1978)

6   BOULLOUD, A. and CHARRIER, I.: 'Current density on the plane electrode of a positive point-to-plane gap'. 6th international conference on *Gas discharges and applications*, Edinburgh, UK, September 8–11, 1980, pp. 110–113

7   BLACK, W.B., *in* BOVEY, F.A., and WINSLOW, F.H. (Eds.): 'Macromolecules: an introduction to polymer science' (Academic Press, 1978), Chap. 7

8   FUJITA, H.: 'An analysis of mechanical stress in solid dielectrics caused by discharge in voids', *IEEE Trans. Electr. Insul.*, 1987, **EI-22**, pp. 277–285

9   ARBAB, M.N., AUCKLAND, D.W., and VARLOW, B.R.: 'Mechanical processes in the long term degradation of solid dielectrics'. Proceedings of IEE 5th international conference on *Diel. Mats., Meas. and Appl.*, Canterbury, UK, June 1988, pp. 109–111

10   PEYROUS, R., COXON, P., and MORUZZI, J.: 'Mass spectra of ionic species created by dc corona discharges in air'. Proceedings of IEE 7th international conference on *Gas discharges*, London, UK, September 1982

11   PEYROUS, R., and MILLOT, R.M.: 'Gaseous products created by dc corona discharges in air or oxygen fed point to plane gap'. Proceedings of IEE 7th international conference on *Gas discharges*, London, UK, September 1982

12   ALLEN, N.L., COXEN, P., PEYROUS, R., and TEISSEYRE, Y.: 'A note on the creation of condensation nuclei by negative discharges in air at low pressure', *J. Phys. D, Appl. Phys.*, 1981, **14**, pp. L207–90

13   SONE, M., MITSUI, H., and TAKAOKA, K.: 'Clustering of water molecules in several kinds of gases'. Conference on *Electrical insulation and discharge phenomena*, Pocono Manor, Penn., USA, October 28–31, 1990, pp. 696–701

14   YAGI, S., and TANAKA, M.: 'Mechanism of ozone generation in air-fed ozonisers', *J. Phys. D, Appl. Phys.*, 1979, **12**, pp. 1509–1520

15   SHIELDS, A.J., and KEMP, I.J.: 'Degradation and breakdown of mica under partial discharge stressing', *IEE Proc., Sci. Meas. Technol.*, 2000, **147**, (3), pp. 105–109

16   SHIELDS, A.J., and KEMP, I.J.: 'Degradation and breakdown of mica under partial discharge stressing: transverse discharges', *IEE Proc., Sci. Meas. Technol.*, 2000, **147**, (5), pp. 256–260

17 HEPBURN, D.M., KEMP, I.J., SHIELDS, A.J., and COOPER, J.: 'Degradation of epoxy resin by partial discharges', *IEE Proc., Sci. Meas. Technol.*, 2000, **147**, (3), pp. 97–104

18 HEPBURN, D.M., KEMP, I.J., RICHARDSON, R.T.R., and SHIELDS, A.J.: 'Role of electrode material in partial discharge chemistry'. Proceedings of 5th international conference on *Cond. and breakdown in solid dielectrics*, Leicester, UK, July 1995

19 DISSADO, L.A., and FOTHERGILL, J.C.: 'Electrical degradation and breakdown in polymers', in STEVENS, G.C. (Ed.): (Peter Peregrinus Ltd, 1992)

20 MAYOUX, C.J.: 'Influence of low energy ions in the degradation of polyethylene by partial discharges', *European Polym. J.*, 1973, **9**, pp. 1069–1075

21 MAYOUX, C.J.: 'Corona discharge and the ageing process of an insulator', *IEEE Trans. Electr. Insul.*, 1977, **EI-12**, pp. 153–158

22 HARROLD, R.T., EMERY, F.T., MURPHY, F.J., and DRINKUT, S.A.: 'Radio frequency sensing of incipient arcing faults within large turbine generators', *IEEE Trans. PAS*, 1979, **98**, (3), pp. 1167–1173

23 STONE, G.: 'Measuring partial discharge in operating equipment', *IEEE Electr. Insul. Mag.*, 1991, **7**, (4), pp. 9–20

24 KREUGER, F.H.: 'Recognition of discharges', *Electra*, 1970, **11**, pp. 69–91

25 GULSKI, E.: 'Digital analysis of partial discharges', *IEEE Trans. Dielectr. Electr. Insul.*, 1995, **2**, (5), pp. 822–837

26 BOGGS, S.A., and STONE, G.C.: 'Fundamental limitations in the measurement of corona and PD', *IEEE Trans. Electr. Insul.*, 1982, **17**, (2), pp. 143–150

27 BARTNIKAS, R.: 'A commentary on pd measurement and detection', *IEEE Trans. Electr. Insul.*, 1987, **22**, (5), pp. 629–653

28 BOGGS, S.A.: 'Partial discharge: overview and signal generation', *IEEE Electr. Insul. Mag.*, 1990, **6**, (4), pp. 33–42

29 KREUGER, F.H., GULSKI, E., and KRIVDA, A.: 'Classification of pd', *IEEE Trans. Electr. Insul.*, 1993, **EI-28**, (6), pp. 917–931

30 GULSKI, E., and KRIVDA, A.: 'Influence of ageing on the classification of partial discharges in high voltage components', *IEEE Trans. Dielectr. Electr. Insul.*, 1995, **2**, pp. 676–684

31 ZAENGL, W.S., OSVATH, P., and WEBER, H.J.: 'Correlation between band-width of pd detectors and its inherent integration error'. Proceedings of IEEE international symposium on *Electr. Insul.*, Washington DC, USA, June 1986

32 ZAENGL, W.S., LEHMANN, K., and ALBEIZ, M.: 'Conventional pd measurement techniques used for complex hv apparatus', *IEEE Trans. Electr. Insul.*, 1992, **EI-27**, (1)

33 ZAENGL, W.S., and LEHMANN, K.: 'A critique of present calibration procedures of partial discharge measurements', *IEEE Trans. Electr. Insul.*, 1993, **EI-28**, (6), pp. 1043–1049

34 TWIEL, M.M., STEWART, B.G., and KEMP, I.J.: 'An investigation into the effects of parasitic circuit inductance on partial discharge detection'. Proceedings 29th EIC/EMCW Expo., Cincinnati, USA, October 2001

35 BRYDEN, J., KEMP, I.J., NESBITT, A., CHAMPION, J.V., DODD, S., and RICHARDSON, Z.: 'Correlations among tree growth and the measurable parameters of partial discharge activity'. Proceedings of 11th international symposium on *HV Eng.*, **4**, (79.524), London, UK, August 1999

36 PATSCH, R., HOOF, M., and REUTER, C.: 'Pulse-sequence-analysis, a promising diagnostic tool'. Proceedings of 8th international symposium on *High Voltage Eng.*, Yokohama, Japan, August 1993

37 PATSCH, R., and HOOF, M.: 'The influence of space charge and gas pressure during electrical tree initiation and growth'. Proceedings of 4th international conference on *Props. and Appl, of Diel. Mats.*, Brisbane, Australia, July 1994

38 HOOF, M., and PATSCH, R.: 'Pulse-sequence-analysis: a new method for investigating the physics of pd-induced ageing', *IEE. Proc., Sci. Meas. Technol.*, 1995, **142**, (1), pp. 95–101

39 HOOF, M., and PATSCH, R.: 'Voltage-difference analysis: a tool for partial discharge source identification'. Proceedings of IEEE international symposium on *Electr. Insul.*, Montreal, Canada, June 1996

40 WILSON, A., JACKSON, R.J., and WANG, N.: 'Discharge detecting techniques for stator windings', *IEE Proc. B, Electr. Power Appl.*, 1985, **132**, (5), pp. 234–244

41 KEMP, I.J., GUPTA, B.K., and STONE, G.C.: 'Calibration difficulties associated with partial discharge detectors in rotating machine applications'. Proceedings of EEIC/ICWA conference on *Electr. Insul.*, Chicago, USA, 1987

42 VAILLANCOURT, G.H., MALEWSKI, R., and TRAIN, D.: 'Comparison of three techniques of partial discharge measurements in power transformers', *IEEE Trans. Power Appar. Syst.*, 1985, **PAS-104**, (4), pp. 900–909

43 WOOD, J.W., SEDDING, H.G., HOGG, W.K., KEMP, I.J., and ZHU,: 'Partial discharges in HV machines; initial considerations for a PD specification', *IEE Proc. A, Sci. Meas. Technol.*, 1993, **140**, (5), pp. 409–416

44 KOPF, U., and FESER, K.: 'Rejection of narrow-band noise and repetitive pulses in on-site pd measurements', *IEEE Trans. Dielectr. Electr. Insul.*, 1995, **2**, (6), pp. 1180–1191

45 BORSI, H., GOCKENBACH, E., and WENZEL, D.: 'Separation of partial discharge from pulse-shaped noise signals with the help of neural networks', *IEE Proc., Sci. Meas. Technol.*, 1995, **142**, (1), pp. 69–74

46 LOUIS, A.K., MAASS, P., and RIEDER, A.: 'Wavelets: theory and applications' (John Wiley & Sons, 1997)

47 DAUBECHIES, I.: 'The wavelet transform, time-frequency localization and signal analysis', *IEEE Trans. Inf. Theory*, 1990, **36**, (5), pp. 961–1005

48 MALLAT, S.G.: 'A theory for multiresolution signal decomposition: the wavelet representation', *IEEE Trans. Pattern Anal. Mach. Intell.*, 1989, **11**, (7), pp. 674–693

49 ANGRISANI, L., DAPONTE, P., LUPO, G., PETRARCA, C., and VITELLI, M.: 'Analysis of ultrawide-band detected partial discharges by means of a multiresolution digital signal-processing method', *Measurement*, 2000, **27**, pp. 207–221

50   MA, X., ZHOU, C., and KEMP, I.J.: 'DSP based partial discharge character-isation by wavelet analysis'. Proceedings of 19th international symposium on *Discharges and electrical insulation in vacuum* (ISDEIV2000), Xi'an, China, 18–22 September 2000

51   MA, X., ZHOU, C., and KEMP, I.J.: 'Investigation into the use of wavelet theory for partial discharge pulse extraction in electrically noisy environments'. Proceedings of IEE 8th international conference on *Dielectr. Mats. Meas. and Appl.*, Edinburgh, UK, 17–21 September 2000

52   SHIM, SORAGHAN, J.J., and SIEW, W.H.: 'Detection of pd utilising digital signal processing methods, part 3: open loop noise reduction', *IEEE Electr. Insul. Mag.*, 2001, **17**, (1), pp. 6–13

53   MA, X., ZHOU, C., and KEMP, I.J.: 'Wavelet for pd data compression'. IEEE conference on *Electrical insulation and dielectric phenomena* (CEIDP 2001), Kitchener, Canada, October 2001

54   SIMONS, J.S.: 'Diagnostic testing of high-voltage machine insulation – a review of ten years' experience in the field', *IEE Proc. B, Electr. Power Appl.*, 1980, **127**, (3), pp. 139–154

55   PAKALA, W.E., and CHARTIER, V.L.: 'Radio noise measurement on overhead power lines from 2.4 to 800kV', *IEEE Trans. Power Appar. Syst.*, 1971, **PAS-90**, pp. 1155–1165

56   SAWADA, Y., FUKUSHIMA, M., YASUI, M., KIMOTO, I., and NAITO, K.: 'A laboratory study of RI, TIV, and AN of insulators strings under contaminated condition', *IEEE Trans. Power Appar. Syst.*, 1974, **PAS-93**, pp. 712–719

57   BERADELLI, P.D., CORTINA, R., and SFORZINI, M.: 'Laboratory investiga-tion of insulators in different ambient conditions', *IEEE Trans. Power Appar. Syst.*, 1973, **PAS-92**, pp. 14–24

58   IEEE RADIO NOISE AND CORONA SUBCOMMITTEE REPORT, RI limits Task Force, Working Group No. 3: 'Review of technical considerations on limits to interference from power lines and stations', *IEEE Trans. Power Appar. Syst.*, 1980, **99**, (1), pp. 35–379

59   KANNUS, K., LEHTIO, A., and LAKERVI, E.: 'Radio and TV interference caused by public 2.4 kV distribution networks', *IEEE Trans. Power Deliv.*, 1991, **6**, (4), pp. 1856–1861

60   STEWART, B.G., HEPBURN, D.M., KEMP, I.J., NESBITT, A., and WATSON, J.: 'Detection and characterisation of partial discharge activity on outdoor high voltage insulating structures by RF antenna measurement tech-niques'. Proceedings of international symposium on *High-voltage engineering*, London, UK, August 1999

61   BROWN, P.: 'Non-intrusive partial discharge measurements in high voltage switchgear'. IEE Colloquium on *Monitors and condition assessment equipment*, December 1996

62   LUNDGAARD, L.E.: 'Partial discharge, Pt XIII: acoustic partial discharge detection – fundamental consideration', *IEEE Electr. Insul. Mag.*, 1992, **8**, (4), pp. 25–31

63   LUNDGAARD, L.E.: 'Partial discharge, Pt XIV: acoustic partial discharge detection – practical application', *IEEE Electr. Insul. Mag.*, 1992, **8**, (5), pp. 34–43

64   JONES, S.L.: 'The detection of partial discharge activity within power transformers using computer aided acoustic emission techniques'. Proceedings of IEEE international symposium on *Electr. Insul.*, 1990, Toronto, Canada

65   LINDNER, M., ELSTEIN, S., LINDNER, P., and TOPAZ, J.M.: 'Daylight corona discharge imager'. Proceedings of international symposium on *High voltage Eng.*, 4.349–4.352, London, UK, August 1999

66   HALSTEAD, W.D.: 'A thermodynamic assessment of the formation of gaseous hydrocarbons in faulty transformers', *J. Inst. Petroleum*, 1973, **59**

67   ROGERS, R.R.: 'Concepts used in the development of the IEEE and IEC codes for the interpretation of incipient faults in power transformers by dissolved gas in oil analysis'. IEEE Winter Meeting, 1978

68   IEC 599: 'Guide for the interpretation of the analysis of gases in transformers and other oil-filled electrical equipment in service', 1978

69   IEC 60599: 'Mineral oil-impregnated electrical equipment in service – Guide to the interpretation of dissolved and free gas analysis', second edn., 1999

70   DUVAL, M.: 'Dissolved gas analysis: it can save your transformer', *IEEE Electr. Insul. Mag.*, 1989, **8**, (6), pp. 22–27

71   SHROFF, D.H., and STANNETT, A.W.: 'A review of paper ageing in power transformers'. *IEE Proc. C, Gener. Transm. Distrib.*, 1985, **132**, (6), pp. 312–319

72   UNSWORTH, J., and MITCHELL, F.: 'Degradation of electrical insulating paper monitored with high performance liquid chromatography', *IEEE Trans. Electr. Insul.*, 1990, **EI-25**, (4), pp. 737–746

73   EMSLEY, A.M., XIAO, X., HEYWOOD, R.J., and ALI, M.: 'Degradation of cellulosic insulation in power transformers. Part 2: Formation of furan products in insulating oil', *IEE Proc., Sci. Meas. Technol.*, 2000, **147**, (3), pp. 110–114

74   CARBALLIERA, M.: 'HPLC contribution to transformer survey during service or heat run tests'. IEE Coll., Publ. 184, 1991

75   MACKINLAY, R.: 'Managing an ageing underground high-voltage cable network', *Power Eng. J.*, 1990, **4**, pp. 271–277

76   KREUGER, F.H., WEZELENBURG, M.G., WIENER, A.G., and SONNEVELD, W.A.: 'Partial discharge Pr XVIII: Errors in the location of partial discharges in high voltage solid dielectric cables', *IEEE Electr. Insul. Mag.*, 1993, **9**, (6), pp. 15–23

77   BAUMGARTNER, R., FRUTH, B., LANZ, W., and PETTERSON, K.: 'Partial discharge, part IX: PD in gas-insulated substations – 1', *IEEE Electr. Insul. Mag.*, 1991, **7**, (6), pp. 5–13

78   BAUMGARTNER, R., FRUTH, B., LANZ, W., and PETTERSON, K.: 'Partial discharge – Part X: PD in gas-insulated sub stations – measurement and practical considerations', *IEEE Electr. Insul. Mag.*, 1992, **8**, (1), pp. 16–27

79   CHU, F.Y.: 'SF_6 decomposition in gas insulated equipment', *IEEE Trans. Electr. Insul.*, 1986, **EI-21**, pp. 693–726

80   GRAYBILL, H.Q., CRONIN, J.C., and FIELD, E.J.: 'Testing of gas insulated substitutions and transmission systems', *IEEE Trans. Power Appar. Syst.*, 1974, **PAS-93**, pp. 404–413

81   LUNDGAARD, L.E., RUNDE, M., and SKYBERG, B.: 'Acoustic diagnoses of gas insulated substations – a theoretical and experimental basis', *IEEE Trans. Power Deliv.*, 1990, **PD-S**, (4), pp. 1751–1759

82   LUNDGAARD, L.E., TANGEN, G., SKYBERG, B., and FAGUSTAD, K.: 'Acoustic diagnoses of GIS: field experience and expert systems', *IEEE Trans. Power Deliv.*, 1992, **PD-7**, (1), pp. 287–294

83   PEARSON, J.S., FARISH, O., and HAMPTON, B.F. *et al.*: 'Partial discharge diagnostics for gas insulated substations', *IEEE Trans. Dielectr. Electr. Insul.*, 1995, **2**, (5), pp. 893–905

84   JUDD, M.D., FARISH, O., and HAMPTON, B.F.: 'Excitation of UHF signals by partial discharges in GIS', *IEEE Trans. Dielectr. and Electr. Insul.*, 1996, **3**, (2), pp. 213–228

85   CIGRE Task Force 15/33.03.05: 'Partial discharge detection system for GIS: sensitivity verification for the UHF method and the acoustic method', *Electra*, 1999, (183), pp. 74–87

Chapter 5

ZnO surge arresters

A. Haddad

5.1 Introduction

High voltage systems are often subject to transient overvoltages of internal or external origin. The resultant surges travel along the transmission line and can cause damage to unprotected terminal equipment. Corona losses and the earth return path can attenuate and distort the surges, but the magnitude of the surge may still exceed the insulation level of the equipment. Surge arresters provide a limitation of the overvoltage to a chosen protective level. The superiority of the recently developed zinc oxide (ZnO) material over earlier silicon carbide (SiC) renewed interest and boosted the use of surge arrester protection.

The ideal surge arrester would be one that would start conduction at a voltage level at some margin above its rated voltage, hold that voltage level with as little variation as possible for the duration of the overvoltage surge and cease conduction as soon as the voltage across the arrester returns to a value close to the rated voltage. Such an arrester would therefore conduct only that current required to reduce the surge voltage to the arrester protective level, and absorb the energy that is associated with the overvoltage. The basic non-linear formula that relates voltage V and current I in a surge arrester is given by:

$$I = kV^{\alpha} \tag{5.1}$$

where k is a constant and α is the coefficient of non-linearity.

Surge arresters are used for protection of power system equipment against surge overvoltages because they offer low protection levels and permit the reduction of insulation levels, which has a substantial effect on the cost of high voltage equipment. Early overvoltage protective devices were simple spark gaps capable of supporting essentially no voltage during conduction. Major development steps in the evolution of the arrester technology, which are in the direction of the optimum arrester characteristic, have been made since then.

5.2   Evolution of overvoltage protection practice

The evolution of surge arrester technology has been characterised by both the gradual improvement of the various arrester components, and more importantly by four successive major steps: the simple spark gap, the valve-type arrester, the introduction of active gaps and the gapless metal oxide arrester. The latter is associated with ZnO varistors, and the two former arresters were made with SiC resistors. The introduction of each arrester had an important impact on protection levels and cost of the power system equipment as a whole.

5.2.1   Simple spark gaps

During the first half of the twentieth century, protection of apparatus in electrical power systems was provided by rod gaps (also referred to as coordinating gaps) and a very high withstand voltage for the insulation [1].

The advantage of rod gaps lies in their simplicity and cost. However, they cannot support any voltage during their operation and cannot clear the power frequency follow current, which means that after sparking over, a permanent fault occurs which leads to interruption of supply. Moreover, the time lag to sparkover and the dependence of sparkover on many factors may result in failure of the protective system. The combination of a high protective level with spark gaps still finds application today in low voltage applications, but at high service voltages increasing the withstand voltage of insulation has a marked influence on costs.

At 33 kV systems and lower, it is not recommended to use simple rod gaps because they may be bridged by birds; instead duplex gaps and/or triggered gaps of expulsion type may be employed.

5.2.2   Valve-type arresters

The discovery of the non-linear properties of silicon carbide (SiC) around 1930 made feasible the introduction of valve-type surge arresters to protect power systems against atmospheric discharges. The general design of a conventional arrester consists of plate-type gaps spaced by insulating rings with series non-linear SiC resistors also known as thyrite. The spark gap performs the switching function and the SiC resistor limits the follow current and enables the arrester to reseal. By subdividing the spark gap, it is possible to reseal at higher voltages eliminating steady state energy dissipation. Compared with spark gaps, valve-type arresters have a number of advantages but the protective level remains relatively high. The SiC elements have severe specification standards because most of the voltage is supported by the non-linear resistors and the arc voltage is comparatively negligible; also most of the energy associated with the discharge of a transmission line/cable is absorbed by the SiC non-linear resistors.

5.2.3 Surge arresters with active gaps

A significant improvement of arresters to alleviate problems of utilisation was achieved by the introduction of active gaps (also called current-limiting gaps) [2]. These are characterised by the presence of blast coils that produce a strong magnetic field during the passage of arrester follow current. The arcs across the individual gaps are then elongated and blown towards the edge of the chamber, thus producing an increased voltage drop during the follow current period. The power frequency voltage across the series non-linear resistors is therefore reduced with the consequent reduction of the follow current. The arc voltage opposes the follow current and interrupts it before the working voltage reaches zero. A further component of the active design is the grading system which ensures that the voltage is distributed uniformly between the series spark gaps.

Compared with simple gap arresters, active gap arresters have the following advantages:

(i) arc voltage is of the order of the voltage drop across the non-linear SiC resistor
(ii) protective level is substantially lowered
(iii) the energy absorbed by the resistors decreases; some of it is absorbed by the elongated arc and the other part remains in the transmission line because the follow current is interrupted before voltage zero
(iv) constant voltage during flow of arrester current
(v) the roots of the arc move along the electrode which reduces the change in arcing found in simple plate gap arresters.

5.2.4 Metal oxide surge arresters

The relatively high protective levels provided by SiC arresters became more and more an economical disadvantage with the increase of maximum system voltages. In order to reduce the insulation levels of the apparatus it was, therefore, necessary to try to reduce the protection levels of the surge arresters. Furthermore, the increase of transmission line lengths resulted in an increase of the energy that the arrester had to absorb in case of a line discharge through it. The search for new materials to obtain superior non-linear $V-I$ characteristics led to the discovery of zinc oxide (ZnO) varistors in the late 1960s. The impedance of ZnO varistors at voltages below the rated voltage is so high that the resulting current is in the milliampere range. The direct consequence of this low current consumption was the possibility of constructing surge arresters with no series gaps. The first power system gapless metal oxide surge arresters were completed in the mid 1970s [3, 4]. The absence of the gaps and the extreme non-linearity of the voltage–current characteristic of the material resulted in the following additional changes in the main features of arrester protection:

(i) elimination of grading resistors/capacitors, which reduced further the number of parts used for the arrester construction
(ii) energy absorbed by the non-linear resistor represents only a fraction of the discharge energy of the transmission line because there is no follow current,

and parallel connections of varistors to increase the energy absorption capability of the arrester are now possible [5]

(iii)   lower discharge voltage (residual voltage which appears across the arrester terminals when a discharge current is flowing through it [6, 7])

(iv)   lower protective level (highest discharge voltage that appears between arrester terminals during specific conditions of a discharge operation [6, 7])

(v)   faster switching capability compared with a spark gap time response.

5.2.5   Existing applications of ZnO surge arresters

5.2.5.1   Applications

In addition to the extensive use in low voltage suppressors used in electronic equipment and domestic appliances [8], ZnO material has now largely replaced SiC material in modern surge arresters. Because of their reliability and superiority, ZnO surge arresters are used in various overvoltage protection schemes and in coordination with overcurrent protection [9]. Among these schemes are standard protection of overhead distribution systems [10–13], underground distribution systems [14–17], GIS systems [18–20], HVDC systems and cables [18, 19], EHV series capacitors [21–23], EHV shunt capacitors [24], EHV breakers [25] and rotating electrical machines [26]. Surge arresters, which limit surges entering underground power cable and provide adequate protection at open end of feeders [15, 27], known as riser-pole and open-end arresters, are needed in order to maintain acceptable protective margins on cable systems.

Surge arresters are also used at bonding points of the cable system [16, 17]. They are referred to as sheath voltage limiter (SVL) arresters. The voltage rating is, however, much smaller than that for phase-conductor arresters, for example: 3 and 6 kV arresters are acceptable on 138 kV systems, 9 kV arresters will be needed only for higher fault current or very long sections of cable. The rating of SVL arresters for single bonded systems is usually higher than that for cross bonded systems [16] because of the induced voltage involved. The main duties for SVL arresters are to:

(i)   withstand the sheath induced voltage; this is not onerous because the magnitude is a few tens of volts

(ii)   dissipate short circuit current in case of a line fault

(iii)   be water and humidity proof.

5.2.5.2   Design improvements

Field experience from these various applications has led to some important improvements in design features which were suggested to overcome problems encountered on surge arresters:

(i)   use of shunt gap module to increase tolerance of ZnO arresters to temporary system overvoltage, particularly useful for riser-pole, elbow and oil-immersed arresters [28]

(ii)   use of special thermal shields and ventings for pressure relief to avoid fragmentation which occurs due to thermal shocks following fault currents [29, 30]

(iii) soldering techniques for mechanical strength [18]
(iv) use of ceramic grading capacitors to control voltage distribution along the arrester column [19]
(v) better $E-J$ characteristic of the material to reduce size of the arrester and improve voltage distribution [20]
(vi) special requirements for the working environment and conditions [31, 32].

5.3 Basic properties of ZnO material

5.3.1 Composition and effect of additives

The semiconducting properties of ZnO were very well known [33] well before Matsuoka *et al.* [34] investigated the non-linear voltage–current characteristic of ZnO doped with alkali earth metal oxides. It was found that with appropriate amounts (0.5 mol%) of additives having an ionic radius larger than that of Zn^{+2} (which would not dissolve in the ZnO lattice and would thus form a segregation layer at the ZnO grain boundary), the ceramic exhibits a high coefficient of non-linearity α on its $V-I$ characteristic. Furthermore, it was shown that the new material was exceptionally superior compared with conventionally used SiC varistors [35]. The non-linearity coefficient, α, of early ZnO varistors was in the range 25–50. In contrast, that of SiC varistors was between 2 and 7 [36].

In 1971, Matsuoka [37] disclosed a large number of compositions which produce useful ZnO varistors. A coefficient $\alpha = 50$ was obtained when 99.9 per cent pure ZnO is doped with 1 mol% Sb_2O_3 and 0.5 mol% of Bi_2O_3, CoO, MnO and Cr_2O_3 [38–42]. Adding more dopants has allowed the fabrication of ZnO varistors with large grain size and the reduction of the breakdown voltage of a single grain boundary [43–47]. Paraseodymium oxide rich composition, which does not contain any bismuth, was used to manufacture ZnO varistors with a two-phase microstructure and improve its electrical properties such as energy absorption capability, residual voltage and life performance [48, 49].

It is postulated that the transition metal oxides are involved in the formation of interfacial states and deep bulk traps, both of which contribute to the highly non-linear resistivity property. The grain size, hence the breakdown voltage per each grain boundary, could be controlled with additives such as TiO_2, Sb_2O_3, BeO, Al_2O_3, K_2O and SiC. The electrical stability has been improved by adding ZrO_2, Cr_2O_3 or Nb_2O_5 to bismuth oxide rich compositions. Table 5.1 summarises the effect of various additives on ZnO varistor properties reported in the presently reviewed literature.

5.3.2 Fabrication process

Figure 5.1 shows a diagram of the ZnO surge arrester fabrication process. The basic material used to manufacture metal oxide varistors is pulverised, very finely grained and highly pure (99.9 per cent) ZnO with a particle size of 1–10 μm. Several doping elements and specially prepared powders [43, 50] are added in the form of fine oxide powders. The concentration of the individual components range from

Table 5.1   Effect of additives on ZnO material electrical properties

Additives	Effect on ZnO properties	Reference
• Bi_2O_3 + alkali metals: CaO, CoO, BaO, SrO, MnO	• non-linearity appears • highest α with BaO	34, 37, 40
• K_2O • 3 mol% MeO • seed grains: ZnO + 0.5% $BaCo_3$	• inhibits grain growth • minimum for good non-linearity • large grain size • breakdown voltage reduced • α reduced	52 41 43
• Ga_2O_3	• increases α • large leakage current	47, 102
• ZrO_2	• better energy absorption capability	47
• SiO_2	• low leakage current	47
• Pr_6O_{11} substituted to Bi_2O_3	• 2-phase microstructure (no spinel) • improved $V-I$ characteristic • higher energy absorption capability • better life performance • lower residual voltage	48, 49
• Al_2O_3	• lower residual voltage • grain growth enhancer • large leakage current • high α	44, 48
• Frit glass: 63% PbO, 25% Bi_2O_3, 12% SiO_2	• lower leakage current • same α	50
• TiO_2, Sb_2O_3	• grain growth enhancer • lower potential barrier height • high α	44, 45, 102
• Cr_2O_3	• enhances thermal stability	50
• Nb_2O_5	• α increased to 60 • leakage current decreased • higher energy absorption capability • doubles life performance	46

p.p.m. to per cent [51], and the composition and proportion of additives differ from manufacturer to manufacturer. The resulting powder is ball mixed in deionised water [46, 52] or alcohol [50] for 10–24 hours, after which a homogeneous aqueous slurry with small particles is obtained. The homogeneity of the mixture is of immense importance for the quality of the end product. The slurry is then calcinated at 700°C [37, 41] in a spray drier for a duration of 30 min [43] to 2 h [37]. Homogeneous spheroidal granulates of 100 μm in diameter are obtained [51] and are compacted, in the next production stage, under a pressure of 340–400 kg/cm. The pressed material

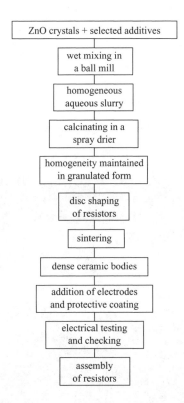

Figure 5.1   Simplified diagram of the manufacturing process of ZnO material

is moulded into disc shaped blocks having a diameter of 3–100 mm and a thickness of 1–30 mm.

The blocks are then sintered in an electric furnace in air or oxygen for one or two hours under constant temperature between 1000–1400°C. This heat treatment has the effect of further densifying the compacted powder into a solid ceramic body. The temperature and duration of the process are very critical for the characteristics of ZnO varistors. It is reported [51] that in this process the adjacent powder particles are united by means of diffusion, and subsequently grow into large grains. In the next stage, the dense ceramic bodies are furnace cooled at a rate of 50–200°C/h at the end of which they are coated with a collar material (glass, epoxy) on the peripheral surface to protect them from the environment and prevent flashover. Metallic electrodes (silver, gold, copper, aluminium) are also deposited on each face of the blocks in the last production stage. Finally, before assembling, routine checks are performed to verify the resistor electrical properties

5.3.3   Microstructure

The grain size, hence the breakdown voltage per each grain boundary, could be controlled with additives such as TiO_2, Sb_2O_3, BeO, Al_2O_3, K_2O and SiC.

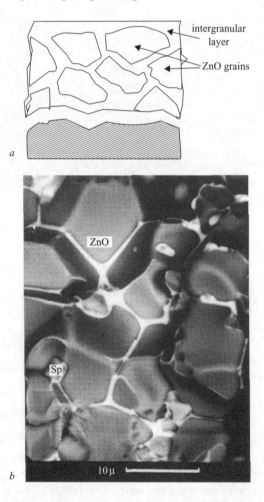

Figure 5.2

    a   Schematic representation of ZnO material microstructure

    b   Typical SEM micrograph of ZnO material. (Courtesy of GREUTER, F., PERKINS, R., ROSSINELLI, M., and SCHMUCKLE, F.: 'The metal-oxide resistor at the heart of modern surge arresters', *ABB Rev.* Surge Arresters, 1989, **1**)

The fabrication process results in a polycrystalline material composed of semiconducting ZnO grains surrounded by a thin layer of intergranular materials (Figure 5.2*a*). The microstructure has been examined by means of several optical techniques such as x-ray diffraction, electron microprobe, scanning electron microscopy (SEM) and transmission electron microscopy (TEM) [41, 42, 45, 47, 48, 50, 52]. The resulting photomicrographs as reproduced on Figure 5.2*b* reveal that the microstructure of ZnO varistors is made up of minute ZnO crystals or grains, approximately 5 to 20 μm in size. These grains are separated physically from each other by thin layers of intergranular materials of approximately 0.2 mm in thickness. The nature and composition of these layers depends on the additives used for doping ZnO. Very sensitive measurements [40, 42] showed that the intergranular layer can be as small as 2 Å.

Spinel particles of $Zn_7Sb_2O_{12}$ which are thought to play a passive electrical role, were also observed [39, 41, 48] in the intergranular phase and their origin was traced back to the presence of Sb_2O_3 additive in the ZnO composition during the sintering process. In addition to the spinel crystallites, a pyrochlore phase ($Bi_2\{Zn_{4/3}Sb_{2/3}\}O_6$) was identified in the intergranular layer and a bismuth rich phase at the triple point which is formed with three or more ZnO grains [47].

5.3.4 Conduction mechanism in ZnO varistors

There are two distinct regions of conduction: the prebreakdown region where the varistor behaves almost in a linear fashion (the current is thermally activated [39, 47, 53] which is described by the Schottky emission over the potential barrier [54]) and the breakdown region where high non-linearity appears and more complex conduction processes take place. An overview of these studies is given below.

5.3.4.1 Progressive evolution of the breakdown conduction theory

Since the early studies [55], the origin of the non-linearity of conductivity has been attributed to the regions situated in the ZnO grain boundaries. This was soon confirmed by thin film experiments [52] and photoconductivity measurements [39]. Several models describing the conduction phenomenon have been developed to give an account of the non-linear electrical properties of zinc oxide varistors (Table 5.2). On the other hand, there are other known mechanisms which have been ruled out due to incompatibility with ZnO properties. The diffusion theory has been excluded [55] since the current in ZnO is strongly voltage dependent. The avalanche breakdown mechanism has also been rejected [27, 39, 53, 55] because it results in a positive temperature coefficient; ZnO is known to exhibit a strong negative temperature coefficient.

The first model of the ZnO conduction mechanism was based on space charge limited current (SCLC) [55]. Although yielding high coefficients of non-linearity α [56], this model is inappropriate to ZnO varistors because of the passive role of the intergranular layer in determining the breakdown voltage [39, 53]. Subsequently, a simple tunnelling process [39, 47] and a two-step transport mechanism [52] that consisted of steps of electron hopping [57] and electron tunnelling were suggested. None of these models could account for the high α observed in ZnO varistors.

Most of the accepted models assume a common description of the band structure in the vicinity of the intergranular layer region. The potential barrier is assimilated to an assembly of two back-to-back Schottky junctions. The Schottky emission model [27] was extended to include thin intergranular layers, electron traps in the intergranular layer [41] and to consider the depletion regions formed in the ZnO grains [58, 59]. The parallel currents in the bismuth rich intergranular layer are important when the varistor

Table 5.2   The various models considered to explain the conduction mechanism in ZnO varistors

Model	Feature	Reference
Diffusion theory	ZnO current strongly dependent on voltage (rejected)	55
Avalanche theory	ZnO temperature coefficient is negative (rejected)	39, 55
SCLC theory	high α but intergranular layer is electrically inactive (rejected)	53, 55, 56
Simple tunnelling	ZnO capacitance is function of voltage and low α (rejected)	41, 43, 47, 52
Electron hopping and tunnelling	low α (rejected)	52, 57
Schottky emission	linear conduction: suitable for prebreakdown regime	39, 53, 54
Double Schottky barrier (DSB) without intergranular layer	linear conduction	27
Parallel currents	important in low current regime	60
DSB with thin layer	low α	41
DSB with thin layer and electrons traps	better α	41
Depletion layer and holes and tunnelling	high α but does not account for negative capacitance	53, 59, 61
Depletion layer and holes	high α	56, 61, 62, 63
Depletion layer and recombination of holes	high α and improved results	58

is operated in the low conduction regime [60]. A further improvement of the depletion layer model is achieved by introducing a process of creation, by hot electrons [56] and impact ionisation [58, 59, 61], of holes [62, 63] in the depletion layers which are inversely polarised compared with the double Schottky barrier. Under the effect of the field, the created holes are collected at the interface and narrow the potential hill at the intergranular interface and favour tunnelling [53]. However, tunnelling was rejected and the presence of holes was associated at the interface with lowering of the potential barrier height [60, 64]. As a result, a sharp increase of current for a very small increase in applied voltage is obtained. A further step in the formulation of the conduction mechanism takes into account recombination of the created holes [58].

5.3.4.2   Description of the high conduction model

Consider the situation where two ZnO grains meet to form a grain boundary. The thickness of the intergranular layer is assumed very thin, so that it can be neglected. Gap states are then formed at the interface and trap a negative charge. This charge is screened by the ionised donors' ZnO. As a result, depletion layers appear on both sides of the barrier. The resulting double Schottky potential barrier has a height of the order 1–3.2 eV and the total spatial extent of the potential barrier is about 0.2 μm, representing the depletion layer width $(X_L + X_R)$, which is very large compared with the thickness of the intergranular layer (2 Å). The form of the potential barrier depends on the interface density of states, the donor density and the applied voltage V.

The main features of this model are:

(i)   parabolically decaying potentials on both sides of the interface
(ii)   asymmetric distortion of the barrier under an applied voltage V
(iii) decay of the barrier height with voltage
(iv)   at large applied voltages, a very large potential drop ($\sim$5 eV) within very short distances ($\sim$0.1 mm); therefore, very large electric fields ($\sim$1 MV/cm) are reached in such conditions.

The last feature is of particular importance since under such magnitudes of electric field, generation of hot electrons process takes place [56, 58]. Under a very high field, electrons are injected over the potential barrier into the positively biased depletion region. The strong electric field easily accelerates the particles to attain high kinetic energy. Minority carriers (holes) are then created by impact ionisation and are swept towards the interface. As a result, the barrier height is lowered which increases the current sharply. Further increase of current is obtained if recombination of the swept holes with trapped electrons at the interface is considered.

5.4   Thermal performance of ZnO surge arresters

5.4.1   Background

The performance of surge arresters in power systems is determined on the one hand by the electrical and thermal properties of the varistors, and on the other by the design and installation of the arrester.

Compared with SiC gapped arresters, ZnO gapless metal oxide arresters offer a protection closer to the ideal. However, because these arresters contain no gaps, a leakage current flows through the material at working voltages which causes power losses and heating of the ZnO elements. This can be dangerous to the stability of the arrester, particularly in the low conduction regime where the V–I characteristic of ZnO material is very sensitive to temperature. Badly dimensioned arresters are exposed to the risk of thermal runaway. A further practical problem relating to power system usage concerns pollution of arresters, which can cause abnormal voltage and current distribution in the arrester in both steady state and transient conditions.

5.4.2   Heat dissipation capability and thermal stability of ZnO surge arresters

In surge arresters, the ZnO valve elements are generally located in an environment consisting of gas or solids which will limit cooling of the valve elements when submitted to their continuous operating voltage. As can be seen on Figure 5.3, for the

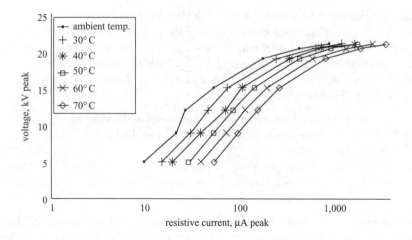

Figure 5.3   Temperature dependence of ZnO voltage current characteristic

same applied voltage level, any temperature rise would increase the current because of the high sensitivity to temperature of the $V-I$ characteristic in the low conduction regime. A sort of positive feedback will occur and may cause damage to the arrester if the material heat is not dissipated. For this reason, thermal stability is one of the most important application criteria for metal oxide surge arresters. It is frequently analysed with the help of a heat loss input heat balance diagram [65–74].

The thermal stability of ZnO surge arresters is affected by ambient temperature and heat dissipation capability [65, 66], impulse degradation [67, 68] and ageing [69–71]. To obtain thermal stability, the electrical power dissipation in the element must be balanced against heat output to the environment. Near the thermal equilibrium, it is possible to express the thermal dissipation capacity Q of a surge arrester as:

$$Q = C_T(T - T_a) \tag{5.2}$$

where T is the temperature of ZnO valve elements, T_a the ambient temperature and C_T the thermal dissipation factor. The heat generation, P, which is voltage and material composition dependent, may be approximated by:

$$P = Ae^{-(W_c/kT)} \tag{5.3}$$

where W_c is the activation energy, $k = 0.86 \times 10$ eV/K (Boltzmann constant), T is the temperature of the material and A depends on the applied voltage level and the physical dimensions of the valve elements.

The above curves are shown schematically in Figure 5.4. For an ambient temperature T_a and an applied voltage V, the two curves intersect at two points X and Y. The lower point X is at a stable operating temperature T_x and is referred to as the lower stability point. The upper point Y is also at a stable operating temperature T_y and is called the upper limit stability point. At these temperatures, the power input equals the power output.

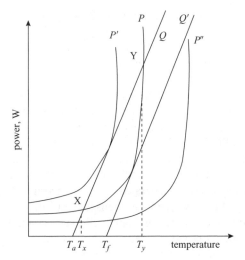

Figure 5.4   Thermal stability diagram for ZnO arresters

    Q heat loss at ambient temperature T_a
    Q' heat loss at ambient temperature T_f
    P power (heat) dissipation at voltage V_1
    P' power (heat) dissipation at voltage V_2 with $(V_2 > V_1)$
    P'' power (heat) dissipation at voltage V_3 with $(V_3 < V_1)$

The valve element temperature always settles at the stable operating point, as long as the initial valve temperature does not exceed the instability threshold. Following any temperature rise up to T_y, produced for example by the absorption of an overvoltage energy, the valve element state will revert to the lower stability point X. For temperatures higher than T_y, the valve element would have an ever increasing power dissipation which would be much higher than the heat dissipation capacity of the arrester. Therefore, when the arrester is operated above point Y, the arrester is no longer able to dissipate the heat introduced into the ZnO valve elements. As a consequence, the temperature will increase and thermal runaway will occur. The arrester is designed for steady state operation at temperature T_x. The energy difference between points X and Y is called the allowable surge energy. It can be seen, in Figure 5.4, that there is a maximum temperature range (T_y-T_x) for the ambient temperature T_a, beyond which the valve element would be unstable. It is noted that this range is not a unique value, but is highly dependent upon the ambient temperature and on the applied voltage. This is a simplified representation which does not include various limiting factors, such as transient changes in the rate at which the energy is delivered and dissipated.

For the ambient temperature T_a, the maximum energy discharge capacity $E_d(T_a)$ is given by [69, 70]:

$$E_d(T_a) = mC_{Tm}(T_y - T_x) \tag{5.4}$$

where C_{Tm} is the heat capacity per unit mass 0.54 J/g/°C and m is the mass of the valve element.

It should be noted, however, that the energy capacity of an arrester may be lower in practice due to thermal shock or puncture. The larger the difference between T_y and T_x, the larger the capacity of the arrester. If the surge energy does not exceed the allowable value, then the arrester temperature will return to T_x. However, the time taken to return to stable conditions is important. Up to one hour may be needed for a surge arrester to recover its initial conditions after absorption of two rectangular surges of 600 A amplitude and 2 ms duration [75]. In addition, the upper stability point is the steady state and not the transient limit, which is higher [71, 72]. For one ZnO product, it was found [76] that the allowable temperature rise due to surge current is 110°C if the arrester is operated at an alternating voltage which produces 1 mA current through the valve elements. Consequently, Figure 5.4 can only be used to evaluate thermal instability arising from steady state conditions, where there has been a sufficient amount of time allowed to stabilise both valve element and arrester housing temperature, and is of limited use when thermal instability due to sudden energy input, switching or lightning surges or even TOV is considered [71].

5.4.3   Thermal runaway

The term thermal runaway as defined by both existing IEC [77] and JEC standards [78] is used to describe a situation culminating in failure if the ZnO arrester is operated above the upper stability point Y. However, there are a number of factors which influence the position of point Y.

5.4.3.1   The ambient temperature

Increasing the ambient temperature lowers the dissipation curve Q and hence decreases the maximum energy discharge capacity [66, 70]. For an ambient temperature of T_f, the upper and lower stability points X and Y coincide with each other and there is zero energy capacity.

5.4.3.2   Ageing and degradation of the arrester

An ageing characteristic that leads to increased power dissipation also decreases the maximum energy capacity [65, 66, 70]. However, an ageing characteristic, which shows decreased power dissipation, increases the maximum energy discharge capacity [70]. In the first case, which is more realistic, the heat generation P is expressed as [67, 69]:

$$P = P_0(1 + h\sqrt{t}) \tag{5.5}$$

where P_0 is the heat generation at the initial stage, h is constant and t is time.

The degradation in metal oxide surge arrester elements increases the leakage current and hence the heat generation P. The effect of ageing is to shift the stability point X to higher temperatures, therefore, degrading it more. Consequently, the thermal runaway threshold is shifted towards lower temperatures [69, 70], and the margin T_y–T_x becomes smaller. The ageing subsequent to the application of voltage will make the heat generation curve *P* move to *P'* which has a new operating point. This evolution of the operating point on the straight line continues until the stability points X and Y are at the same position.

5.4.3.3 Applied voltage level

Thermal runaway of metal oxide elements at high AC stress is caused by a positive feedback process due to the temperature dependence of the resistive component of leakage current. Moreover, the power–voltage characteristic is highly non-linear so that a small increase in voltage will result in a considerably higher power. The effect of this is to shift curve *P* of Figure 5.4 upwards. This displacement will be of similar effect as that described for the ageing effect.

5.4.3.4 Heat dissipation capability of the housing

From the heat dissipation point of view, a small unit is better than a complete arrester. The main reason for this is related to the heat flow along an arrester column. Heat dissipation is reduced at the centre of a long column [68]. As a consequence, the temperature of the middle section of an arrester is higher than those at the ends [65, 79, 80]. Up to six times improvement can be achieved if cooling metal fins are used [66].

Improving the heat dissipation capability of the arrester means that the curve *Q* on Figure 5.4 will be lifted to higher values, hence the thermal property of the arrester is improved. For this reason, conducting tests on pro-rated units or model units must ensure that the unit reproduces the heat dissipation and thermal characteristics of the complete arrester to which the tests are to be related [68].

5.4.4 Thermal runaway critical condition

The critical condition of thermal runaway is evaluated analytically based on the temperature dependence and the heat loss of the arrester [65]. As shown above, thermal runaway occurs when the lower stability point X moves to the upper stability point Y, i.e. the heat dissipation curve *Q* is tangential to the generation curve *P*. At that instant, we can write the critical conditions,

$$P = Q \tag{5.6}$$

and

$$\frac{\partial P}{\partial T} = \frac{\partial Q}{\partial T} \tag{5.7}$$

Using Equations (5.2) and (5.3), the critical values of temperature T_{cr} and power P_{cr} [65] are:

$$T_{cr} = T_a \left(1 + \frac{kT_a}{W_c^2} (W_c + 2kT_a) \right) \tag{5.8}$$

and

$$P = 4\frac{C_d}{D} (T_{cr} - T_a) \tag{5.9}$$

where C_d is the heat dissipation coefficient W/cm/K and D is the diameter of elements.

5.4.5   Dynamic stability of ZnO surge arresters

The energy capacity of a surge arrester may also be affected by thermal stability limitations, because the surge amplitude, duration and energy have critical values above which thermal runaway is triggered. Nishikawa *et al.* [76] quoted an allowable surge energy density of 350 J/cm for an ambient temperature of 40°C and a maximum 2 ms duration for a 1.25 p.u. temporary overvoltage.

5.4.6   Simulation of thermal characteristics of ZnO surge arresters

Thermal stability tests are very expensive and need careful measurements with highly sensitive equipment [74]. It is therefore desirable to have a simple circuit model that will reproduce the thermal properties of the arrester. Lat [71] proposed an electrical analogue model of the thermodynamic behaviour of the arrester assembly. The electrical equivalence is based on representing power flow as a current and temperature as a voltage. The circuit as shown in Figure 5.5 takes account of the electrical power

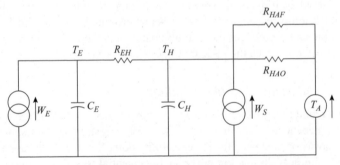

C_E, C_H : thermal capacities of the valve element
and adjacent housing, respectively
R_{EH} : thermal resistance from element to external
surface of the housing
R_{HAO} : thermal resistance from housing to ambient,
radiation and natural convection components
R_{HAF} : thermal resistance from housing to ambient,
forced convection temperature
T_E : valve element temperature
T_H : housing temperature
T_A : ambient temperature
W_E : electrical power input to valve element
W_S : heat input due to solar radiation

Figure 5.5   *Thermal model of a ZnO arrester. (Courtesy of LAT, M.V.: 'Thermal properties of metal oxide surge arresters', IEEE Trans. PAS, 1983, PAS-102, (7), pp. 2194–2202, ©2003 IEEE)*

input, the solar radiation and the ambient temperature. It is claimed that this circuit is valid for steady state and transient behaviour, giving results which agree closely with experimental data [71, 81]. Another circuit [82], which considers the same input components, is a ladder network whose element characteristics are determined experimentally [83]. An important factor, the heat dissipation transfer resistance can be studied using the three heat transfer mechanisms involving radiation, conduction and convection [82].

5.5 Degradation and ageing of ZnO surge arresters

The terms varistor degradation and varistor stability are often used to describe the electrical state of a varistor relative to its past or future state when under the influence of external stress. The degree of degradation is a good indication of varistor reliability and is usually used for predicting the life span of ZnO surge arresters.

Severe electrical degradation causes a large increase in leakage current at low voltages. The degradation process affects mainly the varistor leakage and prebreakdown conduction and not the behaviour of the varistor at voltages higher than the breakdown voltage.

Most degradation studies have investigated the changes in the voltage level which cause a current of 1 mA to flow through the ZnO elements at ambient temperature, after removing the applied degrading stress. Others [84, 85] have used the thermally simulated currents (TSC) method. The TSC currents are known to be associated with stored charge within the varistor which is released at a constant heating rate e.g. 4°C/min, after a biasing direct voltage has been removed. Measurement of the TSC spectrum gives useful information on the amount of trapped carriers, the polarisation of dipoles and space charge polarisation caused by ion migration which are all dependent on the degree of degradation [64, 84, 86].

Another method [86] uses changes incurred to the potential barrier height. It is now established [64, 86] that the barrier height, which determines the amplitude of current, is decreased after each ageing cycle (continuous applied voltage at high temperature) and after applying a fast high current surge (4/10 μs, 250 kA). The consequence of a decreasing barrier height is the increase in leakage current. The degree of degradation has also been evaluated through partial discharge detection [87].

5.5.1 Differences between degradation and thermal runaway

It is useful to note two differences between normal degradation and the thermal runaway condition [66]:

(i) In degradation mode the gradual increase of current with time is different from the thermal runaway increase. It has a slower rate of change and shows little dependence upon temperature. The rate of change of current due to degradation for most products was found [65, 69, 88] proportional to the square root of time:

$$\Delta i \propto \sqrt{\Delta t} \qquad (5.10)$$

whereas this change is in exponential form during the thermal runaway process [65]:

$$\Delta i \propto e^{\sqrt{\Delta t}} \tag{5.11}$$

(ii)   The change in the $V-I$ characteristic, which is caused by degradation, is preserved even after removing the stress. In thermal runaway, however, recovery may be possible after cooling down.

5.5.2   Factors affecting rate of degradation

A number of factors have been identified to affect the rate of ZnO degradation.

5.5.2.1   Composition and fabrication process

The physical explanation of degradation and ageing is related to electromigration of oxygen ions perpendicular to the grain boundaries which lowers the potential barrier height [86]. Reference 64 gives a comprehensive summary of the various physical explanations of the degradation process. The amount of additives [84, 89] and the heat treatment [85], with which ZnO is produced, affect the non-linearity coefficient of the material. Products with high coefficients degrade more because of the relatively high currents which flow in the material. Furthermore, it is observed that the long term leakage current characteristic depends on the composition and manufacturing process of the elements even if they have similar $V-I$ characteristics at the initial stage.

5.5.2.2   Homogeneity

Evidence of non-uniformity of ZnO material has been shown by means of dot electrodes [90, 91] and an infra-red radiation thermocamera [91]. Heterogeneous conductivity leads to inhomogeneous energy dissipation. The resulting local overheating may then lead to an irreversible change of varistor properties after surge stress, because of localised paths of current.

5.5.2.3   Ambient temperature

Since the current is thermally activated in the low conduction region, there is a strong dependence of the $V-I$ characteristic on the temperature, so that high temperatures cause high currents to flow, and this process is known to accelerate the ageing of the sample. As an example [81], continuous operation under 135°C for one day degrades the ZnO arrester by a decrease of 12 per cent in the voltage level, V_{1mA}, which produces a current of 1 mA through the arrester.

5.5.2.4   Working voltage

The highly non-linear characteristic of ZnO has permitted the reduction of the protective level. However, continuous application of the working voltage progressively degrades the ZnO elements and may end up in thermal runaway after some operating time [66]. The rate of degradation is largely affected by the magnitude [68, 92] and type (direct or alternating) [89] of the applied voltage. Testing ZnO at higher voltages and temperatures results in accelerated ageing.

5.5.2.5   Design and physical arrangements of ZnO surge arresters

The main features that could affect the degradation rate are the heat dissipation capability of the complete housing [66, 68, 93] and the voltage distribution along the arrester column. As will be discussed later, the voltage distribution along an arrester column is distorted by stray capacitances, and may enhance the non-uniform temperature distribution, which in turn will accelerate the degradation of the arrester. Besides, extensive laboratory measurements [69, 70] showed that degradation could affect the collar material (rim) severely, which may overshadow the ageing of the ZnO body. New organic materials are, however, more efficient as far as corona [94] and long term stress [70] are concerned.

5.5.2.6   Pollution of surge arresters

The effect of pollution is related to that of the dry bands which cause partial discharges and corona effects. These types of electrical stress cause surface erosion and localised damage to the collar material protecting the ZnO blocks [94]. The partial discharges cause a drop in oxygen content of the gas surrounding ZnO blocks together with simultaneous formation of corrosive gases and this was found to accelerate ZnO degradation [95]. Moreover, a white deposit of silicone, whose origin is not explained, is observed [96] in contaminated arresters where corona activity is present.

5.5.2.7   Environment

The working environment of ZnO protective devices, which may be gaseous (air, SF_6) or liquid (transformer oils) and solid (sidewall), plays an important role in the life span of the material. ZnO elements in SF_6 are degraded more than those in air [69]. However, adding to SF_6 a small amount (3 per cent) of oxygen, the presence of which seems to play an important role in the chemical process, ensures less degradation [95]. Water vapour and moisture presence also cause an increased leakage current at low temperatures [95, 97]. The relatively high working temperature of ZnO valve elements in transformer oils ($\sim 95°C$) degrades these arresters at a faster rate. For an applied voltage $V = 0.7\ V_{1mA}$, the arrester life has been estimated to be as short as 0.36 years [98].

5.5.2.8   Surge absorption capability

In general, just after the application of a surge, there is a measurable degradation of ZnO electrical properties. Three types of overvoltages are of interest; switching surges, temporary overvoltages and lightning surges:

(i)   *Switching surges* Following the absorption of a high energy surge, the cooling conditions of the arrester have very little effect [90]. Experiments [4, 66, 68] showed that degradation is more severe under high current surge than high energy surge. Degradation resulting from energy surge absorption is therefore negligible in practice and has almost no effect on the life of the elements. However, high energy surge applied to ZnO elements exhibiting localised spots of conduction could melt the material [66, 99]. Switching surge degradation is pronounced under unidirectional surges but can be reduced to some extent by applying surges of opposite polarity [67]. Only 1 per cent degradation was observed [68] on the direct voltage which produces 1 mA (V_{1mA}), after the discharge of positive and negative switching surges. After absorbing two rectangular surges of 600 A and 2 ms duration followed by power frequency, less than −4 per cent maximum variation in voltage, V_{1mA}, was observed [4], and the ZnO element could return to its initial working conditions of voltage and temperature after just one hour. As can be seen on Figure 5.6a, $V_{1\,mA}$ undergoes as little as 5 per cent change when the arrester is subjected to 1000 switching surges of 600 A magnitude and 2 ms duration [4].

(ii) *Temporary overvoltage (TOV)* No change is measured after TOV absorption [68]. Durability of MOV against TOV depends mainly on global structural uniformity [100]. Temperature effects, however, may lead to thermal runaway if the surge parameters are higher than specific allowable values given in manufacturers' specifications (Figure 5.6b) [76].

(iii) *Lightning discharge* Degradation due to discharge of the surge energy largely depends on the peak value of the discharge current rather than the discharge energy. The presence of AC stress does not enhance nor restore the surge degradation. A few high current surges degrade ZnO arresters more than many high energy surges of lower current; two (8/20 μs) surges of 65 kA amplitude caused the leakage current to increase by 50 per cent [68]. Figure 5.7a shows the changes in V_{1mA} as a function of the number of applied surges for two current amplitudes (40 kA and 100 kA) of a fast surge (4/10 μs) [4]. The allowable number of surges decreases when the amplitude of the surge increases; e.g. a life of 50 years allows 130 surges of 50 kA or only 30 surges of 100 kA [89]. Figure 5.7b shows the overall effect of fast surge absorption on the $V–I$ curve. As may be observed in the figure, the relative degradation is more severe for the lower voltages of the $V–I$ characteristic.

5.5.3 Destruction mechanism

Ultrasonic methods that are sensitive for detection of physical defects such as small voids or cracks, have been used to check the physical condition of overstressed ZnO elements [91]. There was no microstructure difference which could be revealed between new and degraded (aged) blocks when scanning electron micrography (SEM) was used [69]. However, when a high energy such as those of excessive direct or alternating voltage levels or a very long surge overvoltage (>100 μs) is absorbed, a puncture will occur in the ZnO element [99]. Such failure may be caused by current concentration in non-uniform elements. At 820°C, the main additive Bi_2O_3 melts. ZnO grains along the localised current path may then melt as a result of high temperature.

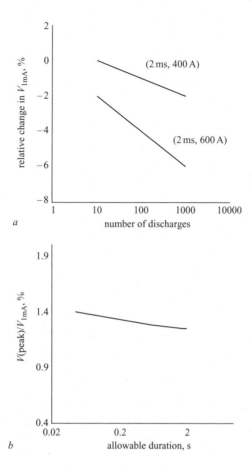

Figure 5.6

   a Degradation of ZnO surge arresters after absorption of switching surges. (Courtesy of NISHIKAWA, H., SUKEHARA, M., and YOSHIDA, T., *et al.*: 'Thyristor valve gapless arrester element lifetime performance and thermal stability', *IEEE Trans. PAS*, 1985, **PAS-104**, (7), pp. 1883–1888, ©2003 IEEE)

   b Degradation of ZnO surge arresters following temporary over-voltages. (Courtesy of NISHIKAWA, H., SUKEHARA, M., and YOSHIDA, T., *et al.*: 'Thyristor valve gapless arrester element lifetime performance and thermal stability', *IEEE Trans. PAS*, 1985, **PAS-104**, (7), pp. 1883–1888, ©2003 IEEE)

In the case of high current fast surges (time to crest less than 50 μs), a cracking has been observed. The heat, which is generated within a very short time, may not transfer sufficiently quickly to other parts of the element. This thermal expansion stress is transformed into a mechanical one and produces the crack. This localised destruction may be healed by cutting the material around the defect [100].

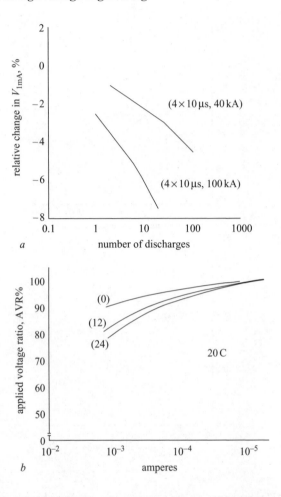

Figure 5.7

    a Degradation of ZnO surge arresters caused by lightning impulses. (Courtesy of HAHN, E.E.: 'Some electrical properties of zinc oxide semiconductor', *J. Appl. Phys.*, 1951, **22**, (7), pp. 855–863)

    b Changes in *V–I* characteristic following surge absorption. (Courtesy of NISHIKAWA, H., SUKEHARA, M., and YOSHIDA, T., *et al.*: 'Thyristor valve gapless arrester element lifetime performance and thermal stability', *IEEE Trans. PAS*, 1985, **PAS-104**, (7), pp. 1883–1888, ©2003 IEEE)

5.6   Life estimation of ZnO surge arresters

5.6.1   Long term accelerated ageing tests

As described in the previous sections, the continuous applied voltage on ZnO surge arresters during their service duty causes a permanent flow of current which affects the arrester qualities in the long term. As a consequence of this degradation, the life span during which the arrester will provide efficient protection will be limited. In order to assess this life span, long term accelerated ageing tests have been performed [4, 65, 66, 69, 70, 75, 76, 84, 88, 89, 92, 96]. These tests are normally run for long periods (approx. 1500 hours) under a constant elevated voltage level ($\sim$ knee of conduction) and at a constant temperature which is considerably higher than that anticipated in actual use (70–150°C). The current I or power loss P is monitored versus time t until a given value is reached. This value is often referred to as the end of life criterion. Various criteria have been adopted:

(i)   when power loss reaches a critical value such as 1 W/cm of length [96]
(ii) when the leakage current reaches a fixed value, e.g. 100 mA [76]
(iii) when power loss reaches a value that is double the initial value measured at the beginning of the test [70, 84, 88].

During these long term tests, it has been observed that the P–t and I–t curves go through three stages: (i) a stable period of working life where the arrester does not show any change in its characteristics; then comes (ii) a degradation period where P and I increase with time; and after this degradation stage thermal runaway is preeminent; (iii) a sudden sharp increase of I and P takes place.

It was suggested [66] that the end of life criterion should be chosen at the beginning of the degradation period. Figure 5.8 shows schematically the various shapes measured during long term tests.

The difference in the long term accelerated ageing curves is essentially due to the difference in composition and manufacturing process of the tested ZnO products [70, 89]. Dissimilar shapes have been recorded for ZnO products showing identical initial V–I characteristic but having different compositions [89]. Another important factor, which could play a key role in the curve shape, is the nature of the collar material. It is known that the collar material can overshadow the characteristics of the ZnO material when it is of unsuitable nature [69, 70].

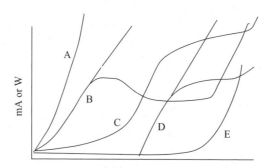

Figure 5.8   Measured typical shapes of current/power versus time curves during long term accelerated ageing tests

5.6.2   *Dakin–Arrhenius plots of life span*

The theory developed by Dakin [101] has been widely used for ZnO surge arrester life prediction [4, 65, 66, 69, 70, 76]. On the basis that the ageing process of ZnO is a chemical reaction, its duration may be described by the Arrhenius' equation:

$$\log \xi = \frac{A}{T} + B \qquad\qquad (5.12)$$

where ξ is the life span, A, B are constants depending on the material and stress and T is the absolute temperature.

The time taken for power or current to reach the end of life criterion is a measure of the tested ZnO element lifetime, ξ, at the test applied voltage and temperature. If lifetimes obtained in this way at various temperatures are plotted on a log scale versus the reciprocal of the test absolute temperature, Dakin–Arrhenius life plots can be obtained [65, 69]. Extrapolation to lower temperatures allows us to obtain the life span of the ZnO surge arresters under normal working conditions. Figure 5.9 shows a typical example of life plots. Most studies [75, 96, 97] have indicated that the measured current and power dissipation data agree with the Dakin–Arrhenius equation. However, there is a third component to consider in P and I equations; that is the number of surges that the arrester absorbs during its working service life [67]. Subsequently, however, it has been clearly shown [70] that the above equation could be validated for ZnO only within a limited range of temperature (115–175°C) since

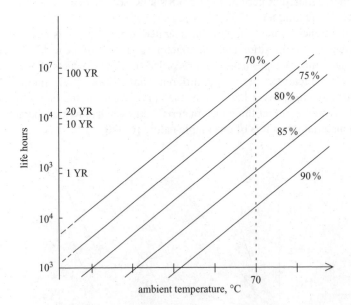

Figure 5.9   *Typical Dakin–Arrhenius life plots. (Courtesy of BRONIKOWSKI, R.J., and DuPONT, J.P.: 'Development and testing of MOVE arrester elements', IEEE Trans. PAS, 1982, **PAS-101**, (6), pp. 1638–1643, ©2003 IEEE)*

the data for 70°C were significantly lower than those expected by extrapolation of the data from higher temperatures.

5.6.3 Alternative methods of life estimation

The extrapolation to lower temperatures has always been treated with caution [84], and other methods [65, 67, 69, 70, 102] of estimating life have been adopted:

(i)   Kirkby *et al.* [70] proposed that the energy capacity determined from the thermal stability diagram would be of more value to the user than a predicted uncertain life.

(ii) Tominaga *et al.* [65] and Vicaud [69] suggested an analytical model in which it is assumed that the power P and current I are proportional to the square root of time:

$$P = P_0(a\sqrt{t} + b) \tag{5.13}$$

$$I = I_0(c\sqrt{t} + d) \tag{5.14}$$

where P, I are power and current at time t, P_0, I_0 are power and current at start of test, b, d are constants and a, c are determined from the Dakin–Arrhenius law:

$$a = a_0 e^{-(W_c/kT)} \tag{5.15}$$

$$c = c_0 e^{-(W_c/kT)} \tag{5.16}$$

with k Boltzmann constant $= 0.86 \times 10^{-4}$ eV/K, W_c activation energy ~ 0.5 eV for ZnO and a_0 and c_0 constants.

If the end of life criterion is defined, the combination of the above equations easily yields the life span. However, experience showed that the development of P–t or I–t could deviate from a square root law and exhibit a more complex behaviour.

5.7 Test procedures for the characterisation of ZnO arrester

Modelling of ZnO surge arresters uses experimental data acquired over a wide range of current and voltage amplitudes and types. Such tests include accurate recording of the voltage and current traces as well as the measurement of the V–I characteristic. There are two distinct regions of conduction in ZnO arresters

5.7.1 Prebreakdown regime of conduction: AC and DC tests

The low current region of the V–I curve is determined from direct or alternating voltage tests. Current magnitudes up to 10 mA can be used without damage to the ZnO material. However, currents above 20 mA can cause too much heat dissipation which can lead to irreversible degradation of the arrester. A laboratory circuit for testing ZnO surge arresters with alternating and direct voltages incorporates suitable power sources and adequate transducers, with simple dividers and shunt resistors giving sufficient accuracy for the determination of the V–I curve. For 50 Hz alternating voltages, the

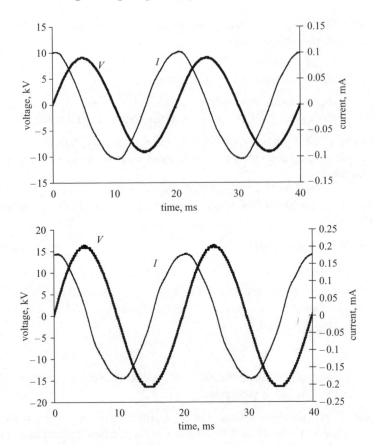

Figure 5.10   Typical AC records of voltage and current obtained in the prebreakdown region

current waveshape changes significantly around the knee of conduction as illustrated in Figures 5.10 and 5.11 and in order to compare DC and AC characteristics, only the peak resistive component of the AC current is used in the AC $V–I$ curve.

5.7.2   Breakdown regime of conduction and up-turn region: impulse tests

In constructing the $V–I$ curve, switching impulses of up to 2.5 kA can be used since they are within the energy requirements of the arresters. In the up-turn region with high magnitude currents up to 100 kA, fast or lightning impulses are commonly applied depending on arrester class and type. These fast impulses have lower energy content compared with switching impulses, hence their suitability for high current amplitude testing. Figure 5.12 shows a typical laboratory arrangement for testing ZnO arresters with impulse currents. With suitable setting of the current impulse generator, the test loop should be as short as possible to minimise stray inductance effects. The voltage measurement requires transducers with a fast response time (usually less than 20 ns)

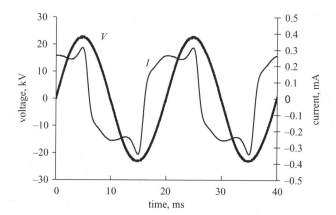

Figure 5.11   Typical AC records of voltage and current obtained in the breakdown region (around the knee of conduction)

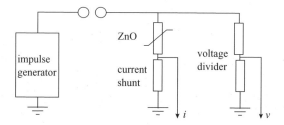

Figure 5.12   Simple layout for impulse testing of ZnO surge arresters

and no overshoot on their transfer characteristic. The transducer should be located very close to the arrester to avoid induced spikes [103]. Current measurement also requires specialised tubular current shunts or fast response current transformers. Figure 5.13 shows some typical fast impulse voltage and current records obtained on a 15 kV rated surge arrester.

Figure 5.14 shows a typical measured $V–I$ curve for polymeric surge arresters using the methods described above. As can be seen, the $V–I$ characteristic is highly non-linear in the breakdown region above 10 mA.

5.7.3   Voltage distribution along arrester columns

5.7.3.1   Effect of stray capacitances

At working power frequency voltages, the resistive component of current is very small compared with the capacitive component. The stray capacitances then play an important role in determining the voltage distribution [104]. The voltage distribution along an arrester is not uniform because of the effect of the stray capacitances to both the high and low voltage electrodes. This non-uniformity should be kept to a minimum in order to avoid damage to the upper parts of the arresters which are inevitably more

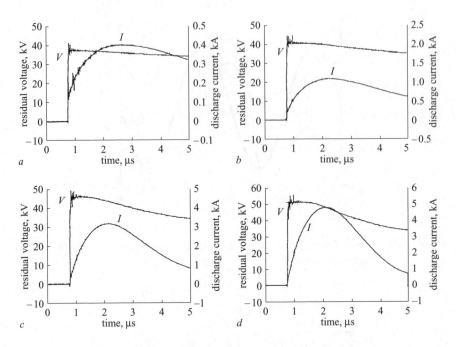

Figure 5.13   Typical voltage and current traces obtained with a fast impulse generator

stressed (Figure 5.15). In contrast, in the high conduction regime, the effect of stray capacitances is secondary and the distribution becomes more uniform because of the low ZnO resistance.

5.7.3.2   Effect of pollution

In addition to the stray capacitances, pollution of the arrester housing is a key factor in the performance of the arrester. The electric field uniformity along the arrester may be improved by means of grading rings at the top of the arrester and at the flanges for tall column arresters which are used in EHV. The effect of stray capacitances is further reduced if the base of the arrester (low voltage side) is raised above ground [105–109]. Additional grading resistors and/or capacitors are also used in particular applications but the cost and size of the arrester are increased. Besides, the effectiveness of the protection offered by the arrester will depend on the grading elements' reliability [20].

5.7.3.3   Voltage distribution measurements methods

Measurement of the voltage distribution along an arrester can be achieved in the same way as for insulator strings. A number of techniques have been used such as:

(i)   sparking gaps (rods, spheres) in which the voltage is determined from a known breakdown voltage of the gaps

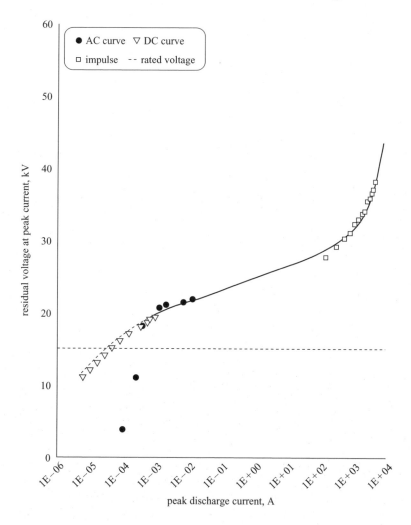

Figure 5.14   Measured V–I curve over the whole range of current for 15 kV rated surge arrester

(ii)   voltage divider methods which use potentiometers or microcapacitances
(iii)   electrostatic probes; for which the voltage is calculated from the calibration curve of the probe
(iv)   the neon tube method where the voltage is determined from the known discharge voltage of the neon tube [110].

All the above cited methods use metallic links to the point where the voltage is to be measured. This results in the disturbance of the equipotential distribution in the vicinity of the measuring point. Therefore, regardless of the method accuracy, the

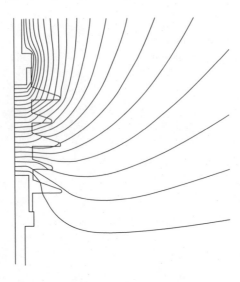

Figure 5.15   Computed equipotential distribution on a polymeric surge arrester

true voltage that would exist without the presence of the measuring device, is slightly different from the measured value.

Recently, two new optical methods have been developed [111, 112], which are characterised by the advantage of having an electrically insulated input and output, and hence no extra distortion of the field distribution is introduced. Moreover, these methods cause a minimum electromagnetic interference and consume very little energy.

(i)   One method [105, 112], which allows measurement of current at any point along the arrester column, is an optoelectronic technique in which a sensor comprising a light emitting diode (LED) is inserted between two adjacent ZnO elements so that when current flows through the elements, the LED is also subjected to this current.

(ii)   The other method [112, 113] measures the field at any point between the high and low voltage electrodes. It uses the Pockels' effect principle. A light beam, which can be generated either by a laser or an LED and sent through a light waveguide (LWG), is linearly polarised before it is goes through the Pockels' device. The refraction index of Pockels' devices is electric field sensitive. As a result of this property, the two orthogonal components of the light electromagnetic wave travel with different velocities inside the Pockels' device. The output light beam is then elliptically polarised proportional to the field strength. An analyser detects the phase shift before the beam is transmitted to the photo detector which converts the light into electricity. From the calibration curve, a very accurate field measurement can be achieved without field distortion. This method can be used for field strengths up to 40 kV/cm [113].

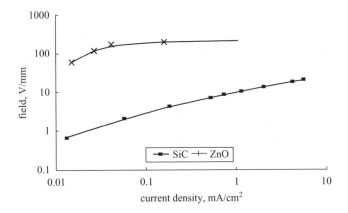

Figure 5.16   E–J characteristics of ZnO and SiC materials

5.8   Characteristics of ZnO surge arresters

5.8.1   Background

Since 1968 when Matsushita Electric Co. Ltd announced the new technology of ZnO varistors [34], many investigators have been involved in studying the properties of ZnO in order to understand and formulate the complex process of ZnO conduction. The importance of ZnO arresters is a consequence of their voltage–current $(V–I)$ characteristic, which is far closer to the ideal than that of SiC, the gapped arresters which they have largely replaced. Figure 5.16 shows the measured field current density $(E–J)$ characteristics of both ZnO and SiC materials.

It is not surprising, in view of the extreme non-linearity of its conduction process, that the ZnO characteristic is a function of temperature and depends on the type of the applied voltage, especially in the low conduction region. Furthermore, a large dependence on frequency is observed. These features represent a departure from the ideal, and may have important implications for ageing of arrester elements and change of arrester performance.

5.8.2   Frequency response of ZnO material

Early measurements [39, 52, 114] on the frequency response of ZnO material ceramics were achieved using a General Radio capacitance bridge, and because of the very small size of the tested samples, the applied voltage was merely 1 V r.m.s. In the frequency range $10–10^5$ Hz, the calculated relative permittivity from capacitance measurements is unexpectedly high (1000–1600) considering that the relative permittivity for ZnO is in the range $\varepsilon_r = 8$ to 10, and for the additives is in the range $\varepsilon_r = 16$ to 25. This can be partially accounted for by the microstructure of ZnO material. As used in surge arresters, the material consists of ZnO grains (approximately 25 μm) of high conductivity separated by a very thin layer (<100 Å) of additives which controls the non-linear characteristics. Therefore, the true dielectric thickness

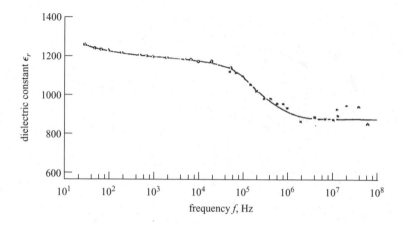

*Figure 5.17   Permittivity dependence on frequency in ZnO material. (Courtesy of LEVINSON, L.M., and PHILIPP, H.R.: 'Long time polarization currents in metal oxide varistors', J. Appl. Phys., 1976, **47**, (7), pp. 3177–3181)*

of a ZnO block is very small compared with the physical thickness. Microscopic photography measurements showed that the ratio of sample thickness to dielectric thickness was about 1000. The direct effect of this reduced thickness would be the increase of the observed capacitance. Consequently, the permittivity of the material would not necessarily be as high as the calculated values from the experimental data. The nature of the additives is clearly shown to influence the measured values of both capacitance and dissipation factor [50, 115]. A more plausible explanation is based on the existence of depletion layers in the ZnO grains adjacent to the intergranular layer and the trapping of electrons at the interface. The net effect of the former is the rise of the material capacitance [38].

Extensive characterisation work [38, 42, 50, 115, 116] used bridges, Q-meters and transmission line techniques over a wide range of frequency (30–10^8 Hz). In order to validate the application of dielectric theory, the measurements were performed at very low stresses, well below the knee of conduction, where the conduction in ZnO material is linear. A decreasing dielectric constant with increasing frequency was measured (Figure 5.17). Measurements at higher frequencies (1 MHz–1 GHz) on a sample 0.162 cm thick and 0.305 cm in diameter confirmed a relative permittivity for the material of about 900, and microwave transmission line techniques revealed an inductance value of about 0.35 nH associated with the body of the ZnO sample. The dielectric constant measured with polarisation current techniques [116] shows a decrease with increasing frequency and a strong temperature dependence at very low frequencies below 100 Hz.

The dissipation factor, however, was found first to decrease with increasing frequency, to a minimum between 1–10 kHz, and then increased to reach a peak at around 300 kHz. Finally, it decreased for higher frequencies. The observed peak is reminiscent of a typical broadened Debye resonance [117]. In this range of frequency, tan δ

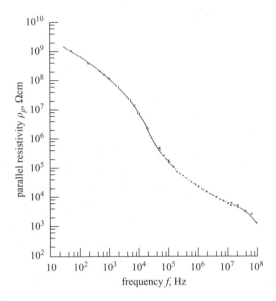

Figure 5.18 Dependence of ZnO resistivity on frequency. (Courtesy of LEVINSON, L.M., and PHILIPP, H.R.: 'Long time polarization currents in metal oxide varistors', J. Appl. Phys., 1976, **47**, (7), pp. 3177–3181)

varied only by a factor of ten. The immediate consequence of this behaviour would be a strong frequency dependence of the resistance in a parallel R–C circuit representation. Accordingly, the equivalent parallel resistivity decreases with increasing frequency (Figure 5.18). At low frequencies, the resistivity was mainly attributed to the resistance of the intergranular layer, which is very high compared with the grains resistance. At higher frequencies, however, it was supposed that the intergranular resistance fell to the low limiting value representing the grains resistance.

A peak in loss angle accompanied by a fall in permittivity is a common dielectric behaviour but the Maxwell–Wagner model [118], which is used to explain the dielectric behaviour of inhomogeneous solids and polycrystalline semiconductors, fails to account for the decreasing parallel resistivity with increasing frequency. The highly disordered intergranular layer and the existence of interface states and electron traps are thought to be the cause of the model failure. The loss angle peak can be interpreted as being caused by electron trapping [119].

Although of great importance, the above-published data were obtained from experiments performed in order to examine the basic physics of these materials in which the samples studied were of very small size (thickness = 2 mm, diameter = 0.3–2 cm) and the voltages were very low (up to 10 V). Consequently, fewer problems were encountered in generating the voltages and measuring the physical characteristics. The dangers of extrapolating and scaling the properties of such non-linear materials are clear, since different phenomena may appear in large samples

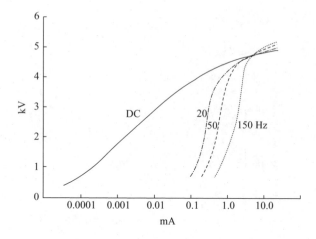

Figure 5.19   Frequency dependence of ZnO V−I curve

and with very high voltages which would be more characteristic of elements used in surge arresters in power systems.

The first high voltage variable frequency test results on ZnO elements [120, 121], which used frequencies between 30 and 10^6 Hz, showed a rapidly falling dissipation factor up to 1 kHz and a decreasing capacitance with increasing frequency in the high frequency range above 10 kHz. Subsequent research [132] showed the dependence of the V−I curve (Figure 5.19) and power consumption in ZnO surge arrester blocks. Such data yielded decreasing material resistance, capacitance and dissipation factor with increasing frequency.

5.8.3   Impulse response

There is now an extensive published literature on the response of the ZnO material and complete surge arresters to impulse currents of different magnitudes and shapes. Unlike in the prebreakdown region, the resistive current dominates for impulse currents in the kiloampere range.

5.8.3.1   Initial voltage overshoot

One peculiar observation in the ZnO fast transient response, which was observed by many investigators, is the initial spike/overshoot measured on the residual voltage (Figure 5.20). Careful studies linked this phenomenon to circuit inductance [120], the impulse rate of rise [122, 123], the impulse amplitude [124], the nature and amount of additives [125], the preinjection of carriers (using a double impulse technique) [125, 126], the value of the non-linearity coefficient and the difference in the rate of charge accumulation at the electron and hole traps located at the interface of the grain and intergranular layer [127]. This type of overshoot is, however, not observable on SiC material [127, 128]. This leads to the attribution of the overshoot to ZnO material,

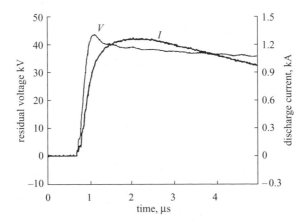

Figure 5.20   Measured spike on ZnO surge arrester residual voltage

its limited capability to switch fast suggesting that it takes a finite time to turn on the material from a low to a high conduction regime.

It should be emphasised, however, that this time to turn on is not conclusively established. Subnanosecond tests [129] on small ZnO varistors showed no detectable time delay in the conduction process, and high voltage tests [128] claimed a time delay of 3 ns. One major difference between these two sets of conflicting results is the size of test and measurement loops. Minimisation of these loops and adoption of coaxial field-free techniques established that the overshoot; hence the delay to turn on is caused by circuit arrangements [130]. Figure 5.21 illustrates how the voltage spike recorded with a parallel divider is avoided with the coaxial measurement.

5.8.3.2   Effect of discharge current impulse shape and magnitude

Laboratory tests have shown that faster discharge current impulses produce higher peaks of residual voltage but the *V–I* curve constructed from the voltage at the instant of peak current versus peak current is not affected. This may indicate that the resistance of the material is not greatly influenced by the current rate of rise. The higher magnitudes of current, however, have a faster rate of rise for a given test circuit. It is suggested that discrete current paths form through the material using the lowest potential barriers. With increasing current magnitude, the number of paths increases with some of them getting shorter. Parallel branches having resistances and inductances in series can represent such a process [131].

5.8.4   Combined stress response

The improved understanding of the material behaviour and complete arrester performance has allowed better characterisation of ZnO surge arresters and more widespread application. Various types of voltage including direct, alternating, variable frequency, impulse voltages and their combinations have been used to test ZnO arresters. Each set of tests has revealed an important property of the material: frequency dependence

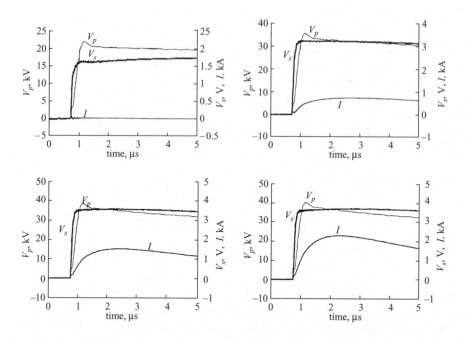

Figure 5.21   Initial voltage overshoot (V_p) on ZnO arresters caused by measurement errors. No overshoot on coaxial measurement (V_s)

of the $V-I$ curve [132]; crossover of the different $V-I$ characteristics [133]; over-shoot on the residual voltage [134, 120, 122–127]; finite time to turn-on [128]. Mixed (AC plus DC) voltage tests provide an insight into three important aspects of ZnO characteristics.

5.8.4.1   Mixed (AC + DC) characteristics

ZnO surge arresters are used in HVDC systems [121, 135] where, because of the con-verters used in such systems, high frequency voltages are generated. As a result, a mixed voltage in which an alternating voltage is superimposed upon a direct voltage appears on the network. A comparison of both the current–voltage and power-dissipation–voltage characteristic curves for direct and alternating voltages, respectively, show a crossover point near the knee of high conduction (Figure 5.22). This crossover is also frequency dependent, and its physical cause has not been explained, although it appears to be associated with the dynamic response of the material:

(i)   *Low voltage capacitance measurements*   Although the above introduction estab-lishes engineering interest in mixed voltage tests, hitherto such a procedure has been used mainly in physical studies, where it is a widely used laboratory method for measuring capacitance [42, 44, 115, 49, 136]. Here, a very small alternating voltage is superimposed on a larger direct voltage. The AC level is

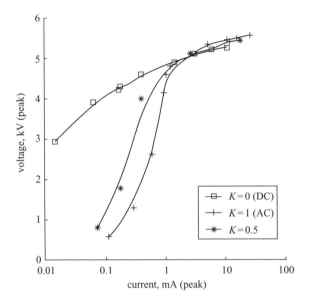

Figure 5.22   Crossover of DC and AC characteristics

then kept constant while the direct voltage is varied [44, 136, 137]. The importance of these $C-V_{dc}$ curves lies in the close link between the capacitance and the physical behaviour of the material. $C-V_{dc}$ curves are far more sensitive to microstructural fluctuations than $I-V$ curves [44], and have been used to investigate many of the physical properties of ZnO material, such as the density of interface states [42, 52], donor concentration [49, 115] and change in depletion layer.

(ii) *Crossover of DC and AC characteristics* The crossover was first attributed to a significant contribution of dielectric loss under alternating applied voltage in the prebreakdown (low current) region [48, 58, 70, 121, 132, 133, 139], and a time lag which was observable in the current in the breakdown region. However, according to the measured data, less than 200 ms are needed for the current to reach its steady state value, so that this explanation involving such a short response time would not account for any crossover at 50 Hz. Another explanation relates to polarisation currents which are much higher than conduction currents, in contrast to the breakdown region where conduction prevails [116, 140, 141]. It was predicted that lower conduction currents would be obtained at higher frequencies for the same applied voltage, because the time lag would prevent the current from reaching its static value. However, the polarisation currents are more important at very low temperature and the order of magnitude of the ratio of polarisation currents to DC equilibrium current is less than that of the ratio of AC to DC values. Polarisation phenomena can thus give only a partial explanation of the crossover.

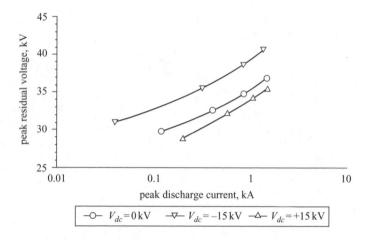

Figure 5.23   *Typical effect of DC prestress (trapped charge) on the protective characteristic of a 15 kV-rated ZnO surge arrester*

5.8.4.2   Mixed (DC + impulse) characteristics and effect of trapped charge

In AC power systems, under normal service conditions, surge arresters on isolated lines and cables can experience a direct voltage stress arising from trapped charge. If followed by a switching surge of reverse polarity, severe requirements will be imposed on the surge arresters. Combinations of direct and surge voltages can also occur in HVDC systems and due to circuit breaker restrikes during switching of capacitor banks or unloaded transmission lines where trapped charges are present. Circuit breaker restrikes can cause high rate-of-rise overvoltages of peak values of 2 to 3 p.u. following rapid polarity reversal, and may cause up to 8 p.u. especially for some types of oil circuit breakers [142]. This condition may be particularly onerous where the surge is opposite in polarity to the preexisting working voltage [143–146]. For a given level of peak discharge current through the arrester, a higher level of residual voltage of upto ten per cent is found to appear across an arrester when the polarity of the applied impulse opposes that of the direct voltage (Figure 5.23). This effect is especially marked in the peak residual voltage which is affected by the increased di/dt. Experimental tests have shown [147] that the effect increases with decreasing arrester protection level.

5.8.4.3   Mixed (AC + impulse) characteristics and effect of working voltage level

Transient high amplitude surges commonly occur in high voltage power systems, due to either external or internal influences, such as lightning strikes or switching operations. For most systems, these conditions are likely to occur when the system is already operating at its normal alternating working voltage. Laboratory experiments [148] revealed that for an impulse voltage superimposed on the power frequency

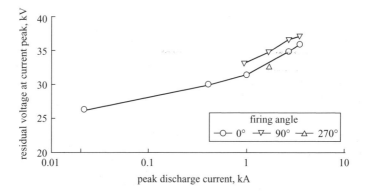

Figure 5.24 Typical effect of AC prestress (working voltage) on the protective characteristic of a 15 kV-rated ZnO surge arrester

voltage, it was found that firing at peaks with opposite polarity impulses produces the highest residual voltage levels (Figure 5.24).

In the conditioning test recommended by IEC 99-4:1991 [77] for the operating duty test, it is specified that four groups of five lightning nominal current impulses (8/20 μs) shall be applied at intervals of 50–60 s between impulses and 25 to 30 minutes between groups. These impulses should be applied superimposed on an elevated system voltage determined by ageing tests. The firing angle is specified to be 15 to 45 electrical degrees from the zero crossing. The American Standard ANSI/IEEE C62-11: 1987 [149] specifies a firing angle of 60° before the crest voltage. It is recommended that the impulse should be of same polarity as the alternating voltage at the instant of firing. This IEC recommendation regarding the firing angle in the duty tests was originally specified for gapped surge arresters and was to ensure satisfactory firing of the series gap. However, since modern ZnO surge arresters do not have any series spark gaps, it may be more appropriate to include a firing angle that produces higher residual voltages.

5.8.4.4 Multiple impulse response

Between 60 and 70 per cent of all lightning ground flashes contain more than one stroke. A typical flash will consist of three or four strikes with time intervals between strokes of 20 to 200 ms. In addition, induced lightning surges on transmission lines are known to produce a transient overvoltage with two peaks of reverse polarity. A rapid series of transients may also occur during certain switching transients.

Extensive laboratory tests have shown that the cumulative nature of multiple impulses forms a significant part of the ZnO material degradation process, especially the insulating walls of the ZnO elements [150, 151]. Such deterioration was related to energy dissipation constraints. However, the $V–I$ characteristic of the arrester is not significantly affected by multiple impulses of low energy [152, 153].

5.8.5   *Equivalent circuit of ZnO material*

5.8.5.1   Review

The main objective in an equivalent circuit is to be able to represent and to model the ZnO surge arrester under different conditions of voltage, temperature and frequency. Furthermore, equivalent circuit representations should be simplified as far as possible, consistent with the arrester element characteristics, to facilitate their application. The proliferation of different equivalent circuits has arisen in response to data progressively acquired from new measurements. Unexpected behaviour and properties of the ZnO element such as temperature dependence, time lag of current growth, polarisation currents, frequency dependence and fast transient response characteristics have helped to refine the characterisation of the material.

The simplest model is an $R–C$ parallel circuit where R is defined to be highly non-linear with increasing voltage; a similar non-linear behaviour, although of less degree, is assigned to the capacitance. Such a circuit can be useful for engineering modelling of arrester behaviour [82, 83]. However, as will be demonstrated in this section, this simple equivalent circuit, even with non-linear elements, is ultimately non-representative of the complete behaviour of the ZnO arrester. Experimental data have shown that a more complex equivalent circuit is needed in order to take into account the various parameters which affect the arrester characteristics.

Starting with the idealised microstructure of the block consisting of cubic conducting ZnO grains surrounded (coated) by a segregation layer (the intergranular layer) which is responsible for the non-linear behaviour of the material [37, 39, 115], early equivalent circuits consisted of a small resistance representing the ZnO grains in series with an inductance and a parallel $R–C$ circuit representing the voltage and frequency-dependent resistance and capacitance of the intergranular layer (Figure 5.25a) [37]. The capacitance of ZnO grains is sufficiently small to be neglected at normal working voltages [39, 114, 120]. The series inductance represents the inductance of the arrester body and accounts for the response to steep currents [154]. A slightly different equivalent circuit, which consists of a resistance R_g representing the ZnO grains in series with two parallel $R–C$ circuits, was suggested in Reference 37. One $R–C$ circuit was used to represent the intergranular material component and the other to account for the interface between ZnO grains and the intergranular layer. Other circuits represented the ZnO blocks by three parallel branches [155], a resistance R which is essentially frequency independent and is given by the $V–I$ characteristic, a capacitance C which is weakly dependent on the applied voltage, temperature and frequency, and a resistance Z which is nearly voltage independent but strongly dependent on frequency. It is specified that at low voltages losses are determined by Z but at higher voltages losses are governed by R. Other earlier proposed circuits [140, 141], accounting for the measured polarisation currents proposed [55], consisted of:

(i)   an inductance associated with the varistor body itself
(ii)   a voltage and temperature-dependent resistance to represent losses due to conduction
(iii)   a voltage, temperature and frequency-dependent resistance to represent polarisation with losses

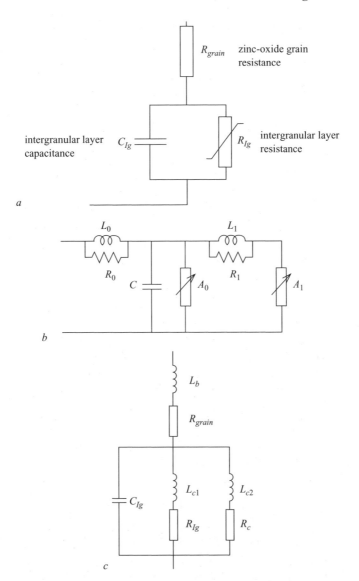

Figure 5.25

　a　Simple equivalent circuit for ZnO material
　b　Equivalent circuit for fast impulses. (Courtesy of IEEE working
　　　group 3.4.11: 'Modelling of metal oxide surge arresters', *Trans-
　　　action on Power Delivery*, 1992, **7**, (1), pp. 302–309, ©2003
　　　IEEE)
　c　Equivalent circuit to account for the dynamic conduction at high-
　　　magnitude fast impulses

(iv)   a frequency and temperature-dependent capacitance also associated with polarisation with losses
(v)   a capacitance to account for polarisation without losses.

An equivalent circuit based on laboratory test data included the series grain resistance and inductance and contains a three-branch parallel circuit [128]:

(i)   a non-linear resistance which is a function of temperature
(ii)   a voltage and temperature dependent capacitance which is also affected by the rate of change of the voltage
(iii) a turn on element, which will account for the dynamic charge distribution at the grain boundary; this is a function of voltage rate of rise of voltage and the time constant *t* to reach the equilibrium of electrons and holes at the grain boundary.

Recent research on zinc oxide surge arresters has concentrated on the very fast transient characterisation [156, 157], and two main equivalent circuits were proposed. In addition to the above elements, these circuits include parallel branches (Figures 5.25*b* and 5.25*c*) with inductances to account for the fast transient measurement data.

5.8.5.2   Determination of the equivalent circuit main components

Low voltage tests and/or computation [158] can be used to estimate the parallel capacitance C_{ig}, and high current impulse data can be used to calculate the ZnO grain R_g and intergranular layer R_{ig} resistances. The voltage V_{ig} across the intergranular layer resistance is related to current by:

$$\frac{V_{ig}}{V_B} = \left(\frac{I}{I_B}\right)^{\beta} \tag{5.17}$$

where $\beta = 1/\alpha$, V_B and I_B are the base values.
    The intergranular layer resistance is:

$$R_{ig} = \frac{V_{ig}}{I} = \frac{V_B}{I_B}\left(\frac{I}{I_B}\right)^{\beta-1} = R_B\left(\frac{I}{I_B}\right)^{\beta-1} \tag{5.18}$$

This yields the resistive voltage current behaviour for the arrester:

$$V = I(R_g + R_{ig}) = I\left(R_g + R_B\left(\frac{I}{I_B}\right)^{\beta-1}\right) \tag{5.19}$$

which gives

$$V = I(R_g + kI^{\beta-1}) \tag{5.20}$$

with

$$k = \frac{R_B}{I_B^{\beta-1}}$$

Such an expression is readily applicable [159] in computer packages such as Spice.

5.9   Monitoring of ZnO surge arresters

Zinc oxide surge arresters are designed to last a useful lifetime of at least 20 to 30 years. Over this period, the arrester is expected to absorb a large number of surges and limit the voltage to a safe level, cope with the harsh environment and withstand temporary overvoltages and system voltage fluctuations. Long term accelerated ageing tests, which are conducted with elevated applied voltage and temperature, have indicated that when the resistive component of current and the power consumption in a ZnO surge arrester increases to more than twice the initial value, the arrester has degraded to a stage equivalent to the end of its useful life. This criterion is used in many condition monitoring techniques for surge arresters. Most of these techniques require the measurement of applied voltage and leakage current through the surge arrester.

One technique to assess arrester health is to perform offline laboratory tests such as DC or AC current injection to monitor [159] the changes in the arrester's V–I characteristic over a wide range of current (0.5 μA to 1 mA). This method, however, is not cost effective as it requires outages and careful transport arrangements.

Other more cost-effective methods are based upon online monitoring which involves leakage current analysis [160–162]. These techniques use harmonic analysis of the total leakage current to obtain the third harmonic of the resistive current. Subsequent scaling allows reconstruction of the resistive current level. Voltage measurement in some of these techniques is based on a capacitive probe pick up. The analysis compensates for voltage harmonic content as well as influence of adjacent phases. Errors of more than 30 per cent were observed with these techniques, which occur because at system voltages, the arrester leakage current is predominantly capacitive. The resistive component is generally less than five per cent of the total leakage current. Considering this small ratio and the highly non-linear conduction in ZnO material, accurate discrimination of the resistive component is not easy. Also, the third harmonic is a small fraction of the resistive component which may impose a demanding resolution limit for the transducers used in arrester leakage current measurements. Furthermore, any harmonic contents in the voltage will give rise to complex current waveshapes. In addition, most existing techniques for current discrimination assume a constant linear capacitance behaviour of the ZnO surge arrester, which is inconsistent with the measured non-linear ZnO capacitance.

The ZnO element may be represented by a parallel RC circuit, both R and C being non-linear, and the current can be resolved into conduction and displacement components. A conventional laboratory method of obtaining the two components of current uses a constant loss-free high voltage capacitor to compensate for the capacitive current [17, 32, 55, 87, 120]. This compensation technique and/or the standard Schering bridge method may be used to estimate the ZnO parameters. The additional need in these methods for a high voltage capacitor may be avoided with the attenuator compensation technique [163, 164] or by means of special electronic circuits [91, 165]. However, these methods do not allow for the ZnO capacitance being voltage dependent. Another method with similar assumptions [24] uses Fourier analysis to resolve the measured current into in-phase and quadrature components. For a detailed description of leakage current measurement and diagnostic indicators of ZnO surge arresters, refer to BS EN 60099-5:1997 or its equivalent EN60099-5:1996 including amendment A1:1999 [166].

Recently, a point-on-wave method [167], which requires voltage and current traces, has identified variations during the voltage cycle not only in the equivalent resistance of the sample, but also in its capacitance. The method is based on the expression of the average power and assumes a single valued voltage conduction current characteristic.

The total current for an *RC* parallel equivalent circuit is:

$$I_t(t) = I_c(t) + I_r(t) = C\frac{dV(t)}{dt} + I_r(t) \tag{5.21}$$

where I_c and I_r are the capacitive and resistive components of the total leakage current I_t.

It was shown [167] that the instantaneous resistive current magnitude at two instants t_1 and t_2 corresponding to the same level of voltage on the cycle can be expressed as:

$$I_r(t_1) = I_r(t_2) = \left(\frac{1}{2}(I_t(t_1) + I_t(t_2))\right) - \left(\frac{1}{2}(I_t(t_1) - I_t(t_2))\right)$$

$$\times \left(\frac{dV(t_1)/dt + dV(t_2)/dt}{dV(t_1)/dt - dV(t_2)/dt}\right) \tag{5.22}$$

This equation allows the calculation of the resistive current around the cycle for any waveshape. If the voltage is sinusoidal with no harmonic content or with only odd harmonics without phase shift, it can be shown that:

$$I_r(t_1) = I_r(t_2) = \tfrac{1}{2}(I_t(t_1) + I_t(t_2)) \tag{5.23}$$

Figure 5.26 shows typical examples obtained with this discrimination technique on a 15 kV surge arrester in the low and high conduction regimes. As can be seen, both the resistive and capacitive components show non-linearities.

5.10   Standards and application guidelines

This section is a general summary of important aspects of the existing international standards (IEC60099 and IEEE/ANSI C62.110).

5.10.1   Standard definitions of important parameters

The following standard definitions are used for both the testing and selection of zinc oxide surge arresters.

5.10.1.1   Continuous operating voltage (U_c)

U_c is the maximum permissible value of a sinusoidal power frequency voltage which may be continuously applied across the arrester terminals. The normal system voltage

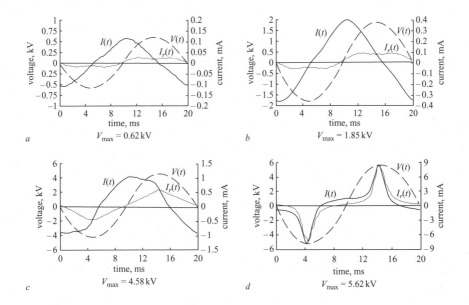

Figure 5.26   Discriminated leakage current components using the point-on-wave technique

should not exceed this value. Some manufacturers use the terms maximum continuous operating voltage (MCOV) and continuous operating voltage (COV), the former being an upper limit for the arrester permissible continuous voltage.

5.10.1.2   Continuous current (I_c)

I_c is the continuous r.m.s. or peak magnitude current which flows through the arrester when the continuous operating voltage, U_c is applied across its terminals.

5.10.1.3   Rated voltage (U_r)

U_r is the maximum permissible short duration r.m.s. value of power frequency voltage between the arrester terminals. In IEC60099-4:1991 [77], the rated voltage is defined as the 10 s power frequency voltage used in the operating duty test involving high current and long duration impulses. However, it should be noted that some national standards use a number of impulses at nominal current with superimposed power frequency voltage in order to determine the rated voltage which may produce different values from those recommended by IEC. The standards bodies are now considering these discrepancies. Table 5.3 reproduces the recommended rated voltages [77].

5.10.1.4   Discharge current

The discharge current is used to describe the impulse current which flows through the arrester under test.

Table 5.3   Recommended values for ZnO surge arrester rated voltages [77]

Range of rated voltage (kV r.m.s.)	Steps of rated voltage (kV r.m.s.)
<3	under consideration
3–30	1
30–54	3
54–96	6
96–288	12
288–396	18
396–756	24

5.10.1.5   Switching current impulse

The switching current impulse is the peak magnitude of discharge current having a recommended standard impulse shape, a virtual front time greater than 30 μs but less than 100 μs and a virtual time to half value on the tail of roughly twice the virtual front time.

5.10.1.6   Nominal discharge current (I_n)

I_n is the peak magnitude of a standard lightning current impulse (8/20) that an arrester is rated for and used for its classification. It is considered as the main parameter for determining the protective characteristic and the energy absorption capability of an arrester.

5.10.1.7   High current impulse

The high current impulse is the peak magnitude of discharge current having a 4/10 impulse shape which is used to test the stability of an arrester if subjected to direct lightning strokes.

5.10.1.8   Residual voltage (U_{res})

U_{res} is the peak value of the impulse voltage that appears between the terminals of an arrester during the flow of discharge current. Some countries still use the term discharge voltage to describe the residual voltage.

5.10.2   Classification of ZnO surge arresters

5.10.2.1   Line discharge class

ZnO surge arresters with 10 kA and 20 kA nominal discharge currents are given a standard class number (1 to 5) according to their energy absorption capability. Table 5.4 gives the specified classes of IEC99-4:1991 [77].

Table 5.4 *ZnO surge arrester classes as defined by IEC60099-4:1991 [77] using the line discharge test*

Arrester classification	Line discharge class	Surge impedance of the line (Ω)	Virtual duration of peak (μs)	Charging voltage (kV DC)
10 000 A	1	4.9 U_r	2000	3.2 U_r
10 000 A	2	2.4 U_r	2000	3.2 U_r
10 000 A	3	1.3 U_r	2400	2.8 U_r
20 000 A	4	0.8 U_r	2800	2.6 U_r
20 000 A	5	0.5 U_r	3200	2.4 U_r

Table 5.5 *Peak currents for switching impulse residual voltage tests [77]*

Arrester classification	Peak currents (A)
20 000 A, line discharge classes 4 and 5	500 and 2000
10 000 A, line discharge class 3	250 and 1000
10 000 A, line discharge classes 1 and 2	125 and 500

5.10.2.2 Pressure relief class

The pressure relief class is determined by the capability of an arrester to withstand internal fault currents without violent shattering of the housing; this is particularly applicable to porcelain housed surge arresters.

5.10.3 Other important arrester characteristics

5.10.3.1 Protective characteristic and protective levels of an arrester

The protective characteristic of an arrester includes three main protective levels:

(i) Slow front overvoltages (switching impulse protective level): the maximum residual voltage at the specified switching impulse current. Table 5.5 reproduces the standard recommended peak magnitudes for the switching currents to determine the switching impulse protective level.

(ii) Fast front overvoltages (lightning impulse current, 8/20, protective level): the maximum residual voltage at nominal discharge current.

(iii) Very fast transients (steep current protective level): the residual voltage for steep currents whose front times are between 0.9 and 1 μs and tail times no longer than 20 μs. A reliable measurement procedure for the precise determination of the residual voltage under steep current is under consideration. The technique should take into account inductive effect errors as described in section 5.7.2.

5.10.3.2   Power frequency withstand voltage versus time characteristic of an arrester

Corresponds to the maximum duration for which the arrester can withstand a given level of continuous power frequency voltage without sustaining any damage or thermal instability. This characteristic is usually above the temporary overvoltage characteristic.

5.10.3.3   Pollution withstand characteristics

The pollution performance of an arrester housing is determined according to relevant IEC507:1991 [168] and IEC815:1986 [169] standards. In addition, the arrester has to withstand temperature gradients and internal partial discharges which may appear on the arrester column due to surface pollution. An extensive review of pollution characteristics of zinc oxide surge arresters is given in Reference 170.

5.10.4   Standard tests

The current standards specify the following tests (for a detailed description of these tests, refer to IEC 99-4:1991 [77]):

(i)   type or design tests, which are made on a new arrester design to establish its performance and compliance with the relevant standard; these tests will be required on subsequent arresters with a similar design

(ii)   routine tests are made on each arrester to ensure that the product meets the design specifications

(iii)   acceptance tests, which are agreed between the manufacturer and the purchaser.

In the operating duty cycle, correct selection of the test sample depends on the correct selection of the reference voltage and current. The reference current, I_{ref}, of an arrester is defined as the peak value of the resistive component of the total current flowing through the arrester. The reference current ranges between 0.05 mA and 1.0 mA per square centimeter of disc area for single column arresters. The voltage obtained at the reference current is defined as the reference voltage, U_{ref}.

5.10.5   Recommended arrester identification

IEC60099-5:1999 recommends that '...metal oxide surge arresters shall be identified by the following minimum information which shall appear on a nameplate permanently attached to the arrester:

• continuous operating voltage
• rated voltage

- rated frequency, if other than one of the standard frequencies
- nominal discharge current
- pressure relief rated current in kA r.m.s. (for arresters fitted with pressure relief devices)
- the manufacturer's name or trade mark, type and identification of the complete arrester
- identification of the assembling position of the unit (for multiunit arresters only)
- the year of the manufacture
- serial number (at least for arresters with rated voltage above 60 kV).

5.11   Selection of gapless metal oxide surge arresters

Figure 5.27 reproduces the procedure of arrester selection specified by the standards [166]. During the selection process of a metal oxide surge arrester for a particular application, the following arrester parameters need to be determined: the continuous operating voltage, the rated voltage, the nominal discharge current and the protective characteristic, the line discharge class, the pressure relief class, the pollution withstand and live washing capabilities and the mechanical properties [166]. The following are particularly important.

(i)   *Continuous operating voltage U_c*: must be higher than the maximum attainable power frequency system voltage including any harmonic content. It is proposed to use a safety factor of 1.05 to account for harmonic content. The continuous operating voltage level to be used depends on the system earthing and the earth fault clearing arrangements.

(ii)   *Rated voltage*: is determined from the temporary overvoltage levels where the arrester is to be installed. The power frequency voltage characteristic of the arrester should be higher than the temporary overvoltage characteristic of the system. The standards recommend considering temporary overvoltages due to earth fault overvoltages, load rejections, resonance effects and Ferranti effect. Table 5.6 reports current rating for the whole range of standard rated voltages.

(iii)   *Nominal discharge current*: is normally selected according to the lightning discharge current through the arrester. Table 5.7 summarises the recommended values of arrester nominal discharge current for the range of voltages in use in power systems.

(iv)   *Energy absorption capability*: zinc oxide surge arresters must be able to absorb the energy due to a variety of transient overvoltages arising from closing or reclosing long lines; disconnection of capacitor banks or cables with restriking circuit breakers; and lightning. Formulae for the calculation of the energy absorbed by the arrester for the different overvoltages are given in Reference 166.

(v)   *Coordination of ZnO arresters*: ZnO surge arresters are now used as the primary overvoltage protection of power systems equipment such as transformers.

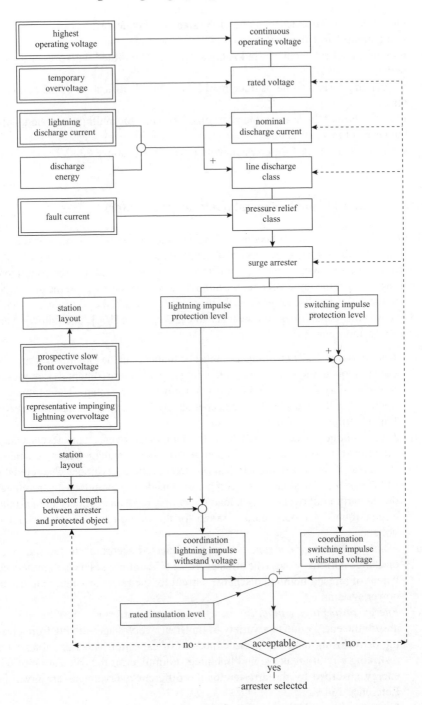

Figure 5.27   Recommended procedure for the selection of ZnO surge arresters[1]

Table 5.6   Arrester classification and rated voltages [166]

	Standard nominal discharge current				
	20 000 A	10 000 A	5000 A	2500 A	1500 A
rated voltage U_r (kV r.m.s.)	$360 < U_r \le 756$	$3 \le U_r \le 360$	$U_r \le 132$	$U_r \le 36$	this low voltage range is now under consideration

Table 5.7   Practical examples of arrester ratings (extracted from [166])

System voltage level	Recommended nominal discharge current (kA)
up to 72 kV: short lines (less than 5 km)	5
up to 72 kV: low flash density and effectively shielded lines	5
up to 72 kV: high flash density and earth resistance	10
above 72 kV and up to 245 kV	10
above 245 kV and up to 420 kV	10
above 420 kV	20

The transformer bushings are normally protected by air gaps which need to be coordinated with the surge arrester. At distribution voltages, some of the arresters may be in vulnerable positions which requires careful examination of the gap/arrester assembly. It has been shown that the gap distance may be adjusted to act as a relief device for the arrester when an unexpectedly high magnitude surge hits the system [171].

(vi)   *Multicolumn surge arresters*: the energy dissipation capability of ZnO surge arresters depends on their surface area; larger ZnO blocks are required for larger amounts of energy. However, there is a block size limit that can be manufactured economically with a guaranteed good homogeneity and uniform current distribution. For this reason, multicolumn surge arresters are preferred. Usually, the matching of the parallel columns is a difficult aspect of the arrester construction since a good balance of the current sharing must be achieved. The normal practice is to match the voltage–current characteristics of individual columns at two points; usually at 1 mA and at the nominal discharge current. Matching at other points is not certain and the characteristics can be up to six per cent different for matched columns [172]. Such imbalance could lead to premature failure following service degradation, which can be caused by leakage current, pollution and other weather effects, solar radiation and ambient temperature differences.

5.12   Location and protective distance of surge arresters

The selection process described in the previous section allows specifying the best choice of a ZnO surge arrester to protect against prospective overvoltages at a given site/location. However, the surge arrester will offer the optimised protection level only if it is installed directly across the terminals of the equipment to be protected. Unfortunately, in practice, it is not always possible to locate the arrester close to the equipment, and an inevitable separation distance will be required. Protection against switching overvoltages is normally guaranteed for most applications since travelling wave effects can be neglected on short distances, and the switching protective level of the arrester will appear across the equipment terminals. For fast transient overvoltages, however, conductor inductance and travelling wave effects can cause significant voltage differences between the arrester and the equipment. The following three main approaches were adopted to predict the effectiveness of a surge arrester protection at a given location.

5.12.1   Effect of distance on protective level

When a surge arrester of total length a (including its leads) is located a distance d away from the protected equipment, estimation of the voltage level, V_{equ}, at the equipment for a given overvoltage steepness S and a propagation speed u can be achieved using the following simplified expression:

$$V_{equ} = V_a + \frac{2 \times (d + a) \times S}{u} \tag{5.24}$$

where V_a is the residual voltage of the surge arrester, and d is the distance between the arrester and the equipment including the lengths of connection leads.

Equation (5.24) indicates that the voltage level at the equipment will increase with separation distance and steepness of the incoming surge. This method, however, should be used with some caution since it does not take into account the characteristics of the equipment to be protected, e.g. equivalent capacitance in the case of transformers or shunt reactors. Computer simulations of such systems indicate that this simplified approach exaggerates the increase in protective level (Figure 5.28).

5.12.2   Calculation of separation distance

IEEE Standard C62.22-19:1991 [173] recommends a calculation procedure of the maximum allowable separation distance d between the equipment to be protected and the surge arrester connection point for which an adequate overvoltage protection would still be provided. The following empirical equation was derived from extensive

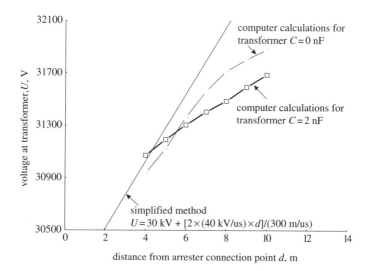

Figure 5.28   *Effect of distance on protective characteristic of ZnO surge arrester (15 kV-rated surge arrester, $S = 40$ kV/µs, $V_{res} = 30$ kV, $u = 3 \times 10^8$ m/s)*

EMTP simulations of single transformer substations with system voltages ranging from 69 kV to 765 kV:

$$d \leq \left(\frac{0.1155\, u V_{res}}{S}\right) \left(\frac{0.957\, V_{BIL} - V_{res}}{2.92\, V_{res} - 0.957\, V_{BIL}}\right) \tag{5.25}$$

where V_{BIL} is the basic lightning impulse insulation level (BIL) of the transformer (kV), u is the surge velocity (m/µs), $V_{res} = V_a + L_a\, dI/dt$ is the residual voltage across the arresters and its leads (kV) and S is the steepness of the incoming surge at the surge arrester connection point (kV/µs). L_a is the inductance of the arrester branch of length a.

5.12.3   Calculation of arrester protective zones

IEC 99-5:1996 [166] recommends the technique proposed in IEC 71-2:1996 [174] for estimating the coordination lightning impulse withstand voltage, and proposes the calculation of the protective zone, $L_p = d + a$, of the arrester using:

$$d = \frac{N}{A}\left[\left(\frac{V_{BIL}}{1.15} - V_a\right)\right](L_{sp} - L_f) - a \tag{5.26}$$

where N is the number of lines connected to the substation (usually $N = 1$ or 2), A is a factor describing the lightning performance of the overhead line connected to the substation (see Table 5.8 reproduced from Reference 174), V_{BIL} is the rated lightning impulse withstand voltage, V_a is the lightning impulse level protection level of the surge arrester, L_{sp} is the span length on the incoming overhead line and L_f is the

Table 5.8   Factor A for various overhead lines [166, 174]

Type of line	Factor A (kV)
distribution lines (phase-to-phase flashovers)	
• with earthed cross-arms (flashover to earth at low voltage)	900
• wood–pole lines (flashover to earth at high voltage)	2 700
transmission lines (single-phase flashover to earth)	
• single conductor	4 500
• double conductor bundle	7 000
• four conductor bundle	11 000
• six and eight conductor bundle	17 000

length of the overhead line section with outage rate equal to acceptable failure rate. For more details on these factors see Reference 166.

5.13   Note

[1] The author thanks the International Electrotechnical Commission (IEC) for permission to reproduce: Figure 5.166 from its International Standard IEC 60099-5 Edition 1.1 2000-03 Surge arresters – Part 5: Selection and application recommendations. All such extracts are copyright of IEC, Geneva, Switzerland. All rights reserved. Further information on the IEC is available from www.iec.ch. IEC has no responsibility for the placement and context in which the extracts and contents are reproduced by the author; nor is IEC in any way responsible for the other content or accuracy therein.

5.14   References

1   KAUFMANN, R.H., and HALBERG, M.N.: 'System overvoltages – causes and protective measures', *in* BEEMAN, D. (Ed.): 'Industrial power system handbook' (McGraw-Hill Book Co., London, 1955, 1st edn.), pp. 278–336

2   BURGER, U.: 'Surge arresters with spark gaps', *in* RAGALLER, K. (Ed.): 'Surges in high voltage networks' (Plenum Press, New York, 1980), pp. 251–282

3   SAKSHAUG, E.C., KRESGE, J.S., and MISKE, Jr. S.A.: 'A new concept in station arrester design', *IEEE Trans. PAS*, 1977, **PAS-96**, (2), pp. 647–656

4   KOBAYASHI, M., MUZINO, M., AIZAWA, T., HAYASHI, M., and MITANI, K.: 'Development of zinc oxide non-linear resistors and their applications to gapless surge arresters', *IEEE Trans. PAS*, 1978, **PAS-97**, (4), pp. 1149–1158

5 BARGIGIA, A., MAZZA, G., CARRARA, G., and DE NEGRIS, M.: 'Evolution of long duration current impulse withstand test on surge arresters for HVAC systems', *IEEE Trans. Power Deliv.*, 1986, **PWRD-1**, (4), pp. 173–183

6 ABB HIGH VOLTAGE TECHNOLOGIES LTD: 'Dimensioning, testing and application of metal oxide surge arresters in medium voltage networks'. Manufacturer's booklet, 1984

7 See general literature and websites of surge arrester manufacturers: Bowthorpe EMP, Ohio Brass, Raychem, Siemens, etc.

8 STANDLER, R.B.: 'Protection of electronic circuits from overvoltages' (John Wiley & Sons, New York, 1989)

9 MUENCH, F.J., and DUPONT, J.P.: 'Coordination of MOV type lightning arresters and current limiting fuses', *IEEE Trans. Power Deliv.*, 1990, **5**, (2), pp. 966–971

10 LAT, M.V., and KORTSCHINSKI, J.: 'Distribution arrester research', *IEEE Trans. PAS*, 1981, **PAS-100**, (7), pp. 3496–3505

11 SHIH, C.H., HAYES, R.M., NICHOLS, D.K., KOCH, R.E., TIMOSHENKO, J.A., and ANDERSON, J.G.: 'Application of special arresters on 138 kV lines of Appalachian power company', *IEEE Trans. PAS*, 1985, **PAS-104**, (10), pp. 2857–2863

12 KERSHAW, S.S., GAIBROIS, G.L., and STAMP, K.B.: 'Applying metal-oxide surge arresters on distribution systems', *IEEE Trans. Power Deliv.*, 1989, **4**, (1), pp. 301–303

13 HILEMAN, A.R., ROGUIN, J., and WECK, K.H.: 'Metal oxide surge arresters in AC systems, part V: protection performance of metal oxide surge arresters', *Electra*, 1991, (133), pp. 132–144

14 BURKE, J.J., SAKSHAUG, E.C., and SMITH, S.L.: 'The application of gapless arresters on underground distribution systems', *IEEE Trans. PAS*, 1981, **PAS-100**, (3), pp. 1234–1243

15 NIEBUHR, W.D.: 'Protection of underground systems using metal-oxide surge arresters', *IEEE Trans. Ind. Appl.*, 1982, **IA-18**, (2), pp. 188–191

16 REID, W.E., SECREST, S.R., HASSLER, S.P., and POTTER, M.E.: 'MOV arrester protection of shield interrupts on 138 kV extruded dielectric cables', *IEEE Trans. PAS*, 1984, **PAS-103**, (11), pp. 3334–3341

17 PARMIGIANI, B., QUAGGIA, D., ELLI, E., and FRANCHINA, S.: 'Zinc oxide sheath voltage limiter for hv end ehv power cables: field experience and laboratory tests', *IEEE Trans. Power Deliv.*, 1986, **PWRD-1**, (1), pp. 164–170

18 POTTER, M.E., and SOKOLY, T.O.: 'Development of metal oxide varistors for gas-insulated surge arresters', *IEEE Trans. PAS*, 1982, **PAS-101**, (7), pp. 2217–2220

19 MILLER, D.B., REUTHER, R.W., NOURAI, A., and TAHILIANI, V.: 'Development of gapless metal oxide arresters for 362 kV gas insulated substations', *IEEE Trans. PAS*, 1982, **PAS-101**, (7), pp. 2178–2186

20 IMATAKI, M., UJITA, K., FUJIWARA, Y., ISHIBE, S., and NITTA, T.: 'Advanced metal oxide surge arrester for gas insulated switchgear (GIS)', *IEEE Trans. PAS*, 1984, **103**, (10), pp. 2990–2998

21   STEINSTROM, L., LINDBERG, P., and SAMUELSSON, J.: 'Testing procedure for metal oxide varistors protecting EHV series capacitors', *IEEE Trans. Power Deliv.*, 1988, **3**, (2), pp. 568–583

22   HAMANN, J.R., MISKE, S.A., JOHNSON, I.B., and COURTS, A.L.: 'A zinc oxide varistor protective system for series capacitors', *IEEE Trans. PAS*, 1981, **PAS-100**, (3), pp. 929–937

23   COURTS, A.L., HINGORANI, N.G., and STEMLER, G.E.: 'A new series capacitor protection scheme using non-linear resistors', *IEEE Trans. PAS*, 1978, **PAS-97**, (4), pp. 1042–1052

24   McGRANAGHAN, M.F., REID, W.E., LAW, S.W., and GRESHAM, D.W.: 'Overvoltage protection of shunt capacitor banks using MOV arresters', *IEEE Trans. PAS*, 1984, **PAS-103**, (8), pp. 2326–2336

25   RIBEIRO, J.R., and McCALLUM, M.E.: 'An application of metal oxide surge arresters in the elimination of need for reclosing resistors in EHV breakers', *IEEE Trans. Power Deliv.*, 1989, **4**, (1), pp. 282–291

26   HARDER, J.E.: 'Metal oxide arrester ratings for rotating machine protection', *IEEE Trans. PAS*, 1985, **PAS-104**, (9), pp. 2446–2451

27   NIEBUHR, W.D.: 'Application of metal-oxide-varistor surge arresters on distribution systems', *IEEE Trans. PAS*, 1982, **PAS-101**, (6), pp. 1711–1715

28   WESTROM, A.C., and SCOTT, W.L.: 'Increasing the reliability of MOV arresters', *IEEE Trans. Power Deliv.*, 1990, **5**, (4), pp. 1823–1829

29   KOCH, R.E., TIMOSHENKO, J.A., ANDERSON, J.G., and SHIH, C.H.: 'Design of zinc oxide transmission line arresters for application on 138 kV towers', *IEEE Trans. PAS*, 1985, **PAS-104**, (10), pp. 2675–2680

30   KOCH, R.E., and SONGSTER, H.J.: 'Development of a non-fragmenting distribution surge arrester', *IEEE Trans. PAS*, 1984, **PAS-103**, (11), pp. 3342–3352

31   HARUMOTO, Y., TSUDA, Y., KOBAYASHI, M., MIZUNO, M., and AOKI, T.: 'Evaluation for application of built-in type zinc gapless surge arresters for power system equipments', *IEEE Trans. Power Deliv.*, 1987, **PWRD-2**, (3), pp. 750–757

32   VOGLER, L., SMITH, C.T., and TAYLOR, T.F.: 'Some aspects of design and performance of metal oxide surge arresters'. IEE conference on *Lightning protection*, 1984, pp. 107–113

33   HAHN, E.E.: 'Some electrical properties of zinc oxide semiconductor', *J. Appl. Phys.*, 1951, **22**, (7), pp. 855–863

34   MATSUOKA, M., MATSUYAMA, T., and IIDA, Y.: 'Voltage nonlinearity of zinc oxide ceramics doped with alkali earth metal oxide', *Jpn. J. Appl. Phys.*, 1969, **8**, pp. 1275–1276

35   MASUYAMA, T., and MATSUOKA, M.: 'Current dependence of voltage nonlinearity in SiC varistors', *Jpn. J. Appl. Phys.*, 1968, **7**, p. 1294

36   LEVINE, J.D.: 'Theory of varistor electronic properties', *CRC Crit. Rev. Solid State Sci.*, 1975, **5**, pp. 597–608

37   MATSUOKA, M.: 'Nonohmic properties of zinc oxide ceramics', *Jpn. J. Appl. Phys.*, 1971, **10**, (6), pp. 737–746

38 LEVINSON, L.M., and PHILIPP, H.R.: 'ZnO varistors for transient protection', *IEEE Trans. Parts Hybrids Packag.*, 1977, **PHP-13**, (4), pp. 338–343

39 LEVINSON, L.M., and PHILIPP, H.R.: 'The physics of metal oxide varistors', *J. Appl. Phys.*, 1975, **46**, (3), pp. 1332–1341

40 PHILIPP, H.R., and LEVINSON, L.M.: 'Low-temperature electrical studies on metal-oxide varistors, a clue to conduction mechanisms', *J. Appl. Phys.*, 1977, **48**, (4), pp. 1621–1627

41 BERNASCONI, J., KLEIN, H.P., KNECHT, B., and STRASSLER, S.: 'Investigation of various models for metal oxide varistors', *J. Electron. Mater.*, 1976, **5**, (5), pp. 473–495

42 ROSSINELLI, M., BLATTER, G., and GREUTER, F.: 'Grain boundary properties of ZnO varistors', *Proc. Br. Ceram. Soc.*, 1986, pp. 1–17

43 EDA, K.: 'Grain growth control in ZnO varistors using seed grains', *J. Appl. Phys.*, 1983, **54**, (2), pp. 1095–1099

44 BOWEN, L.J., and AVELLA, F.J.: 'Microstructure, electrical properties and failure prediction in low clamping voltage zinc oxide varistors', *J. Appl. Phys.*, 1983, **54**, (5), pp. 2764–2772

45 TRONTELJ, M., KOLAR, D., and KRASEVEC, V.: 'Properties of ZnO ceramics containing TiO_2', *Proc. Br. Ceram. Soc.*, 1986, pp. 143–151

46 ASOKAN, T., NAGABHUSHANA, G.R., and IYENGAR, G.N.K.: 'Improvement of non-linear characteristics of multicomponent ZnO-based ceramics containing Nb_2O_5', *IEEE Trans. Electr. Insul.*, 1988, **23**, (2), pp. 279–287

47 SWEETANA, A., GUPTA, T., CARLSON, W., GREKILA, R., KUNKLE, N., and OSTERHOUT, J.: 'Gapless surge arresters for power systems applications, vol. 1: development of 500 and 1200 kV arresters'. EPRI EL 3166, **1**, Project 657-1, final report, September 1983

48 MUKAE, K., TSUDA, K., and SHIGA, S.: 'Zinc oxide-praseodymium oxide elements for surge arresters', *IEEE Trans. Power Deliv.*, 1988, **3**, (2), pp. 591–597

49 MUKAE, K., TSUDA, K., and NAGASAWA, I.: 'Capacitance vs. voltage characteristics of ZnO varistors', *J. Appl. Phys.*, 1979, **50**, (6), pp. 4475–4476

50 BI-CHIOU CHIOU, and JIH, F.W.: 'Studies on ZnO-doped varistors with conventional pellet form and thick film form', *Proc. Br. Ceram. Soc.*, 1986, pp. 129–141

51 GREUTER, F., PERKINS, R., ROSSINELLI, M., and SCHMUCKLE, F.: 'The metal-oxide resistor at the heart of modern surge arresters'. *ABB Rev. Surge Arresters*, 1989, **1**

52 MORRIS, W.G.: 'Electrical properties of ZnO-Bi_2O_3 ceramics', *J. Am. Ceram. Soc.*, 1973, **56**, (7), pp. 360–364

53 MAHAN, G.D., LEVINSON, L.M., and PHILIPP, H.R.: 'Theory of conduction in ZnO varistors', *J. Appl. Phys.*, 1979, **50**, (4), pp. 2799–2812

54 MENTH, A., STREIT, P., and KNECHT, B.: 'Present state and development potential of solid state varistors', *in* RAGALLER, K. (Ed.): 'Surges in high voltage network' (Plenum Press, New York, 1980), pp. 283–298

55   TOMINAGA, S., SHIBUYA, Y., FUJIWARA, Y., and NITTA, T.: 'Electrical properties of zinc oxide valve element for a surge arrester'. IEEE Power Engineering Society Summer Meeting, paper A78 595-1, Los Angeles, CA, USA, 16–21 July 1978, pp. 1–8

56   BLATTER, G., and GREUTER, F.: 'Carrier transport through grain boundaries in semiconductors', *Phys. Rev. B*, 1986, **33**, (6), pp. 3952–3966

57   JONSCHER, A.K.: 'Hopping losses in polarisable dielectric media', *Nature*, 1974, **250**, pp. 191–193

58   MANUEL, P.: 'Modelisation des mecanismes de conduction non lineaires dans les varistances ZnO', *Rev. Phys. Appl.*, 1987, **22**, (9), pp. 971–983

59   SATO, K., TAKADA, Y., and MAEKAWA, H.: 'Electrical conduction of ZnO varistors under continuous DC stress', *Jpn. J. Appl. Phys.*, 1980, **19**, (5), pp. 909–917

60   SUZUOKI, Y., OHKI, A., MIZUTANI, T., and IEDA, M.: 'Electrical properties of ZnO-Bi$_2$O$_3$ thin-film varistors', *J. Phys. D, Appl. Phys.*, 1987, **20**, pp. 511–517

61   BLATTER, G., and GREUTER, F.: 'Carrier transport through grain boundaries in semiconductors', *Phys. Rev. B*, 1986, **33**, (6), pp. 3952–3966

62   PIKE, G.E., and SEAGER, C.H.: 'The DC voltage dependence of semiconductor grain boundary resistance', *J. Appl. Phys.*, 1979, **50**, (5), pp. 3414–3422

63   PIKE, G.E.: 'Grain boundaries in semiconductors', *in* PIKE, SEAGER and LEAMY, (Eds): 'Grain boundaries in semiconductors'. Proceedings of the Annual Meeting Mat. Res. Proc., 1981, **5**, pp. 369–379

64   EDA, K.: 'Zinc oxide varistors', *IEEE Electr. Insul. Mag.*, 1989, **5**, (6), pp. 28–41

65   TOMINAGA, S., SHIBUYA, Y., FUJIWARA, Y., IMATAKI, M., and NITTA, T.: 'Stability and long term degradation of metal oxide surge arresters', *IEEE Trans. PAS*, 1980, **PAS-99**, (4), pp. 1548–1556

66   MIZUNO, M., HAYASHI, M., and MITANI, K.: 'Thermal stability and life of the gapless surge arrester', *IEEE Trans. PAS*, 1981, **PAS-100**, (5), pp. 2664–2671

67   FUJIWARA, Y., SHIBUYA, Y., IMATAKI, M., and NITTA, T.: 'Evaluation of surge degradation of metal oxide surge arresters', *IEEE Trans. PAS*, 1982, **PAS-101**, (4), pp. 978–985

68   KAN, M., NISHIWAKI, S., SATO, T., KOJIMA, S., and YANABU, S.: 'Surge discharge capability and thermal stability of a metal oxide surge arrester', *IEEE Trans. PAS*, 1983, **PAS-102**, (2), pp. 282–289

69   VICAUD, A.: 'AC voltage ageing of zinc-oxide ceramics', *IEEE Trans. Power Syst.*, 1986, **PWRD-1**, (2), pp. 49–58

70   KIRBY, P., ERVEN, C.C., and NIGOL, O.: 'Long term stability and energy discharge capacity of metal oxide valve elements', *IEEE Trans. Power Deliv.*, 1988, **3**, (4), pp. 1656–1665

71   LAT, M.V.: 'Thermal properties of metal oxide surge arresters', *IEEE Trans. PAS*, 1983, **PAS-102**, (7), pp. 2194–2202

72 YEANFANG, W., YUYI, Z., and YUGIN, L.: 'Study of thermal stability of metal oxide surge arrester'. Fifth international symposium on *High voltage engineering*, paper 83.01, Braunschweig, FRG, August 1987

73 KIRKBY, P.: 'The long term stability of valve elements used in metal oxide surge arresters'. CEA report 199 T 425, July 1986

74 ST-JEAN, G., and PETIT, A.: 'Metal oxide surge arrester operating limits defined by a temperature-margin concept', *IEEE Trans. Power Deliv.*, 1990, **5**, (2), pp. 627–633

75 TOMINAGA, S., AZUMI, K., NITTA, T., NAGAI, N., IMATAKI, M., and KUWAHARA, H.: 'Reliability and application of metal oxide surge arresters for power systems', *IEEE Trans. PAS*, 1979, **PAS-98**, (3), pp. 805–812

76 NISHIKAWA, H., SUKEHARA, M., and YOSHIDA, T., *et al.*: 'Thyristor valve gapless arrester element lifetime performance and thermal stability', *IEEE Trans. PAS*, 1985, **PAS-104**, (7), pp. 1883–1888

77 IEC99-4:1991 incorporating amendment No.1:1998: 'Surge arresters, part 4: metal oxide surge arresters without gaps for AC systems'. IEC standard, 1998

78 JEC-217: 'Metal oxide surge arresters'. Standard of the Japanese Electrotechnical Committee, published by the IEEJ, 1984

79 RICHTER, B.: 'Investigation of MO surge arrester under constant high voltage stress by application of potential-free measuring devices'. Fifth international symposium on *High voltage engineering*, paper 82.10, Braunschweig, FRG, 1987

80 RICHTER, B., and HEINRICH, B.: 'Investigation of the stresses of MO surge arresters for high voltage application'. Sixth international symposium on *High voltage engineering*, paper 26.02, New Orleans, USA, 1989

81 YUFANG, W., YUJING, L., YUNYI, Z., and LIBO, H.: 'Study of thermal stability of 500 kV MOA prorated section'. Sixth international symposium on *High voltage engineering*, paper 26.10, New Orleans, USA, 1989

82 HINRICHSEN, V., and PEISER, R.: 'Simulation of the electrical performance and thermal behaviour of metal oxide surge arresters under AC stress'. Sixth international symposium on *High voltage engineering*, paper 26.04, Braunschwieg, FGR, 1987

83 HINRICHSEN, V., and PEISER, R.: 'Simulation of the AC performance of gapless ZnO arresters'. Fifth international symposium on *High voltage engineering*, paper 82.09, Braunschweig, FGR, 1987

84 PHILIPP, H.R., and LEVINSON, L.M.: 'Degradation phenomenon in zinc oxide varistors: a review', *Advances in Ceramics*, 1984, **7**, pp. 1–21

85 NAWATA, M., KAWAMURA, H., and IEDA, M.: 'Studies on degradation mechanism of zinc oxide ceramics varistor by thermally simulated current'. Fifth international symposium on *High voltage engineering*, paper 82.06, Braunschweig, FRG, 1987

86 STUCKI, F., BRUESCH, P., and GREUTER, F.: 'Electron spectroscopic studies of electrically active grain boundaries in ZnO', *Surf. Sci.*, 1987, **189/190**, pp. 294–299

87   DE MORAIS, H.M., DA COSTA, E.G., and NOWAKI, K.: 'Ageing of zinc oxide blocks through partial discharges'. Fifth international symposium on *High voltage engineering*, paper 83.05, Braunschweig, FRG, 1987

88   SAKSHAUG, E.C., KRESGE, J.S., MARK, D.A., and KARADY, G.G.: 'Contamination and hot wash performance of zinc oxide station arresters', *IEEE Trans. PAS*, 1982, **PAS-101**, (5), pp. 1095–1104

89   OYAMA, M., OHSHIMA, I., HONDA, M., YAMASHITA, M., and KOJIMA, S.: 'Life performance of zinc-oxide elements under DC voltage', *IEEE Trans. PAS*, 1982, **PAS-101**, (6), pp. 1363–1368

90   ZANDER, W.: 'Influence of material heterogeneity on the ageing of ZnO arrester material'. Fifth international symposium on *High voltage engineering*, paper 82.05, Braunschweig, FRG, 1987

91   MIZUKOSHI, A., OZAWA, J., SHIRAKAWA, S., and NAKANO, K.: 'Influence of uniformity on energy absorption capabilities of zinc oxide elements as applied in arresters', *IEEE Trans. PAS*, 1983, **PAS-102**, (5), pp. 1384–1390

92   BRONIKOWSKI, R.J., and DUPONT, J.P.: 'Development and testing of MOVE arrester elements', *IEEE Trans. PAS*, 1982, **PAS-101**, (6), pp. 1638–1643

93   NISHIWAKI, S., KIMURA, H., SATOH, T., MIZOGUSHI, H., and YANABU, S.: 'Study of thermal runaway/equivalent prorated model of a ZnO surge arrester', *IEEE Trans. PAS*, 1984, **PAS-103**, (2), pp. 413–421

94   VONG, N.M., MILLER, R., RYDER, D.M., DOONE, R.M., and SPARROW, L.: 'Deterioration in the performance of zinc oxide surge arrester elements due to partial discharges'. Sixth international symposium on *High voltage engineering*, paper 26.22, New Orleans, USA, 1989

95   KNOBLOCH, H.: 'The influence of the surrounding medium on the service behaviour of metal oxide varistors'. Sixth international symposium on *High voltage engineering*, paper 26.07, New Orleans, USA, 1989

96   DEM'YANENKO, K.B., and SERGEEV, A.S.: 'Stability of highly non-linear zinc oxide resistors under the prolonged action of a commercial frequency potential', *Elektrotekhnika*, 1984, **55**, (9), pp. 46–51

97   SAKSHAUG, E.C., BURKE, J.J., and KRESGE, J.S.: 'Metal oxide arresters on distribution systems fundamental considerations', *IEEE Trans. Power Deliv.*, 1989, **4**, (4), pp. 2076–2089

98   YUFANG, W., YUJIN, L., and YUYI, Z.: 'Working stability of MOV in the oil'. Sixth international symposium on *High voltage engineering*, paper 41.02, New Orleans, USA, 1989

99   EDA, K.: 'Destruction mechanism of ZnO varistors due to high currents', *J. Appl. Phys.*, 1984, **56**, (10), pp. 2948–2955

100 SHUFANG, L., YUJIN, L., and YUYI, Z.: 'Energy absorption characteristic of metal oxide elements under transient overvoltage'. Fifth international symposium on *High voltage engineering*, paper 83.02, Braunschweig, FRG, 1987

101 DAKIN, T.W.: 'Electrical insulation deterioration treated as a chemical rate phenomenon', *AIEE Trans.*, 1948, **67**, pp. 113–122

102 CARLSON, W.G., GUPTA T.K., and SWEETANA, A.: 'A procedure for estimating the lifetime of gapless metal oxide surge arresters for AC application', *IEEE Trans. Power Syst.*, 1986, **PWRD-1**, (2), pp. 67–74

103 HADDAD, A., NAYLOR, P., METWALLY, I., GERMAN, D.M., and WATERS, R.T.: 'An improved non-inductive impulse voltage measurement technique for ZnO surge arresters', *IEEE Trans. Power Deliv.*, 1995, **10**, (2), pp. 778–784

104 KIZEBETTER, V.E., SEGEEV, A.S., and FIRSOV, A.V.: 'Calculation of the distribution of voltage and current along the elements of an overvoltage limiter with contaminated and wet covering', *Elektrotekhnika*, 1987, **58**, (3), pp. 21–27

105 PEIBAI, Z., GUOHANG, W., HUIMING, M., JIALU, S., CIZHANG, F., and NAIXIANG, M.: 'Analysis of the potential distribution of gapless surge arrester'. Sixth international symposium on *High voltage engineering*, paper 26.09, New Orleans, LA, USA, 1989

106 BENZAOUA, F., HADDAD, A., ROWLANDS, A.R., and WATERS, R.T.: 'A circuit model for polluted ZnO polymeric surge arresters'. 10th international symposium on *High voltage engineering (ISH)*, Montreal, Canada, 1997, **3**, pp. 317–320

107 ROUSSEAU, A., GOURMET, P., NIKLASCH, H., and BURET, F.: 'A new use of pollution tests on ZnO arresters'. Sixth international symposium on *High voltage engineering*, paper 27.17, New Orleans, LA, USA, 1989

108 LENK, D.W.: 'An examination of the pollution performance of gapped and gapless metal oxide station class surge arresters', *IEEE Trans. PAS*, 1984, **PAS-103**, (2), pp. 337–345

109 LENK, D.W., STOCKUM, F.R., and GRIMES, D.E.: 'A new approach to distribution arrester design', *IEEE Trans. Power Deliv.*, 1988, **3**, (2), pp. 584–590

110 OYAMA, M., OHSHIMA, I., HONDA, M., YAMASHITA, M., and KOJIMA, S.: 'Analytical and experimental approach to the voltage distribution on gapless zinc-oxide surge arresters', *IEEE Trans. PAS*, 1981, **PAS-100**, (11), pp. 4621–4627

111 KOJIMA, S., OYAMA, M., and YAMASHITA.: 'Potential distributions of metal oxide surge arresters under various environmental conditions', *IEEE Trans. Power Deliv.*, 1988, **3**, (3), pp. 984–989

112 RICHTER, B.: 'Application of fibre optics to the measurement of leakage current in metal oxide surge arresters'. Fourth international symposium on *High voltage engineering*, paper 64.05, Athens, Greece, 1983

113 HIDAKA, K., and KOUNO, T.: 'Simultaneous measurements of two orthogonal components of electric field using a Pockel device', *Rev. Sci. Instrum.*, 1989, **60**, (7), pp. 1252–1257

114 LEVINSON, L.M., and PHILIPP, H.R.: 'High frequency and high-currents in metal-oxide varistors', *J. Appl. Phys.*, 1976, **47**, pp. 3117–3121

115 MATSUURA, M., and YAMAOKI, H.: 'Dielectric dispersion and equivalent circuit in nonohmic ZnO ceramics', *Jpn. J. Appl. Phys.*, 1977, **16**, (7), pp. 1261–1262

116   LEVINSON, L.M., and PHILIPP, H.R.: 'Long time polarization currents in metal oxide varistors', *J. Appl. Phys.*, 1976, **47**, (7), pp. 3177–3181

117   LAGRANGE, A.: 'Les varistances a base d'oxide de zinc: des elements de protection a l'etat solide en plein development', *Rev. Gen. Electr.*, 1986, (9), pp. 9–18

118   DANIEL, V.: 'Dielectric relaxation' (Academic Press, London and New York, 1967)

119   CORDARO, J.F., MAY, J.E., and SHIM, Y.: 'Bulk electron traps in zinc oxide varistors', *J. Appl. Phys.*, 1986, **60**, (12), pp. 4186–4190

120   ELLI, E., WARNOTS, E., and NASCIMENTO, L.F.M.: 'Laboratory tests to characterize metal-oxide surge arresters', *Energ. Elettr.*, 1986, (2), pp. 81–88

121   HORIUCHI, S., ICHIKAWA, F., MIZUKOSHI, A., KURITA, K., and SHIRAKAWA, S.: 'Power dissipation characteristics of zinc-oxide arresters for HVDC systems', *IEEE Trans. Power Deliv.*, 1988, **3**, (4), pp. 1666–1671

122   DANG, C., PARNELL, T.M., and PRICE, P.J.: 'The response of metal oxide surge arresters to steep fronted current impulses', *IEEE Trans. Power Deliv.*, 1986, **PWRD-1**, (1), pp. 157–163

123   RICHTER, B., KRAUSE, C., and MEPPELINK, J.: 'Measurement of the U–I characteristic of MO-resistors at current impulses of different wave shapes and peak values'. Fifth international symposium on *High voltage engineering*, paper 82.03, Braunschweig, Germany, 1987

124   BREILMANN, W.: 'Protective characteristic of complete zinc-oxide arrester and of single elements for fast surges'. Fifth international symposium on *High voltage engineering*, paper 82.04, Braunschweig, Germany, 1987

125   EDA, K.: 'Transient conduction phenomena in non-ohmic zinc oxide ceramics', *J. Appl. Phys.*, 1979, **50**, (6), pp. 4436–4442

126   MODINE, F.A., and WHEELER, R.B.: 'Fast pulse response of zinc-oxide varistors', *J. Appl. Phys.*, 1987, **61**, (8), pp. 3093–3098

127   TUA, P.F., ROSSINELLI, M., and GREUTER, F.: 'Transient response of electrically active grain boundaries in polycrystalline semiconductors', *Phys. Scr.*, 1988, **38**, (3), pp. 491–497

128   SCHMIDT, W., MEPPELINK, J., RICHTER, B., FESER, K., KEHL, L., and QIU, D.: 'Behaviour of MO-surge arrester blocks to fast transients', *IEEE Trans. Power Deliv.*, 1989, **4**, (1), pp. 292–300

129   PHILIPP, H.R., and LEVINSON, L.M.: 'ZnO varistors for protection against nuclear electromagnetic pulses', *J. Appl. Phys.*, 1981, **52**, (2), pp. 1083–1090

130   HADDAD, A., NAYLOR, P., METWALLY, I., GERMAN, D.M., and WATERS, R.T.: 'An improved non-inductive impulse voltage measurement technique for ZnO surge arresters', *IEEE Trans. Power Deliv.*, 1995, **10**, (2), pp. 778–784

131   HADDAD, A., and NAYLOR, P.: 'Dynamic impulse conduction in ZnO arresters'. 11th international symposium on *High voltage engineering (ISH)*, London, UK, 1999, **2**, pp. 254–257

132 HADDAD, A., FUENTES-ROSADO, J., GERMAN, D.M., and WATERS, R.T.: 'Characterisation of ZnO surge arrester elements with direct and power frequency voltages', *IEE Proc.*, 1990, **137**, (5), pp. 269–279

133 HADDAD, A., ELAYYAN, H.S.B., GERMAN, D.M., and WATERS, R.T.: 'ZnO surge arrester elements with mixed direct and 50 Hz voltages', *IEE Proc. A, Sci. Meas. Instrum. Manage. Educ.*, 1991, **138**, (5), pp. 265–272

134 TREGUIER, J.P.: 'Equipment de telecommunications: varistances a base d'oxyde de zinc avec d'autres composants pour un meilleur niveau de protection', *Rev. Gen. Electr.*, 1986, (9), pp. 24–31

135 EKSTROM, A.: 'Application guide for metal oxide arresters without gaps for HVDC converter stations'. CIGRE Working Group 33/14-05, June 1988

136 PIKE, G.E.: 'Electronic properties of ZnO varistors, a new model', *in* 'Grain boundaries in semiconductors'. Annual Meeting Material Research Society Proceedings, Boston MA, 1981, **5**, pp. 369–379

137 PIKE, G.E., GOURLEY, P.L., and KURTZ, S.R.: 'Impact ionisation near GaAs grain boundaries', *Appl. Phys. Lett.*, 1989, **43**, (10), pp. 939–941

138 EMTAGE, P.R.: 'The physics of zinc oxide varistors', *J. Appl. Phys.*, 1977, **48**, (10), pp. 4372–4384

139 KORN, S.R., MARTZLOFF, F.D., MAY, J., and WOLFF, B.L.: 'Transient voltage suppression', *in* SMITH, M.C., and McCORMICK, M.D. (Eds): 'Manual of semiconductors' (Production Dept. (G.E.C.), 1982)

140 HONG-YANG ZHENG: 'Study on the power loss characteristics of ZnO arrester elements used in H.V.D.C. converter station'. Fifth international symposium on *High voltage engineering*, paper 83.03, Braunschweig, FRG, 1987

141 YU-JIN LIANG, YU-YI ZHAO, and HONG-YAN ZENG: 'Survey of low current characteristics of metal oxide varistor'. Fifth international symposium on *High voltage engineering*, paper 82.11, Braunschweig, FRG, 1987

142 STOKES, A.D.: 'High voltage line-dropping switching stresses', *IEE Proc. C, Gener. Transn. Distrib.*, 1992, **139**, (1), pp. 21–26

143 CHI, H.: 'The transient behaviour of ZnO varistors under DC service voltage'. Seventh international symposium on *High voltage engineering*, paper 85.01, Dresden, Germany, 1991

144 HADDAD, A., METWALLY, I., NAYLOR, P., TONG, Y.K., and WATERS, R.T.: 'Impulse response of ZnO arresters with pre-existing direct voltage stress'. Eighth international symposium on *High voltage engineering*, paper 77.04, Yokohama, Japan, 1993, pp. 417–420

145 HADDAD, A., GERMAN, D.M., METWALLY, I., NAYLOR, P., GRIFFITHS, H., and WATERS, R.T.: 'The transient response of ZnO surge arresters to sudden voltage reversal'. 28th universities *Power engineering* conference, Staffordshire University, UK, 1993, pp. 482–485

146 NAYLOR, P., HADDAD, A., METWALLY, I., TONG, Y.K., GERMAN, D.M., and WATERS, R.T.: 'Trapped charge effects on the transient response of surge arresters'. 29th universities *Power engineering* conference, Galway, Ireland, 1994, pp. 441–444

147   HADDAD, A., NAYLOR, P., TONG, Y.K., MARLEY, W.A., GERMAN, D.M., and WATERS R.T.: 'Direct voltage and trapped charge effects on the protective characteristic of ZnO surge arresters', *IEE Proc., Sci. Meas. Technol.*, 1995, **142**, (6), pp. 442–448

148   HADDAD, A., NAYLOR, P., TONG, Y.K., METWALLY, I., GERMAN, D.M., and WATERS R.T.: 'Trapped-charge and point-on-wave effects on the protective characteristic of surge arresters'. 9th international symposium on *High voltage engineering (ISH)*, Graz, Austria, 1995, **4**, pp. 7558-1 to 7550-4

149   ANSI/IEEE C62.11:1987: 'IEEE Standard for metal-oxide surge arresters for ac power circuits'. American National Standard, 1987

150   SARGENT, R.A., DUNLOP, G.L., and DARVENIZA, M.: 'Effects of multiple impulse currents on the microstructure and electrical properties of metal oxide varistors', *IEEE Trans. Electr. Insul.*, 1992, **27**, (3), pp. 586–592

151   DARVENIZA, M., MERCER, D.R., and TUMMA, L.R.: 'Effects of multiple stroke lightning impulse currents on ZnO arresters'. CIRED, 1993, pp. 2.25.1–2.25.5

152   HADDAD, A., ABDUL-MALEK, Z., GERMAN, D.M., and WATERS, R.T.: 'Double impulse testing of ZnO surge arresters'. 23rd international conference on *Lightning conference (ICLP)*, Florence, Italy, 1996, **II**, pp. 696–702

153   HADDAD, A., and ABDUL-MALEK, Z.: 'A new rotating spark gap', *Rev. Sci. Instrum.*, 1997, **68**, (12), pp. 1–5

154   OZAWA, J., OOSHI, K., SHIRAKAWA, S., NAKANO, K., MUZUKOSHI, A., and MARUYAMA, S.: 'Fast transient response and its improvement of metal oxide surge arresters for GIS'. Sixth international symposium on *High voltage engineering*, paper 26.03, New Orleans, USA, August 1989

155   KNECHT, B.: 'Solid-state arresters', *in* RAGALLER, K. (Ed.): 'Surges in high voltage networks' (Plenum Press, New York, 1980) pp. 299–321

156   IEEE working group 3.4.11: 'Modelling of metal oxide surge arresters', *IEEE Trans. Power Deliv.*, 1992, **7**, (1), pp. 302–309

157   KIM, I., FUNABASHI, T., SASAKI, H., HAGIWARA, T., and KOBAYASHI, M.: 'Study of ZnO arrester model for steep front wave', *IEEE Trans. Power Deliv.*, 1996, **11**, (2), pp. 835–841

158   HADDAD, A., and NAYLOR, P.: 'Finite element computation of capacitance networks in multiple electrode systems: application to ZnO surge arresters', *IEE Proc., Sci. Meas. Technol.*, 1998, **145**, (4), pp. 129–135

159   HADDAD, A., and NAYLOR, P.: 'Dynamic impulse conduction in ZnO arresters', 11th international symposium on *High voltage engineering (ISH)*, London, UK, 1999, **2**, pp. 254–257

160   BREDER, H., and COLLINS, T.: 'Supervision of gapless zinc oxide surge arresters', IEE Conf. Publ. (236), London, 1984

161   SHIRAKAWA, S., ENDO, F., and KITAJIMA, H., *et al.*: 'Maintenance of surge arrester by portable arrester leakage current detector,' *IEEE Trans. Power Deliv.*, 1988, **3**, (3), pp. 998–1003

162 LINDQUIST, J., STRENSTROM, L., SCHEI, A., and HANSEN, B.: 'New method for measurement of the resistive leakage currents of metal oxide surge arresters in service', *IEEE Trans. Power Deliv.*, 1990, **5**, (4), pp. 1811–1819

163 DEEP, G.S., and NOWACKI, K.: 'Measurement of resistive leakage current of metal-oxide surge arresters'. Fourth international symposium on *High voltage engineering*, paper 64.07, Athens, Greece, 5–9 September 1983

164 NAIDU, S.R., and SRIVASTAVA, K.D.: 'The validity of circuits for measuring the resistive leakage current of zinc oxide arresters'. Fifth international symposium on *High voltage engineering*, paper 82.12, Braunschweig, FRG, 24–28 August 1987

165 VICAUD, A., and EGUIAZABAL, D.: 'Characterisation electrique sous faible courant de ceramique ZnO pour parafoudres'. Societe des electriciens et des electroniciens (SEE) conference on *Varistance a base d'oxide de zinc*, paper B-11, pp. 85–87, Ecole Superieure d'Electricite, Gif-sur-Yvette, France, 6 March 1986

166 IEC 60099-5:1996 incorporating amendments 1:1999: 'Surge arresters, part 5: selection and application recommendation'. IEC Standard, 1996

167 SPELLMAN, C., and HADDAD, A.: 'A technique for on-line condition monitoring of ZnO surge arresters'. 10th international symposium on *High voltage engineering (ISH)*, Montreal, Canada, 1997, **4**, pp. 151–154

168 IEC507:1991: 'Artificial pollution tests on high-voltage insulators to be used on AC systems'. IEC Standard, 1991

169 IEC815:1986: 'Guide for the selection of insulators inrespect of polluted conditions'. IEC Standard, 1986

170 CIGRE Task Force 33.04.01: 'Polluted insulators: a review of current knowledge'. Brochure 158, 2000, pp. 134–185

171 HADDAD, A., ABDUL-MALEK, Z., GERMAN, D.M., and WATERS, R.T.: 'Coordination of spark gap protection with zinc oxide surge arresters', *IEE Proc., Gener., Trans. Distrib.*, 2001, **148**, (1), pp. 21–28

172 ABDUL-MALEK, Z., HADDAD, A., GERMAN, D.M., and WATERS, R.T.: 'Current sharing in multi-column surge arresters'. 10th international symposium on *High voltage engineering (ISH)*, Montreal, Canada, 1997, **5**, pp. 145–148

173 ANSI/IEEE C62.22: 1991: 'IEEE guide for the application of metal-oxide surge arresters for alternating-current systems'. IEEE Standard, 1991

174 IEC 71-2:1996: 'Insulation co-ordination, part 2: application guide'. IEC Standard, 1996

Chapter 6

Insulators for outdoor applications

D.A. Swift

6.1 Introduction

The product developments that have taken place since John Looms published, in 1988, his excellent book [1] that covered much of this subject have mainly concerned polymeric materials – especially silicone rubber. Although the principal modifications made to porcelain and glass insulators over this period are those required for high voltage DC applications, much more has – nonetheless – been learned about the performance of such insulators under AC energisation. The discussed proposals of some years ago for ultra high voltage transmission systems resulted in the production of very high mechanical strength insulators of the cap and pin design. However, as such proposals generally failed to come to fruition, at least one insulator manufacturer now has a product that is ready, should a market become available in the future.

The design, the specification for a certain usage and the dimensioning of the insulator for that usage in an outdoor environment are usually dominated by the need to take account of wetted pollution on its surface. Often, the consequential problem is that of flashover across this polluted surface when the leakage current exceeds a certain value. However, some potentially dangerous defects in the structure of such insulators have been known to occur at values of leakage current well below that of the critical flashover limit. The associated mechanisms are quite involved and, although studied in some detail already, are still not fully understood.

Essentially, the troublesome pollution comes in two forms. The soluble components – e.g. salts from the sea and industrial gases that result in weak acids being formed – produce an electrolyte when dissolved in water. This water is provided by condensation from humid air, fog, mist and drizzle. Rain is usually helpful, but not always so – especially when heavy. Generally, it washes the pollution off the surface before it reaches the troublesome magnitude. The exceptions to this general situation have needed certain insulator profiles to be used – or palliatives to be adopted – to minimise the probability of flashover under such heavy wetting conditions.

Paradoxically, insulators in some desert regions are severely affected. This occurs because the sand can contain a very high concentration of salt. For example, in parts of Tunisia chemical analyses have widely confirmed the aggressiveness of desert sand because it may contain up to 18 per cent of salt (Znaidi, private communication). Presumably such situations arise because, at some time in the past, such land was below sea level.

A further point worthy of general mention is pollution flashover caused by snow and ice. Here, the main problem lies with the difficulty in standardising an artificial pollution test. As the salt fog test [1] took many years to be accepted internationally, it can be reasoned that only young people have sufficient time available in their working lifetime to attempt such an enterprise if they wish to reap benefit – in personal recognition – of their research efforts!

Before proceeding to explain the specific details of leakage current, it is useful to mention the state of the polluted surface from the point of view of its propensity to become wetted. The usually used terms are hydrophilic – i.e. surface readily wets so that a film of water is produced – and hydrophobic – i.e. water resides as discrete beads. This difference is due to the magnitude of the free surface energy; the lower value being for the hydrophobic case.

Many thousands of scientific papers, covering both practical and theoretical issues, have been published on this subject. The rate of publication continues to increase as the years go by. At the International Symposium on High Voltage Engineering, which is a biennial event, at least ten percent of the papers – often being at least five hundred in total – cover this subject. One wonders why is this so? The reason is related to the strategic importance of power transmission to the wellbeing of all industrial nations and the complex issues involved in trying to explain the flashover and some longer lasting failure processes of the insulating components of this system. A major review of polluted insulators, fairly recently published as a brochure [2] by the International Conference on Large High Voltage Electric Systems (CIGRE, which is the acronym of the French version of this title) has been prepared by a taskforce of 15 experts from 12 countries. This CIGRE publication comprises 185 pages that summarise research published in 382 references. The same international committee is currently working on a companion brochure dealing with guidelines for the selection and dimensioning of insulators for use in polluted environments [3] – hopefully, to be published in early 2005. Their research findings are currently being used by the International Electrotechnical Commission (IEC) to revise an old document concerning polluted insulators [4] – probably not to be completed until at least 2006. Although these new CIGRE and IEC documents deal essentially with the same subject, the aim of the IEC publication is to achieve a more rule of thumb approach than that being adopted by CIGRE.

6.2   Role of insulators

Insulators for an overhead power line (OHL), in many substation applications and on the overhead electrification systems of railways, must, primarily, support the conductors. Also important, as already mentioned, is the need to avoid frequent flashover events from occurring.

Although the total mechanical failure of such an insulator is, fortunately, a rare event, its occurrence may be very serious. For example, should a vertical insulator of an OHL (often referred to as a suspension unit) break, then its conductor could be supported by the insulators of the neighbouring support structures (often called towers) at either side. Then, it is possible that this conductor could be reenergised but with little ground clearance!

The consequences of a flashover vary from being annoying to being very costly. For example, the damage resulting from the external flashover of the insulating housing of a high power circuit breaker during a synchronising operation, when the voltage across the polluted surface can increase to twice the normal value, could be extremely large.

6.3 Material properties

Table 6.1 provides a summary of the directly relevant properties – in a generic way – of the mechanical properties of materials that are currently in use for outdoor insulators. These values will be referred to later as various aspects of designs and problems are discussed.

Because silicone rubber now seems to be preferred for insulators that are employed in highly polluted environments, their properties with respect to hydrophobicity warrant a special mention. High temperature vulcanised (HTV) silicone rubber, which is now the usual variant of this polymer for constructing insulators, is commonly composed of polydimethysiloxane and alumina trihydrate (ATH) filler. This filler adsorbs the heat from electrical discharges and thereby prevents surface tracking from taking place. Many researchers think that the hydrophobicity of silicone rubber insulators is due to the low molecular weight (LMW) silicone chains, comprising methyl groups attached to silicone atoms, that have diffused to the surface. Upon breaking this attachment (e.g. by electrical discharges), the hydrophobicity of this surface is lost until it becomes recovered by a new lot of LMW. As ATH affects the mobility of this oil-like substance, it plays an important part in the recovery rate of hydrophobicity.

Table 6.1   Mechanical properties of the dielectric components of high voltage insulators [2]

	Units	Porcelain	Glass	Polymer*	RBGF**
Tensile strength	MPa	30–100	100–120	20–35	1300–1600
Compressive strength	MPa	240–820	210–300	80–170	700–750

* silicone and carbon based
** resin-bonded glass fibre used for the core of a polymeric insulator

That is, the different formulations used by the various manufacturers may result in different recovery rates. When the silicone rubber surface becomes polluted, it is variously thought that, in time, these LMWs cover the pollutants by a process due to evaporation or diffusion, or are related to the capillary effect (i.e. surface tension). Therefore, the size and shape of the solid particles (e.g. spherical, flakes) may well affect the rate of this transfer. Another complicating factor is that LMWs come in two forms: one linear, the other cyclic. The former seems to be more effective in covering the pollutants. Obviously, the available quantity of LMW is important. In newly produced HTV silicone rubber, it is about three per cent by mass. Therefore, a pessimistic assessment is that the rate and the amount of LMW diffusion from the bulk to the surface of the silicone rubber, and hence onto the pollutants, decreases as the material ages under service stress and weathering. However, there is a well supported postulate that the quantity of LMWs is maintained because the bulk material tries to achieve thermodynamic equilibrium. That is, silicone rubber is a dynamic material and so the effects of ageing have to be assessed from this viewpoint. A silicone rubber coating is gaining increasing use as a way of enhancing the pollution flashover performance of porcelain insulators. Typically, such coating materials are made from room temperature vulcanised (RTV) silicone rubber. In its newly produced state, the LMW content is about five per cent by mass, but it is of the cyclic form.

Some comments concerning the resin-bonded glass fibres (RBGF) used in the core of a polymeric insulator are also necessary. Although this structure has enormous tensile strength when compared with that of glass and porcelain (Table 6.1) a somewhat disturbing type of failure – usually called brittle fracture – has occurred in a few cases. This defect appears as a very sharp break in the cross-section of the rod as though it had been sliced with a knife. An elegant explanation [5], which still remains to be proved, involves the replacement of the aluminium and calcium ions in the glass with hydrogen ions from an acid. Various acids have been suggested including oxalic acid – resulting from electrical discharges within the structure that produce reactive oxygen and oxides of nitrogen that combine with polymer carbon. In this ion exchange reaction, the proton is smaller than the replaced metal ion thereby inducing a tensile stress in the surface of the glass that causes spontaneous cracking. Such fractures have been shown to happen readily on non-coated fibres of standard E-type glass. Although numerous attempts have been made in the laboratory to produce this type of failure using commercially produced RBGF rod and realistic acid concentrations, none is known to have succeeded. Therefore, it is tempting to conclude that there is no problem until the glass in the fibre is stripped of its coating of epoxy resin. Perhaps a weakness has been built into a few insulators during the manufacturing stage. A French/Swiss research team has fairly recently proposed [6] that incorrect curing of the resin may cause such a weakness.

6.4   Examples of design

In general terms, the various shapes of insulators can be divided into two groups: discs and cylinders. Of the former, the cap and pin designs for OHL constructions are the most numerous. The latter group embraces such designs as longrods, of both the porcelain and the polymeric types, posts and barrels.

6.4.1 Cap and pin insulators

For areas in which there is little pollution the standard profile, Figure 6.1*a*, is adequate. When, however, pollution is a major problem then two measures are adopted. Either the leakage path length is increased – by having substantially deeper ribs – as shown in the anti-fog profile of Figure 6.1*b*, or the surface is made very smooth to minimise the catch and retention of pollutants – as illustrated in Figure 6.1*c* by the aerodynamic

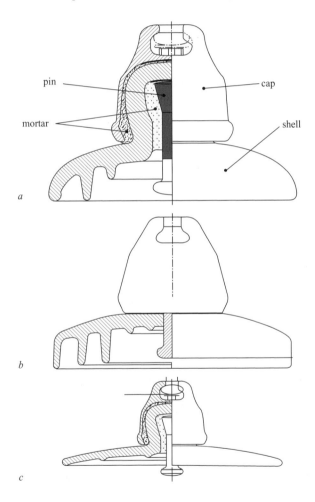

Figure 6.1   *Typical examples of the profile of a cap and pin insulator*
    a   standard design
    b   anti-fog design
    c   aerodynamic design

profile. Often, the anti-fog design is used in coastal and heavy industrial regions and the aerodynamic design is employed in deserts.

The advantage, from the mechanical viewpoint, of the cap and pin design is that it employs the insulating material – glass or porcelain – in its compressive mode, because these materials are very strong (see Table 6.1) under this type of mechanical stress. This desirable arrangement to support the axial load is obtained by the conical shape of the part of the steel pin embedded within the mortar that fixes it to a wraparound cast iron cap.

The seemingly obvious shortcoming of the cap and pin insulator is the large amount of metal – mainly the caps – that occupies the axial length of a string of such units. However, there are two compensating aspects. One is the substantial amount of leakage path length of ceramic that can be obtained per unit axial length, and the other is associated with the small radial dimension of this insulating section around the pin cavity – which can account for an appreciable amount of the surface resistance. Also, having some metal in the path of a propagating arc root can be beneficial due to its ability to arrest such movement [7].

6.4.1.1   Some problems

From time to time, defective insulators have been found on operational lines that have necessitated changes in design. In other cases, such defects suggest that quality control in manufacturing may need to be improved. Sometimes insulators are in service for many years before faults reveal themselves – thereby making the reasoned cause very difficult to prove. These problems have puzzling features, which have stimulated much research, but aspects still warrant further study.

6.4.1.2   Glass shattering under DC energisation

When insulator designs that have low failure rates on AC lines were installed on DC lines, the shattering of the glass shell occurred much more frequently than was anticipated. An obvious difference between AC and DC energisation is the unidirectional flow of ion current in the latter. For soda-lime glass, ionic conductivity depends mainly on temperature, but it is also a function of the amount and the type of the alkali that is present in the silica network. Further, impurity inclusions can cause electric stress enhancement in the glass – as they often do in other dielectrics – thereby leading to a concentration of the ions. Although various mechanisms, based on these phenomena, have been proposed for causing this high shatter rate under DC enegisation, none has been universally accepted. However, the technological remedial approach that has been shown to be highly successful is to improve the quality of both the raw material and the manufacturing process. Such improvements have achieved a high resistivity toughened glass [8]; such comparative resistivity values are shown as a function of temperature in Figure 6.2.

An aspect that is worthy of a brief mention is that of thermal runaway. It is readily apparent from Figure 6.2 that glass is more conducting as its temperature increases. Therefore, if this temperature increase is caused by ionic conduction, a positive feedback situation arises – resulting in a continuously increasing ionic current for a

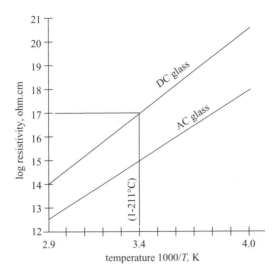

Figure 6.2   Resistivity–temperature relationship for glass insulators developed for AC and DC lines [8]

constant electric field. Laboratory experiments have shown that this can occur with a somewhat simple unit – constructed from impure glass. When, however, such an insulator is in a string the situation may be different. If the other insulators are of very high resistivity glass, these units control the current. That is, the electric field within the more conducting glass is not constant but decreases as the glass becomes hotter – thereby resulting in a negative feedback condition. Obviously, if all the units in the string were identical and of impure glass, then positive feedback would result. Although it is not very likely that all such units would be identical they could, nevertheless, be near enough for this latter situation to arise in some cases. This may be the explanation of the glass shattering on the AC-type cap and pin insulators when they were energised on the DC lines.

6.4.1.3   Corrosion under DC energisation

Electrolytic corrosion of insulators on DC lines is a well researched problem [9]. The positive electrode dissolves gradually into the electrolyte, produced by wetted pollution, due to the flow of leakage current. For ferrous material, such corrosion occurs through the following ionising reaction:

$$Fe \rightarrow Fe^{2+} + 2e^- \tag{6.1}$$

The loss of weight $W(z)$ of the positive electrode due to a flow of current $I(z)$ for duration T was found to follow approximately Faraday's Law:

$$W(z) = I(z)TZ \tag{6.2}$$

where Z is the electrochemical equivalent (i.e. the mass of ion taken into the solution by a current of one ampere flowing for one second).

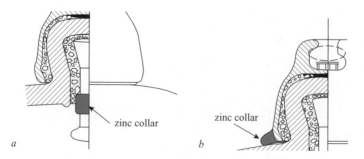

Figure 6.3   *Use of sacrificial zinc to protect the metal parts of cap and pin insulators for DC applications*

  a on the pin
  b on the cap

From data reported for DC lines in Japan, the annual quantity of charge (i.e. $I(z)T$) involved in such a process has been estimated as 200–300 C for inland regions and 500 C for coastal ones. For a very severe region in New Zealand, the corresponding value was 1500–3000 C.

From the practical viewpoint, the pin is the component that suffers the most. When its metal is substantially reduced, the insulator string can no longer support the conductor.

The method successfully developed to provide protection of the steel pin – in addition to normal galvanising – was to use a zinc sleeve as a sacrificial electrode [9]. This sleeve is fused around the pin in the position shown in Figure 6.3*a*. A corresponding method of protecting the cap [8] – by bonding a pure zinc collar around its rim – is shown in Figure 6.3*b*.

6.4.1.4   Radial cracks in porcelain under AC energisation

During the refurbishment of some UK lines, a considerable number of 400 kV suspension insulators were found to be defective [10]. In one of the two major problems, radial cracks had developed in the part of the shell external to the head region. Most of these faulty units were at the high voltage end of the string. Subsequent investigations showed that the corrosion of the part of the pin embedded within the mortar – for about 10 mm in from the air surface, see Figure 6.4 – was the cause [11]. The corrosion products resulted in an increase in the radial dimension of the pin, thereby causing a tensile hoop stress to develop in the porcelain. Interestingly, the corrosion of the pin external to the mortar was much less than that found for DC insulators.

An assessment of an infra-red method for detecting such radial cracks in situ [12] showed that – under humid conditions – the caps of sound units at the high voltage end of the string were a few degrees warmer than the caps of the defective insulators. Follow-up research revealed that adjacent dry band discharges produced this heating and that similar discharges occurred around the pin cavity. By simulating this latter

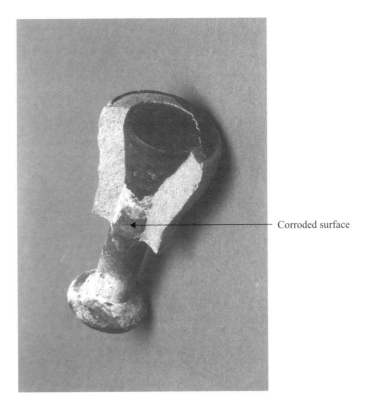

Corroded surface

Figure 6.4   Corroded pin (with some mortar removed)

condition in the laboratory, using a construction in which the pin surface was covered with a collection of electrodes, the current distribution was found to be that shown in Figure 6.5 [unpublished findings of the author].

The comparison of Figures 6.4 and 6.5 strongly suggests that the leakage current flow through the mortar causes, first, corrosion of the galvanised layer and then of the underlying steel. Findings from complementary metallurgical experiments support this conclusion. However, many aspects of this problem remain to be explained properly – including the part played by the stray capacitance current [13] in causing the concentration of the cracked units at the high voltage end of the string.

As zinc corrodes less quickly than does steel, the design of the replacement insulator includes a collar – which is similar to that shown in Figure 6.3*a* for the DC case.

6.4.1.5   Head cracks in porcelain under AC energisation

In contrast to the radial cracks, the second major problem noted on some of the UK lines involved cracks that were hidden beneath the cap. For complete insulators, such

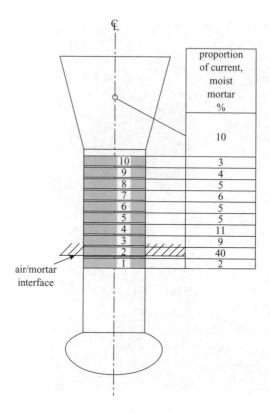

Figure 6.5   Diagrammatic representation of a pin fitted with electrodes, and the corresponding current distribution for a dry-band discharge terminating on the mortar surface adjacent to the porcelain

cracks can only be detected electrically. In the worst case, one third of the units at the high voltage end of the suspension string were defective.

Upon removing the caps, typical examples of the cracks and the puncture path through the porcelain were found to be those shown in Figure 6.6. From breaking open a substantial number of supposedly sound units, a few partially completed puncture paths were found. These varied from being very short to being ones that had nearly developed through the complete thickness of the porcelain.

The findings of subsequent research [14] indicate that incipient defects, that were built into the porcelain at the manufacturing stage, can extend in length under an appreciable voltage while in service. The occurrence of such an excess voltage at the high voltage end of the string is most likely to happen as polluted insulator surfaces start to become moistened – when, for example, $RH > 70$ per cent. Prior to this stage, the electrostatic voltage distribution along the string applies and so – because of stray capacitance – the voltage drop, ΔV, across the high voltage end units is appreciably greater – e.g. by factor 2 – than that across the units in the middle of the string. For a resistance $R(u)$ of the surface of a unit, the power dissipated in that surface

puncture path

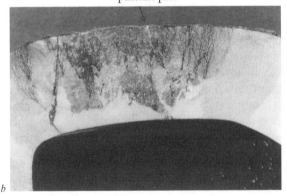

Figure 6.6   *Cracks and puncture path in the head of the porcelain shell*

 a view of top of the head
 b view of the face of the cracked porcelain

is $\Delta V^2/R(u)$. Therefore, if all the units in the string have the same value of $R(u)$ at the start of this transition from the electrostatic to the resistive voltage distribution, the power available to effect counteracting drying is greatest for the high voltage end units. As a consequence, the value of $R(u)$ for the high voltage end insulators becomes larger than that of the others in the string and so the resistive voltage distribution readjusts to enhance the value of ΔV across these units. Sometimes the increase in ΔV^2 is substantial and so positive feedback applies for a while. By this process, the value of ΔV across a high voltage end unit continues to increase until it becomes a large portion of the total voltage across the string. However, this situation may only be short lived because moistening often becomes sufficiently large to swamp this differential drying process.

Generally, head cracks are more disruptive than radial ones. Should an insulator string become so severely polluted that a flashover occurs, the flow of such a large current through a crack in the head of the porcelain shell will cause a very large rise in pressure. Then the cap usually splits. Occasionally it does not; instead, the whole contents of the insulator are ejected like shot from a canon. In this situation, it seems very likely that the current path was essentially along the axis of the insulator. In both cases, the insulator string breaks. In contrast, an arc within a radial crack in the shell outside the cap can very rapidly escape into free air.

It would thus seem wise that, from time to time, sample units – especially those from the high voltage end of the string – should be removed and subjected to a high voltage test, particularly for those lines in highly polluted areas.

6.4.2   Longrods

The obvious way to minimise the quantity of metal along the axial length of the insulator is seen in Figure 6.7 for the longrod design. For the porcelain version, there is the additional advantage that any corrosion products will subject this material to a compressive stress – which it is well able to withstand (see Table 6.1). Of course, the rod section of this design has to carry a tensile load and these diagrams clearly show the advantage of the glass fibre version (again, see Table 6.1) in that the polymeric insulator is much slimmer – and hence much lighter – than the porcelain one.

Both versions of this design can have alternate long and short sheds. Although the leakage path length is obviously less than that of corresponding designs having all sheds as large as possible, it is an optimum arrangement to achieve the benefit of increasing the spacing between the longer sheds. This larger axial spacing is useful when a substantial amount of wetting occurs – e.g. heavy rain, live washing – because the run-off stream of water has a chance to break up before contact is made with the surface below.

6.4.3   Posts

The multiple cone version of this insulator, Figure 6.8a, is worthy of a brief mention because again it shows an important difference between a glass and a porcelain construction. In some cases, such porcelain posts have developed radial cracks (Figure 6.8b), which drastically reduce the path length of the leakage current and thereby the pollution flashover strength (Figure 6.8c). The reduced amount is approximately equal to the proportion of the post so affected (Figure 6.8d) – as seen by the close correlation between the experimental and theoretical curves (unpublished work of the author). These cracks are probably caused by the volume growth of the mortar that bonds these cones together. Such mortar growth is known to cause radial cracks to develop in the porcelain shells of some cap and pin insulators [15]. The corresponding shattering of glass cones does not seem to occur. This may be due to the much

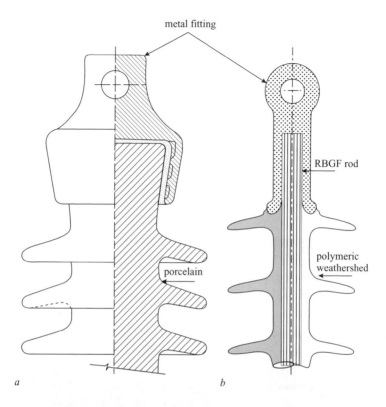

Figure 6.7   *Examples of longrods*

    a   ceramic version (based on [2])
    b   polymeric version (courtesy of Sediver's 'Mechanical application
        guide for composite line post insulators', 1991)

greater tensile strength of toughened glass to that of porcelain. Another difference is that Portland cement is used for the porcelain version whereas Fondu cement is employed for glass ones.

It is worthy of note, that much of this failure phenomenon with the porcelain-type of multiple post is similar to the radial cracking in porcelain cap and pin insulators.

As a consequence of the cracking of these porcelain cones, the solid core porcelain post was developed. As per the longrod line insulator, this design avoids putting porcelain into a tensile mode should there be any metal corrosion, or cement, growth.

A design of a polymeric post is shown in Figure 6.9 [16]. The general arrangement is depicted in Figure 6.9*a*, and the ability to include a fibre optic link [1] within such a construction is illustrated in Figure 6.9*b*.

A particular advantage of having a RBGC core lies both with its high mechanical strength and its flexibility. Both of these features enable such posts to withstand more readily than their porcelain equivalent the short circuit forces that occur under

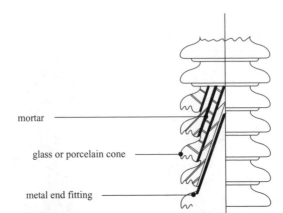

mortar

glass or porcelain cone

metal end fitting

a

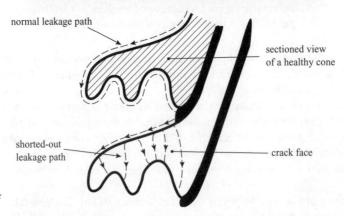

b

normal leakage path

sectioned view
of a healthy cone

shorted-out
leakage path

crack face

c

Figure 6.8   *Multiple cone post*
   a   general construction
   b   typical example of a radial crack in a porcelain cone
   c   probable conducting path across the face of the crack

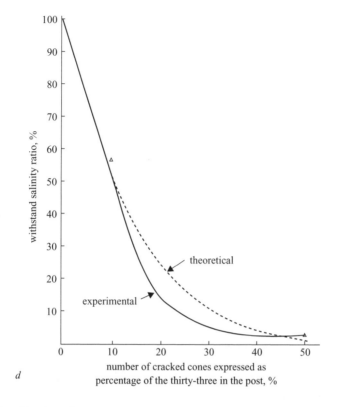

d

Figure 6.8   Continued

 d withstand salinity in artificial salt fog of a 400 kV post comprising
      various numbers of cracked cones; expressed as percentage ratio
      of that of a completely sound post

high current fault conditions and the violent vibrations that happen under seismic
activity.

6.4.4   Barrels

Such hollow insulators have a number of important applications; e.g. weather housing
of bushings, surge arrester enclosures, circuit breaker interrupter heads, cable sealing
ends. Sometimes they are pressurised to greater than 50 kPa gauge. The various
standards relating to porcelain, annealed glass and toughened glass – for use with
voltages greater than 1000 V – have, in recent years, been incorporated into a single
document: that is, IEC 62155.

   The design that is now attracting much interest is the polymeric unit, especially
ones having silicone rubber sheds bonded onto a glass fibre tube – for example,
see Figure 6.10. Special features are that such hollow insulators can be made into

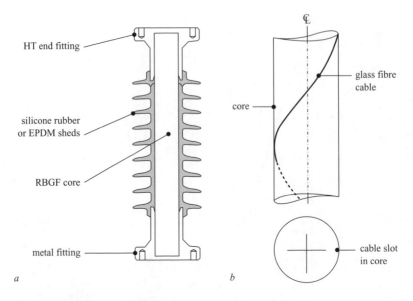

HT end fitting

glass fibre
cable

core

silicone rubber
or EPDM sheds

RBGF core

metal fitting

cable slot
in core

a

b

Figure 6.9   Example of a polymeric post
    a   general arrangement
    b   inclusion of a light link

long lengths with large diameters. Also, such insulators have an excellent pollution
flashover performance.

The resin bonded glass fibre tube is constructed on a mandrel, as per standard
practice. In one design of bushing manufacture [16] the silicone rubber sheds are
applied to this tube. The end fittings are attached while the silicone rubber is still
pliable and this complete assembly is then vulcanised in an autoclave. Final machining
is carried out when this assembly has cooled down. Such hollow core insulators can
currently be constructed with a maximum internal diameter of 600 mm and an overall
maximum length of 6 m. The significance of this additional size over that of the
corresponding porcelain insulator will be discussed briefly later with regard to circuit
breaker interrupter heads (section 6.6.3).

6.5   Flashover mechanisms

Relative to the breakdown of the corresponding length of an unbounded air gap, the
flashover of a polluted insulator is usually a slow process that takes place at a fairly
low value of average electric field.

6.5.1   Surface wettability

A good measure of the wetting of an insulating material is the contact angle, θ, of a
sessile water drop residing on its surface (Figure 6.11). This angle is controlled by

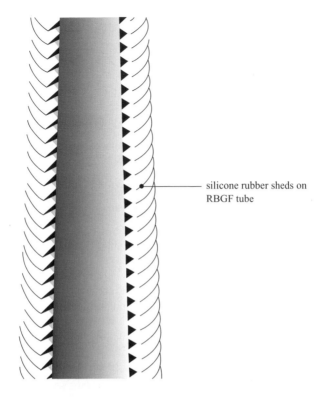

silicone rubber sheds on
RBGF tube

Figure 6.10   Example of a polymeric hollow core insulator

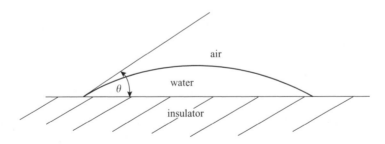

air

water

θ

insulator

Figure 6.11   Sessile water drop on an insulator surface

three coefficients of interfacial tension, λ, as related by the Young–Dupre equation:

$$\lambda_{wa} \cos\theta = \lambda_{ia} - \lambda_{iw} \tag{6.3}$$

where subscripts a, i and w refer to air, insulator and water, respectively.

For a clean and new surface, θ is about 30° for ceramic and about 100° for silicone rubber. However, should this surface become coated with a layer of pollution – albeit only dust – then this angle becomes essentially zero. The difference between these

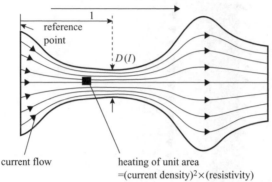

electric field = (current density) × (resistivity)

reference point

$D(I)$

current flow

heating of unit area = (current density)^2 × (resistivity)

Figure 6.12   Developed surface of an arbitrary hydrophilic insulator that is polluted and wetted

two types of material is that, for such a polluted surface, θ always stays zero for ceramic (i.e. the hydrophilic case) but it will, in time, return to a substantial value for silicone rubber (i.e. the hydrophobic one).

6.5.2   Hydrophilic case

When a film of polluted water covers the complete surface (see Figure 6.12) of an energised insulator, this electrolyte conducts a leakage current. Joule heating causes some of the water to evaporate. The corresponding power density is $j^2\rho$, where j is the surface current density and ρ is the surface resistivity. In a region where this power density is a maximum or the thickness of the water is less than a critical value, ρ increases with time and the electric field E in the electrolyte at this point also increases; that is, because $E = j\rho$. The electric field in the air immediately above this point has approximately the same value. Once the ionisation level in this air is reached, a discharge occurs. Such a region is usually referred to as a dry band, but strictly speaking it need not be completely dry. This discharge can be a spark, a glow or an arc – depending upon its duration and the magnitude of the current it conducts. However, if a pollution flashover is to take place it has to be an arc whose roots propagate over the surface of the electrolyte and along the length of the insulator. On a vertically mounted string of cap and pin insulators, such discharges originate around the pin cavity – where j is a maximum – and around the cap – where ρ is often high due to the washing effect of rain.

From Kirchoff's law, the voltage V and current I relationship of an arc plus electrolyte on such an insulator surface (Figure 6.13) can be expressed as:

$$V = IR + V(e) + If(\rho)f(X) \tag{6.4}$$

R is the resistance of the arc; $V(e)$ is the sum of the voltage drops at the junctions between the arc and electrolyte (i.e. anode phenomenon – about 200 volts and cathode

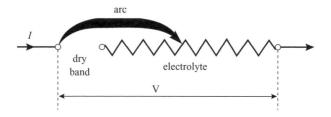

Figure 6.13   *Equivalent circuit for an arc spanning a dry band plus a portion of the neighbouring electrolyte*

one – about 700 volts); $f(\rho)$ is a function of ρ, the value of which varies along the length of the electrolyte – principally because of enhanced heating around the arc's root. Note: ρ decreases as the electrolyte's temperature increases until saturation of the solution occurs; thereafter, it increases again as evaporation of the water continues further. $f(X)$ is a function of the arc length, X, the sum of the electrolyte's length plus that of the dry band – together, designated L – and the effective size of the arc's root on the electrolyte's surface; obviously, this size can have two values – one for the anode, and the other for the cathode.

From Suit's arc equation [17]:

$$R = I^{-(1+a)} \int_0^X A(x)\, dx \qquad (6.5)$$

where a is a constant and $A(x)$ has a value at position x from the beginning of the arc that depends upon the proximity of that part of the arc to the electrolyte's surface, the amount of water on that surface and the extent that thermalisation has taken place within the arc. To a first approximation, A can be considered to have an effective value – i.e. a weighted average – that applies for the whole length of the arc.

Because the resistance of the arc decreases as its current increases, its length extends when in so doing its current increases. That is, the criterion for the propagation of the arc's root to effect this increase in length can be stated as:

$$\frac{dI}{dX} > 0 \qquad (6.6)$$

For the AC case, the flashover process usually occurs over a few half cycles and so the reignition of the arc after current zero needs to be taken into account. This constraint can be specified [18] as a reignition voltage $V(r)$ of the form:

$$V(r) = NXI^{-n} \qquad (6.7)$$

Again to a first approximation, effective values for the constants n and N relating to the surface and shape of a practical insulator – rather than a gap in free air – should be used.

The various mathematical models that have been adopted to solve these equations for simple insulator shapes – i.e. the function $f(X)$ is then simple – have been well reviewed by Rizk [19].

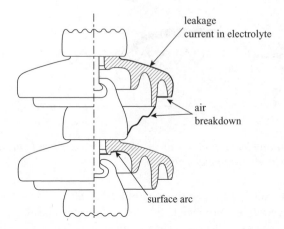

Figure 6.14   *Illustration of how the breakdown of various air gaps on a practical insulator can lead to the shortening of an arc that is starting to flash over that surface*

For the more complex insulator shapes, e.g. like the anti-fog design shown in Figure 6.1, the first additional aspect that needs to be taken into account is the form factor, F, of its surface. F can be readily obtained by considering the developed view, Figure 6.12, of that surface to become:

$$F = \pi^{-1} \int_0^L \frac{dl}{D(l)} \tag{6.8}$$

In this equation, $D(l)$ is the diameter of the insulator at leakage position l from the reference point and L is now the total leakage path of the insulator.

The second additional aspect that needs to be taken into account for practical situations is that the arc does not necessarily hug the surface of the insulator over its entire length – an example is illustrated in Figure 6.14. The sparkover of the air gaps between sheds and the ribs of a shed can effectively shorten the arc's length, and buoyancy forces can result in either its shortening or its lengthening, depending upon the part of the insulator's surface under consideration. Mathematically, this can be represented by stating that the arc's length $X(a)$ is related to the corresponding spanned length $X(i)$ of the insulator by:

$$X(a) = kX(i) \tag{6.9}$$

where $k < 1$ for the shortening case and $k > 1$ for the lengthening one.

Experience indicates that once the critical conditions have been reached, flashover is likely to occur even though an appreciable length of the insulator remains to be spanned by the arc. Therefore, up to this stage the resistance of the electrolyte around the arc root is much smaller than that of the rest of the electrolyte's length that still remains to be spanned and so – for all practical purposes – ρ can be taken to have a constant value and $f(X)$ can be considered to be of the form:

$$f(X) = F\left(1 - \frac{X}{L}\right) \tag{6.10}$$

The solution to obtain the flashover criterion is straightforward for the DC case. The value of the flashover voltage $V(\text{FO, DC})$ is determined as being:

$$V(\text{FO, DC}) \approx (\rho F)^{a/1+a}(kAL)^{1/1+a} \tag{6.11}$$

In contrast, the rigorous solution for the AC case is quite involved [19]. However, a method of obtaining a simple relationship for an approximate criterion is to make the following assumptions [20]. The first one is that $n = a$; the physical argument being that both constants are strongly related to the temperature of the arc. The second assumption is to make V equal to $V(r)$ in Equation (6.4), thereby obtaining I in terms of X; that is, the least value of the supply voltage is that required to cause reignition. The third assumption, and the one not so readily justified, is that the critical value of X is about $2L/3$. This value is that of the DC case when the electrolyte is a long rectangular shape. Furthermore, for the AC case, it is consistent with the value obtained by making a more rigorous analysis. With these assumptions, the flashover voltage for the AC case becomes:

$$V(\text{FO, AC}) \approx 0.7N(\rho F/2(N - A))^{n/1+n}(kL)^{1/1+n} \tag{6.12}$$

In this equation, $V(\text{FO})$ represents a value somewhat less – maybe only slightly so – than that of the peak of the waveform.

By comparing the calculated values with the measured ones, the effective quantities for the various constants can be determined. Those for a and n can be obtained from simple bench-top experiments. Data obtained in artificial pollution tests made on the simpler designs of post and longrod insulators can be used to find the quantities of A and N. The corresponding values for k are best derived from the testing of insulators that have the anti-fog profile. Ballpark values are $a = n = 0.5$, $A = 10$ volts per mm square root amps and $N = 100$ volts per mm square root amps. For anti-fog cap and pin insulators, the value for k is found [21] to vary between 0.4 and 1; it depends upon the pollution severity – smaller values relate to higher resistivities – and whether AC or DC energised – smaller values apply to the DC case.

An examination of Equations (6.11) and (6.12), using the parameter values quoted above, readily shows that $V(\text{FO, DC})$ is much lower than $V(\text{FO, AC})$ for the same level of pollution severity. That is, the lack of the reignition process makes DC insulators substantially longer than those for the corresponding AC case if their flashover is to be avoided.

For the AC case involving flashover in less than one half cycle – i.e. on a short insulator – then, obviously, reignition does not apply and so the process is more akin to that of the DC situation. For this reason, linearity between AC flashover voltage and insulator length does not apply over the complete range of length. This means that reliable information about the pollution flashover performance under AC energisation is best obtained by testing fullscale insulators at, or close to, the relevant operating voltage. From the theoretical viewpoint, this restriction does not apply to the DC case – which is an aspect that needs to be considered further by experimentation.

Because such pollution flashover tests on high voltage insulators are both costly and time consuming, the ability to calculate – with a reasonable degree of accuracy – the flashover voltage pollution severity relationship would be very beneficial. Therefore, further theoretical studies directed towards this end are highly warranted.

6.5.3   Hydrophobic case

When the water drops remain discrete, which is the distinctive feature of this case, the flashover path involves the intervening air gaps [22]. Here the water drop plays two major roles. In one, it acts as a stress enhancer, due to its much larger permittivity than that of both the surrounding air and the insulating material on which it resides. In the other, it reduces the length of the air path across the dry part of the insulator. In this second role, the flashover path is through the drop when the water is highly conducting or across its surface when it is not so. A contributing feature is the deformability of the drop under the influence of the electric field, such that stress enhancement becomes greater and the air path between the drops in the direction of the electric field becomes smaller. Sometimes the drops can coalesce to form rivulets. Also, the migration of soluble pollutants from the seemingly dry part of the surface to the drops is another important factor [23].

Consequently, the flashover mechanism of a hydrophobic surface is appreciably different from that of a hydrophilic one. For one thing, it develops faster. More important from the practical viewpoint, the average electric stress is substantially greater than that of the corresponding hydrophilic surface – but, unfortunately, no authoritative method is yet available to enable this difference to be calculated.

6.5.4   Ice and snow conditions

The processes responsible for the flashover of ice-covered insulators and those on which snow has settled are not yet fully understood. Nonetheless, the following observations apply.

A flashover caused by ice accretions usually occurs when a water film forms on that ice during its thawing stage. This water film has a substantially lower resistance than that part of the insulator on which ice was either not present or has fallen away. Most of the voltage is then impressed across the latter – i.e. it is somewhat similar to the dry band situation described earlier. Depending upon the relative lengths of these two parts, the arcs that are produced across the more highly stressed sections can extend to produce a total electrical failure. The probability for such a flashover to happen may be significantly increased when fog, drizzle or rain is also present at the critical moment. Pollution on the insulator's surface prior to the formation of this ice also reduces the insulator's strength.

The quantity of snow that covers an insulator is usually the greatest when it is horizontally mounted. The density of this snow and the resistance of the melted water are the main parameters that affect the value of the leakage current. The parts of the snow that have a high current density are likely to melt more quickly due to Joule heating. This snow may then drop away from the insulator thereby leading to a non-uniform voltage distribution along its length. Depending on the resistance of the remaining snow and the length of the insulator it does not cover, arcs may bridge the latter and thereby provide the precursors of a possible flashover event.

6.6 Electrical characteristics

Although the largest population of insulators that are subjected to severe pollution is – by far – on operational lines, the published information concerning their flashover performance is very limited. A reason is, possibly, that the Electricity Supply Industry does not wish to be embarrassed when their failure statistics are higher than they had designed for in the quest to provide a very reliable supply at minimum cost. However, some information does become known via networking sources – especially within technical meetings, such as CIGRE Working Group 33.04.13 – thereby aiding research thinking.

The research findings on high voltage insulators have been obtained using natural pollution testing stations and artificial pollution test chambers. Testing the same insulator using both of these facilities provides the most thorough investigation. The severity of the pollution can be expressed in a number of ways. The main ones are:

(i)   the density of the pollutants on the insulator surface
(ii)   the surface conductivity of those pollutants when adequately wetted
(iii) the quantity of the polluting medium used in the relevant artificial test.

An example of the latter is the amount of sodium chloride per unit volume of water used to produce a salt fog.

For many years, only the quantity of the soluble component of the pollution that settled on an insulator's surface in service was considered important [4], it often being expressed as the equivalent salt deposit density (ESDD) – where the salt is sodium chloride. However, it has recently been recommended [3] that the corresponding non-soluble deposit density (NSDD) should also be measured, because this non-soluble component helps to retain surface water. With silicone rubber insulator, the non-soluble material plays another important part in that it temporarily masks the water-repellent properties of the rubber. Both the type and the density of this deposit need to be known to assess the consequences of having such an insulator so covered.

The advantage of the natural pollution investigations is that they fully replicate the service conditions. They show both the extent to which an insulator collects and retains the pollution from a certain environment plus the ability of this energised insulator to withstand this pollution deposit when wetted under fog, mist and, drizzle. The disadvantages are that they take some years to conduct, they are costly when comprehensively done and the results apply directly only to that type and severity of pollution.

The attractions of the artificial pollution test are:

(i)   the severity of the pollution can be varied over a wide range
(ii)   it is performed fairly quickly

(iii)   it is cheaper than the making of the corresponding test conducted under natural conditions.

However, with regard to this last point, the test source is costly because of the need to provide a current of some amperes – often more than ten – with negligible voltage regulation. The corresponding principal disadvantage from the practical viewpoint is that the test shows only the ability of the insulator to cope electrically with a controlled amount of wetted pollution. Consequently, care must be taken in interpreting the findings. For example, if the flashover voltage pollution severity curves for an anti-fog type of insulator and a corresponding aerodynamic one are shown in the same diagram, then a cursory inspection suggests that the pollution flashover of the latter is much inferior to that of the former. The more discerning observer realises, of course, that – from the practical viewpoint – these two curves should be substantially separated along the pollution severity axis to take account of the much smaller amount of pollutants that the aerodynamic insulator catches in service.

Major research programmes – most employing artificial pollution but a substantial number using natural pollution – have involved the testing of more than 120 types of ceramic insulator and nearly 30 types of polymeric ones. A comprehensive record of the findings is provided in the CIGRE review [2] and a summary of the main features is provided in the rest of this section.

6.6.1   Performance under natural pollution

Under natural pollution, the flashover statistics of a test insulator are obtained over a number of years relative to those of a control one. Also, the flashover frequency can be determined by slightly changing the length of the insulator for the same test voltage. Some good examples for marine pollution have been obtained from the results obtained at the Brighton Insulator Testing Station (BITS) on the south coast of England. Upon the closure of this long operating station, much of the test equipment was relocated about 55 miles to the east at Dungeness (to become DITS). The practice adopted by the operators of BITS – that is, Central Electricity Research Laboratories (CERL) of the Central Electricity Generating Board – was to publish many of its findings [1]. In contrast, the tests conducted by the current owners of DITS – National Grid Company – were made under contract conditions, thereby making their general dissemination much more difficult. Regrettably this site is currently mothballed awaiting the emergence of a new owner.

The results from these test stations can be expressed in the form of a figure of merit (FOM). This FOM is defined as the length, L_c of the control insulator divided by the length L_t of the test insulator for flashovers to occur across both of them under the same voltage and environmental conditions. That is:

$$\text{FOM} = \frac{L_c}{L_t} \tag{6.13}$$

Note: the length in this method is the axial value between fixing points. The insulator with the largest FOM value has the best pollution flashover performance. That is, the axial length of this insulator is equivalent to that of the control one divided by FOM.

Table 6.2   *Figure of merit (FOM) values of some porcelain cap and pin insulators when AC energised at BITS [2]*

Insulator type	Orientation*	FOM
Reference	vertical	1.00
Standard	horizontal	0.79
Anti-fog	horizontal	1.35
Anti-fog	angled	1.11

* horizontal is about 75° to the vertical; angled is about 45° to the vertical

Table 6.3   *Figure of merit (FOM) of some polymeric insulators; AC energisation at BITS [2]*

Insulator type	Orientation*	FOM
Silicone rubber	horizontal	>1.53
Silicone rubber	vertical	>1.53
EPDM	horizontal	1.28
EPDM	vertical	1.12
EPR	horizontal	1.19
EPR	vertical	1.17

* horizontal is about 75° to the vertical

Some values of FOM for strings of cap and pin insulators under various orientations that had been tested at 275 kV and 400 kV are shown in Table 6.2. These are compared relative to that for a vertical string of reference insulators – known as CERL Reference A [2] – which is the large anti-fog design. Table 6.3 gives corresponding FOM values for polymeric insulators tested at 550 kV.

The larger FOM value for a horizontally mounted string of cap and pin insulators, relative to that of a corresponding vertically mounted one, is reasoned to be due to the beneficial effect of rain. That is, with the former both the pin side as well as the cap side are well washed. This finding at BITS is supported from the general performance of the insulators on UK overhead lines. For example, the total leakage path of the tension string on a 400 kV line is about 30 per cent less than that of the suspension string in the same location. Further proof became obvious many years ago when the coastal region of Kent was subjected to about a week of strong dry wind off the North Sea. After this sustained build-up of salt on the insulators, the wind then changed direction to become a prevailing south westerly involving a little mist. Some flashovers involving these tension strings, but not the suspension ones, then occurred. Fortunately, this south westerly wind also soon brought in rain so this problem was fairly short lived.

Of special note: the FOM for the silicone rubber insulators can be expressed only as being greater than 1.53 because no flashover of this type of insulator occurred during the four years of testing in this severe marine environment. In this last respect, the pollution severity of BITS has been assessed as being equivalent to a salinity of 60 kg/m^3 in an artificial salt fog test. This excellent performance of the silicone rubber type is now being confirmed in service. For example, South Africa now has no pollution flashovers [24, 25] on previously troublesome lines (they were then insulated with either ceramic or epoxy resin insulators). Further, China now has about 600 000 silicone rubber insulators in service and, likewise, their pollution flashover performance is reported as being excellent [26].

A further use of the FOM concept is that it enables the flashover statistics at a site to be determined even though the data come from different insulator types. That is, the axial flashover stress of a test insulator is normalised to that of the control insulator by dividing its stress by FOM. By this technique, the statistics of the flashover events at BITS are shown in Figure 6.15 as a cumulative frequency versus normalised axial flashover stress E_n.

The relationship shown in Figure 6.15 can be expressed in terms of the number of flashovers per year $N(FO)$ of this equivalent control insulator. That is:

$$N(FO) = K \left[\frac{E_n - E_o}{E_o} \right]^m \tag{6.14}$$

For an insulator type of similar profile to that shown in Figure 6.1b and of overall diameter of 394 mm, the constants K, m and E_o have values of 16.1, 2.1 and 99 kV/m (based on system voltage), respectively.

This equation indicates that the distribution of flashover events is truncated with respect to axial stress. That is, when E_n is equal to or less than E_o there should be no flashover at this site. This is not strictly true, however, because E_o is the mean

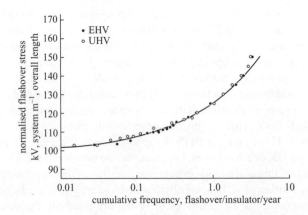

Figure 6.15   Cumulative frequency distribution of 275 kV and 400 kV ceramic insulators, at normalised flashover axial stress, for a test period of about eight years at BITS. (Courtesy of CIGRE's paper 33-01, published in 1982)

value to which must be related a standard deviation. Unfortunately, the data obtained at BITS were not sufficient to determine this standard deviation with an acceptable degree of accuracy. Consequently, this is an area that warrants further research.

6.6.2 *Performance under artificial pollution*

The two types of test standardised by IEC [4] are the salt fog method and the solid layer method; the latter now often referred to as the clean fog test because of the technique used to wet the pollution layer. The former simulates coastal pollution where a thin conductive layer formed by the sea salt covers the insulator's surface. In practice, this layer usually contains little – if any – insoluble matter. The same test also applies to industrial pollution – principally gases that result in the formation of acids – when there is essentially no inert material present on the surface. In contrast, the solid layer (clean fog) method applies when there is an appreciable amount of non-soluble pollution in addition to that forming ions when wetted.

The current versions of these two methods – IEC 507 and IEC 1245 for AC and DC energisations, respectively – were developed for the testing of ceramic insulators. The modification of these tests so as to apply also to polymeric insulators has been considered by CIGRE WG 33. 04 .13. Unfortunately, there is as yet no internationally accepted method to simulate snow and ice although some tests have been developed for research purposes. The development of a test that IEC will ultimately recommend is no mean task. For example, the first one to be produced – the salt fog method – involved collaborative research by the European utilities of CEGB/EdeF/ENEL – lasting many years (perhaps as many as ten!). There are some other noteworthy methods that are currently of non-standardised status; e.g. the dry salt layer (DSL) [2].

The major difficulty with the testing of polymeric insulators – especially those of the silicone rubber variety – is the hydrophobic aspect. The actual insulators tested at BITS were also subjected to the salt fog test. It was then realised that the ranking in performance of the four types of insulator referred to in Table 6.3, was different in this artificial pollution [27] from that found under natural pollution conditions. A number of reasons for this difference have been suggested but, as yet, no explanation seems certain. The temporary loss and then the subsequent recovery of hydrophobicity for the silicone rubber material are of particular concern. Because the dynamic nature of this material has a very large influence on the pollution flashover performance, there is currently some uncertainty that the full potential of silicone rubber for the solving of a number of important problems can be realised. However, some useful results obtained using a solid layer method have been reported [3].

The extent of the change in the flashover strength in a salt fog test of a horizontally mounted porcelain insulator that had been coated with silicone rubber is exemplified in Table 6.4. These results show that over a period involving 11 attempts to obtain a sensible average value of the flashover voltage, FOV, the individual values progressively decreased by factor two. After resting this insulator unenergised for twenty-four hours the initial strength was fully recovered. Obviously, by this procedure the lowest value of the withstand voltage, WSV, could be obtained. However, this is a very pessimistic value to adopt for dimensioning purposes as it came about by having had repeated flashovers – which, hopefully, would never happen in practice.

Table 6.4   Salt fog test on a horizontal porcelain insulator coated with silicone rubber

Voltage* kV	Test number										
	1	2	3	4	5	6	7	8	9	10	11
68										WS	WS
71								WS	FO	FO	FO
74								WS			
77							FO	FO			
80	WS					WS	WS				
83	WS					FO	FO				
86	WS		WS	FO							
89	WS		WS								
92	WS	FO	FO								
95	WS										
104	FO										

* each voltage level was maintained for five minutes
FO is flashover
WS is withstand

On the other hand, it is probably not prudent to assume that the upper safe limit is the first flashover value because constant energisation with severe pollution and long lasting surface discharges may result in the temporary loss of some hydrophobicity. Consequently, much research is required to establish what the meaningful upper value should be.

The situation with the clean fog test may be less serious – but, as yet, this is not fully proven as, however, some aspects of the procedure still remain to be investigated. The main difficulty from the test viewpoint – and one that may also be of concern in some practical situations – is that immediately after a hydrophobic silicone rubber surface is covered with pollution its outer surface is then hydrophilic. Although it is known that water repellent properties are transferred from the silicone rubber to the adhering pollutants, the rate and extent at which this transfer occurs – and the effect on the pollution flashover strength during this transitional period – have not been fully investigated. An example of how the leakage current across a new silicone rubber insulator decreases as a function of elapsed time after an application of slurry – comprising kaolin, sodium chloride, a wetting agent and tap water – is shown in Figure 6.16. In these tests [28], the dry insulator was placed in a clean fog chamber after this elapsed time. Then it was energised and a few minutes later the mist was applied; the contact angle of a water drop placed on the surface of the polluted insulator was measured prior to the start of these tests. These results clearly indicate that the transfer of hydrophobicity from the silicone rubber to the outer layer of the pollution, as manifested by the increase in this contact angle and the decrease in the leakage current, can take an appreciable time to occur. Also, it is very dependent upon the

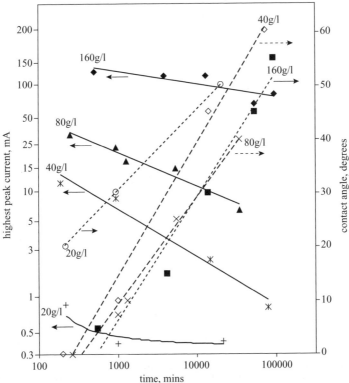

Figure 6.16   *Peak current and contact angle for a silicone rubber insulator as a function of elapsed time from pollution application, at four concentrations of kaolin in the polluting slurry*

concentration of the kaolin in the slurry; the corresponding densities of the soluble and non-soluble components of the pollution – as indicated by ESDD and NSDD, respectively – on the insulator's surface are shown in Table 6.5. As a consequence of this phenomenon, a standardised test will have to specify clearly the time that must pass before such a polluted insulator should be subjected to a withstand test. A complicating feature is that the transfer rate of this water repellent property is also a function of the type of the non-soluble pollutant [29].

6.6.2.1   Effect of soluble component

Generally, the critical stress $E(s)$ at flashover of a ceramic insulator, and pollution severity S, can be expressed in the form:

$$E(s) \propto S^{-p} \tag{6.15}$$

Table 6.5   *ESDD and NSDD measurements for silicone rubber insulator used for investigations into hydrophobicity transfer [29]; results presented in Figure 6.16 apply*

Kaolin concentration g/l	ESDD mg/sq.cm	NSDD mg/sq.cm
20	0.2	0.1
40	0.3	0.3
80	0.3	0.5
160	0.4	1.3

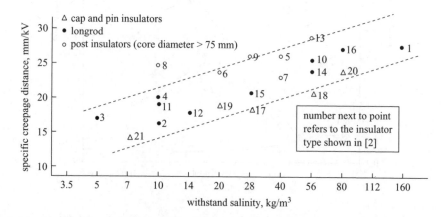

Figure 6.17   *Electric strength of different AC hydrophilic insulators in the salt fog test as a function of pollution severity [2]*

The value of p lies within the range 0.1 to 0.6. Insulators with plain open shedding tend to have the higher values of p, and for many other cases p is often about 0.2. From the mathematical viewpoint, the value of p can be considered as a weighted average of the one for the electrolyte surface (e.g. $p = 0.3$ for brine) and that for the air breakdown ($p = 0$) between parts of the insulator surface.

Some typical relationships of specific creepage length – i.e. the inverse of surface stress – versus pollution severity for various insulator types are shown in Figure 6.17.

6.6.2.2   Effect of non-soluble component

The withstand voltages, WVS, of a cap and pin insulator, a porcelain longrod insulator and of a porcelain post type insulator are shown in Figure 6.18 as a function of NSDD when the non-soluble component is Tonoko material.

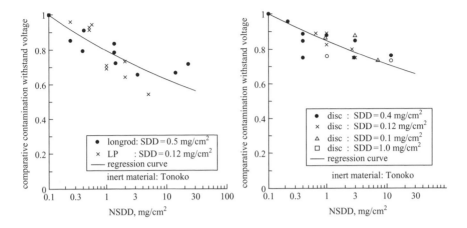

Figure 6.18   *Influence of NSDD on withstand voltage of some hydrophilic insulators [2]*

This relationship can be expressed in the form:

$$WSV \propto (NSDD)^{-0.15} \tag{6.16}$$

which applies for the NSDD range of 0.1–10 mg/cm^2.

6.6.2.3   Influence of average diameter for hydrophilic insulators

Some of the data indicate that at flashover, or withstand, the relationship between specific length *SL*, i.e. the inverse of stress, and average diameter $D(a)$ is of the form:

$$SL \propto D(a)^q \tag{6.17}$$

The best support for Equation (6.17) occurs when the insulators are of the same profile and only the diameter is varied. An excellent example is shown in Figure 6.19 for a DC porcelain housing that had been subjected to a clean fog test with an ESDD of 0.12 mg/cm^2; the value of q is 0.16.

Although this reduction of pollution flashover performance with average diameter is appreciable, field experience indicates that – for cylindrical-type insulators – the pollution collected on the surface of the insulator also decreases as the average diameter increases. A conservative relationship for the relative value of its severity, ESDD(r), – where ESDD(r) = 1 for an average diameter of 115 mm – can be expressed [2] as:

$$ESDD(r) = 2.6D(a)^{-2.1} \tag{6.18}$$

Therefore, for dimensioning an insulator for practical use, both of the aspects covered by Equations (6.17) and (6.18) need to be taken into account.

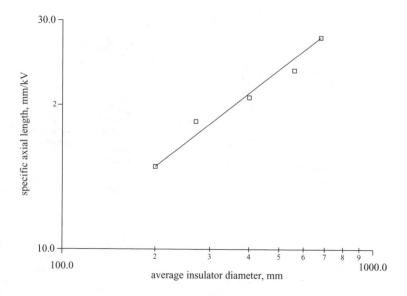

Figure 6.19   Specific axial length versus insulator diameter, for a porcelain insulator under DC energisation in a clean fog test [2]

6.6.3   Interrupter head porcelains

Improvements in the design of the AC circuit breaker, particularly the use of SF_6 in lieu of air, have resulted in the interrupter head being more highly stressed in some cases than has hitherto been the situation. Of particular concern is the head of the horizontally mounted version.

The axial electric stress necessary to cause external flashover of the interrupter porcelain of various designs is shown in Figure 6.20 as a function of the severity of an artificial salt fog. The much inferior performance of the horizontal head (i.e. H values) when compared with that of a corresponding vertically mounted porcelain (i.e. V values) is reasoned to be associated with the way that wetted pollution drains from the insulator's surface [30].

The design of SF_6 circuit breakers has now advanced to the stage that a two-break version is adequate for 400 kV (cf. ten breaks or more for the corresponding air blast breaker). That is, the axial stress on the H1 porcelain for the SF_6 breaker would be about 180 kV r.m.s. per metre during arduous synchronising conditions. The results shown in Figure 6.20 – plus supporting tests conducted at BITS – show that even small amounts of pollution result in flashover for this condition – which is also so even for normal service stressing. The scope for substantially increasing the length of these large diameter porcelains that are suitable for pressurised use is very limited. Two possibilities that warrant study for future development are: (i) porcelain when coated with a hydrophobic material and (ii) other types of barrel construction – e.g. silicone rubber sheds on a resin bonded glass fibre reinforced tube. This latter option is covered in section 6.4.4 and Figure 6.10.

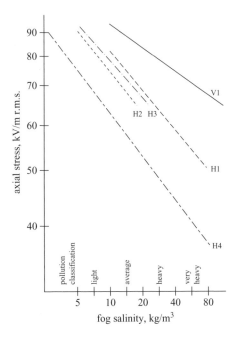

Figure 6.20   *Axial stress to cause flashover against fog salinity for various interrupter porcelains under AC energisation [2]*

6.6.4 *AC versus DC*

Although not many publications contain information on the same – or very similar – insulators that have been subjected to AC and DC energisations for an extensive range of pollution severity, those provided by Electric Power Research Institute (EPRI) for AC [31] and DC [32] are very useful. Data for identical insulators are shown in Figure 6.21, where it is seen that the AC strength (based on peak value) is about 70 per cent greater than that for DC at the same pollution severity condition.

6.6.5 *Transient overvoltages*

A useful comparison is the ratio of the impulse flashover voltage to that for normal stressing. Such information for cap and pin insulators is illustrated in Figure 6.22 as a function of the duration of the impulse waveform (stated as the time for which the voltage is greater than 50 per cent of the peak value). Also included in this diagram are some results for the corresponding temporary AC overvoltage condition (in this case, the duration is the time for which the 50 Hz voltage is applied before flashover occurs).

These and related findings [2] indicate that the pollution flashover problem is associated with the normal stressing condition. Although the magnitude of the transient overvoltage can be considerably greater – especially for the lightning case – than

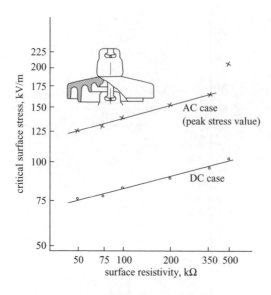

Figure 6.21   Flashover data for identical insulators under AC and DC energisations

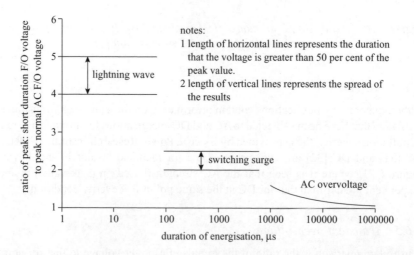

Figure 6.22   Comparison of the short duration stressing for positive polarity impulses with the normal AC stressing strength for a nine-unit string of cap and pin insulators; the surface was very heavily polluted [2]

that of the normal stressing, the slowness of the pollution flashover process is the important feature. That is, the magnitude of this overvoltage has reduced appreciably before the arc can extend by the critical amount over the length of the insulator.

6.6.6 Iced insulator

The flashover voltage of insulators that are covered with ice depends greatly upon the type, the density and the thickness of the ice plus the conductivity of the freezing water, the overall covering of the surface and the number of icicles. Wet grown ice (glaze) with a density of about 0.9 g/cm³ has been found to be more dangerous than other types of atmospheric ice accretions [2]. The maximum withstand stress, expressed per unit leakage length, is shown in Table 6.6 for various insulator types. There are porcelain cap and pin (IEEE and antifog types), EPDM longrod and porcelain post. The influence of freezing water conductivity – as measured at 20°C – on the withstand stress is shown in Figure 6.23. These results are for a six-unit string of standard cap and pin insulators when tested with an air temperature of −12°C. In such tests, the ice thickness, ε, – as recorded on a rotating monitor cylinder – was two centimetres.

Table 6.6 *AC withstand stress for some insulator types covered with ice grown in wet and in dry regions [2]*

Type of ice	Ice density (g/cm³)	Withstand stress (kV/m)*			
		IEEE**	anti-fog	EPDM	post
Glaze with icicles	0.87	70	84	96	90
Rime	<0.3	>148	>146	>168	>197

* maximum withstand voltage for an insulator of 1 m length
** small standard-type insulator

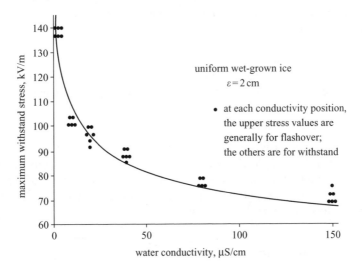

Figure 6.23 *AC withstand stress of a short vertical string of cap and pin insulators as a function of freezing water conductivity [2]*

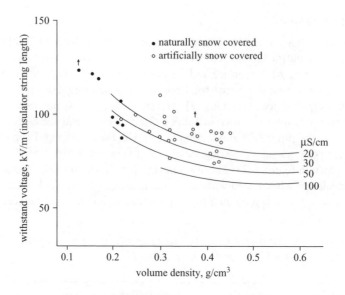

Figure 6.24   AC withstand voltage of a vertical string of porcelain cap and pin insulators as a function of snow density [2]

Some researchers have reported that the withstand voltage of insulators that had their shed spacing completely bridged by artificial icicles was only 60 per cent of that for the case without ice. When the ice thickness is much lower than one millimetre and the insulators are precontaminated, ions from the pollution tend to dominate the total electrical resistivity of the ice. For this latter situation, a flashover stress along the insulator's surface has been found to be as low as 20 kV/m.

When the insulator is energised during the formation stage of the ice, the ice's distribution along the insulator is affected. Often, some sections of the insulator may be free of ice. This condition is particularly the case, in practice, for long insulator strings, and has also been observed in laboratory tests. The cause is reasoned to be due to the heating effect of the surface arcs and/or an increase in the temperature of the air.

6.6.7   Snow on insulators

The AC withstand or flashover voltage decreases with increasing water conductivity from the melted snow, and with increasing snow density up to 0.5 g/cm^3. Thereafter, it remains constant. An example is shown in Figure 6.24 for a string of cap and pin insulators; a few of the results are for a natural snow covering of the insulator but most are for the case where the snow has been artificially produced.

6.7   Selection and dimensioning

Selection and dimensioning is partly dealt with in IEC Publication 815 [4], which has been in existence for more than fifteen years and is based on limited practical experience and some test results. It applies only to conventional porcelain and glass insulators when used for AC applications. Although the approach specified in this document is attractive because of its simplicity, it can lead to poor design.

This IEC 815 concept is based on a specific creepage distance. That is, for a qualitatively defined level of pollution, a corresponding minimum creepage distance is specified for the surface of the glass (or porcelain) along the length of the insulator. The unit of this parameter is millimetres per kilovolt (phase-to-phase) of the maximum value of the system voltage. Just four pollution levels are specified to cover the whole range of typical environments. These pollution levels – and the corresponding specific creepage distances – are called: light, 16 mm/kV; medium, 20 mm/kV; heavy, 25 mm/kV; very heavy, 31 mm/kV. In addition, some limiting geometrical features of the insulator's profile are stated – essentially to avoid too much creepage distance being crammed into a given axial length. Otherwise, this would lead to excessive sparking between sheds and hence short circuiting of the leakage path.

A comprehensive analysis conducted by CIGRE Taskforce 33-13-01 has found the specific creepage distance concept to be wanting in quite a number of aspects. For example, salt fog tests conducted on twenty-seven insulator types of porcelain and glass insulators – comprising different designs and sizes for the cap and pin, longrod insulator, posts and barrels – have found a range, at flashover, from 20.6 to 36.1 mm/kV for the same salinity representing very heavy pollution. This large variation found under very controlled laboratory conditions for vertically mounted insulators, which comply with IEC 815 profile restrictions, is supported by many other results that have been obtained by conducting natural pollution tests. Consequently, it can now be concluded that factors other than creepage path length greatly influence the pollution flashover performance of porcelain and glass insulators under the same environmental and electrical conditions.

In general, the CIGRE review [2] perceives IEC 815 to be weak – and so in need of revision – in the following areas:

(a) performance of polymeric insulators
(b) insulator orientation
(c) extension of applicability to voltages above 525 kV AC system voltage
(d) design for DC applications
(e) insulators with semiconducting glaze
(f) surge arrester housing performance, particularly with reference to polymeric materials
(g) longitudinal breaks in interrupter equipment
(h) radio interference, television interference and audible noise of polluted insulators
(i) effect of altitude
(j) effect of heavy wetting.

From this above list, it can be appreciated that a comprehensive revision of IEC 815 (now called 60815) is a mammoth task. Although it is well under way, the changes proposed to date have yet to be approved by the international vetting authority. Further, many substantial aspects have not yet been dealt with. Therefore, no details are likely be released for quite sometime.

However, the CIGRE document on this subject is nearing completion. It is based on a flow chart similar to that shown in Figure 6.25; the aspects relevant to its various

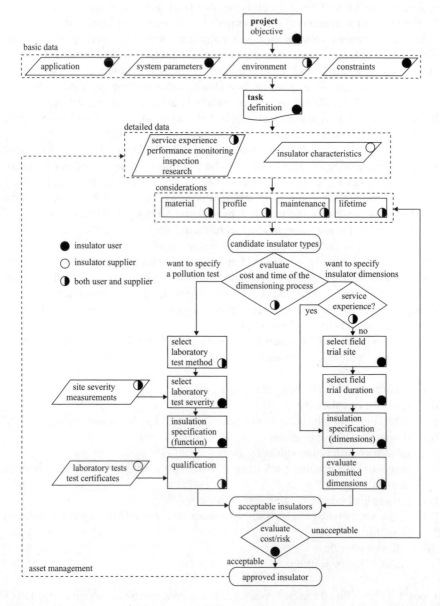

Figure 6.25   *Possible flowchart for the selection and dimensioning of outdoor insulators*

boxes are comprehensively dealt with in the forthcoming Applications Guidelines document [3].

Essentially, this CIGRE approach entails first selecting candidate insulators for the intended installation by using the information contained within their two companion documents [2, 3]. Firstly, the pollution severity to which these insulators will be subjected at this installation has to be determined. Thereafter, the lengths of these candidate insulators are estimated to have an acceptable pollution performance using either relevant service experience or from conducting tests. Obviously, the most thorough testing is carried out in a natural pollution environment that is very similar to that of the intended installation. Unfortunately, this is not always possible due to the required time and the cost of such tests, and so the use of relevant artificial pollution is usually much more likely.

For an AC installation, the aerodynamic catch from air is the principal process by which the insulators are contaminated [1]. In contrast, the electric field around the insulator helps to greatly enhance this catch for a corresponding DC installation.

The determination of the severity of the pollution collected on the surface of a candidate insulator at the site of the installation is the most difficult part of this new procedure. Frequently, a standard insulator will to be used in lieu of candidate insulators for this pollution assessment because, for various reasons – for example, size – it is more convenient to do so. Then, the site pollution severity can be defined as follows.

The site pollution severity (SPS) is the maximum value of either ESDD and NSDD, or leakage current, recorded over an appreciable period of time – i.e. one or more years – on a vertically mounted standard insulator comprising a string of cap and pin units or a porcelain longrod. If rain occurs during this measuring period, the measurements should be repeated at appropriate intervals; SPS is then the largest value recorded during this series of measurements. Note:

(i) even if the largest values of ESDD and NSDD do not occur at the same time, then SPS is – nonetheless – taken as the combination of these largest values
(ii) when there is no natural washing during the measuring period, the maximum value of ESDD and NSDD can be estimated from the plot of deposit density as a function of the logarithm of time.

The pollution deposit density on a certain type of insulator relative to that on a reference insulator (profile similar to that shown in Figure 6.1a, having an overall diameter of 254 mm) is provided in Table 6.7 for vertically installed disc insulators. The corresponding relative value for DC and AC voltages is given in Table 6.8. As horizontally installed insulators – generally – collect less pollution than do vertically mounted ones in the same location, a conservative approach is to use the same information as for the vertical case.

Other aspects of this correlation process are dealt with in the CIGRE document [3].

Prudent managers of high voltage systems will conduct routine inspections, possibly supported with sample tests, of its outdoor insulators over their anticipated long lifetime – e.g. fifty years. It is particularly pertinent that such inspections should be carried out on polymeric insulators to see if any detrimental ageing effects are

Table 6.7   *Ratio of pollution deposit density on an insulator type to that on the standard IEEE type [3]; for the vertical mounting condition*

Insulator type	Desert	Coastal*	Industrial/agricultural	Inland
Aerodynamic	0.4 ± 0.3	0.6 ± 0.3	0.7 ± 0.3	0.6 ± 0.2
Anti-fog	0.8 ± 0.3	0.8 ± 0.3	0.85 ± 0.3	1.08 ± 0.2
Longrod	0.4 ± 0.3	1.6 ± 0.4	0.4 ± 0.3	0.4 ± 0.3

* during rapid pollution build-up (e.g. typhoon or strong wind off the sea)

Table 6.8   *Ratio of pollution deposit density for DC energisation to that for the corresponding AC voltage*

Ratio	Site conditions
1–1.2	natural pollution only, such as sea and desert
1.3–1.9	natural pollution plus industrial pollution whose source is a few kilometres away
2.0–3.0	industrial pollution only, a few kilometres away and considered clean w.r.t. AC voltage

taking place. If so, then this adverse behaviour obviously has to be taken into account when a follow-up selection and dimensioning exercise is carried out.

6.8   Supplements, palliatives and other mitigating measures

Occasionally the pollution flashover performance is not good enough, which may be due to a number of reasons. For example: an unforeseen shortcoming in the original design, an increase in the pollution level – e.g. introduction of new industry into the location – or the strategic importance of the supply facility is raised after its construction. Often, the insulators cannot be changed for longer ones – be it due to cost or installation constraints, such as the size of the support structures – especially when it would reduce the ground clearance of the conductors below a safe level. Fortunately, a whole raft of measures can now be adopted to improve the reliability; some are essentially permanent while others need to be renewed from time to time.

6.8.1   Booster sheds

The booster shed [33] was developed to reduce the incidence of flashover under heavy wetting conditions.

It is constructed from a copolymer of silicone rubber and polyethylene that is radiation cross-linked. Such a polymer has now been shown to have a long and excellent service history in very severe pollution, often saline.

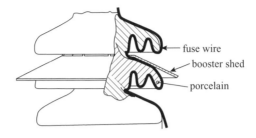

Figure 6.26   An arrangement to initiate a sandwiched arc between a booster shed and a porcelain surface

   The advantage of this supplement is twofold. When it is fitted to every other – or every third – porcelain shed of a vertically mounted insulator, cascading water will break up before such booster sheds are bridged. As a stream of water drops has a much greater resistance than that of a continuous column, the flashover withstand strength is much improved. Second, when an arc extends across the part of the insulator's surface that is covered with this shed, it is more highly cooled – i.e. greater value of $A(x)$ applies in Suit's equation, see section 6.5 – than it would be if the polymer were not present. Therefore, as seen from Equation (6.5), the arc's resistance is correspondingly larger. The value of $A(x)$ increases as the gap between the polymer and the porcelain decreases [33]. However, this gap cannot be made too small; otherwise any propagating discharge would severely damage the adjacent materials. The optimum spacing is about 6 mm. As a consequence of fitting booster sheds, a higher value of pollution severity can be withstood before flashover occurs.
   Should, nonetheless, a flashover take place there is some concern about the resultant damage – especially having seen the consequence of such an arc confined within a cracked head of a cap and pin insulator (see section 6.4.1.1). To investigate this possibility, such a sandwiched arc was initiated using fuse wire, as illustrated in Figure 6.26 (unpublished research of the author). High speed photographs revealed that an arc carrying a current of some kilo amperes rapidly transferred from being so sandwiched to become free to move in the air above the booster shed. The damage was minimal and mainly to the polymeric material. Although extensive blackening of its surface had occurred, it was easily rubbed off to leave essentially sound material. The reason for this fortunate situation lies with the low inertia of the arc, the high buoyancy forces of hot gases and the extensive wriggles of such a sandwiched arc – as witnessed in a complementary experiment involving an arc conducting a current of just one ampere [33].

6.8.2   Shed extender

In contrast to the previous case where the polymeric shed was supported just above the surface of the insulator, the polymeric shed of the extender is bonded to the surface of the shed – see Figure 6.27 – in order to lengthen the leakage path. The adhesive

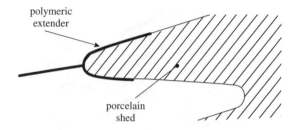

Figure 6.27   A silicone rubber shed extender fitted to a porcelain insulator

used to achieve this bond is a crucial component of this supplement, because surface discharges in any air gaps will damage both the porcelain and the polymer.

Although the maximum increase in the leakage path is obtained by having all these extenders of equal size, there is scope for a long and short shed arrangement to be provided. That is, it helps to cope with polluted cascading water – even though the water run off is far from being in the category of heavy wetting.

Of course, with this construction, there will be a mixture of hydrophobic and hydrophilic surfaces. That is, from the leakage current viewpoint, films and drops of water are now in series. This is another aspect that has so far received little study with regard to the flashover process – and so it becomes a research topic that warrants some investigation.

A fairly recent development [34] that is now coming into service avoids this problem by covering the porcelain with a silicone rubber coating (see sections 6.3 and 6.8.4) after the extender has been added. By this process, a somewhat hybrid hydrophilic–hydrophobic surface has been converted into a fully hydrophobic one.

6.8.3   Shed protector

Should an appreciable part of a porcelain shed be broken, the pollution withstand strength of that insulator is proportionally reduced. Two measures that have been adopted to protect the most vulnerable part of the shed against vandal attack using hard projectiles, such as stones, are shown in Figure 6.28.

In one, Figure 6.28*a*, the complete shed is covered with a polymeric shield – which is essentially a mini booster shed. In the other, Figure 6.28*b*, the circumference of the shed is protected with an impact absorbing polymeric tyre.

6.8.4   Coatings

A long established method to enhance the pollution flashover performance of the porcelain insulator is to coat its surface with a layer of grease, which has two benefits. First, it provides a hydrophobic surface and, second, it encapsulates and/or imparts water repellent properties to the pollutants. The two types of grease are hydrocarbon and silicone. The former is restricted to the temperate zone because of its fairly low sliding temperature, but the latter is used throughout the world. The hydrocarbon version is relatively inexpensive, applied as a substantial layer and can last up to

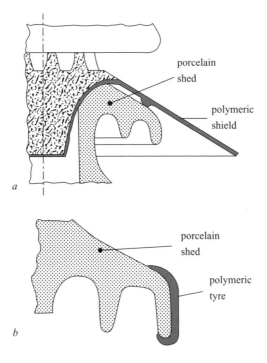

Figure 6.28   *Two examples of shed protectors*
  a   polymeric shield
  b   polymeric tyre

about ten years. Silicone grease is fairly expensive, applied as a thin layer and may have to be replaced after a year. Such greases are normally applied by hand or by using a brush or spray. The disadvantage lies with the removal of dirty grease before a new coating is applied. If this change is not done in good time, trapped electrical discharges that can occur beneath the grease's surface may damage the porcelain. The indication of the need for such a change is the presence of dry band discharges on the surface of the grease.

The good service record of silicone rubber insulators for outdoor applications has stimulated the development of RTV rubber coatings. All known commercially produced variants consist of polydimethylsiloxane polymer, alumina trihydrate – or alternative filler – to increase the tracking and erosion resistance, a catalyst and a cross-linking agent. Such materials are dispersed in either naphtha or trichlorethylene solvent, which acts as the carrier to transfer the RTV rubber to the insulator's surface. End of life is not easy to determine accurately [35]. Service experience with both AC and DC systems has found that the useful life of a RTV coating can vary from less than a year to many years. Basically, when such a coating loses some of its water repellence, it may be washed and/or cleaned using a dry abrasive – such as crushed corncobs or walnut shells – and then recoated provided the existing coating is well bonded to the ceramic surface.

6.8.5   Washing

Because dry pollutants rarely cause leakage current, the obvious way to overcome the flashover problem is to remove them before they reach the critical level – albeit if that means getting them wet to do so!

If the insulator can be deenergised, then washing with tap water is a simple process. However, insulators generally must be kept in service and so live washing – sometimes referred to as hot line washing – has to be undertaken. For this procedure to be conducted without promoting flashover, two conditions are usually essential. The water should be demineralised (resistivity > 50 000 ohm cm). Also, the washing sequence should be such as to avoid the concentration of the dissolved pollutants. That is, for a vertically mounted insulator, the bottom section should be washed clean before the middle section is wetted and finally, the top section is wetted only when the rest of the insulator is clean.

The four methods of live insulator washing that are most often used – which differ mainly in the type of nozzle arrangement that is employed to project the water – are:

  (i)   portable handheld jet nozzles
 (ii)   fixed spray nozzles
(iii)   remote controlled jet nozzles, often automated using robots
(iv)   helicopter mounted nozzles.

The many safety procedures that must be followed have been extensively dealt with in the CIGRE review document [2].

6.8.6   Provisions for ice and snow

For increasing insulator reliability under these special – albeit relatively infrequent – conditions, it is generally necessary to increase the length of the insulators. When ground clearance poses a problem, the usual practice is to use an inverted V insulator mounting arrangement – i.e. two insulators are hung at 45° to the vertical in lieu of one normally suspended. Other measures are dealt with briefly in the CIGRE review document [2].

6.9   Miscellaneous

A number of relevant aspects that do not fit readily into the previous sections are discussed briefly below. These range from physical processes to proposals, speculations, future needs and financial matters.

6.9.1   Cold switch-on and thermal lag

As can be readily reasoned from section 6.5, the pollution flashover strength is at a minimum when the insulator is well wetted – but not sufficiently so for the washing of the surface to have occurred. Such a vulnerable condition can happen when an insulator is reenergised after it has been out of service for some time – e.g. following maintenance/repair or after switch-out overnight for voltage control purposes. It was just such a situation that resulted in the flashover of strings containing some cracked porcelain insulators that were referred to in section 6.4.1.4.

Even with energised insulators, enhanced moistening happens when the insulator's temperature is only a few degrees lower than that of the surrounding air [36]. Generally, this is an early morning situation when the ambient temperature rises quickly – and is especially pertinent in deserts.

Although there is insufficient evidence to quantify this problem accurately, guidance – nonetheless – is provided by considering some results obtained for semiconducting glazed insulators [2]. For example, the flashover voltage of a cold switch-on for a soaking wet condition (0.15 MΩ per suspension unit) was 40 per cent lower than that for the same insulators when only damp (1000 MΩ per unit) and fourfold less than the value for the dry case (15 000 MΩ per unit).

The above quoted conditions apply to hydrophilic surfaces. As yet, there are no corresponding data for the hydrophobic case. Hence, there is a particular need for conducting such a study with the silicone rubber insulator, because its thermal mass and thermal conductivity are appreciably different from those of porcelain.

6.9.2   Semiconducting glaze

A metallic oxide, such as that produced from tin, is doped to provide semiconducting properties. When this material is applied to the surface of a porcelain insulator as a glaze to achieve a continuous current of about 1 mA under normal operating conditions, there is sufficient heat on the insulator's surface to keep it essentially dry in dew or fog.

Such an insulator has a withstand voltage of three to four times that of the corresponding ordinary glazed unit when moistened. Although this insulator type may, therefore, seem a solution to the severe pollution problem – especially that arising from thermal lag – cold switch-on remains a difficulty. However, the latter may be overcome by energising such circuits when thermal equilibrium has become established between the air and the insulator.

Post insulators coated with such a glaze operated very successfully for over ten years at the Brighton Insulator Testing Station but, unfortunately, a type of cap and pin version did not. However, a manufacturer now claims that an improved design has been produced for AC applications by substantially increasing the thickness of this glaze around the pin cavity. These insulators may not be suitable for use on DC systems because of electrolytic corrosion of the glaze.

6.9.3   Live line working

To reduce costs, working on energised OHLs is becoming an increasing practice. If the insulators have to support the weight of the linesmen, their structural integrity is – obviously – of paramount importance. Questions have, therefore, been raised about the possibility, albeit probably rare, of there being the beginning of a brittle fracture in the RBGF rod of the polymeric longrod – see section 6.3. Consequently, there is a need to develop a test for detecting such an eventuality.

A possibility of coping with this uncertainty is to measure the vibrational signature of a polymeric insulator in situ; perhaps by using the method developed for detecting radial cracks in porcelain cap and pin insulators [12]. Basically, a small electro-magnetic vibrator is fixed to the base of the tower and the minute movement of the insulator's surface is measured at ground level by using a laser Doppler vibrometer. By slowly changing the frequency of the vibrator, the resonances of the insulator are determined. From comparing this signature with that of a sound unit, abnormalities can – hopefully – be identified.

6.9.4   Visual annoyance, audible noise and electromagnetic compatibility

Even without causing flashovers, the flow of leakage current across polluted and wetted insulators can be a source of nuisance to people and can interfere with communication systems.

A pollution severity of only one tenth of that necessary for flashovers to happen is sufficient for appreciable gas discharges and audible noise (AN) to occur [2]. Visible discharges can cause annoyance, especially where an OHL passes through a populated area. Although AN does not pose a serious problem under normal surface discharge conditions for either AC or DC systems, single unit sparkovers under DC energisation produce periodic loud bangs that can continue for hours.

The signal to noise ratio for acceptable reception is 20 dB and 35 dB for radio and television, respectively; this difference arises because the eye is more sensitive than the ear to interference. The acceptable upper limits on radio interference (RI) produced by overhead lines and substations are covered generally in CISPR publication 18-2 [37] and for polluted insulators in CISIR publication 18.2, Amendment 1 [38]. For normal surface discharges on AC and DC insulators, RI and television interference (TVI) are not usually severe and do not increase much as the pollution severity gets greater. However, single unit sparkovers on DC insulators can be troublesome. Some small-scale investigations [40] have found that the electromagnetic interfer-ence – under AC energisation – for polluted and wetted silicone rubber insulators was substantially lower than that for two corresponding ceramic insulators – one a short string of glass cap and pin units, the other a porcelain longrod – and that for polymeric insulators made of EPDM and epoxy resin.

Any of these problems can be overcome completely, or be very substantially reduced, by adopting some of the measures covered in the previous section – namely: using greases, silicone rubber coating, semiconducting glaze and by washing.

6.9.5   Electric field distributions

Capacitor-type internal insulation in bushing and current transformers is designed to provide uniform radial and axial electrical stress distributions on the inside of their weather housings. Also, the varistor arrangement of a surge arrester is such as to achieve a similar situation. As can be readily reasoned from section 6.5, an uneven distribution of the pollutants and/or the water will result in a non-uniform voltage dis-tribution along the outside surface of this insulator. Of special note, the heating from the components within this housing can increase the degree of this uneven wetting.

Such differential axial voltage distributions produce a radial electric field. This resultant radial field may cause electrical discharges to occur within this insulator. In extreme situations, the insulator may puncture if the internal stress-relieving measures are not properly designed [40].

6.9.6 Interphase spacers

These insulators are mainly used to prevent mid-span flashovers of overhead lines during conditions of galloping and conductor jumping following ice release. In addition, phase-to-phase spacers may be required to achieve compact line constructions, to reduce phase spacing so as to decrease the magnetic field levels or to improve the aesthetics of the line.

Such spacers are made from either porcelain or polymers. The procedure adopted for obtaining their dimensioning is the same as that outlined in section 6.7 for the line to ground insulation, except that the withstand voltage is $\sqrt{3}$ greater. Further, considering the dire consequences of a phase-to-phase flashover on the operation of a transmission line, an additional safety margin may need to be included.

6.9.7 Compact and low profile lines

Wayleave – sometimes called right of way – requirements are becoming more difficult to obtain in some countries and so there is a growing pressure to make OHLs as narrow as possible. Also, in order to gain permission to erect lines across areas of outstanding natural beauty there is often a need to keep these lines as low as possible. Hence, there is a desire to achieve compact line arrangements, thereby providing scope for the designing of more imaginative insulators/support structures – possibly making some components act as both functions.

One wonders, therefore, if the good pollution flashover performance of silicone rubber could be exploited more than has happened so far. Currently, there is a reluctance to design an insulator around the incorporation of the RTV coating because the latter will probably need to be renewed from time to time. However, if the benefit of its application to the surface of a rigid insulator – e.g. porcelain, epoxy resin, a substantial section of glass fibre rod – outweighs this disadvantage then its use may be justified.

6.9.8 Financial and related matters

Although insulators form a relatively small part of the total cost of a new electricity supply project, the potential consequences of their bad performance can be costly. It is, therefore, important to consider this aspect when selecting insulators for a particular function. One method of approach is to define an operational risk, which is the likelihood of failure – either flashover or mechanical breakage – multiplied by the severity of the consequences of that failure.

In addition to the aspect covered in section 6.7, the general specifications of the minimum technical requirements of insulators – and the tests to verify them – are normally obtained from international and/or national standards, e.g. IEC, IEEE,

British Standards Institute. Further, the user's practice may play its part in this defining process. Insulator costs, even when expressed in relative terms, can vary markedly over a short period of time. Such variations can depend on matters other than technical ones – e.g. if a new manufacturer wishes to break into the market or cheap labour becomes available in an emerging economy.

Even though the electricity supply industry usually expects its insulators to last for thirty plus years, there are special situations in which a much shorter lifetime is still very useful. A good example – drawn from the author's experience – is in the KwaZulu-Natal Province of South Africa. The long and strategic 400 kV OHLs have been reinsulated a number of times in an attempt to reduce the number of serious flashovers caused by pollution. It was estimated that if such events could be eliminated, or substantially reduced, for just seven years by the use of silicone rubber insulators, their cost would be recouped by avoiding the need to live wash (using helicopters) the existing porcelain insulators. It is encouraging to note that the pollution flashover performance of these silicone rubber insulators has been excellent for their nearly ten years now (2004) in service.

Asset management, in its formalised form, has of late become very fashionable in many industries – especially those that have been deregulated. Therefore, it is timely that a paper [41] was presented at the 2002 CIGRE conference that deals with a survey conducted by CIGRE Taskforce 23.18 and to note that this group of experts propose to publish a CIGRE brochure to present their guidelines in 2005. Also, at the thirteenth International Symposium on High Voltage Engineering (held in 2003 at Delft, Holland), the organisers included for the first time a subject entitled 'Asset management of HV equipment: strategies and tools'. Further, such a topic may be discussed at such symposiums in the future. Hopefully, these two respected bodies will provide clear guidelines on how outdoor insulators should be monitored, maintained and replaced when deemed necessary. In the meantime, it is suggested that prudent managers should note the remarks concerning inspections that are included in section 6.7.

6.10   References

1   LOOMS, J.S.T.: 'Insulators for high voltages' (IEE Power Engineering Series, No. 7, Peter Peregrinus Ltd., 1988)
2   CIGRE Taskforce 33.04.01: 'Polluted insulators: review of current knowledge'. CIGRE technical brochure 158, June 2000
3   CIGRE Taskforce 33.13.01 (formally 33.04.01): 'Guidelines for the selection and dimensioning of insulators for outdoor applications: part 1, General principles and the AC case'. CIGRE technical brochure, publication is likely in 2005
4   IEC Publication 815: 'Guide for the selection of insulators in respect of polluted environments'. 1986
5   CHANDLER, H.D., JONES, R.L., and REYNDERS, J.P.: 'Stress corrosion failure on composite longrod insulators'. 4th international symposium on *High voltage engineering (ISH)*, Athens, Greece, paper 23.09, 1983

6   DE TOURILLE, C., PARGAMIN, L., THENVENET, G., PRAT, S., and ṢIAMPIRINGUE, N.: 'Brittle fracture of composite insulators: the new explanation and a field case study'. 12th international symposium on *High voltage engineering (ISH)*, Bangalore, India, 2001, **3**, pp. 683–686

7   SWIFT, D.A.: 'Arresting arc propagation with narrow metal strips', *IEE Proc. A, Phys. Sci. Meas. Instrum. Manage. Educ. Rev.* 1980, **127**, pp. 553–564

8   SEDIVER BROCHURE.: 'HRTG insulators for HVDC applications'. 500A RP 99/MRG, 79 avenue Francois Arago, 92017 Nanterre Cedex, France, December 1999

9   CRABTREE, L.M., MACKEY, K.J., KITO, K., NAITO, K., WATANABE, A., and IRIE, T.: 'Studies on electrolytic corrosion of hardware of DC line insulators', *IEEE Trans.*, 1985, **PAS-104**, (3), pp. 645–654

10 MADDOCK, B.J., ALLNUTT, J.G., and FERGUSION, J.M., *et al.*: 'Some investigations of the ageing of overhead lines'. CIGRE, paper 22.09, 1986

11 CROUCH, A.G., SWIFT, D.A., PARRAUD, R., and DE DECKER, D.: 'Ageing mechanisms of AC energized insulators'. CIGRE, paper 22-203, 1990

12 SWIFT, D.A.: 'Cap and pin insulators: an assessment of techniques for detecting defective porcelain'. International conference on *Revitalising transmission and distribution systems*, IEE Publ., 273, 1987, pp. 1–7

13 SWIFT, D.A., SPELLMAN, C.A., JAMES, G., and HADDAD, A.: 'Stray capacitance of cap and pin insulators in suspension strings: analytical and computer techniques'. 36th universities *Power engineering* conference (UPEC), Swansea, UK, paper 1C-1, 2001

14 BARCLAY, A.L., and SWIFT, D.A.: 'Cap and pin insulators: electrical puncture of porcelain under AC energization'. 5th international conference on *Dielectric materials, measurements and applications*, University of Kent at Canterbury, UK, 27–30 June 1988

15 CHERNEY, E.A., and HOOTON, R.D.: 'Cement growth failure mechanism in porcelain suspension insulators', *IEEE Trans.*, 1987, **PWRD-2**, (1), p. 249

16 MARTIN, R.: 'Silicone rubber station posts and bushings'. Maclean Power Systems brochures, 11411 Addison Street, Franklin Park, 1L 60131-11300, USA, 2002

17 SUITS, C.G.: 'High pressure arcs in common gases in free convection', *Phys. Rev.*, 1939, **55**, pp. 561–567

18 CLAVERIE, P.: 'Predetermination of the behaviour of polluted insulators', *IEEE Trans.*, 1971, **PAS 90**, (4), pp. 1902–1908

19 RIZK, F.A.M.: 'Mathematical models for pollution flashover', *Electra*, 1981, (78), pp. 71–103

20 HOLTZHAUSEN, J.P.: 'A critical evaluation of AC pollution flashover models for HV insulators having hydrophilic surfaces'. Ph.D. dissertation, University of Stellenbosch, 1997

21 HOLTZHAUSEN, J.P., and SWIFT, D.A.: 'The pollution flashover of AC and DC energized cap and pin insulators: role of shortening of the arc'. 11th international symposium on *High voltage engineering (ISH)*, London, UK, 1999, **4**, pp. 333–336

22   SWIFT, D.A.: 'AC flashover mechanism for water droplets on an hydrophobic insulator'. 8th international symposium on *High voltage engineering (ISH)*, Yokohama, Japan, paper 44.09, 1993, pp. 113–116

23   KARADY, G.G., SHAH, M., and BROWN, R.L.: 'Flashover mechanisms of silicone rubber insulators used for outdoor insulation part I and II', *IEEE Trans. Power Deliv.*, 1995, **10**, (4), pp. 1965–1978

24   RAVERA, C.N., BRITTEN, A.C., and SWIFT, D.A.: 'Service experience with polymeric insulators in ESKOM, South Africa'. Conference on *Applications on non-ceramic outdoor insulation*, Paris, France, 7–8 June, 1994

25   RAVERA, C.N., OLIVIER, P.J., BRITTEN, A.C., and SWIFT, D.A.: 'Silicone rubber insulators on ESKOM'S AC transmission lines'. IEE Conf. Publ. 423, *AC and DC power transmission*, 29 April–3 May, 1996

26   LIANG XIDONG, WANG SHAOWU, HUANG LENGCENG, SHEN QUINGE, and CHENG XUEQUI: 'Artificial pollution test and pollution performance of composite insulators'. 11th international symposium on *High voltage engineering (ISH)*, London, UK, 1999, **4**, pp. 337–340

27   HOULGATE, R.G., and SWIFT, D.A.: 'Polymeric insulators: AC flashover voltage under artificial salt-pollution of new and naturally-aged units compared to porcelain'. 6th international symposium on *High voltage engineering (ISH)*, New Orleans, USA, paper 47.03, 1989

28   SWIFT, D.A., SPELLMAN, C.A., and HADDAD, A.: 'Leakage current on silicone rubber insulators and a RTV coating in clean fog: hydrophobicity transfer to pollution'. 12th international symposium on *High voltage engineering* (ISH), Bangalore, India, 2001, **3**, pp. 607–610

29   WANG SHAOWU, LIANG XIDONG, GUAN ZHICHENG, and WANG XUM: 'Hydrophobicity transfer properties of silicone rubber contaminated by different kinds of pollution'. CEIDP, Victoria, Canada, 2000, pp. 373–376

30   HOULGATE, R.G., and SWIFT, D.A.: 'AC circuit breakers: pollution flashover performance of various types of interrupter head', *IEE Proc., Gener. Transm. Distrib.*, 1997, **144**, (1), pp. 50–56

31   Electric Power Research Institute: 'Transmission line reference book, 345kV and above' (Second Edition, 1982)

32   Electric Power Research Institute and Bonneville Power Administration: 'Transmission line reference book: HVDC to ±600 kV', undated

33   ELY, C.H.A., LAMBETH, P.J., LOOMS, J.S.T., and SWIFT, D.A.: 'Discharges over wet, polluted polymers: the booster shed'. CIGRE paper 15-02, 1978

34   Mace Technologies Brochure: 'Insilcure'. PO Box 739, Lonehill 2062, RSA, undated

35   CHERNEY, E.A., and GORUR, R.S.: 'RTV silicone rubber coatings for outdoor insulators', *IEEE Trans. Dielectr. Electr. Insul.*, Oct. 1999, **6**, (5), pp. 605–611

36   ORBIN, D.R.H., and SWIFT, D.A.: 'Surface resistivity of a cool polluted insulator when exposed to warm ambient air'. Proceedings of 4th international conference on *Properties and applications of dielectric materials*, Brisbane, Australia, paper 7244, July 3–8, 1994

37   IEC Publication 437: 'Radio interference tests on high voltage insulators', 1973
38   CISPR Publication 16-1: 'Specification for radio disturbances and immunity apparatus and methods', 1993
39   SWIFT, D.A., and BRITTEN, A.C.: 'Electromagnetic interference from high-voltage insulators: a comparison of hydrophobic and hydrophilic cases'. International conference on *Electromagnetic compatibility*, Rome, Italy, September 17–20, 1996, pp. 846–851
40   MELIK, G.: private communication
41   BARLETT, S.: 'Asset management in a de-regulated environment'. CIGRE paper 23-303, 2002

Chapter 7

Overvoltages and insulation coordination on transmission networks

D.M. German and A. Haddad

7.1 Introduction

In this chapter we review some aspects of insulation coordination and overvoltages on transmission networks. The purpose of insulation coordination is to ensure that the probability of insulation breakdown is limited to an acceptable value and that any breakdown is restricted to self-restoring insulation. It is based on computing the most severe overvoltages occurring on the network and relating these to the breakdown characteristics of the insulation through appropriate margins to obtain withstand voltages for the network components together with the statistical risk of insulation failure. The procedure is summarised in Figure 7.1.

From the network configuration and component data, an appropriate simulation of the network is formulated and used to compute the overvoltages, taking into account the effect of control and protective devices. The breakdown voltage characteristics are based on the type and configuration of the insulation. The coordination procedures relate these two components to give withstand voltages and to quantify the risk of insulation failure.

Coordination is an integral part of the design process, giving criteria that are satisfied through insulation design and through overvoltage control and protection. Although simple in concept, involving comparisons between maximum network over-voltages and minimum insulation breakdown voltages, in practice the process is complex.

Network overvoltages are usually the result of some sudden change in conditions, typically switching operations, lightning strokes or faults. They have a very wide

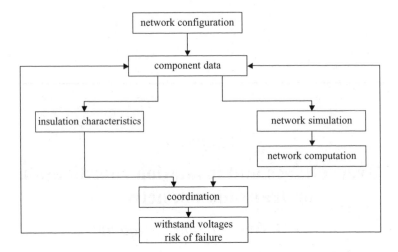

Figure 7.1   Insulation coordination

range of waveshapes and durations, the most important being:

- temporary overvoltages – lightly damped oscillations at supply frequency
- switching overvoltages – damped oscillations at frequencies of <10 kHz
- lightning overvoltages – damped oscillations at frequencies <100 kHz.

The three types of insulation – gaseous, liquid and solid – have different breakdown characteristics, which are dependent on the configuration and environment of the insulation and on the waveshape and duration of the applied voltage. The coordination process relates these various characteristics to the temporary, switching and lightning overvoltages on the network.

Air still provides the major insulation for transmission networks. An important property is that, across the long air gaps necessary for transmission voltages, the breakdown voltage is dependent on the voltage rise time, and is a minimum for the relatively long rise times associated with switching overvoltages. Coordination is therefore based primarily on switching overvoltages, although it is still important to confirm that the performance under both lightning and temporary overvoltages is acceptable.

A second important property of air insulation is that for a given rise time of the applied voltage the actual breakdown value will vary randomly about a mean value. This characteristic is represented by the cumulative probability function $P(U)$ where $P(U = U_b)$ is the probability of breakdown occurring with a crest voltage $U = U_b$ and shown in Figure 7.2.

In computing switching overvoltages, it is often not possible to specify the combination of parameter values that will result in the maximum overvoltage. In particular the circuit breaker may close at any point on the AC cycle and, due to circuit breaker tolerances, the instants at which each pole closes will vary randomly about this point. By computing the switching overvoltages for a large number of switching operations

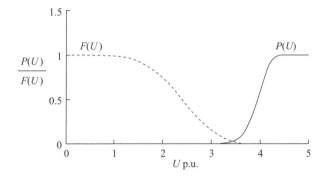

Figure 7.2 Cumulative overvoltage distribution F(U) and breakdown voltage distribution P(U)

with different instants of pole closing a switching overvoltage distribution $F(U)$ is obtained (Figure 7.2), with $F(U = U_e)$ giving the probability of the switching overvoltage exceeding U_e.

From the probability distributions of $F(U)$ and $P(U)$ shown in Figure 7.2, statistical methods are used to specify insulation withstand voltage and risk of failure.

The International Standard IEC 60071: 'Insulation coordination' [1, 2] provides a comprehensive guide to coordinating procedures and specifies coordination factors and standard insulation levels.

Although coordination is based primarily on determining withstand voltages and risks of failure from overvoltage and insulation characteristics, other factors are involved. The economic factor is always important and includes cost comparisons between overvoltage protection, insulation levels, risks of failure and network outages. Another factor of increasing importance is the environmental one and this has resulted in the design of compact lines having both reduced visual impact and transmission corridor width.

Thus, insulation coordination takes into account a range of factors and has parameters that are interdependent. An optimum design therefore involves an iterative and integrated analysis of the network.

7.2 System overvoltages

Increase in demand for electrical energy has led to increase in both the size and capacity of electrical power systems. High voltage AC systems of up to 750 kV are now in operation in several countries; future extension of the working voltage to the megavolt (MV) range is being investigated by many researchers and electrical utilities. Moreover, the advances in solid-state electronics have resulted in the development of high voltage high power thyristors and the expansion of HVDC transmission systems.

However, such advances are inevitably accompanied by technical problems. The cost of equipment and the system as a whole is closely related to the insulation level which has been adopted for the network. The choice of insulation level should take into account both the probability of occurrence and the severity of stress that the system may be subjected to. Therefore, the performance of any protective device inserted into the system should be defined as quantitatively as possible.

Overvoltages represent a major threat to security and continuity of supply. A short summary of the sources of these overvoltages is given in this section. Fortunately, a controlled limitation of overvoltage surges is possible by protective devices. Surge arresters are widely used in electrical power systems, and the superior modern metal oxide surge arresters have renewed interest and widespread use of surge arresters in protection practice (see Chapter 5).

For economic design of equipment and safe operation of power systems, a detailed knowledge of types and sources of overvoltages on power systems is required. Figure 7.3 shows a classification of the different types of overvoltage that are likely to occur on today's modern power systems. As can be seen in the figure, and according to their origin, two main categories of overvoltages can be distinguished: (i) external and (ii) internal overvoltages.

7.2.1   External overvoltages

External overvoltages are generated by sources that are external to the power system network. Their magnitude is essentially independent of the system. Two types of external overvoltage have been identified.

7.2.1.1   Lightning overvoltages

Lightning overvoltages are characterised by very high peak currents and relatively low energy content. They are responsible for nearly half of all short circuits on lines in systems of 300 kV and above [3, 4]. Up to approximately 300 kV, the system insulation has to be designed primarily to withstand lightning surges.

The lightning mechanism is now well understood [5–7] to be due to charged bipolar cumulo-nimbus clouds which induce an opposite charge on the ground. When the local electric field reaches a critical value of about 30 kV/cm, an electric discharge is initiated. A negative polarity discharge advances in steps in a channel known as the stepped leader. When the strongly ionised channel bridges the path between cloud and earth, a very high current discharge takes place; this is known as the return stroke. Subsequent discharges of dart leaders and return strokes may follow along the ionised path at irregular intervals of time for periods of time less than a second. Up to 40 strikes of decreasing magnitude have been observed. It is postulated that these subsequent strikes correspond to discharging deeply located charges inside the cloud (for a more detailed reading on lightning phenomena, refer to Chapter 3).

Lightning discharges may be of positive or negative polarity (according to the polarity of stepped leader tip) and ascending or descending type, i.e. upwards or downwards going discharges. Positive strikes, in general, consist of one long discharge and are observed to discharge currents of very high amplitudes (up to 250 kA), in contrast,

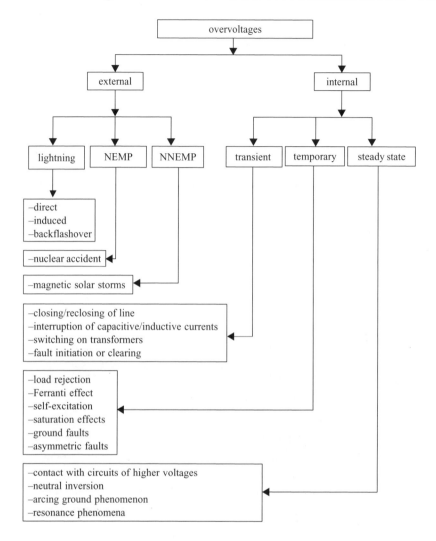

Figure 7.3   Classification of overvoltages

negative strikes are of lower intensity (less than 80 kA). Field observations and measurements have shown that up to 90 per cent of lightning discharges are of negative polarity and descending type. From the point of view of power system overvoltages, the important parameters of the lightning stroke are the current amplitude and shape. Both are statistically distributed. Field measurements [8] have shown that the modal value of current is about 32 kA. Typical rise times are about 2 μs.

Lightning affects overhead lines in three ways, that is (i) a direct strike on a phase conductor or (ii) induced overvoltage if the impact point of lightning is in the vicinity of the line, which is particularly dangerous for low and medium voltage systems.

In addition, (iii) a backflashover may occur due to a rise in potential of the tower top when lightning hits the tower or the earth wire of shielded lines.

The overvoltage, which is determined by the line surge impedance and the discharge current of lightning, is often over one million volts. These surges travel along the line and may cause internal breakdown of insulation of terminal equipment systems. They also generate corona which can be considered as a protective effect. It is the practice that protection of power systems against lightning overvoltages is usually decided on the most probable lightning current shape which might hit the system and the keraunic level or density of lightning flashes in the location area of the system (flashes per kilometer square per year). Up to 0.46 strikes/km/year were recorded and lightning discharges of crest current of approximately 40 kA were found to be the most frequent in the United Kingdom [9]. These figures, although substantially smaller than those of other parts the world [10] (mainly Europe, South America and Africa), show that lightning overvoltages are a serious threat for power systems in the UK.

7.2.1.2   Nuclear and non-nuclear electromagnetic pulses (NEMP, NNEMP)

NEMP are also an external source of power system overvoltages [11]. They are characterised by very short front times (<10 ns) and tens of kiloamperes in amplitude. Their occurrence is fortunately rare and protecting against them is usually not considered in power systems design. An attractive feature of ZnO material is its fast response to limit overvoltages. It has been shown that ZnO material could be used for efficient protection against NEMP [12]. NNEMP have been reported [13] to affect power systems during magnetic storms associated with exceptional solar disturbances.

7.2.2   Internal overvoltages

Unlike external overvoltages, internal overvoltages are principally determined by the system configuration and its parameters. They are mainly due to switching operations and following fault conditions. According to their duration, which ranges from few hundreds of microseconds to several seconds, three main categories of internal overvoltage can be defined.

7.2.2.1   Transient switching surges

Compared with lightning surges, switching surges contain a much higher amount of energy but are of lower amplitude. Several causes have been identified for switching transient overvoltages, a brief account of which will be given here; detailed descriptions can be found elsewhere [14–16]. The following major causes have been reported:

(i)   *Closing and reclosing of a line:* the resistance, capacitance and inductance of transmission lines form a distributed parameter (oscillatory) network [17, 18]. Therefore, any switching operation involves transient phenomena. For example, switching an unloaded line which is open at the other end produces up to 2 p.u. overvoltage, but this value could be even higher if a circuit breaker restrikes on a line with residual voltage. Such voltages are known to exist because of the presence of charges trapped in transmission lines after interruption of supply voltage.

(ii) *Switching of capacitive currents:* this includes capacitor banks, unloaded overhead lines and cables. It involves interruption of small capacitive currents at peak voltage. The recovery voltage may reach 1.5 p.u. across the contacts of the circuit breaker [15]. Half a cycle later, when the voltage of the feeding side attains its maximum, a voltage of 2.5 p.u. is then established across the circuit breaker contacts which may lead to restrike of the circuit breaker.

(iii) *Switching of inductive current:* when interrupting the small inductive currents of unloaded transformers and shunt reactors, current chopping may occur with the current forced to zero before the natural current zero. The high di/dt associated with current chopping results in high induced voltage in the inductive circuit. This type of operation produces overvoltages of 2 to 3 p.u. in modern transformers [15]; however, with transformers loaded with shunt reactors, values up to 5 p.u. may be reached which necessitates the use of surge arresters for protection.

(iv) *Initiation and clearing of system faults:* the most frequent fault on power systems is the short circuit phase-to-earth which is often accompanied by an increase of neutral potential. If the fault occurs at peak voltage, then up to 2.7 p.u. overvoltage can be generated. Reducing the ratio (X_0/X_1), however, will limit significantly the overvoltage; e.g. for a configuration having a ratio equal to 1, no overvoltage has been observed [15]. The single phase fault which causes an asymmetry is therefore the most dangerous fault. The clearing of a fault by circuit breakers, in particular a three-phase fault, will also generate crest overvoltages of up to 3 p.u.

The magnitude of switching surges is affected by the following factors:

(i) line parameters including dimensions, earth resistivity, trapped charges, terminating network and coupled energised circuits
(ii) the circuit breaker performance
(iii) the source network.

Many techniques have been developed [15, 19] and are widely used to reduce the peak value of switching transients to values lower than 2 p.u., which helps economical design of higher voltage systems (400 kV and above). Among these techniques are:

(i) switching resistors
(ii) controlled synchronised closing of circuit breakers
(iii) shunt reactors
(iv) drainage of trapped charges before reclosing by provision of leak resistors.

In addition, protective measures, such as surge protection capacitors and surge arresters, are adopted. Modern metal oxide surge arresters are particularly efficient for such duties because of their fast switching response, energy absorption capability and excellent voltage–current non-linearity. Corona losses and earth return attenuation have, however, little effect on reducing switching surge overvoltages except on long transmission lines.

7.2.2.2   Temporary overvoltages

A definition adopted by many authors [20] describes a temporary overvoltage (TOV) 'as an oscillatory phase-to-ground or phase-to-phase overvoltage of relatively long duration at a given location which is undamped or weakly damped in contrast to switching and lightning overvoltages which are usually highly or very highly damped and of short times'. Temporary overvoltages have also been classified into three groups according to whether the frequency of oscillation is lower, equal to or higher than the working voltage frequency. In relation to operating power system networks, CIGRE WG 33.10 and IEEE Task Force on TOV [21] propose to define temporary overvoltages as 'an overvoltage higher than the highest system voltage and lasting for more than 2 cycles'.

The most frequent causes of temporary overvoltages are Ferranti effect, load rejection, ground faults, saturation effects and ferroresonance. Most of these overvoltages range between 1.2 and 1.5 p.u., but in severe conditions they may reach 2 p.u. [19]. The highest TOV levels are known to occur when long lines are connected to a weak feeding system. Table 7.1 [21] summarises the most important causes of TOV and their properties.

Table 7.1   Summary of important TOV causes and characteristics [21]

Temporary overvoltage phenomena	Important parameters	Overvoltage magnitudes	Typical durations	Methods of control
Common causes				
fault application	fault location system X_0/X_1 ratio fault current magnitude	1.0–1.4 p.u.	2–10 cycles	usually not necessary
load rejection	power flow system short circuit MVA system capacitance machine automatic voltage regulators	1.0–1.6 p.u.	seconds	switched reactors SVS generator controls
line energising	line capacitance system short circuit MVA	1.0–1.2 p.u.	seconds	switched reactors SVS generator controls

Table 7.1   Continued

Temporary overvoltage phenomena	Important parameters	Overvoltage magnitudes	Typical durations	Methods of control
line dropping/ fault clearing	fault conditions line capacitance shunt reactors breaker opening sequence	1.0–1.5 p.u.	<1 second	shunt reactors relaying SVS
reclosing	line capacitance shunt reactors trapped charge levels fault conditions	1.0–1.5 p.u.	seconds	shunt reactors relaying SVS
transformer energising	system short circuit MVA transformer saturation characteristics frequency response characteristics system voltage level	1.0–1.5 p.u.	0–2 seconds	switched reactors SVS harmonic filters breaker closing res.
Special cases parallel line resonance	coupling capacitance between circuits shunt reactor values and saturation line corona losses	1.0–2.0 p.u.	steady state	neutral reactors switched reactors
uneven breaker poles	circuit capacitance shunt reactor values and saturation line corona losses	1.0–2.0 p.u.	steady state	neutral reactors switched reactors
ferroresonance	circuit capacitance transformer saturation transformer characteristics	1.0–1.5 p.u.	steady state	operating procedures
backfeeding	cable or line capacitance system short circuit MVA frequency response characteristics	1.0–2.0 p.u.	seconds	operating procedures shunt motors

Temporary overvoltages due to the Ferranti effect and load rejection are limited by shunt reactors, but because these shunt reactors remain connected to the system under normal working conditions, a problem of reactive power consumption is raised. Thyristor-controlled reactive compensation and shunt reactors with flat magnetising characteristics have also been utilised [20].

7.2.2.3   Steady state overvoltages

Steady state overvoltages are of supply frequency and are sustained for long time periods. They are often related to earthing and neutral arrangements within the system: examples are the neutral displacement in badly designed star-connected voltage transformers, arcing ground phenomena in systems with insulated neutrals and the resonance phenomena appearing as a result of the open-circuiting of one or two phases in a three-phase system which can be initiated by faulty circuit breakers or broken conductors. Most systems are designed so that such overvoltages do not occur.

7.3   Network simulation and analysis

Since switching overvoltages are a major factor in determining insulation levels for transmission networks and as the most severe magnitudes result from line reenergisation, the simulation of the network under these conditions is a prerequisite to overvoltage computations.

The selection of models for the network components is based on:

- the frequency range to be simulated; this depends on the type of overvoltage being computed, with lightning overvoltages requiring a frequency <100 kHz, switching overvoltages <10 kHz and temporary overvoltages <1 kHz
- the extent of the network about the source of the overvoltage requiring a detailed simulation; this again depends on the type of overvoltage and varies from a minimum for lightning to a maximum for temporary overvoltages
- a compromise between model accuracy, complexity and the availability of numerical data for the model parameters.

Transmission networks are usually simulated by circuit models for overvoltage studies and the computations carried out in the time domain.

7.3.1   Transmission lines

Models are based on the transient analysis of the line, a subject extensively covered in the literature [16, 22–24], and three models – the phase model, the 0, α, β model and the modal model – will be summarised.

7.3.1.1 Line parameters

A multiphase line with one conductor/phase and parallel to an earth plane of zero resistivity is described by the following per unit length parameters [16, 17]:

$[L]$ – full matrix of conductor–earth self and mutual inductances
$[P]$ – full matrix of conductor–earth self and mutual potential coefficients
$[C] = [P]^{-1}$ – full matrix of conductor–earth self and mutual capacitances
$[R]$ – diagonal matrix of conductor resistances
ℓ – length of line.

The order of the matrices is equal to the number of phases and $[L]$ and $[C]$ are related by:

$$[L] = (2\pi)^{-1}\mu[G] \tag{7.1}$$

$$[C] = 2\pi\varepsilon[G]^{-1} \tag{7.2}$$

where $[G]$ is obtained from the phase geometry.

7.3.1.2 Phase model

Under transient conditions the line is usually considered as a lossless line with losses represented by resistance networks at each end as shown in Figure 7.4. For a more accurate representation, or to obtain voltages at points along the line, the line may be subdivided and each section represented as in Figure 7.4.

The phase voltages $[V(x,t)]$ and phase currents $[I(x,t)]$ at a point distance x along the line $s'r'$ are related by the lossless line equations [23]:

$$\frac{d^2}{dx^2}[V(x,t)] = [L][C]\frac{d^2}{dt^2}[V(x,t)] \tag{7.3}$$

$$\frac{d^2}{dx^2}[I(x,t)] = [C][L]\frac{d^2}{dt^2}[I(x,t)] \tag{7.4}$$

and from Equations (7.1) and (7.2):

$$[L][C] = [C][L] = \mu\varepsilon[U] \tag{7.5}$$

where $[U]$ is the unit matrix. The phase variables $[V(x,t)]$ in Equation (7.3) and $[I(x,t)]$ in Equation (7.4) are therefore decoupled and a solution in terms of travelling

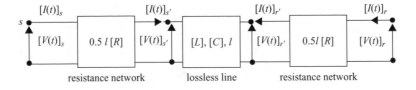

Figure 7.4   Diagram of circuit model for multiphase line

waves of voltage and current readily obtained as [23]:

$$[I(x,t)] = [I(x - vt)]_f + [I(x + vt)]_b \tag{7.6}$$

$$[V(x,t)] = [V(x - vt)]_f + [V(x + vt)]_b = [Z]\{[I(x - vt)]_f - [I(x + vt)]_b\} \tag{7.7}$$

where suffixes f and b refer to waves travelling in directions s' to r' and r' to s', respectively.

The voltage and current waves are related through the phase surge impedance matrix:

$$[Z] = [L]^{1/2}[C]^{-1/2} = \mu^{1/2}\varepsilon^{-1/2}[G] \tag{7.8}$$

and all waves have the same velocity:

$$[v] = [L]^{-1/2}[C]^{-1/2} = \mu^{-1/2}\varepsilon^{-1/2}[U] \tag{7.9}$$

From the travelling wave solutions for the lossless line the voltages and currents at the terminals s' and r' are obtained in a form suitable for computation as [25]:

$$[V(t)]_{s'} = [V(t - T)]_{r'} + [Z]\{[I(t)]_{s'} + [I(t - T)]_{r'}\} \tag{7.10}$$

$$[V(t)]_{r'} = [V(t - T)]_{s'} + [Z]\{[I(t)]_{r'} + [I(t - T)]_{s'}\} \tag{7.11}$$

where, $T = (\mu \cdot \varepsilon)^{1/2} \cdot l$ is the travel time for the line.

These equations define a model for the lossless line and when combined with the conductor resistance can be represented from the line terminals by the circuit shown in Figure 7.5. The line is characterised by the surge impedance $[Z]$, which includes coupling terms between phases, the travel time T and the conductor resistance $\ell[R]$.

7.3.1.3   Earth resistivity

When the earth resistivity $\rho \neq 0$, earth currents penetrate into the earth and their distribution is not readily determined. Numerous studies into earth current have been reported [26–30], and most involve field analysis in the frequency domain assuming a homogeneous earth. Some studies have considered a multilayer earth with different

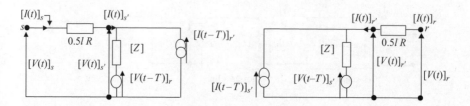

Figure 7.5   Diagram of circuit model for line at s and r (sending and receiving end)

layer resistivities. The overall effect of the current distribution in the earth is to modify both $[L]$ and $[R]$ to equivalent values $[L]_e$ and $[R]_e$ where:

$$[L]_e = [L] + [\Delta L(\omega)] \tag{7.12}$$

$$[R]_e = [R] + [\Delta R(\omega)] \tag{7.13}$$

Power series are widely used to compute $[\Delta R(\omega)]$ and $[\Delta(\omega)]$ [33], with the correction terms evaluated at some representative frequency. This frequency can be approximated by $f = 1/4T$ where T is the travel time for the lossless line.

Equations (7.3) and (7.4) now become:

$$\frac{d^2}{dx^2}[V(x,t)] = [L]_e[C]\frac{d^2}{dt^2}[V(x,t)] \tag{7.14}$$

$$\frac{d^2}{dx^2}[I(x,t)] = [C][L]_e\frac{d^2}{dt^2}[I(x,t)] \tag{7.15}$$

$[L]_e[C]$ and $[C][L]_e$ are non-diagonal matrices and the phase variables in both Equations (7.12) and (7.13) are not decoupled. A simple travelling wave solution is not available and a transformation of variables is necessary to obtain decoupled equations.

7.3.1.4 0, α, β transformation

The 0, 1, 2 or 0, α, β transformations used in fault studies can be used to decouple the line equations [31–33]. The latter is preferred because all the elements in the transformation have real rather than complex values. In addition to analytical advantages, the transformed variables provide a simple insight into the network response. However, to be effective, the arrangement of phase conductors and earth must be assumed to be symmetrical. This requires an approximation, replacing all mutual terms and all diagonal terms by their average values to give, for a three-phase line:

$$[L]_e = \begin{bmatrix} L_S & L_M & L_M \\ L_M & L_S & L_M \\ L_M & L_M & L_S \end{bmatrix} \tag{7.16}$$

$$[C]_e = \begin{bmatrix} C_S & -C_M & -C_M \\ -C_M & C_S & -C_M \\ -C_M & -C_M & C_S \end{bmatrix} \tag{7.17}$$

The values of the elements of the transformation matrix relating the 0, α, β components to the phase variables depend on the convention adopted to define the components and any normalisation procedure used. With the transformation defined by:

$$\begin{bmatrix} V_0 \\ V_\alpha \\ V_\beta \end{bmatrix} = \begin{bmatrix} \dfrac{1}{\sqrt{3}} & \dfrac{1}{\sqrt{3}} & \dfrac{1}{\sqrt{3}} \\ \dfrac{1}{\sqrt{2}} & \dfrac{-1}{\sqrt{2}} & 0 \\ \dfrac{1}{\sqrt{6}} & \dfrac{1}{\sqrt{6}} & \dfrac{-2}{\sqrt{6}} \end{bmatrix} \begin{bmatrix} V_a \\ V_b \\ V_c \end{bmatrix} \qquad (7.18)$$

Equation (7.14) then transforms to:

$$\frac{d^2}{dx^2} \begin{bmatrix} V_0(x,t) \\ V_\alpha(x,t) \\ V_\beta(x,t) \end{bmatrix} = \begin{bmatrix} L_0 & 0 & 0 \\ 0 & L_\alpha & 0 \\ 0 & 0 & L_\beta \end{bmatrix} \begin{bmatrix} C_0 & 0 & 0 \\ 0 & C_\alpha & 0 \\ 0 & 0 & C_\beta \end{bmatrix} \frac{d^2}{dt^2} \begin{bmatrix} V_0(x,t) \\ V_\alpha(x,t) \\ V_\beta(x,t) \end{bmatrix}$$

$$(7.19)$$

where the 0, α, β parameters are given by:

$$L_0 = L_s + 2L_m$$
$$L_\alpha = L\beta = L_s - L_m$$
$$C_0 = C_s - 2C_m$$
$$C_\alpha = C\beta = C_s + C_m$$

Using the same transformation for currents transforms Equation (7.15) to an identical form to Equation (7.19).

The 0, α, β variables are decoupled and the travelling wave solution gives surge impedances:

$$Z_0 = \left(\frac{L_0}{C_0}\right)^{1/2}$$

$$Z_\alpha = Z_\beta = \left(\frac{L_\alpha}{C_\alpha}\right)^{1/2}$$

And travel times:

$$T_0 = (L_0 C_0)^{1/2} \ell$$

$$T_\alpha = T_\beta = (L_\alpha C_\alpha)^{1/2} \ell$$

The ideal line can now be represented by uncoupled 0, α, β networks, each network has terminal equations of the same form as Equations (7.8) and (7.9).

In a similar way the resistance matrix $[R]_e$ now includes off diagonal elements due to the equivalent resistance of the earth return paths, and will also transform to a diagonal matrix with:

$$R_0 = R_s + 2R_m$$

and

$$R_\alpha = R_\beta = R_s - R_m$$

Although there is no coupling between the 0, α, β networks, the terminal conditions can result in an interconnection of the networks. As in symmetrical component analysis of unbalanced faults where the interconnected sequence networks represent the constraints between voltages and currents at the fault points, in a similar way, under sequential switching conditions, the 0, α, β networks are interconnected to satisfy the constraints at the line ends [34]. Thus, if the line is energised from one end by closing phase a with phases b and c open, the constraints between the phase variables at this end are:

$$V_a = E_a, \qquad \text{the source voltage}$$

$$I_b = I_c = 0$$

And between the 0, α, β variables

$$E_a = \frac{V_0}{\sqrt{3}} + \frac{V_\alpha}{\sqrt{2}} + \frac{V_\beta}{\sqrt{6}}$$

$$\sqrt{3}I_0 = \sqrt{2}I_\alpha = \sqrt{6}I_\beta$$

These constraints are satisfied by an interconnection of the 0, α, β networks at the energised end through coupling transformers as shown in Figure 7.6. In a similar way, the interconnected network for other sequential switching combinations can be obtained.

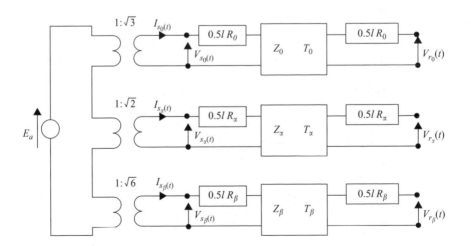

Figure 7.6   Interconnection of 0, α, β network to simulate single pole closing

7.3.1.5 Modal transformation

When the assumption of the symmetry used in Equations (7.14) and (7.15) is not acceptable and a more accurate model is required then modal transformations can be used to transform the variables in Equations (7.3) and (7.4) from the phase to the modal domain and diagonalise the equations [33]. Different transformations are required for voltage and current and are given by:

$$[V'(x,t)]_p = [T]_v [V'(x,t)]_m \qquad (7.20)$$

$$[I'(x,t)]_p = [T]_i [I'(x,t)]_m \qquad (7.21)$$

where $[T]_v$ is the eigenvector matrix of $[L]_e[C]$ and $[T]_i$ the eigenvector matrix of $[C][L]_e$ with suffixes p and m denoting phase and modal variables, respectively.

The lossless line equations then become:

$$\frac{d^2}{dx^2}[V'(x,t)]_m = [\Lambda]\frac{d^2}{dt^2}[V'(x,t)]_m \qquad (7.22)$$

$$\frac{d^2}{dx^2}[I'(x,t)]_m = [\Lambda]\frac{d^2}{dt^2}[I'(x,t)]_m \qquad (7.23)$$

where $[\Lambda]$ is the diagonal eigenvalue matrix of $[L]_e[C]$ and of $[C][L]_e$.

These equations are of the same form as Equation (7.19) and have similar solutions giving:

modal travel time $[T]_m = [\Lambda]^{1/2}$

modal surge impedance $[Z]_m = [\Lambda]^{1/2}$

Again, there is no coupling between modes and Figure 7.7 shows a representation of the model. The eigenvalues and hence the modal velocities are uniquely defined. However, the eigenvectors can be multiplied by any selected non-zero constant without altering their properties and the modal surge impedances are dependent on the particular eigenvector matrix selected.

The lossy networks cascaded at each end of the lossless line may be transformed to modal variables as in Figure 7.7, or left outside the transformation as phase variables.

7.3.1.6 Corona losses

Under transient conditions, corona discharges will occur from the conductor surface when the voltage exceeds the critical corona inception voltage E_c and will establish a space charge around the conductor. As the voltage increases the corona will propagate

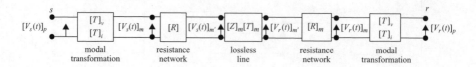

Figure 7.7   Modal simulation of line

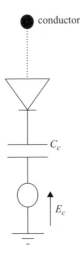

Figure 7.8 *Circuit model for corona on line*
E_c = corona inception voltage
C_c = corona capacitance

further into the space around the conductor until the conductor voltage reaches a maximum and at this point the corona will be suppressed by the space charge. The corona loss represents an energy loss from the line resulting in distortion of the surge front and, for fast-front, low energy surges, significant surge attenuation. This phenomenon may be simulated by the circuit model shown in Figure 7.8 [36]. The corona capacitance C_c can be obtained from impulse voltage tests on the conductors or estimated from the corona inception field and conductor geometry. In using this model, the line must be divided into short cascaded sections and the corona model included at the terminals s' and r' of each section.

7.3.1.7 Bundled conductors

When each phase consists of several conductors in parallel, $[L]$, $[C]$ and $[R]$ may be formulated by considering each conductor rather than each phase and then matrix reduction methods used to obtain equivalent phase matrices [33]. In a similar way, the earth conductor and earth return can be combined. These reductions are frequency dependent and usually evaluated at a representative frequency.

7.3.1.8 Lightning and temporary overvoltages

Although the circuit models are described for the computation of switching overvoltages, they can also be applied to lightning overvoltages except that earth correction parameters are evaluated at a higher frequency. For strikes to earth wires, line spans close to the strike are each represented by a circuit model, and the towers simulated by a surge impedance [35].

For temporary overvoltages with a much lower maximum frequency, pi circuit representations are adequate [23], but a more extensive section of network should be simulated.

7.3.1.9   Trapped charge

When an energised line is isolated there will be a trapped charge left on the line and this will discharge rapidly through any transformer windings still connected to the line, or more slowly through leakage over insulation, and it is necessary to simulate this charge especially for high speed reclosure. Any trapped charge present on reclosure can have a significant effect on switching voltages and is simulated as an initial voltage on the line. This voltage can be computed from the previous line opening sequence or, assuming each pole of the circuit breaker clears at a current zero and interphase capacitance is neglected, then the most severe conditions will be $V_a(0) = V_b(0) = V_{ph}$ and $V_c(0) = -V_{ph}$, where V_{ph} is the crest value of the phase voltage.

7.3.1.10   Summary of line models

The three models for transient analysis of lines are, in order of increasing complexity:

- Phase model: $\rho = 0$, characterised by full surge impedance matrix $[Z]$ and a single travel time T.
- $0, \alpha, \beta$ model: $\rho \neq 0$, symmetrical line characterised by surge impedances Z_0, $Z_\alpha = Z_\beta$ and travel times T_0 and $T_\alpha = T_\beta$. Transformation $0, \alpha, \beta \leftrightarrow a, b, c$ variables defined.
- Modal model: $\rho \neq 0$, asymmetrical line characterised by diagonal surge impedance matrix $[Z]_m$ and diagonal modal travel time matrix $[T]_m$. Transformation modal $\leftrightarrow abc$ computed.

Accuracy of transmission line modelling is limited by uncertainties in parameter values and for general overvoltage studies the $0, \alpha, \beta$ model is adequate.

7.3.2   Cables

Cables may be simulated by similar circuit models to lines with all conductors, core, sheath and earth represented and with cross bonding and earthing taken into account.

7.3.3   Circuit breakers

Circuit breakers are usually simulated by ideal switches with each pole closing at some specified nominal time. In circuit breakers without point-on-voltage wave control the closure can occur with equal probability at any point on the voltage wave, although with pre-arcing it would be biased towards points of maximum voltage. In addition, circuit breaker tolerances will result in a statistical spread of individual pole closing times about the nominal time. To simulate these effects a large number of switching operations are computed with the individual pole closing times varied in a random or systematic manner [33]. Although more accurate models including pre-arcing can be formulated they are not normally necessary for network overvoltage computations.

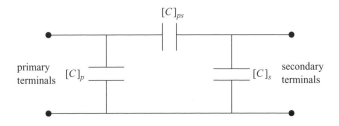

Figure 7.9   Circuit model for transformer capacitance

7.3.4   Transformers

For overvoltage computations, the simplest simulation is an equivalent capacitance network representing the interwinding and self-winding capacitance as shown in Figure 7.9.

When it is relevant to simulate the magnetic couplings between windings the model must account for both the electrical constraints due to the interconnection of the windings and the magnetic constraints imposed by the core construction. A model based on the duality between electric and magnetic circuits meets these requirements. The method is illustrated in Figure 7.10, which shows a three-limb core with one winding on each limb, the windings being connected in star [37]. The isolating transformers are necessary to ensure that both the electrical and the magnetic constraints are satisfied simultaneously. The limb and the yoke inductance are defined by flux-current curves based on the magnetic characteristics of the core material and the core dimensions. The characteristic can include recoil paths within hysteresis loop and are readily simulated in piecewise linear form. The model simulates the varying levels of saturation occurring in different sections of the core, especially during transient conditions. It is readily extended to represent several windings on each limb and to include core losses.

7.3.5   Network reduction

Sections of the network can be replaced by a Thevenin equivalent circuit. Ideally, the frequency response of this circuit should correspond with that of the actual circuit up to an appropriate frequency. Generally, there are insufficient network data available for this and a simple alternative is to use a circuit such as shown in Figure 7.11, where L is the inductance obtained from the fault level at the section terminals and Z the surge impedance looking into the section from these terminals.

7.4   Computed switching overvoltages

The switching overvoltages of greatest relevance are those associated with energising a line terminated in an open circuit, and variables having a significant effect on these

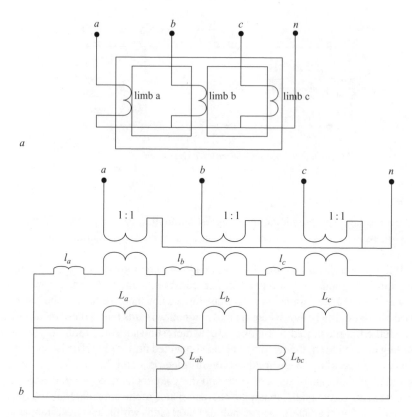

Figure 7.10   Circuit model of three-limb, transformer core with star connected winding

    a   physical arrangement of core and winding
    b   circuit model

    L_a, L_b, L_c = limb inductances
    L_{ab}, L_{bc} = yoke inductances
    l_a, l_b, l_c = leakage inductance

overvoltages are:

- trapped charge on each phase at the instant of energisation
- point on voltage wave at which each phase is energised
- earth resistivity
- line length.

These effects will now be reviewed.

For the simple case of a lossless line with a travel time of $T = 1$ ms, energised from a source voltage $V_s = \cos(\omega t + \theta)$ p.u. with the circuit breaker closed at $t = 0$, Figure 7.12 shows the variation of the maximum overvoltage V_r at the open circuit termination with the switching angle θ. Curve *a* is for zero trapped charge with V_r

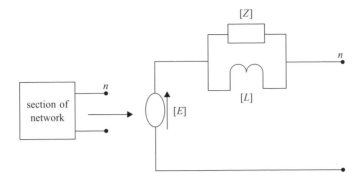

Figure 7.11   Equivalent circuit representing section of network

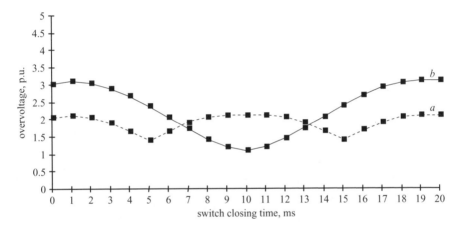

Figure 7.12   Variation of switching overvoltage with instant of switch closing t = 0, when source voltage on phase a is maximum

> *Curve a   zero trapped charge*
> *Curve b   trapped charge = −1 p.u. voltage*

having a maximum value of 2.0 per unit for $\theta = 0°$ and a minimum value of 1.2 p.u. at $\theta = 90°$. Curve *b* shows the effect of a trapped charge represented by an initial voltage of −1.0 p.u. giving an increase in the maximum voltage to 3 p.u. for $\theta = 0°$ with the minimum value being 1.0 p.u. at $\theta = 180°$.

The relative effects of trapped charge, pole closing times and earth resistivity are shown in Figure 7.13. They are based on a 150 km single circuit line and computed using the 0, α, β model and EMTP [33]. The line has four conductors/phase and a single earth wire, and the 0, α, β parameters are given in Table 7.2 with earth resistivity ρ having greatest effect on the 0 parameters. The operating conditions are shown in Table 7.3.

Figure 7.13(i) shows the voltage waveshapes at the line termination for the ideal case of $\rho = 0$, simultaneous pole closing at the instant when $E_{a0} = 1$ p.u.,

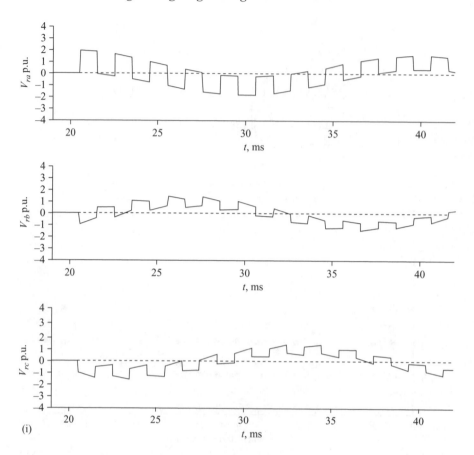

Figure 7.13   *Phase voltages at line termination*
i   simultaneous pole closing, zero trapped charge, $\rho = 0$

$E_{b0} = E_{c0} = -0.5$ p.u. and with zero trapped charge. The phase voltages $V_a(t)$, $V_b(t)$ and $V_c(t)$ are linear combinations of the 0, α, β travelling waves appearing at the termination and with simultaneous pole closing on to a symmetrical source, no zero components are present. The voltages have well defined waveshapes and show clearly the travelling wave components arriving simultaneously. The maximum value of the overvoltage $V_{ra} = 2$ p.u. on phase a.

Figure 7.13(ii) shows the effect of sequential switching. The closing times t_a, t_b and t_c are selected within a pole spread of 5 ms and switch at voltages of $E_{a0} = 1$ p.u., $E_{b0} = 0.87$ p.u. and $E_{c0} = -1$ p.u. The first pole to close, a results in $V_{ra}(t)$ of the same waveshapes and magnitudes as case (i), and produces coupled voltages on phases, b and c. When c and then b close, these coupled voltages present different initial conditions from those seen by phase a resulting in modified waveshapes following closure and with $V_{rc}(t)$ now reaching a maximum value of 2.3 p.u.

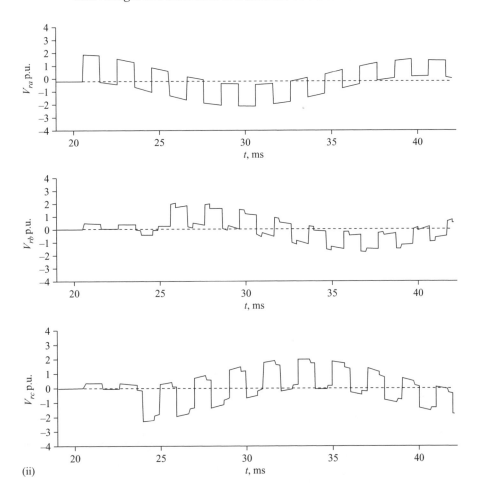

(ii)

Figure 7.13   Continued

   ii   sequential pole closing, zero trapped charge, $\rho = 0$

Figure 7.13(iii) shows the effects of $\rho \neq 0$ resulting in longer travel times for the zero waves. This results in further distortion of the waveshapes and greater attenuation because of the increased value of R_0. The maximum overvoltage is now 2.35 p.u. on phase b.

Figure 7.13(iv) shows the effect of trapped charge represented by $V_{a0} = V_{b0} = -1$ p.u. and $V_{c0} = 1$ p.u. The wave magnitudes are increased with a maximum of 3.78 p.u. appearing on phase b.

Finally, Figure 7.13(v) is as for case (iv) but includes the effect of a surge arrester connected between each phase and earth at the line termination. The arresters have a protective level of 2.38 p.u., cause further distortion of the wave shapes and limit the maximum voltage to 2.07 p.u.

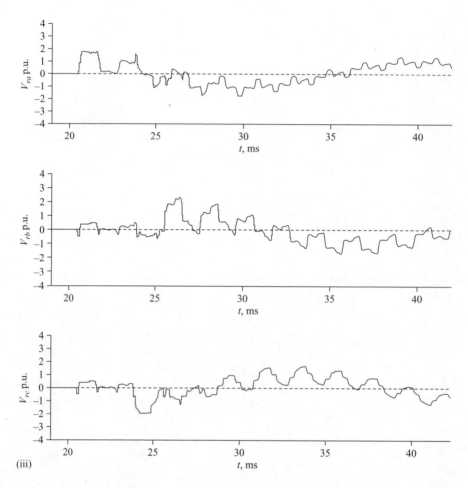

Figure 7.13   Continued

   iii   sequential pole closing, zero trapped charge, $\rho = 100$ Ωm

   Sequential pole closing, earth resistivity and trapped charge can all result in high values of the switching overvoltage, however it is not possible to specify the combination that will result in the most severe overvoltages, hence it is necessary to consider a statistical analysis of the overvoltage distribution.

   As an example the effect of random pole closing on the overvoltages was computed using a statistical switching model with random times computed for each pole and based on a normal distribution about a nominal closing time of 20 ms and a standard deviation of 0.83 ms. Three operating conditions to show the effect of trapped charge and of surge arresters were considered using the data in Table 7.4.

   Figure 7.14 shows the cumulative distribution of the overvoltages $F(U)$ together with the density distribution $f(U)$. These distributions are based on 1000 switching

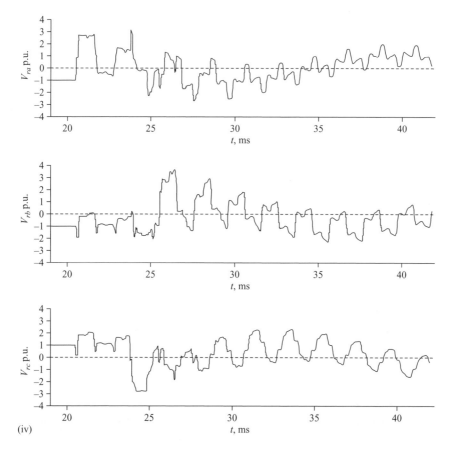

(iv)

Figure 7.13   Continued

   iv   sequential pole closing, trapped charge $-1.0, -1.0, 1.0$ p.u.,
           $\rho = 100\ \Omega\mathrm{m}$

operations taking the highest overvoltage from each switching operation. The mean values of the overvoltages together with the standard deviations are given in Table 7.5.

   The distribution with zero trapped charge is shown in Figure 7.14 case (vi) with overvoltages ranging from 1.75 p.u. to 2.95 p.u. The effect of trapped charge is to move the distribution to a higher value ranging from 2.75 p.u. to 4.90 p.u. but with a similar distribution as shown in Figure 7.14 case (vii). When surge arresters are included at the line termination the overvoltages are limited to the narrow distribution of 2.05 p.u. to 2.25 p.u. seen in Figure 7.14 case (viii).

7.5   Insulation coordination

Although coordination between switching overvoltages and the strength of self-restoring insulation can be based on the maximum value of overvoltage and the

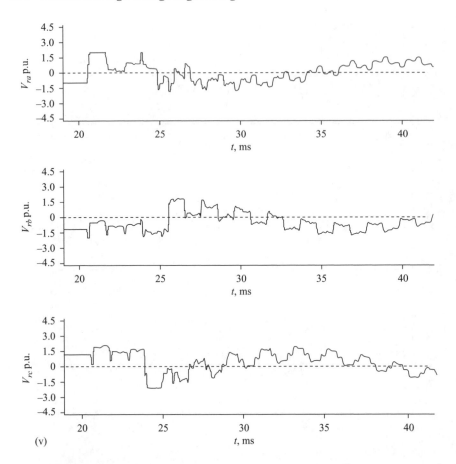

(v)

Figure 7.13   Continued

    v   sequential pole closing, trapped charge −1.0, −1.0, 1.0 p.u., surge arresters included, $\rho = 100\ \Omega$m

minimum value of breakdown voltage, this takes no account that the probability of both occurring simultaneously is very low and it is more realistic to use statistical data for coordination.

IEC 60071: 'Insulation coordination' [1, 2] specifies procedures for the selection of withstand voltages for phase-to-earth, phase-to-phase and longitudinal insulation. A full statistical analysis would consider all variables that may influence the overvoltage distribution, including various configurations of the network. Such an analysis is both time consuming and difficult to optimise and most studies consider one or two configurations. The procedures include correction factors based on operational experience on transmission networks.

Statistical data are obtained from computations of overvoltages and laboratory/field tests on insulation characteristics. It is generally obtained as: $F(U)$ –

Table 7.2   0, α, β line parameters

ρ (Ωm)	Z_0 (Ω)	T_0 (ms)	R_0 (Ω/km)	Z_α (Ω)	T_α (ms)	R_α (Ω/km)
0	442	0.50	0.065	256	0.50	0.043
100	553	0.63	1.6000	259	0.51	0.048

Table 7.3   Operating conditions

Case	ρ (Ωm)	t_a (ms)	t_b (ms)	t_c (ms)	V_{a0} (p.u.)	V_{b0} (p.u.)	V_{c0} (p.u.)
(i)	0	20.0	20.0	20.0	0	0	0
(ii)	0	20.0	25.0	23.3	0	0	0
(iii)	100	20.0	25.0	23.3	0	0	0
(iv)	100	20.0	25.0	23.3	−1	−1	1
(v)*	100	20.0	25.0	23.3	−1	−1	1

* includes surge arrester at line termination

Table 7.4   Operating conditions

Case	ρ (Ωm)	V_{a0} (p.u.)	V_{b0} (p.u.)	V_{c0} (p.u.)
(vi)	100	0	0	0
(vii)	100	−1	−1	1
(viii)*	100	−1	−1	1

* includes surge arresters at line termination

the cumulative distribution of switching overvoltages, where $F(U = U_e)$ is the probability that the maximum switching overvoltage $U_e \geq U$, $f(U)$ is the density distribution where $f(U) = -dF(U)/dU$ and $f(U = U_e)dU$ is the probability that $U_e \leq U \leq U_e + dU$; $P(U)$ – the cumulative distribution of breakdown voltage where $P(U = U_b)$ is the probability of breakdown with a crest voltage $U = U_b$.

The probability of breakdown occurring at voltage U is $f(U)P(U)dU$ and the risk of failure R for all values of U is given by [38, 39]:

$$R = \int_0^\infty f(U)P(U)dU \tag{7.24}$$

Because there is an upper limit to the switching voltage U_{et} and a lower limit to the breakdown voltage U_{b0} the integration limits $0 \rightarrow \infty$ can be replaced by U_{b0} and U_{et}, where $F(U = U_{et}) = 0$ and $P(U = U_{b0}) = 0$.

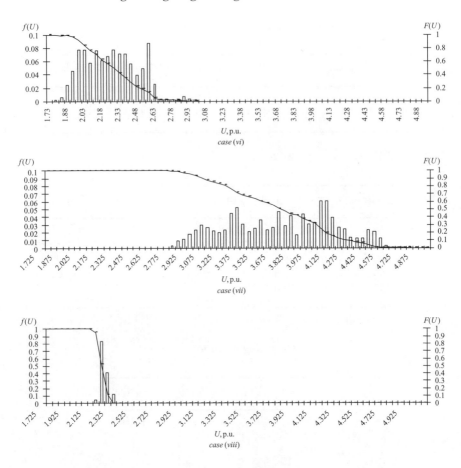

Figure 7.14   *Density and cumulative density distributions of overvoltages*

   case (vi):   zero trapped charge, no arresters
   case (vii):   1.0 p.u. trapped charge, no arresters
   case (viii):   1.0 p.u. trapped charge, arresters at line end

Insulation coordination is then achieved by optimising the overvoltage data through protection and control, and the breakdown voltage data using different insulation levels in order to obtain an acceptable risk of failure. An acceptable risk value is based on satisfactory operational experience on similar networks and is generally within a range of 10^{-1} to 10^{-4} per switching operation [2].

7.5.1 *Analytical expressions for $F(U)$ and $P(U)$*

Although the risk of failure is best estimated from $F(U)$ and $P(U)$, often insufficient data is available and analytical expressions based on limited data are widely used. The expressions should give a good fit over the region U_{b0} to U_{et} and IEC71 recommends

Table 7.5   Overvoltage data

Case	U_{e50} (p.u.)	σ_e (p.u.)
(vi)	2.21	0.231
(vii)	3.64	0.449
(viii)	2.16	0.035

the use of Weibull probability functions of the form:

$$F(U) = 1 - e^{-((U_{et}-U)/\beta)^{\gamma}} \tag{7.25}$$

and

$$P(U) = 1 - e^{-((U-U_{b0})/\beta)^{\gamma}} \tag{7.26}$$

where β are scale parameters and γ shape parameters. These expressions are readily evaluated and differentiated.

$F(U)$ can be readily characterised using the U_{e50} switching overvoltage, where $F(U = U_{e50}) = 0.5$ and σ_e the standard deviation of $F(U)$. Then, with the assumptions [2] that:

$$U_{et} = U_{e50} + 3\sigma_e \tag{7.27}$$

and

$$U_{e2} = U_{e50} + 2.05\sigma_e \tag{7.28}$$

where $F(U = U_{e2}) = 0.02$ – a 2 per cent probability that $U_{e2} \geq U$. Equation (7.25) becomes:

$$F(U) = 1 - e^{-((3\sigma_e+U_{e50}-U)/\beta)^{\gamma}} \tag{7.29}$$

giving:

$$F(U = U_{e50}) = 0.5 = 1 - e^{-(3\sigma_e/\beta)^{\gamma}} \tag{7.30}$$

$$F(U = U_{e2}) = 0.02 = 1 - e^{-(0.95\sigma_e/\beta)^{\gamma}} \tag{7.31}$$

and from Equations (7.30) and (7.31) we obtain: $\gamma = 3.07$ and $\beta = 3.4\sigma_e$.
Equation (7.29) now becomes:

$$F(U) = 1 - e^{-((3\sigma_e+U_{e50}-U)/3.41\sigma_e)^{3.07}}$$

$$F(U) = 1 - 0.509^{(1+(U_{e50}-U)/3\sigma_e)^{3.07}} \tag{7.32}$$

and with acceptable accuracy can be represented by:

$$F(U) = 1 - 0.5^{(1+(U_{e50}-U)/3\sigma_e)^{3}} \tag{7.33}$$

and

$$f(U) = -\frac{\ln(0.5)}{\sigma_e}\left(1 + \frac{U_{e50} - U}{3\sigma_e}\right)^2 0.5^{(1+(U_{e50}-U)/3\sigma_e)^3} \tag{7.34}$$

Similarly $P(U)$ is characterised using U_{b50} and σ_b, and with the assumptions that:

$$U_{b0} = U_{b50} - 4\sigma_b$$

$$U_{b16} = U_{b50} - \sigma_b$$

where $P(U = U_{b16}) = 0.16$; Equation (7.26) becomes:

$$P(U) = 1 - 0.5^{(1+(U-U_{b50}/4\sigma_b))^5} \tag{7.35}$$

These expressions can then be used to estimate the risk of failure of the insulation.

7.5.2 Risk of failure

Using the data in Table 7.5 and Equations (7.33) and (7.34) the distributions $F(U)$ and $f(U)$ for cases (vi), (vii) and (viii) are obtained and are shown in Figure 7.15, as a, b and c. For each case a distribution $P(U)$ is obtained from Equation (7.35), using $U_{b50} = 2.8$ p.u. and $\sigma_b = 0.168$ and is also shown in Figure 7.15. The risk of failure is then evaluated as:

case (vi), no trapped charge, $R = 0.02$
case (vii), trapped charge, $R = 0.94$
case (viii), trapped charge, surge arrester, $R = 5 \times 10^{-06}$

The risk of failure for case (vi) is rather high, whereas for case (vii) the risk is unacceptably high and for case (viii) the insulation level too high. For each case the distribution $P(U)$ can be optimised to give an acceptable risk value.

7.5.3 Simplified method

A simplified method is available [2] which assumes that $F(U)$ and $P(U)$ are each defined by one value:

U_{e2} – the statistical overvoltage where $F(U = U_{e2}) = 0.02$, i.e. a 2 per cent probability of being exceeded
U_{10} – the statistical withstand voltage where $P(U = U_{b10}) = 0.10$, i.e. a 10 per cent probability of breakdown and 90 per cent probability of withstand.

Coordination is achieved through the statistical coordination factor K_{es}, where:

$$U_{b10} = K_{es}U_{e2}$$

K_{es} is selected to meet a required risk of failure from data for typical transmission networks. Figure 7.16 shows typical variations of the risk of failure with the statistical coordination factor K_{es} [2]. Although the curve will be modified by the shape of the

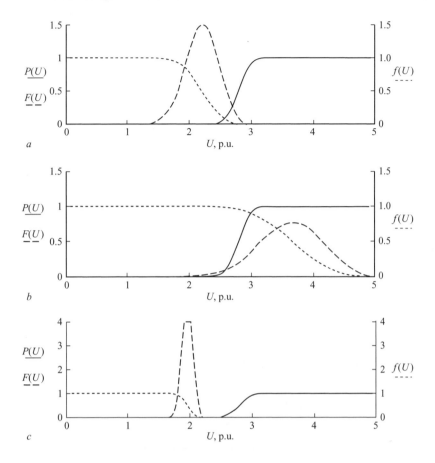

Figure 7.15   *Weibull representation of f(U), F(U) and P(U)*

    a   case (vi), no trapped charge, no arrester, $R = 0.02$
    b   case (vii), trapped charge, no arrester, $R = 0.94$
    c   case (viii), trapped charge, arrester, $R = 5.4\text{E-06}$

overvoltage K_{cs} has a low sensitivity to changing R. Table 7.6 shows the statistical withstand voltage obtained for $R = 0.005$.

7.5.4   Withstand voltage

Having optimised $F(U)$ and $P(U)$ to obtain an acceptable risk of failure, then U_{b10}, the statistical breakdown voltage, is available. $P(U)$ is based on laboratory/field tests where the conditions may differ from those on the actual network. Factors such as manufacturing tolerances, environmental conditions and ageing can modify $P(U)$. These factors are accounted for through a safety factor K_s to give the withstand voltage $U_{rw} = K_s U_{b10}$, where typically $K_s = 1.05$ [2].

    A further correction for altitude may be necessary (see Chapter 2).

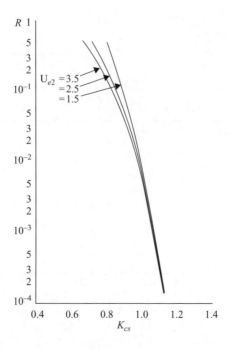

Figure 7.16   *Variation of risk of failure R with statistical coordination factor* K_{cs} *[2]*[1]

Table 7.6   *Statistical   withstand voltages*

Case	U_{e2} (p.u.)	K_{es}	U_{b10} (p.u.)
(vi)	2.67	1.04	2.78
(vii)	4.54	1.04	4.72
(viii)	2.23	1.04	2.32

7.6   Compact transmission lines

One area where both insulation coordination and environmental factors have had a combined impact is that of compact lines. In the normal development of transmission networks load growth and relocation of load centres will at some time necessitate a reinforcement of the network involving additional lines/substations at present voltage levels or at a higher voltage level. Such developments are often opposed because of the high level of public concern about their effect on the environment, including visual impact, land requirement and feared health risks in electric and magnetic fields. This has led to the design of compact lines of reduced dimensions thus lessening both the visual impact and land requirement, but still providing the reliability of conventional networks, and also allowing their upgrading [40, 41]. The reductions have been achieved through:

- improvements in overvoltage protection and control through using modern surge arresters
- developments in insulation including polymeric insulation
- changes in design such as insulating crossarms on towers.

The principal gains obtained using compact lines are:

- maximisation of power transfer through available transmission corridor
- minimisation of environmental impact.

7.6.1 Insulation

Composite insulators made of polymeric materials such as EPDM and silicone rubber are used extensively on compact networks. They have a high mechanical strength under tensile, compressive and bending loads and are lighter than porcelain insulators. Their performance under polluted conditions is superior due to their hydrophobicity and this is highly important on compact systems where reduced clearances can increase their sensitivity to pollution (for further reading, refer to Chapter 6).

Insulating crossarms are widely used to reduce tower dimensions and typically consist of a glass fibre core with silicon rubbers sheaths and sheds. They are usually mounted in a horizontal V or line post arrangements as shown in Figure 7.17 [45] and prevent lateral movement of the conductor at the tower with some designs allowing limited longitudinal movement to relieve torsional stress. A 400 kV line with insulating crossarms and requiring a transmission corridor width of 19 m compared with 30 m for a conventional line is described in Reference 42.

Insulating interphase spacers are sometimes necessary along compact line spans to limit conductor movement, and consist of silicone rubber sheaths moulded to an epoxy reinforced glass fibre rod [43]. The spacers allow further compaction of the line but increase the risk of phase-to-phase breakdown under polluted conditions.

7.6.2 Surge arresters

Surge arresters provide the protection necessary to limit switching overvoltages at line terminations to less than 2.0 p.u. Arresters are usually connected phase-to-earth and are less effective against phase-to-phase overvoltages. Phase-to-phase arresters may be required on compact lines and can be integral with interphase spacers. Phase-to-phase arresters have been considered as a feasible alternative to resistance switching [44].

7.6.3 Comparison between compact and conventional network

Table 7.7 is based on the data taken from Reference 1 and shows standard values of impulse withstand voltages and minimum air clearances.

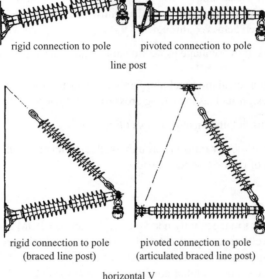

rigid connection to pole   pivoted connection to pole

line post

rigid connection to pole   pivoted connection to pole
(braced line post)   (articulated braced line post)

horizontal V

*Figure 7.17   Typical insulating crossarm [45] (courtesy of paper 400-04 from the
Leningrad CIGRE symposium, 3–5 June, 1991)*

Table 7.7   Standard insulation levels

Nominal voltage (kV r.m.s.)	Switching impulse withstand voltage (kV pk)	Lightning impulse withstand voltage (kVpk)	Minimum clearance conductor-structure (m)
275	750	950	1.6
275	850	1050	1.8
400	850	1175	1.8
400	950	1300	2.2
400	1050	1425	2.6

Switching impulse withstand voltage on 400 kV networks are typically at the
upper limits of 1050 kV. With appropriate overvoltage protection this can be reduced
to 850 kV, reducing the minimum clearances from 2.6 to 1.8 m. Compact networks
take advantage of these reductions to provide a network with a similar risk of failure
factor to that of a conventional design.

A comparison between 275 kV line with a switching impulse level of 850 kV
and a compact 400 kV line with a switching impulse level of 850 kV indicates that
a conventional 275 kV transmission corridor could accommodate a compact 400 kV
line with a doubling in transmission capacity.

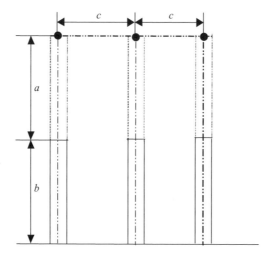

Figure 7.18   Busbar arrangement

Table 7.8   Busbar dimensions

	Conventional substation	Compact substation
SIWL (kVpk)	1050	850
Mini ph earth clearance (m)	2.6	1.8
Mini ph ph clearance (m)	3.6	2.6
Insulator height 'a' (m)	4.0	3.2
Support height 'b' (m)	3.0	3.0
Bubsar centres 'c' (m)	4.6	3.6

Comparable gains are feasible with compact air insulated substations as is demonstrated by the simple busbar arrangements shown in Figure 7.18. Table 7.8 shows the principal dimensions for both networks allowing the same margins above the minimum clearances. The data indicate a reduction of 11 per cent in height and 21 per cent in width.

The effect of compaction on the profile of the electric and magnetic fields for the above geometry is shown in Figure 7.19 and is based on a symmetrical busbar voltage of 400 kV and symmetrical currents of 2 kA/phase and assumes zero earth resistivity. Compaction has resulted in maximum values for both the electric and magnetic fields being some ten per cent higher than those for the conventional layout. The magnetic field is dependent upon the earth resistivity ρ_e and profiles are shown for $\rho_e = 0$ and $\rho_e = \infty$.

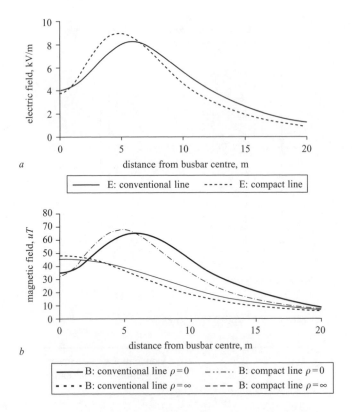

Figure 7.19   *Electric and magnetic field profiles under conventional and compact busbars*

   a   electric field profiles
   b   magnetic field profiles

These simple comparisons show advantages for compact networks, however these will be modified because compact lines are usually designed for a particular situation and there is limited standardisation, this often introduces cost penalties.

7.7   Acknowledgement

The authors wish to thank Niamat Ullah for his help in obtaining and preparing the computation results.

7.8   Note

[1] Figures from British Standards reproduced with the permission of BSI under licence number 2003SK/0157. British Standards can be obtained from BSI Customer Services, 389 Chiswick High Road, London W4 4AL (Tel +44 (0) 20 8996 9001).

7.9 References

1 IEC 60071-1: 1993: 'Insulation co-ordination part 1: definitions, principles and rules'

2 IEC 60071-1: 1993: 'Insulation co-ordination part 2: applications guide'

3 RAGALLER, K., BURGER, U., and RUOSS, E.: 'Introduction and survey in high voltage networks', *in* RAGALLER, K. (Ed.): 'Surges in high-voltage networks' (Plenum Press, London, 1980), pp. 1–24

4 BERGER, K.: 'Lightning surges', *in* RAGALLER, K. (Ed.): 'Surges in high voltage networks' (Plenum Press, London, 1980), pp. 25–62

5 GOLDE, R.H.: 'Lightning, vol. 1: physics of lightning' (Academic Press, London, 1977)

6 GOLDE, R.H.: 'Lightning, vol. 2: lightning protection' (Academic Press, London, 1977)

7 GARY, C.: 'La foudre', *in* EYROLLES, E.D.F. (Ed.): 'Les properties dielectriques de l'air et les tres hautes tensions' (Paris, 1984), pp. 92–229

8 ANDERSON, R.B., ERIKSSON, A.J., KRONINGER, H., MEAL, D.V., and SMITH, M.A.: 'Lightning and thunderstorm parameters'. Proceedings of the IEE conference on *Lightning and power systems*, Conf. Publ. 236, 1984, pp. 57–66

9 ERSKINE, A.: 'Lightning flash current magnitudes and ground flash density in the UK'. Proceedings of the 24th *UPEC*, Belfast, 1989, pp. 565–568

10 GALLAGHER, T.J., and PEARMAIN, A.J.: 'High voltage – measurement, testing and design' (John Wiley & Sons, 1983)

11 AGUET, M., and IANOVICI, M.: 'Phenomenes transitoires en haute tension', *in* 'Traite d'Electricite', Vol. XXII Haute Tension, E.P.F., GIORGI Lauzanne, 1982, pp. 67–109

12 LEVINSON, L.M., and PHILIPP, H.R.: 'ZnO varistors for transient protection', *IEEE Trans. on Parts Hybrids Packag.*, 1977, **PHP-13**, (4), pp. 338–343

13 SMITH, J.: 'Geomagnetic'. Proceedings of the 25th *UPEC*, Aberdeen, 1990, pp. 358–360

14 LE ROY, G.: 'Les contraintes internes', *in* EYROLLES, E.D.F. (Ed.): 'Les proprieties dielectriques de l'air et les tres hautes tensions' (Paris, 1984), pp. 1–90

15 ERCHE, M.: 'Switching surges', *in* RAGALLER, K. (Ed.): 'Surges in high-voltage networks' (Plenum Press, London, 1980), pp. 63–97

16 BICKFORD, J.P.: 'Electric power transmission systems' (Peter Peregrinus Ltd, London, 1976)

17 EATON, J.R.: 'Electric power transmission systems' (Prentice-Hall, Inc, Englewood Cliffs, 1972)

18 CSUROS, L.: 'Overvoltage protection', *in* The Electricity Council (Ed.): 'Power system protection 2 systems and models' (Peter Peregrinus Ltd, Stevenage, 1981), pp. 276–316

19 DIESENDORF, W.: 'Temporary overvoltages', *in* 'High-voltage electric power systems' (Butterworths, London, 1974)

20   GLAVITSCH, H.: 'Temporary overvoltages', *in* RAGALLER, K. (Ed.): 'Surges in high voltage networks' (Plenum Press, London, 1980), pp. 131–163

21   CIGRE working Group 33.10 and IEEE Task force on TOV: 'Temporary overvoltages causes, effects and evaluation', CIGRE conference 1990 session, paper 33-210, 1990

22   DOMMEL, H.W., and MEYER, W.S.: 'Computation of electromagnetic transients', *Proc. IEEE*, 1974, **62**, pp. 983–993

23   GREENWOOD, A.: 'Electrical transients in power systems' (John Wiley & Son, Chichester, 1991)

24   CHOWDHURI, P.: 'Electromagnetic transients in power systems' (John Wiley & Son, Chichester, 1996)

25   DOMMEL, H.W.: 'Digital computer solutions of electromagnetic transients in single and multiphase networks', *IEEE Trans.*, 1969, **PAS-88**, pp. 388–389

26   CARSON, J.R.: 'Wave propagation in overhead wires with ground return', *Bell Syst. Tech. J.*, 1926, **5**, pp. 539–554

27   NAKAGAWA, M., and IWAMOTO, K.: 'Earth return impedance for the multilayer case', *IEEE Trans.*, 1976, **PAS-95**, pp. 671–676

28   DERI, A., TEVAN, C., SEMLYEN, A., and CASTANHEIRA, A.: 'The complex ground return plane, a simplified model for homogeneous and multi-layer earth return', *IEEE Trans.*, 1981, **PAS-10**, pp. 3686–3693

29   WEDEPOHL, L.M., and EFTHYMIADIS, A.E.: 'Wave propagation in transmission lines over lossy ground: a new complete field solution', *Proc. IEEE*, 1978, **125**, pp. 505–510

30   MARTI, J.R.: 'Accurate modelling of frequency dependent transmission lines in electromagnetic transient simulations', *IEEE Trans.*, 1982, **PAS-101**, pp. 147–157

31   CLARKE, E.: 'Circuit analysis of a.c. power systems' (John Wiley & Son, Chichester, vol. 1, 1950)

32   TAVARES, M.C., PISSOLATO, J., and PORTELA C.M.: 'Mode domain multiphase transmission line model-ore in transient studies', *IEEE Trans.*, 1999, **P-D 14**, pp. 1533–1540

33   DOMMEL, H.W.: 'Electromagnetic transients program reference manuel (EMTP theory book)'. Bonneville Power Administration, Oregon, 1986

34   BRANDEO FARIA, J.A., and HILDEMARO BRICERO, J.: 'On the modal analysis of asymetrical three-phase transmission lines using standard transformations', *IEEE Trans.*, 1997, **PWRD-12**, pp. 1760–1765

35   CHISHOLM, W.A., and CHOU, Y.L.: 'Travel time of transmission towers', *IEEE Trans.*, 1985, **PAS-104**, pp. 2922–2928

36   AL-TAI, M.A., ELAYAN, H.S.B., GERMAN, D.M., HADDAD, A., HARID, N., and WATERS, R.T.: 'The simulation of corona on transmission lines', *IEEE Trans.*, 1989, **PD-4**, pp. 1360–1369

37   GERMAN, D.M., and DAVIES, A.E.: 'The simulation of transformer feeders following switching operations', *IEEE Trans.*, 1982, **PAS-100**, pp. 4510–4514

38   JONES, B., and WATERS, R.T.: 'Air insulations at large spacings', *IEE Proc.*, 1978, **125**, pp. 1152–1176

39 ELOVAORA, J.: 'Risk of failure determination of overhead line phase-to-earth insulation under switching surges', *Electra*, 1978, (56), pp. 69–87

40 ELAHI, H., PANELI, J., STEWART, J.R., and PUENTE, H.R.: 'Substation voltage uprating – design and experience', *IEEE Trans.*, 1991, **PD-6**, pp. 1049–1057

41 TONG, Y.K., BENHACKE, R.H., CLARK, A.M., SPARROW, L.T., WHITE, H.B., and OPASCHASTAT, P.: '500kV compact line designs for the greater Bangkok area'. CIGRE international session, group 22, 1998

42 AMMAN, M., and PAPAILOU K.O.: 'Long term experience with silicon composite insulators in the h.v. lines of "Energie Overt Suisse"'. Mid power conference, Cyprus, 1998

43 KARADY, C.G., SCHNEIDER, H.M., and HALL, J.F.: 'Utilisation of compact insulators for compacting transmission lines', CIGRE symposium, *Compacting overhead transmission lines*, June 1991, Leningrad, paper 400-05

44 RIBEIRO, J.R., LAMBERT, S.R., and WILSON, D.D.: 'Protection of compact transmission lines with metal oxide arresters'. CIGRE symposium , *Compacting overhead transmission lines*, June 1991, Leningrad, paper 400-06

45 PARIS, L., and PARGAMIN, L.: 'Application of composite insulators for overhead transmission lines'. CIGRE symposium, *Compacting overhead transmission lines*, June 1991, Leningrad, paper 400-04

Chapter 8

Earthing

H. Griffiths and N. Pilling

8.1 Introduction

Three-phase power systems are earthed by connecting one or more selected neutral points to buried earth electrode systems. Such earths are referred to as system earths. At electrical installations, all non-live conductive metallic parts are interconnected and also earthed to protect people against electric shock, and in this role, the earth is referred to as a protective earth. Under normal conditions, there is only a residual current or no current at all in the earth path. However, very high magnitudes of current return to source via the earth path under fault conditions. The earth also conducts lightning currents and the current path may involve part of a power system either directly or by induction. The earthing system, or part of it, may therefore also be specifically designed to act as a lightning protective earth [1].

The earth is a poor conductor and, therefore, when it carries high magnitude current, a large potential gradient will result and the earthing system will exhibit an earth potential rise (EPR). Earth potential rise is defined in the recently published CENELEC document HD 637 S1:1999 [1] as the voltage between an earthing system and reference earth. In the UK, this quantity has previously been referred to as the rise of earth potential (ROEP) or the earth electrode potential. In the US, it is referred to as the ground potential rise (GPR). Here, this quantity will be referred to as the earth potential rise (EPR).

Soil and rock resistivity may vary considerably from region to region, and it is rarely constant either vertically or horizontally in the area of interest around an electrical installation. This variability makes the construction of earth models for the prediction of earth potential rise a very difficult task. The magnitude of power frequency earth fault currents can range from a few kA up to 20–30 kA, and earth impedances of high voltage substations may lie in the range from 0.05 Ω to over 1 Ω. Although higher fault current magnitudes are generally associated with lower magnitude earth impedances, earth potential rises can be as high as several tens of kV. Consequently, there is a potential risk of electrocution to people in the vicinity of power systems during earth faults, and damage to equipment may also occur unless measures are taken to limit the earth potential rise and/or to control potential differences in critical places. Lightning transients can also generate currents of several tens of kA in the earthing system, and this requires the power system to be protected against overvoltages. The discharge of transient current into earth may also present an electrocution hazard but the tolerable limits are less well defined compared with power frequency currents.

In the past, earthing systems were designed to achieve earth resistances below a specified value or on a particular density of buried conductor. Current practice, however, dictates that such systems are designed to control potential differences within and around the electrical installation. These potential differences are specifically referred to as step and touch voltages. For example, in 1992, the UK Electricity Association introduced a new earthing standard (EA-T.S.41-24) [2], with specific step and touch safe voltage levels, replacing Engineering Recommendation S.5/1 [3]. In North America, the change to quantify safe voltages occurred earlier [4]. In some standards, consideration is also given to the maximum earth potential rise of the earthing system. The extent of the rise of potential on the ground surface around a substation can be described in terms of a hot zone, and this is used to identify whether third parties in the vicinity of the installation are affected.

Nowadays, there is great awareness of the safety issues concerning the earthing of electrical installations. Some recent developments have also highlighted the importance of this subject and the new challenges that face earthing system designers:

a   *Restricted land area*: replacing or upgrading substations in urban areas restricts the options available to control earth potential differences.
b   *Urban encroachment*: urban encroachment of transmission lines has raised concerns about earth potential rise at transmission line tower bases.
c   *Mobile communication base stations*: the rapid growth in mobile telecommunications has resulted in power transmission line towers hosting GSM (global system for mobile communication) base stations. Such installations bring the transmission system into close proximity to the distribution systems supplying the base stations, and introduce challenges for insulating and protecting against faults on the high voltage side.
d   *Modernisation of railway electrification systems*: new electrification schemes involve siting transmission substations in close proximity to the railway network (typically 400 kV/25 kV) where, traditionally, the earthing systems of the respective systems were segregated.
e   *Windfarms*: the design of earthing systems for windfarms requires special consideration because of the distributed nature of the electrical network and the absence of a low impedance earth at the power system interface.

This chapter aims to outline the current state of the subject of earthing covering both power frequency and transient aspects. To begin with, the main earthing system components at transmission and distribution levels are described and different methods of earthing the neutral are considered. The next section is devoted to earth resistivity, and conduction mechanisms are described together with site investigation techniques for determining earth structure and resistivity. In reviewing power frequency performance, the procedure for designing safe earthing installations is explained based on information contained in the latest standards and published literature. This procedure has a number of stages which include calculations to determine (i) the earth impedance of different earthing system components, (ii) earth fault current magnitude and the proportion of this current that will flow through the earthing system and (iii) step and touch voltages and their maximum acceptable values. Each of these stages can be carried out with different degrees of complexity and detail depending on the level of accuracy required. Accordingly, earthing system design may be carried out entirely using analytical formulae using a homogeneous earth model or may require the use of specialised earthing software and complex earth models. A section is devoted to the description of the measurement of earthing system impedance and potentials which are important for the validation of the performance of installed earthing systems. To conclude the treatment of power frequency aspects, a new approach to earthing system design is introduced which uses risk management techniques to identify priorities for investment to improve safety.

The final section of this chapter deals with the transient behaviour of earthing systems. A summary of current earthing design recommendations from standards is provided and the subject of soil ionisation under high current magnitudes is considered in some detail. Circuit models and simulation software for evaluating the high frequency and transient performance of earthing systems are also described.

8.2   Earthing system components and system earthing methods

8.2.1   Transmission system

The 400 kV/275 kV transmission system operated by the National Grid in the UK comprises over 250 substations interconnected by 15 000 km of double circuit overhead lines, supported by about 26 000 transmission towers [5]. Approximately 200 faults per annum occur on the transmission system, of which almost 90 per cent can be classified as earth faults [6]. Although few in number, such faults produce current magnitudes of between 4 kA and 35 kA flowing through substation earthing systems, depending on the level of generation, the system configuration and the type and location of the fault [7]. Normally, faults are cleared by main protection systems in less than 160 ms, the precise speed depending on the technology of the circuit breaker. Backup protection would normally provide clearance within 500 ms [8].

At transmission level, earthing systems consist of two main components: main earth grids [2] and extended earth electrode systems formed by transmission tower lines. The main earth grid of an outdoor transmission substation will cover a substantial area, typically 30 000 m^2, although indoor substations can be much more compact with less scope for a substantial earth grid. The earth wire of transmission tower lines also forms an important part of the transmission earthing system in addition to its lightning shielding role. The earth wire interconnects the latticed steel towers and is terminated at the substation earth grids at the ends of each line. In this way, an extended earthing system is formed by virtue of the connections made to earth by the tower footings. The presence of the earth wire, therefore, reduces the earth impedance of the system at the main substation or indeed at any point along the transmission line. As a result, any rise in earth potential is transferred over a much larger area. The impedance of the extended earthing system formed by the transmission line earth wire and towers is commonly referred to as the chain impedance. The earthing systems of transmission substations directly supplying distribution systems (typically at 132 kV in the UK) are also connected to the lower voltage shielded tower line circuits. In addition to lowering the earth impedance of the system, a shielded circuit system also results in a significant reduction of the proportion of earth fault current flowing back to the source through earth. This is due to the direct conductive path formed by the shield and also the inductive mutual coupling between the faulted phase conductor and the shield.

In urban areas, the transmission substation may be situated close to subtransmission and distribution substations. Usually, the earthing systems of these substations will be bonded together. At such sites, the overall earth impedance can be very low due to the effect of the extended earthing systems of the lower voltage systems. However, although this interconnection reduces the earth impedance of the transmission system, potentials developed due to high magnitude fault currents on the higher voltage system will be transferred over wide areas to the lower voltage earthing systems.

8.2.2   Distribution system

In the UK, the distribution system uses voltages from 132 kV down to 6 kV. It comprises approximately 180 000 km of overhead lines and 133 000 km of underground cables. It is estimated that there are approximately 5000 substations operating from 132 kV to 20 kV and about 170 000 substations distributing power from 11 kV to the low voltage system [9].

On the distribution system, the fault rate is significantly higher compared with the transmission system. With the considerably greater route length of the distribution system, it is not surprising that there are significantly more faults occurring at this level of the system. It is estimated that there may be 25 000 faults per annum on the HV distribution system, 70 per cent of which involve a connection to earth [10, 11]. Typically, earth fault current magnitudes are 25 kA at 132 kV, 1–2 kA at 33 kV and 11 kV substations. Below 66 kV, the system is designed to restrict earth fault current magnitudes. Although fault current magnitudes on the distribution system are generally much lower compared with the transmission system, fault clearance times are considerably longer. As a general rule, the fault clearance time will increase the lower the distribution voltage level. On 11 kV systems employing IDMT protection, fault clearance times may exceed 1 s. It is important to establish accurate fault clearance times because the tolerable level of current passing through the human body depends on the shock exposure time, which is normally assumed equal to the fault duration. The longer the fault, the lower the tolerable body limit.

The 132 kV shielded double circuit system in the UK originally functioned as a transmission grid. Over a period of time, it evolved into a distribution system as generation displaced to the Super Grid at 275 kV and 400 kV [12]. These days, new circuits at 132 kV are often constructed using unearthed wood pole single circuits and the vast majority of overhead line construction at 66 kV and below has always been unearthed. Hence, distribution substations connected to overhead line circuits may not always have the benefit of an extended earthing system.

On the other hand, most distribution circuits in urban areas will consist of underground cables. The metallic sheaths of these cables form an extended earthing system by virtue of their connection to earthing points along the circuit. This system can be highly interconnected and cover a large area providing low earth impedance values. Also, similar to the action of the shielding wire on overhead lines, the sheath of a cable provides an alternative path for the earth fault current to return to source. The mutual coupling between the core and sheath of a cable is considerably greater than that between a phase conductor and earthwire of an overhead line, and therefore this mutual effect is much greater for cables.

Older underground cables were constructed with lead sheaths and covered by an insulating layer of bitumen-impregnated hessian (PILCSWA – paper-insulated lead sheathed steel wire armoured). It has been found that, over a period of time, this type of insulation degrades leaving the lead sheath of the cable effectively in direct contact with soil. Such cables, therefore, provide an additional fortuitous earth connection [13]. These PILCSWA cables are steadily being replaced by plastic insulated cables, which will result in a general increase in earth impedances for such urban systems.

8.2.3   Methods of system earthing–treatment of neutral

8.2.3.1   Background

In the early period of development of three-phase power systems, it was common practice to isolate neutral points and operate the system in an unearthed state. In the UK, the majority of the high voltage systems were operated this way until 1912 [14, 15], and in Germany this was the case until 1917 [16]. However, as power systems grew in size, problems with this method of operation emerged because the magnitude of earth fault current in an unearthed system increases with the phase-to-earth capacitance of the network. Above a certain current threshold, persistent intermittent arcing will occur during the fault resulting in damage to equipment close to the arcing fault. Also, damage can occur to other parts of the network as a result of high magnitude overvoltages that are developed [16–18]. Therefore, the permanent single-phase-to-earth fault on such systems becomes unmanageable and fast fault detection and isolation, or alternatively a method of suppressing the arc, are required [16].

Nowadays, some parts of high voltage networks still operate unearthed because it can be advantageous, under certain circumstances, to continue operating in the presence of an earth fault [19–21]. However, the most commonly recommended practice is to earth at least one neutral point of the network, and there are important advantages in operating the system in this way. The main advantage is that the power system is safer because earth faults are also much easier to detect, to clear and to locate [21].

Also, there is a reduction in overvoltage magnitude during earth faults. The Electricity Safety, Quality and Continuity Regulations 2002 state 'A generator or distributor shall, in respect of any high voltage network which he owns or operates, ensure that the network is connected with earth at, or as near as is reasonably practicable to the source of voltage. ...' [22]. The most convenient way of making this connection is to use the neutral formed by the star point of the supply transformer. For delta-connected equipment, an earthing transformer may be employed for this purpose [21].

According to how the neutral is connected to earth, earthed systems are categorised as either solidly earthed or impedance earthed. Impedance earthed systems can be classified as resistance, reactance or resonant type.

8.2.3.2   Solid earthing

Solid earthing describes the connection of the system neutral point directly to the installation earth. This results in high earth fault levels but minimises system overvoltages.

8.2.3.3   Resistance earthing

With resistance earthing, a transformer neutral point and earth are connected together through a resistor. On systems at and above 11 kV a liquid earthing resistor (LER) is normally employed. The resistance value is chosen such that earth fault current levels are reduced considerably, typically to 1000 A at 11 kV. However, although fault current flow is restricted, sustained phase voltages during the earth fault are not increased substantially above those for solidly earthed systems. Transient overvoltages are also contained to within acceptable levels. Occasionally, high resistance earthing is used to limit overvoltages on previously ungrounded systems [21].

8.2.3.4   Reactance earthing

With this method of system earthing, a reactor is placed between the neutral and earth. When reactance earthing is used, it is recommended that the earth fault level should not be reduced to below 60 per cent of the three-phase fault level, in order to avoid high transient overvoltages [21]. Therefore, reactance earthing is not considered as an alternative to resistance earthing. Reactance earthing can be employed in systems that have an earth fault level exceeding the three-phase fault level per phase [21].

8.2.3.5   Resonant earthing

An alternative method of neutral earthing is achieved by inserting a reactor between the neutral point and earth such that it compensates the phase to earth capacitance of the system operating under the earth fault condition. This method is known as resonant earthing [23] or Petersen coil earthing, named after its founder. It is also sometimes referred to as arc suppression coil earthing or ground fault neutraliser earthing. The device, the arc suppression coil, is therefore also known as a Petersen coil or a ground fault neutraliser. Resonant earthing offers two main advantages. First, during a single-phase-to-earth fault, arcing is suppressed automatically without the need for current interruption. Second, if required, the network can be operated with the fault present because the capacitive earth fault current is compensated. A disadvantage of this method is that the transient and steady state voltages on healthy phases during the earth fault are greater than those on low impedance or solidly earthed systems. Also, the detection and location of an earth fault is more difficult because conventional overcurrent protection cannot be applied.

8.2.4 Application of different system earthing methods

In the UK power system, solid earthing has been adopted on networks operating at 66 kV up to 400 kV. At the transmission level (275 kV and 400 kV), the power system is also multiple earthed because it operates as a meshed system [24]. This practice reduces transformer capital costs, where graded insulation can be used and surge arresters of lower temporary overvoltage rating can be employed [11, 25–27]. At voltage levels from 6.6 kV to 66 kV, the system is normally radial configured, and therefore it is earthed at a single point corresponding to the neutral or derived neutral of the supply transformers. On networks at 33 kV and below, low resistance earthing is used.

A system is classified as effectively earthed when the single-phase-to-earth fault current is at least 60 per cent of the three-phase short circuit. This is achieved when the zero sequence resistance is not greater than the positive sequence reactance ($R_0 \leq X_1$) and the zero sequence reactance is not more than three times the positive sequence reactance ($X_0 \leq 3X_1$) [21]. The main advantages of effectively earthed systems are that arcing faults are very rare, overcurrent earth fault relays can be used to detect and isolate the faulty circuit and the transient and steady state neutral and phase voltages with respect to earth are suppressed [16].

8.3 Earth resistivity and measurement techniques

8.3.1 Conduction mechanisms and resistivity

Electrical conduction in earth is predominantly electrolytic conduction in the solutions of water-bearing rocks and soils. Metallic conduction, electronic semiconduction and solid electrolytic conduction can occur but only when specific native metals and minerals are present [28]. Accordingly, the resistivity of soil or rock normally depends on:

a the degree of porosity or fracturing of the material
b the type of electrolyte
c temperature.

Figure 8.1 indicates the effect of variations in salt, moisture and temperature on resistivity [31].

8.3.2 Resistivity data of soils and rocks

At power frequency, the series impedance of the conductors forming a concentrated earth electrode is small, and when the soil has fully consolidated around the installed

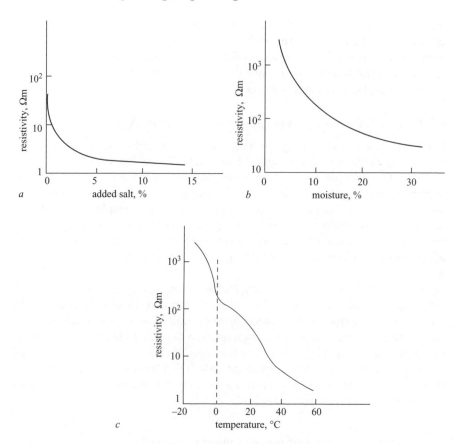

Figure 8.1   *Variation of resistivity with salt (a), moisture (b) and temperature (c)*
(reproduced from IEEE Std.81 1983, IEEE Copyright 1983 and IEEE
Std.80 2000. All rights reserved [31])

system the contact resistance between electrode and the soil can also be neglected.
Therefore, it is the resistivity of the surrounding earth that determines the earth resistance of the electrode. The resistivity and earth structure will also determine the
potential distribution on the surface of the ground around the electrode, and it may be
necessary to construct very accurate earth models in order to predict step and touch
voltages around an installation.

Figure 8.2 [28] shows the wide range of resistivity for different soil and rock
types. As can be seen in the figure, resistivity can vary from about 10 Ωm (clays) to
10^6 Ωm (granites), and, in general, resistivity increases with the age of the geological
formation [29]. Also noticeable is that the wide resistivity ranges for nearly all types
of soil and rock overlap. This illustrates the difficulty in estimating earth resistivity
from a geological classification. Accordingly, it is recommended that practical investigations be carried out on site in order to measure directly the earth resistivity for
each specific electrical installation [30]. However, the variable qualities of natural

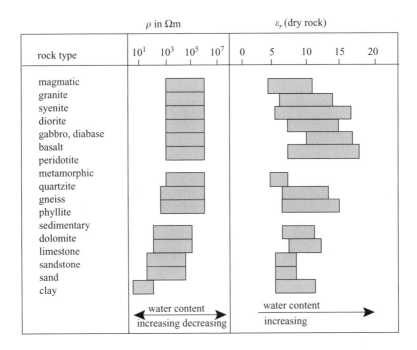

Figure 8.2   Mean value ranges of resistivity and permittivity for different rock types (reproduced from [28])

soil and rock make this task difficult. As expected, variations in resistivity with depth are common. Lateral variations can also be significant even across one site, although this effect is sometimes neglected [31]. In addition, seasonal variations in water content will also result in changes in earth resistivity and temperature variations at the soil surface may also have an effect.

8.3.3   Site investigation and measurement techniques of earth resistivity and structure

It is important to determine an accurate earth resistivity model as a basis for designing an earthing system, and there are a number of different but complimentary methods and sources of information available for this purpose. Many of these, such as geological maps, borehole data, seismic testing and ground penetrating radar, are useful for identifying physical boundaries in the earth. However, they do not quantify accurately earth resistivity. On the other hand, only earth resistivity measurements carried out on the ground surface may not be sufficiently adequate to define the boundaries of a earth model uniquely. Earth resistivity surveying used in conjunction with one or more of the methods for identifying earth region boundaries may offer a much greater degree of confidence in the derived earth models.

8.3.3.1 Geological maps

Geological maps are available in solid and drift formats. Solid maps describe the underlying bedrock, and drift maps detail superficial deposits. The British Geological Survey produced a series of maps for the UK, including 1:625000 scale maps, showing the geology of the UK landmass. A comprehensive series of 1:50000 scale solid and drift maps is also available with supporting literature [32]. Although these maps are two-dimensional, data for certain vertical cross-sections are provided which gives the depths of different layers.

8.3.3.2 Borehole data

At electrical installations requiring detailed civil engineering works, borehole samples can provide information about changes in the earth structure and type with depth. Although this type of information is useful for defining layer boundaries in earth analysis, measurement of the resistivity of the removed earth sections is not recommended because the compaction and moisture content of the samples will be significantly affected by the extraction process and may no longer be representative [31].

8.3.3.3 Seismic surveying

To perform a seismic survey, an acoustic wave is generated on the ground surface (e.g. using a sledgehammer) and the reflected waves from earth layer boundaries are measured using geophones placed in a line away from the source. By analysing the travel times and amplitudes of the returning waveforms, estimates of the thickness and density of the subsurface layers can be obtained. An estimation of the depth to bedrock or the level of the water table is particularly useful for earthing investigations [33, 34].

8.3.3.4 Ground penetrating radar

A radar set is used to generate short bursts of VHF electromagnetic waves in the frequency range 35 MHz to 900 MHz. Reflections of these waves are produced from boundaries between materials with contrasting resistivity and/or dielectric properties. By estimating the velocity of propagation, it is possible to determine the depth to these boundaries. Using this technology, depths of investigation may be limited to tens of metres in dry sands, and as little as 1 m in wet clay [34].

8.3.3.5 Earth resistivity measurements

The most widely recommended and commonly used earth resistivity technique in earthing investigations is the Wenner method [30, 35–38]. This method uses four equally spaced electrodes, as shown in Figure 8.3 and its characteristic feature is a constant spacing a between adjacent electrodes. Current is circulated between the outer two electrodes and the potential difference is measured between the inner two electrodes. A DC instrument is used to energise the circuit but in order to overcome the effects of telluric currents, electrochemically-produced currents and interference from DC power sources, a periodic reversal in the direction of the current is necessary.

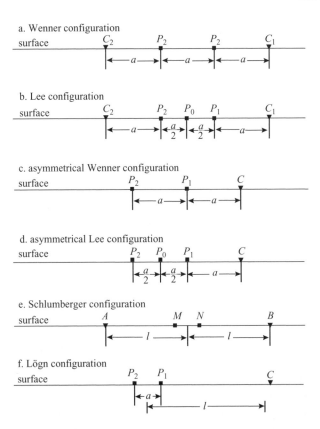

Figure 8.3   Various soil resistivity test electrode configurations (reproduced from [38])

Connection of the voltage terminals can also be delayed until any switching transients have decayed. Telluric currents flow in the earth over large areas with changing magnitude and direction and varying periods. They correspond to regular and irregular changes in the geomagnetic field and are particularly high during magnetic storms. Electrochemically-produced currents can be much higher in magnitude than telluric currents and are produced, for example, by differential oxidation processes. Lee records the use of instruments such as the Megger in earth resistivity measurement as far back as 1928 [38], and a modern version is used extensively in the UK nowadays [39]. A number of alternative earth resistivity instruments are available from different manufacturers including those specifically designed for geophysical exploration. The specifications of these instruments vary considerably.

Earth resistivity measurements taken in the vicinity of existing electrical instal-lations may be significantly affected if test transects are close to the buried earth grid or the routes of the substation's underground cable circuits. The presence of buried metallic objects would, as expected, result in a general underestimation of the resistivity of the native ground. However, such distortion in the apparent resistivity readings may not be easily recognisable and could be interpreted as a variation due to the natural heterogeneity of the earth. Therefore, it is advisable to identify the location and avoid measurements around cable circuit routes. If the orientation of an earth resistivity transect is varied around a centre point, the presence of underground metallic structures may be indicated by changing trends in the resulting apparent resistivity curves [31, 40].

(i) The Wenner method

For any given spacing a, the ratio of voltage to current will yield a specific value of resistance. If the current electrodes are assumed as point sources and the resistivity, ρ, to be homogeneous, it can be shown by applying the principle of superposition that the resistivity is related to the measured resistance, R, by the following formula [38]:

$$\rho_A = 2\pi a \frac{V}{I} = 2\pi a R \tag{8.1}$$

Even if the earth is heterogeneous, the quotient of V/I can be obtained and will be related to a quantity that is dimensionally equivalent to resistivity. This quantity can be considered as an index indicative, in a complex way, of the different resistivities present beneath the surface. This index is known as apparent resistivity. If a number of readings are taken for different constant interelectrode spacings, a, an apparent resistivity curve can be plotted as a function of a. The curve can be compared against theoretically obtained curves to derive approximate earth models, and the apparent resistivity will usually lie within the range of actual material resistivities. As electrode spacing is increased, the current path will involve a larger volume of earth and encompass deeper earth strata. Therefore, the apparent resistivity as a function of probe spacing provides an indication of the change in resistivity of the earth as a function of depth. However, the depth rule, which states that 'the electrode spacing is equal to the depth of penetration', should be applied with caution because the correspondence between spacing and depth is only approximate. There are a number of variants of the Wenner configuration and these are also shown in Figure 8.3 [38]. A distinguishing feature of these alternative methods is that the potential measurement is made across a much smaller distance and this has two implications:

a   The voltage measurement at small distances can be approximated to a measure of the electric field at that location and it is, therefore, easier to detect geological changes. In contrast, if the potential is measured across a large distance, effects of geological anomalies tend to be attenuated.

b   The measurement of such small voltage magnitudes with sufficient accuracy may be difficult in practice either because of limitations in the resolution of the instrumentation or due to high levels of background noise in the measured signal.

Taking into account these considerations, the Wenner method is well suited to applications requiring reasonable accuracy but where the detection of local heterogeneity is not of primary concern.

(ii) Other methods

The Schlumberger–Palmer method is another configuration that uses unequally spaced probes. For this arrangement, the spacing between the potential electrodes is increased such that they are much closer to the current electrodes [31]. In this way, a higher potential difference is obtained compared with other arrangements and this can be beneficial when the resolution of the resistivity meter is a limitation. With both the Schlumberger and Wenner configurations, the requirement to investigate the resistivity of deeper earth for large area substations introduces practical difficulties. For such tests, a very large interelectrode separation is required which necessitates long test leads and a relatively unobstructed path along the test route. In order to overcome these difficulties, dipole techniques were used to survey to greater depths, up to several kilometres, by separating the current and voltage circuits, which are referred to as dipoles. Measurements carried out at a very large spacing (330 km) indicated an upper limit to the resistivity of deep strata in the order of 10^4 Ωm [41].

(iii) Resistivity surveying techniques

In order to obtain information on earth resistivity changes with depth, the centre point of the test electrode setup is kept fixed and the interelectrode spacing is varied. This is known as vertical profiling. If the whole configuration is moved along a traverse at a fixed spacing, it is referred to as horizontal profiling [38].

8.3.3.6 Transient electromagnetic technique

In earthing applications, the depth of interest for determining resistivity may vary from a few metres to several hundred metres depending on the size of the electrode system. The transient electromagnetic technique (TEM) offers a practical method for obtaining earth resistivity data at significant depths. In this technique, a large closed transmitter loop is laid out on the earth's surface. A DC current is established in the loop which is abruptly interrupted. This sudden change induces currents in the earth which decay according to the resistivity and structure of the earth. The decaying currents produce a magnetic field which is measured by a receiver on the earth's surface. By analysing the nature of the transient decay, it is possible to determine equivalent resistivity models [42].

8.3.3.7 Earth resistivity maps

Numerous earth resistivity measurements were carried out in 'carefully selected parts of the UK with a view to ascertaining the range of values which are representative of each of the principle geological formations' [43]. These measurements were taken around the time when the first 132 kV national grid network was established, for the purpose of calculating telephone interference. The apparent resistivity values at a spacing of 500 feet were used to categorise the resistivity of the solid geology. As a result, two electrical resistivity maps were produced based on solid geological maps; one covering England and Wales [44] (c.f. Geological map of the UK (South) [45]) and the other of Southern Scotland [46] and these are reproduced in Figures 8.4*a* and 8.4*b*. It is stated that the maps are not suitable for quantifying the resistivity of

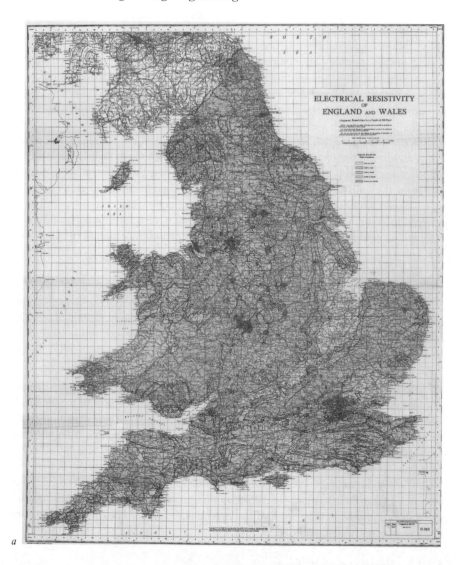

Figure 8.4a   Electrical resistivity map of England and Wales (reproduced from both a 1934 Ordnance Survey Map and [44])

subsurface layers. These maps need updating as good progress has been made since their establishment in terms of measurement techniques and analysis of data.

8.3.3.8   Formation of earth models

(i) Apparent resistivity curves

The recommended practice for earth resistivity investigations in earthing system design is to carry out vertical profiling at a number of positions around the

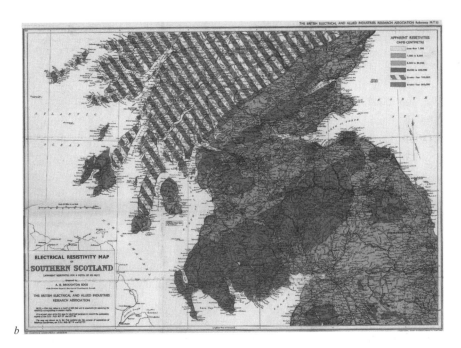

b

Figure 8.4b   *Electrical resistivity map of Southern Scotland (reproduced from © John Bartholomew, ERA, IPR [46])*

installation [30]. Test results of vertical profiles taken at different sites are shown in Figure 8.5 [47]. Curve A of Figure 8.5 illustrates a case of when the apparent resistivity varies little with spacing which indicates reasonably homogeneous earth conditions. However, all other curves indicate that changes occur in apparent resistivity as a result of the non-uniformities in the earth. Curve E represents a two-layer earth with the top layer consisting of a higher resistivity than the lower layer. By contrast, curve D shows a low resistivity near the surface and suggests very high resistivity at depth. The reflection coefficient k is used to describe the contrast in resistivity between the upper layer (ρ_1) and the lower layer (ρ_2):

$$k = \frac{\rho_2 - \rho_1}{\rho_1 + \rho_2} \qquad (8.2)$$

Curves B and C are characteristic of three-layer earth structures: low resistivity near the surface, an intermediate layer of higher resistivity and lower resistivity at depth. Habberjam [48] refers to such three-layer curves as K-type curves, which is one of four different classifications of three-layer earth models. Apparent resistivity curves may be obtained which indicate earth structures consisting of more than three layers. However, as the complexity of the earth structure increases, it becomes more difficult to construct reliable models.

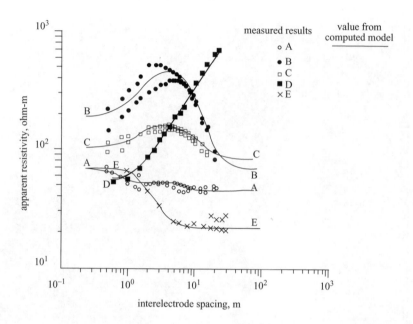

Figure 8.5   Apparent resistivity versus interelectrode spacing (reproduced from [47])

(ii) Simple models

Some earth structures may be very complex and practical testing can yield apparent resistivity curves with many maxima and minima. It may therefore be difficult to establish a suitable earth model for power system earthing applications and the published literature contains many different approaches to this problem. The simplest method that can be applied is to obtain an effective or equivalent value of earth resistivity from the apparent resistivity data [36, 49–51]. The selection of a single value resistivity is convenient because many analytical formulae for computing earth resistance and ground surface potentials are based on homogeneous earth assumptions and some earthing standards still employ such formulae [36, 52]. Recent studies [53] have shown that although it may be possible to obtain acceptable homogeneous equivalent earth models for the calculation of earth resistance of concentrated earth structures, their use for the prediction of earth surface potentials is questionable. Therefore, a homogeneous model would only be acceptable if site measurements indicate a small variation in resistivity over a wide range of interelectrode spacing. Recent measurements [54] of apparent resistivity (Wenner method – vertical profiling) at over 80 sites around the UK showed that significant variations occur over interelectrode spacing in the range from a few metres up to several tens of metres.

(iii) Standard models

To deal with more complex earth structures, the IEEE Std.80 standard recommends the formation of a two-layer earth model or if necessary a multilayer earth model according to the number of maxima and minima on the apparent resistivity curve. A number of techniques for obtaining such models are described [30]. Dawalibi and Barbeito [55] analysed practical earth grids and concluded that non-uniform earth requires the use of multilayer earth models to predict earth grid performance. They also noted the significant effect of vertical discontinuities and stressed the importance of obtaining resistivity readings of sufficiently wide spacing in order to establish the asymptotic value of earth resistivity at depths of the order of the earthing system dimensions.

(iv) Two-dimensional/three-dimensional models

In two or multilayer models, it is assumed that the resistivity changes only with depth and that there are no lateral variations. These are sometimes referred to as one-dimensional models as shown in Figure 8.6. If vertical profiling is combined with horizontal profiling along one line, it is possible to obtain a two-dimensional model. If the measurements are extended such that the horizontal profiling is carried out in two orthogonal directions, a three-dimensional model may be obtained. Typically, one-dimensional measurements may require 20 readings and three-dimensional surveys would require thousands. In order to make such measurements feasible, large numbers of electrodes, multicore cables and automatic switching systems interfaced with computerised data acquisition systems are required. The data from a two-dimensional survey can be plotted using the pseudosection contouring method such that the horizontal location of the plot point corresponds to the mid-point of the particular electrode configuration. This gives an approximate picture of the subsurface earth in a similar way to the apparent resistivity curves from one-dimensional surveys. To analyse data from such detailed surveys, an inversion procedure is required which constructs a model to fit the measured data. Commercial measurement systems with inversion software are available to analyse both two- and three-dimensional measurements [56].

8.4 Power frequency performance of earthing systems

8.4.1 Standards recommendations

The determination of the performance of an earthing system under power frequency earth fault conditions, for the purpose of developing a safe earthing design, requires a number of aspects to be investigated. The main procedure outlined in various standards is the same and is shown below, although the detailed methodology, estimation of parameters and safety criteria differ to some extent:

(i)   measurement of earth resistivity and formulation of earth model
(ii)   calculation of overall earth impedance including main earth grid resistance and impedance of extended earth electrodes using the model developed in (i)
(iii) earth fault current calculation for various fault locations
(iv)   fault current distribution calculation to determine the proportion of fault current passing through the earth

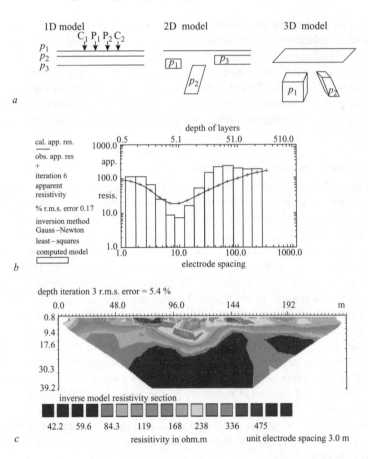

Figure 8.6

    a   three different models used in the interpretation of resistivity measurements

    b   a typical 1D model used in the interpretation of resistivity sounding data (reproduced from [56])

    c   an example of a 2D model

(v)   estimation of earth potential rise and the safety hazard voltages

(vi) specification of safety limits and determination of the safety level of the earthing design

(vii) measurements at installation to verify calculated impedances and potentials and, subsequently, to ensure continued integrity.

Apart from satisfying electrical safety, the earthing system should also have sufficient mechanical strength and be corrosion resistant, and the system should have adequate thermal capability for carrying the maximum fault current. These aspects are dealt with in detail in most earthing standards.

8.4.2 Earth impedance

8.4.2.1 Calculation of earth resistance of concentrated earth electrodes

Analytical expressions of earth resistance for simple forms of earth electrodes are well established [57, 58]. Formulae for various geometries including the sphere, rod, plate, strip, disk, ring, grid, grid and rods are specified in many standards [30, 36, 52]. All these expressions for earth resistance assume homogeneous earth.

The resistance of a hemispherical electrode of radius r in earth of resistivity ρ is given by:

$$R = \frac{\rho}{2\pi r} \tag{8.3}$$

This expression encapsulates the two main features of the power frequency behaviour common to all concentrated earths; i.e. that earth resistance is proportional to earth resistivity and inversely proportional to the extent of the electrode. Based on a realistic range of earth resistivity (section 8.3), the earth resistance of an earth grid may therefore vary from several tens of ohms for small grids to as little as 0.01 Ω for large systems in low resistivity earth. Unless the earthing system is very extensive or situated in very low resistivity earth, the series impedance of the conductors forming the earthing system is very small in comparison with the earth resistance at power frequency. Accordingly, in most cases, it can be assumed that all metallic points on the electrode are approximately at the same potential.

Recently, there have been considerable developments in techniques for evaluating the earth resistance of practical systems. Work has focused on improvements and extensions of the analytical formulae for simple electrode systems, in particular, to account for two-layer horizontal earth models [59–61]. Also, computer programs have been developed which model earth electrodes as systems of interconnected cylindrical conductor segments [4, 62]. Features of more advanced programs include the ability to model more accurately the asymmetry of practical earthing system designs. Such computer programs are now commonly employed to design grids and are used in conjunction with earthing standards.

New approaches to the modelling of earthing systems include the application of the boundary element method [63–65]. Using this method, it is possible to analyse earthing systems that are located in earth which exhibits both horizontal and vertical variations in resistivity and which may also include local inhomogeneities.

8.4.2.2 Chain impedance of extended earthing systems of tower lines

The impedance of the extended earthing system formed by a transmission line earth wire, commonly referred to as the chain impedance, depends on a number of factors. These are the size and type of the earth wire, the size and design of the tower footings and the nature of the earth along the route of the line [66]. Figure 8.7 shows the 50 Hz earth impedance of a tower line for four different earth resistivity conditions as a function of line length. As can be observed, the earth impedance seen from the sending end decreases as the line length increases. For a given earth resistivity, there is a particular length of line, known as the effective length, which results in a minimum

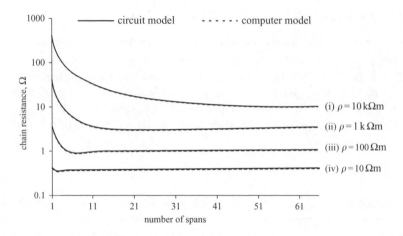

Figure 8.7   Chain impedance of a shielded transmission line as a function of line length [67]

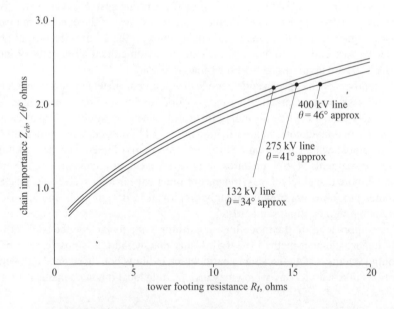

Figure 8.8   Chain impedance and phase angle for long lines as a function of tower footing resistance (reproduced from [36], courtesy of Energy Networks Association (ENA))

value equal to the characteristic impedance [67]. The chain impedance is commonly expressed for an infinite half line as a function of tower footing resistance, which is assumed constant along the route of the line. Figure 8.8 shows typical values of chain impedance (Z_{ch}) for different line construction types [36]. The chain impedance of

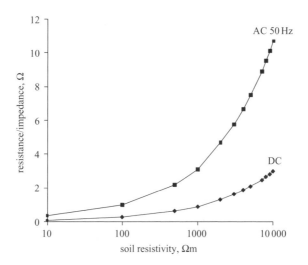

Figure 8.9   *Comparison between DC chain resistance and AC (50 Hz) chain impedance for a 400 kV tower line (reproduced from [67])*

an infinite half line can also be estimated using the following formula [52]:

$$Z_{ch} = \sqrt{Z_s R_t} \tag{8.4}$$

where Z_s, the longitudinal self-impedance per span, can be calculated using the Carson–Clem formula or by using the complex image method [68, 69].

The tower footing resistance R_t can be estimated from standard earth resistance formulae (see section 8.4.2.1).

Figure 8.9 shows both the DC chain resistance and AC chain impedance of a line as functions of earth resistivity corresponding to the asymptotic values of the impedance/resistance length curves. The curves indicate a significant difference between the DC and the AC case that increases with earth resistivity. It is important to appreciate this relationship because sometimes the AC chain impedance is estimated from DC tests.

More detailed circuit models of the earthing system formed by a transmission line can be established to account for the resistances of individual tower footings. However, in practice, such data are often not available.

8.4.2.3   Earth impedance of extended systems with sheathed cables

Urban distribution systems, in particular those at the lower end of the high voltage range (typically 11 kV), comprise extensive interconnected networks of underground cables. The sheaths of these cables interconnect the main substation earth grid to other substation earths and provide an extended earthing system. In addition, for older construction types of cable employing lead sheaths and hessian servings, direct contact is established between the soil and the lead sheath over a period of time. As a result, such cables can be considered as horizontal extensions to the grid [36].

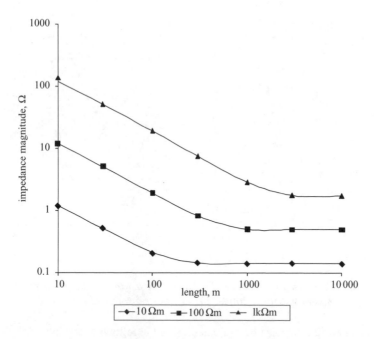

Figure 8.10   Earth impedance of an 11 kV 185 mm² uninsulated cable sheath as a function of length for different earth resistivities [47]

Figure 8.10 shows the earth impedance magnitude of an uninsulated sheath as a function of cable length. The graphs show that sheath earth impedance is proportional to earth resistivity and, for a given resistivity, the impedance decreases in magnitude up to a certain length. Beyond this length, referred to as the effective length, the impedance is practically constant. The effective length can extend to about 1 km in 100 Ωm earth [47]. Plastic insulated cables do not provide this fortuitous earth effect and, consequently, the gradual replacement of older cables would, over a period of time, increase the earth impedance of urban systems. Detailed analyses of the earthing contribution of underground cable circuits can be found elsewhere [70–73].

8.4.3   Interactions between fault currents and earthing systems

8.4.3.1   Fault current magnitude

Fault current calculations are primarily required to specify the interrupting capacity of circuit breakers at various points on the system, to determine the short time thermal rating of equipment and to enable the setting of protective relays. These calculations are also required to determine the current magnitude that will flow through earthing systems under fault, which in turn determines the earth potential rise.

Techniques for calculating both symmetrical and total asymmetrical fault currents are described in IEEE/ANSI [74–77] standards and in IEC909 [78]. These standards employ quasi steady state equations and a number of simplifying empirical factors [79–81]. Recent studies have shown that such approaches produce highly conservative results when compared with detailed dynamic fault current calculation techniques [82–84]. The need to employ more detailed fault calculation algorithms has been identified [85, 86].

For an accurate assessment of earth potential rise, it is necessary to account for the dynamic components in the fault current waveform. These dynamics include the DC offset and the decaying AC component that is present if the fault is near to a generator. In contrast to standard fault current calculations, which require the determination of current magnitude at a particular point in time, in earthing system assessments the cumulative effect of the current is required to determine the electrical shock severity. Accordingly, the r.m.s. equivalent value of the total asymmetrical earth fault current over the fault period should be calculated. In one standard [30], in order to represent the most severe condition, it is recommended that the maximum DC offset is used and that the AC component is assumed to remain constant at its initial subtransient value. For any given earthing system under study, different earth fault locations should be examined because these can have a considerable effect on the earth return current magnitude. For example, at substations, studies should examine faults on both the high and low voltage sides and faults external to the substation should also be considered [2, 30].

8.4.3.2 Fault current distribution

When an earth fault occurs at a substation, the current returns through the earth along the path of least impedance and enters the earth grid connected to the neutral at the source end to complete the circuit. If the earthing system of the source and faulted substation are interconnected, either by the earth wire of a shielded transmission line or the sheath of an underground cable, an additional return path is established for the fault current. Also, as a result of the mutual coupling between the faulted phase conductor and the parallel earth conductor, this additional metallic return path can carry a considerable proportion of the total fault current. In this way, the earth return current is reduced and, hence, the rise of earth potential at the substation. Generally, to determine the earth return current at the fault point as a proportion of the total earth fault current, it is not essential to model the entire system. For practical purposes, the relevant part of the network is that electrically near to the fault and the use of a two-substation model is sufficiently accurate. This arrangement is shown in Figure 8.11 where the faulted substation and equivalent source substation are interconnected by a single shielded circuit. For this circuit, the earth return current, I_{gr}, can be solved in terms of the total fault current, the impedance parameters of the interconnecting circuit and the substation earth impedances:

$$I_{gr} = I_f \frac{z_c - z_{mp,c}}{z_c + ((Z_{eA} + Z_{eB})/l_{AB})} \qquad (8.5)$$

where I_f is the phase conductor fault current, z_c is the self-impedance/unit length of the shield conductor, $z_{mp,c}$ is the mutual impedance/unit length between the phase and shield conductors, Z_{eA} and Z_{eB} are the earthing system impedances of the sending end and receiving end substations, respectively, and l_{AB} is the circuit length.

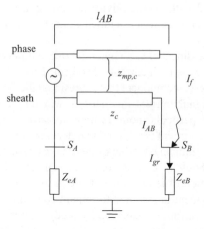

Figure 8.11   Fault current distribution between sheath and ground return for a two-substation arrangement employing a three-core cable (reproduced from [87])

This expression shows that when the circuit is relatively short, as is generally the case for cable circuits, the earth return current magnitude is strongly influenced by the values of the substation grid impedance. Accordingly, when considering cable circuits, it is necessary to evaluate fault current distribution case-by-case [87]. Earthing standards provide both nomograms and analytical relations for a number of two-substation arrangements employing three-core and single-core paper-insulated cables [36, 52]. On the other hand, shielded overhead lines are generally quite long and the magnitudes of the substation earth impedances are sufficiently low in comparison to the system longitudinal impedances not to affect current distribution. This is why, in the case of overhead lines, a shielding factor can be applied which is determined only by the line parameters for a particular tower line type [36]. For a more detailed analysis of fault current distribution, a number of different analytical methods have been developed considering both overhead lines [88, 89] and cable circuits [90, 91]. Specialised circuit analysis software for analysing complex systems is also available [4].

8.4.4   Measurement of earth impedance and potentials

For new installations, the measurement of the earth resistance or impedance of an earthing installation confirms whether the system performs according to its designed value. For existing installations the measurement is, to some extent, a method of establishing the continued integrity of the earthing system.

Measurements are made difficult due to a number of factors including stray currents, the settling of the earth grid after installation and distortion of ground surface potentials by the presence of buried metallic objects. In the case of large area grounding systems or earth grid connected to extended earthing systems, the reactive component may be significant, and, therefore, it is imperative to carry out AC rather than DC testing. For earthing systems with low impedance, the mutual coupling effect between test leads becomes a serious issue. Low magnitude current testing, the preferred option for safe and non-invasive measurements, is made difficult by the presence of relatively high levels of power frequency background noise.

8.4.4.1   The fall-of-potential (FOP) earth resistance/impedance measurement technique

The most representative measurement of the earth impedance of an installation is the staged fault test. This test produces realistic fault current magnitudes and, by using a remote voltage reference, the rise of earth potential and hence the earth impedance can be calculated. As an alternative to the staged fault, one circuit feeding a substation can be taken out of service and used as a test circuit. The staged fault test or the out-of-service circuit test is seldom performed due to economic penalties and system operational constraints. For this reason, it is necessary to measure the earth impedance of an energised substation without any interruption to the supply.

(i) Fall-of-potential set up

The measurement is achieved by injecting current into the earthing system using a temporary test circuit. This type of test is generically referred to as the fall-of-potential (FOP) method [92], although many variants of this technique have been developed. The FOP method is established as the most suitable method for measuring the earth resistance of high voltage substation earthing systems and is preferred to techniques such as the two-point or three-point methods [93]. The principle of the FOP method is shown in Figure 8.12. A test current is injected into the earth electrode under test and returns via an auxiliary current electrode at some distance *C* away. The difference in potential between the test electrode and an auxiliary potential electrode is measured at successive points, identified a distance *P* along a straight line close to the current carrying lead. Such an arrangement is referred to as an in-line FOP test. The ratio of voltage to current, known as the apparent resistance, is plotted against the distance *P* to form an earth resistance curve shown in Figure 8.13 [94]. If the distance *C* is

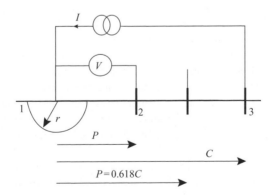

Figure 8.12   Fall-of-potential test electrode arrangement

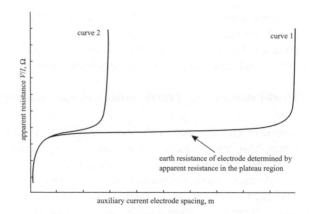

Figure 8.13   Earth resistance curves with and without plateau regions

sufficiently large in comparison with the dimensions of the test electrode (approximately ten times the maximum dimension of the test earthing system), the apparent resistance curve (curve 1) has a definite plateau. A plateau is formed only when the resistance zones of the electrode under test and the auxiliary electrode do not overlap. The value in this region of the plateau yields the earth resistance of the electrode under test. If C is comparatively small, there is no plateau region in the apparent resistance curve (curve 2) and estimating the earth resistance from such a curve is more difficult. To address this problem, Curdts [95] developed an analytical model of the FOP method. The model was based on a hemispherical shaped test electrode of radius r in uniform earth, and it was shown that the apparent resistance at any one distance P is equal to:

$$R = \frac{\rho}{2\pi} \left[\frac{1}{r} - \frac{1}{P} - \frac{1}{C-r} + \frac{1}{C-P} \right] \qquad (8.6)$$

With the approximation that r is small compared with C, the solution of Equation (8.6) which yields the earth resistance of a hemispherical electrode $R = \rho/2\pi r$ is given by the condition $P = 0.618C$. This method of analysing the fall-of-potential curve is therefore known as the 61.8 per cent rule (it is interesting to note that the 61.8 per cent rule has a numerical counterpart in pure geometry, known as the golden rule). The 61.8 per cent rule offers a technique for estimating the earth resistance of large area electrodes when it is impractical to set the auxiliary electrode far enough away such as to obtain a plateau region in the apparent resistance curve. Nevertheless, it is recommended that C should still be about ten times r. If the test electrode and the auxiliary electrode were of the same size, as expected, the correct position to place the probe would be at 50 per cent.

One of the practical difficulties in applying the 61.8 per cent rule is the requirement to measure the distances C and P from the electrical centre of the earthing system. Tagg developed a number of techniques which addressed this issue culminating in the slope method to enable the earth resistance to be determined without requiring the position of the electrical centre of the grid to be known [96–98]. The slope method assumes a distance x between the chosen point of reference, e.g. the substation fence, and the true electrical centre of the earthing system under test. The test requires three apparent resistance readings to be taken corresponding to 20, 40 and 60 per cent of C. Using these values of resistance, x can be evaluated by solving three simultaneous equations and the required auxiliary potential electrode location and corresponding resistance is determined using the 61.8 per cent rule. Tagg also considered that the slope method would provide an indication whether there was sufficient distance to the auxiliary current electrode.

(ii) Effect of non-uniform earth on FOP measurements

The effect of earth non-uniformity on the fall-of-potential technique has been investigated for two-layer earths [35, 96, 99]. It was demonstrated that a two-layer earth causes a shift in the auxiliary potential electrode location yielding the earth resistance of the electrode under test i.e. the 61.8 per cent method does not apply. For positive reflection coefficients (high resistivity lower layer), the auxiliary potential electrode would need to be situated beyond the 61.8 per cent position, towards the auxiliary current electrode. Recent work by Ma *et al.* [100] has identified the need to carry out accurate earth resistivity measurements in conjunction with FOP tests in order to construct accurate earth models to assist in the interpretation of the FOP curve.

(iii) Effect of test lead mutual coupling on FOP measurements

The potential difference measured in AC fall-of-potential tests consists of two components: the actual voltage difference between the earthing system under test and the auxiliary potential electrode, and the induced potential due to alternating current flow in the current test loop.

The effect of mutual coupling between current and voltage test leads can be considerable and, for an in-line FOP arrangement in particular, there can be a significant overestimation of the impedance value of the system under test.

Measurements on large area earthing systems are more susceptible to the effects of mutual coupling for two reasons: (i) the value of earth impedance to be measured is quite low, typically less than 1 Ω, (ii) large area earthing systems, by virtue of their large resistance zones, require long test lead distances.

Typically, mutual impedance can be about 0.5 Ω/km but the precise value is dependent on the test lead spacing [36], and it is recommended that this effect should always be accounted for when measuring earthing systems with a resistance of less than 1 Ω [31]. Recent work by Ma and Dawalibi [101] examined the mutual coupling effect under different conditions and confirmed that when the operating frequency is high or the earthing system is very large, i.e. having low earth resistance, measurement results are severely affected. The influence of mutual coupling on earth impedance measurements of shielded transmission lines, for different test lead configurations, is quantified by Harid *et al.* [102].

If the earthing system under test has extended earths such as tower lines, there may also be mutual coupling between the extended earth and the potential measuring circuit. The mutual coupling effect can be calculated using Carson's formulae for infinitely long conductors [103] or simplified versions such as the Carson–Clem formulae [68]. Velazquez *et al.* [104] and White and Rogers [69, 105], using the complex image method [106], developed equations for calculating the coupling effect between finite length parallel and angled conductors.

As would be expected, the mutual coupling reaches a maximum when the auxiliary potential test lead is laid in parallel with the auxiliary current test lead, and the coupling reduces as the spacing between the circuits is increased, and falls to a minimum when the auxiliary potential test lead is at 90° to the auxiliary current test lead. If the angle between test leads is greater than 90°, the measured impedance will be lower due to the mutual coupling. This means that the so-called 180° test, where the potential lead is laid out in the opposite direction to the current lead, results in underestimation of the earthing impedance of the test system. For large area earthing systems, it is recommended that the earth impedance be measured using angled test lead arrangements, preferably at 90°.

(iv) Practical testing considerations

Measurements are much more difficult on low impedance earthing systems due to the presence of background noise and mutual coupling effects. Background noise arises from unbalanced loading on the three-phase system, the presence of harmonics or any other mechanism that causes a current to circulate through the grid. This noise can be quantified by the standing voltage present on the earthing system with respect to a remote earth and magnitudes are typically in the range from 100 mV to several tens of volts. The background noise can be measured using a frequency-selective voltmeter or a spectrum analyser. Figure 8.14 shows an example of voltage noise measured on the earthing system of a 400 kV substation over a range of frequencies. The background noise may also fluctuate according to changing system conditions, for example at substations supplying electrified railway systems. Figure 8.15 shows the background voltage measured on the earthing system of a 400/25 kV railway supply substation as a train was passing. An assessment of the background noise conditions on an earthing system will enable selection of the minimum required magnitude of the test current in order to achieve an adequate signal to noise ratio. As an alternative, the earth impedance of high voltage energised substations can be measured at a frequency away from the power frequency in order to reduce or eliminate the effects of noise [107, 108]. Figure 8.16 shows earth impedances at frequencies either side of 50 Hz measured at a 275 kV substation using an impedance measurement system (IMS). The required 50 Hz impedance is obtained by interpolation [94].

(v) Test instrumentation

So-called composite instruments designed to measure earth resistance of concentrated earths are not recommended for measuring systems of impedance of less than 1 Ω. Larger earthing systems with extended earths will exhibit reactance as well as resistance and may require the separate measurement of current and voltage using specialised instrumentation [2, 94].

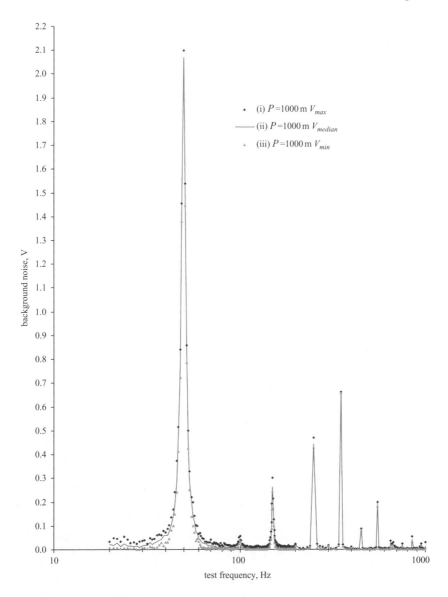

Figure 8.14 *Background noise as a function of frequency measured at a 400 kV transmission substation*

8.4.5 *Maintenance and integrity testing of earthing systems*

The importance of regular testing (e.g. every five or six years) to prove the continued integrity and performance of an earthing system is recognised in many earthing standards [2, 30]. As a part of this exercise, a detailed up-to-date site plan of the earthing system is required [1]. Earthing systems should be inspected and tested to

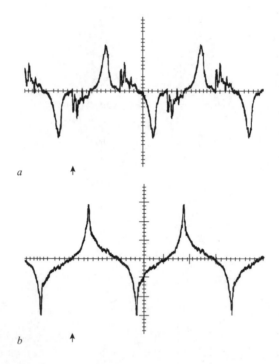

Figure 8.15   Background noise present on railway earthing system
  a   remote-track voltage: 2 V/div, 5 ms/div, train passing
  b   remote-track voltage: 4 V/div, 5 ms/div, train passing

ensure that all joints and connections are sound and secure. Joint resistances should be measured and proven to be of negligible value. In particular, earthing and bonding connections to plant and earth mats should be checked and neutral to earth connections verified. If great reliance is placed on extended earthing systems or interconnection with other substations, such connections should also be verified. The integrity test is typically carried out using a variable voltage source capable of delivering currents up to 300 A [1]. Various methods [1, 109] for evaluating the integrity of the earthing system include:

- radio frequency injection to establish conductor routes or breaks in conductors
- comparing voltage drops across different parts of a grid
- comparing current sharing through grid
- comparing earth impedances measured from different points on the grid
- point-to-point continuity measurements
- joint resistance measurements.

It should be appreciated that where an earth grid is defective, measurements may be hazardous.

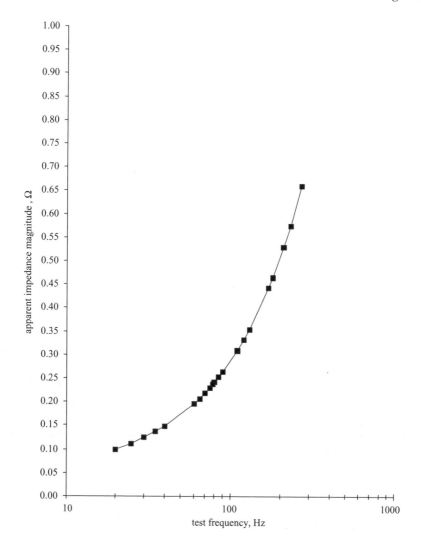

Figure 8.16   *Earth impedance magnitude against frequency measured at a 275 kV*
substation [94]

8.4.6   Special installations

A number of additional considerations for earthing design apply to the following
special installations.

8.4.6.1   GIS substation

Due to the much smaller area occupied by GIS installations and the fast transients
that are generated under switching operations, special measures are necessary for the
design of earthing systems for such installations [30].

8.4.6.2 Earthing systems for electric railways

In the UK, Engineering Recommendation P24 [110] specifies the earthing arrangements for AC traction supplies from 25 kV to 132 kV. However, recent developments in upgrading the electrified rail system in the UK have resulted in the siting of 400/25 kV substations in close proximity to the rail system. This means that detailed evaluations are required in order to investigate the effect of faults; and in particular, the effect of a fault on the 400 kV side on the rail earthing system. A recent system-wide analysis of safety hazards arising from earth faults on a railway supply system is reported by Natarajan *et al.* [111]. The safety threshold limits applicable to coupling into telecommunication systems from AC electric power and AC traction installations is provided in ITU standard K.33 [112].

8.4.6.3 Cellular phone systems on transmission line towers

The increased use in recent years of high voltage towers for hosting GSM transmitter stations has required special earthing arrangements at the tower bases. Such installations bring into close proximity the high voltage system supported by the tower (e.g. 400 kV) and the lower voltage supply system of the mobile station (e.g. 11 kV). A new engineering recommendation (G78) [113] addresses the design of such systems.

8.4.6.4 Wind farms

Land-based wind farms are often located in remote elevated areas with high earth resistivity. The substation that interconnects the wind farm to the main power system is usually quite compact, and, therefore, the substation earth grid may not provide a sufficiently low earth resistance. In such cases, the earthing system can be enhanced by providing a series of buried earth interconnections between the wind turbines extending over the whole area of the farm. This not only lowers the earth impedance at the main substation but also at each individual turbine. Earthing systems may be required at the base of each turbine to control step and touch voltages and for lightning protection [114].

8.5 Electrocution hazards and safety issues

8.5.1 Step and touch potentials

8.5.1.1 Background

Once the earth impedance seen from the point of fault and the proportion of fault current flowing through the earth path have been calculated, the rise of earth potential of the earthing system with respect to remote earth can be estimated. Potential differences across points at locations in and around the installation can also be determined to establish whether a human body would suffer electrocution if it bridged these points. Such potential differences are classified as touch, step and transfer potentials, and examples of them are illustrated in Figure 8.17 [1]. With reference to this figure, the earthing system and any metal structure connected to it will experience the full rise of earth potential U_E. The figure also illustrates the fall of potential on the ground

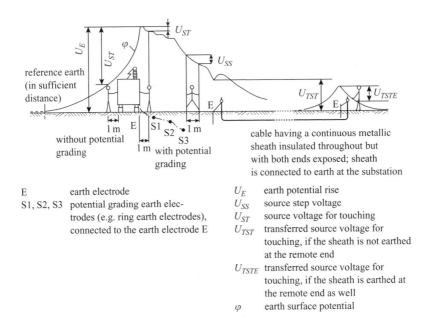

E   earth electrode
S1, S2, S3 potential grading earth elec-
            trodes (e.g. ring earth electrodes),
            connected to the earth electrode E

U_E earth potential rise
U_{SS} source step voltage
U_{ST} source voltage for touching
U_{TST} transferred source voltage for
          touching, if the sheath is not earthed
          at the remote end
U_{TSTE} transferred source voltage for
          touching, if the sheath is earthed at
          the remote end as well
φ earth surface potential

Figure 8.17   Illustration of earth surface potential profile at a substation under earth fault conditions and resulting hazard voltages (reproduced from [1])

surface with distance from the earthing system, which tends to zero at some distant point. As a result, a touch potential, U_{ST}, is experienced if a person standing on the ground also touches metalwork connected to the earthing system. Similarly, if a person stands on the ground in a direction such as to bridge a potential difference, they are said to experience a step voltage, U_{SS}. Transferred potentials describe the potential difference which could be bridged by human contact as a result of either:

a   an insulated conductor connected to the earthing system under fault extending out into a region of zero or low potential (U_{TST} or U_{TSTE})
b   an insulated conductor connected at a remote point of low potential reaching into the substation area of high potential but not bonded to the earthing system.

The magnitudes of these potentials at a given installation and for a fault current magnitude will depend on the location of the person and what they are touching. When making an assessment of safety at an installation, it is not practical to consider every prospective step and touch situation, and, therefore, normally an assessment is made of the maximum touch and step voltages that occur.

Step voltages are normally less hazardous than touch voltages for two reasons:

(i)   the human body can tolerate higher voltages across the foot-to-foot current path (step) compared to the hand-to-feet path (touch). In particular, less current flows through the path of the heart for the foot-to-foot electrocution scenario, and this is expressed in terms of a heart current factor; also, if additional resistances are taken into account (e.g. footwear, surface chippings), the current is limited to a greater extent by the series connection of these additional resistances [115]

(ii) for any given position, the step voltage is lower than the prospective touch voltage.

This generality can lead to the step voltage being neglected as a hazard in its own right. For example, EA Technical Specification 41-24 [2] states, 'if the earthing system is safe against touch potential it will be inherently safe against step potential'. However, IEEE Std.80 [30] recognises that even if safety is achieved within the substation through the control of potential gradients, it may still be possible for step voltage hazards to be present outside the substation. Therefore, it is recommended that step voltages are also computed and compared. IEEE Std.80 (section 19.2) states that 'the best assurance that a substation is safe would come from actual field tests of step and touch voltages'. However, it is noted 'because of the expense, few utilities are likely to make these tests as a routine practice'.

8.5.1.2   Establishing the locations of maximum values of touch and step voltage

The maximum surface potential gradients around an earthing grid normally occur along a diagonal line from the centre extending through a corner point of the earth grid. Inside the grid, potential minima will occur at the surface of the ground near the central point of each mesh as shown in Figure 8.18 for a four-mesh grid. The maximum prospective touch voltages within the area of the grid will correspond to a person standing centrally inside a mesh while touching earthed metalwork, and the location of the highest touch voltage for a symmetrical grid corresponds to the centre of the corner mesh. The touch voltage at this position is referred to as the mesh voltage. IEEE Std.80 notes that the actual worst-case mesh voltage occurs slightly off-centre towards the corner of the grid. Using the IEEE Std.80 standard, the required mesh spacing for the grid can be determined to ensure that the mesh voltage is below the tolerable touch voltage limit. Earthing design standards in the UK follow a different approach where particular attention is given to the touch voltage established when a person stands one metre diagonally out from the corner of the grid [2, 52] while no specific calculation of the mesh voltage is made. The mesh voltage may be greater or less than the corner touch voltage depending on the mesh density and whether earth rods are used at the periphery of the grid [30]. However, provided no earthed metalwork can be touched while standing outside the grid, i.e. if the grid fully covers the area occupied by plant, then the corner touch voltage hazard should not arise. Accordingly, it should only be necessary to calculate the mesh voltage. The location of the worst-case step voltage is accepted by both UK and US standards to be the potential difference across the ground surface one metre diagonally out from the corner of the grid.

8.5.1.3   Magnitude of touch, mesh and step voltages

Until quite recently, touch, mesh and step voltages were calculated using approximate analytical expressions, which incorporated geometrical and empirical correction

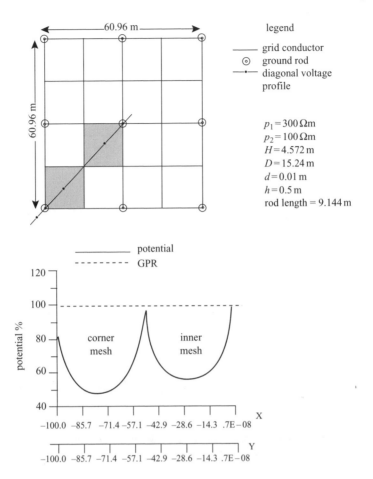

Figure 8.18   *Diagonal voltage profile for a 16-mesh square grid with nine rods in a two layer soil (from IEEE Std.81 1983, IEEE Copyright 1983 and IEEE Std.80 2000. All rights reserved [30])*

factors. Examples of such expressions are available in a number of earthing design standards [2, 30, 52]. More recently, computer programs have become available for computing earth grid resistance and surface potentials around the grid [62, 116, 117]. The most recent edition of IEEE Std.80 [30] outlines the reasons to justifying the use of such programs in carrying out earth design including:

a   a complex earth structure requiring multilayer representation
b   non-uniform grid geometry and the presence of other buried conductors or metalwork
c   a more specific examination of step and touch danger points.

UK standards currently do not take account of computer-based earthing design programs. As an alternative to carrying out detailed estimations of prospective step

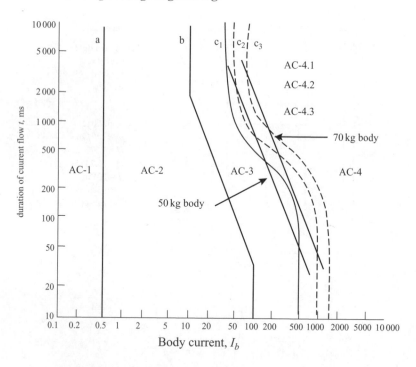

Figure 8.19   Tolerable body current as a function of shock duration – comparison of IEC limits (curved lines) with IEEE Std.80 limits (straight lines) [115, 30]

and touch voltage, the CENELEC document also allows for the estimation of touch voltages by direct measurement [1].

8.5.2   Computation of tolerable voltages

Once the prospective hazardous voltages arising under earth fault conditions have been computed for a given design of earthing system, it is necessary to determine whether or not these exceed recognised safety thresholds known as tolerable voltages. Ultimately, it is the level of current through the human body that determines whether there is a risk of electrocution, and well defined limits apply. However, the tolerable voltage for a particular shock scenario depends not only on the level of tolerable current but also on the values of resistance that are present in addition to the human body resistance. Accordingly, a suitable accidental circuit is also required in order to obtain values of tolerable voltages.

8.5.2.1   Tolerable body currents

Electricity flowing through the human body can be perceived and can cause muscle contraction for currents of a few tens of milliamperes. However, when the current reaches a magnitude of the order of 50 mA, there is a risk of ventricular fibrillation. It is the fibrillation limit that forms the basis of most recommended earthing safety calculations [88]. The fibrillation threshold is not only dependent on current magnitude but also on the current time duration, current path and frequency. Figure 8.19, reproduced from IEC479 and based on work carried out by Biegelmeier and Lee [118], shows the tolerable r.m.s. human body current at power frequency for a range of times for the hand-to-feet body path. The variation in fibrillation threshold for different body current paths is described by the heart current factor [115] and it is interesting to note that the hand-to-hand shock scenario is less onerous than hand-to-foot. The different curves shown in Figure 8.19 correspond to the probability of ventricular fibrillation occurring. For current–time values to the left of curve c1, the risk of ventricular fibrillation is negligible, and curves c2 and c3 delineate the 5 per cent and 50 per cent probability boundaries. Some earthing standards [1, 52, 119] adopt curve c2, although UK industry earthing standard EA TS 41-24 [2] uses curve c1 as the basis of safety calculations. The American earthing standard IEEE Std.80 [30], on the other hand, adopts a different specification for tolerable body current based on the findings of Dalziel [120, 121]. In this case, an empirical formula is used to describe the tolerable current I_b, which applies to 99.5 per cent of a population; this is plotted in Figure 8.19 for comparison:

$$I_b = \frac{k}{\sqrt{t_s}} \tag{8.7}$$

where t_s is the shock duration and k is a constant related to body weight.

For a body weight of 50 kg and 70 kg, $k = 0.116$ and 0.157, respectively. In IEEE Std.80, it is suggested that the 70 kg figure may be applied if this is the average population weight within the area of the electrical installation. However, it is considered that the 50 kg figure is more suitable when applied to areas outside controlled environments (e.g. outside a substation fence).

Whichever current curve is used, the time duration is related to the speed of the circuit protection. The estimation of shock duration may require careful judgement on systems that employ fast acting main but slower backup protection and also circuits using fast autoreclosure.

8.5.2.2 Accidental earth circuit

Once a suitable permissible body current has been selected, the permissible touch voltage can be determined from the accidental shock circuit (Figure 8.20) applicable to either the step or touch voltage scenarios. This circuit comprises: (i) the body impedance Z_b (a resistance is sufficient at power frequency), (ii) additional resistances such as footwear resistance, and (iii) the resistance to earth of the standing point or the feet earth resistance. With reference to the circuit of Figure 8.20, it is important to distinguish between the touch voltage (U_T) and the source voltage for touching or the prospective touch voltage (U_{ST}). U_T is the part of the earth potential rise due to the earth fault which can be picked up by a person, assuming the current is flowing via the human body from hand to feet, and U_{ST} is the voltage which appears during an earth fault between the contact point and earth when these parts are not

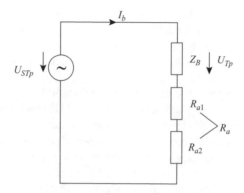

Figure 8.20   *Simple equivalent circuit for the touching shock scenario [1]*

U_{STp} voltage difference acting as a source in the touching circuit with a limited value that guarantees the safety of a person when using additional known resistances (e.g. footwear, standing surface insulating material)

Z_b total body impedance

I_b current flowing through the human body

U_{Tp} permissible touch voltage, the voltage across the human body

R_a additional resistance ($R_a = R_{a1} + R_{a2}$)

   R_{a1} for example resistance of the footwear

   R_{a2} resistance to earth of the standing point

ρ_s resistivity of the ground near the surface in an installation (in Ωm)

t_F fault duration

being touched [1]. It is noticeable that there are considerable differences between standards in the values adopted for each of the resistance components, which are described in more detail in the following sections.

(i) Human body resistance

The resistance of the human body is assumed to be equal to a constant 1 kΩ in both UK and American standards [2, 52, 115]. However, the new CENELEC standard [1] adopts a more precise voltage-dependent resistance model based on IEC data. The characteristics of this model are shown in Figure 8.21, and it can be seen that the IEC model yields higher values of resistance than the constant resistance model at lower voltages [1, 115].

(ii) Footwear resistance

In IEEE Std.80 [30], footwear resistance is neglected. UK standards [2, 52], on the other hand, assume 4 kΩ per shoe and the ITU-T K33 standard allows for a footwear resistance of between 250 Ω and 3 MΩ per shoe depending on the shoe type and condition [122].

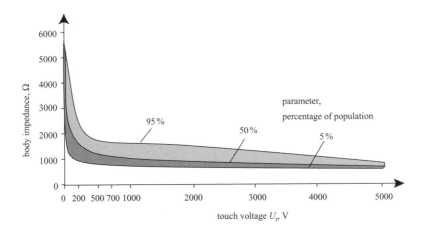

Figure 8.21   Statistical values of human body impedance [1] (copyright ©1994, IEC, Geneva, Switzerland. www.iec.ch)

(iii) Resistance to earth of standing point

In a more accurate model, the resistance to earth of the standing point should be considered as a component of a Thévenin equivalent resistance across the two points that the human body bridges [30, 123, 124]. However, in most standards, a simplified approach is adopted that assumes an additional resistance due only to the earth resistance of the feet. The resistance of the human foot R_f is calculated by representing it as a metallic disc of equivalent area and is given by:

$$R_f = 3\rho \tag{8.8}$$

If the earth surface resistivity ρ is artificially increased, the tolerable touch or step voltage will be higher. Accordingly, it is standard practice in most outdoor substations to add a thin surface layer of high resistivity material, such as gravel. A number of different analytical techniques and computer models are available to account for this effect, and it should be noted that the addition of a surface layer of different resistivity will also alter the surface potential distribution [30].

8.5.2.3   Typical tolerable touch voltages

From the preceding sections, it can be seen that there are considerable differences between standards in both the definition of tolerable body current and also the configuration and parameter values of the accidental circuit. Tolerable touch voltages are therefore very dependent on the model chosen and examples of ranges of tolerable voltages are shown in Figure 8.22.

8.5.2.4   Transferred potential limits

A metallic circuit, connected to the main grid, and extending outside the area of the substation, can transmit transferred potentials. The danger normally arises from a

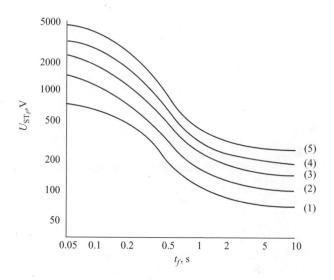

Figure 8.22   Examples of tolerable touch voltage curves for different assumed additional resistances in the accidental circuit [1]

 (1) without additional resistances
 (2) $R_a = 750\ \Omega$ $(R_{a1} = 0\ \Omega, \rho_s = 500\ \Omega m)$
 (3) $R_a = 1750\ \Omega$ $(R_{a1} = 1000\ \Omega, \rho_s = 500\ \Omega m)$
 (4) $R_a = 2500\ \Omega$ $(R_{a1} = 1000\ \Omega, \rho_s = 1000\ \Omega m)$
 (5) $R_a = 4000\ \Omega$ $(R_{a1} = 1000\ \Omega, \rho_s = 2000\ \Omega m)$
 $R_{a1} = 1000\ \Omega$ represents an average value for old and wet shoes

remote touch voltage scenario. The transferred potential will usually be much higher in magnitude than local step and touch voltages, and can even exceed the rise of potential of the main substation [30].

 In the UK, limits are applied to such transferred potentials. For these situations, which assume direct contact between the full rise of earth potential and remote earth, safe limits of 430 V and 650 V are designated depending on the reliability and protection speed of the circuits connected to the substation [2]. These limits derive from a set of CCITT directives (Comitte Consultatif International Telephonique et Telegraphique) [68]. In these directives, the 430 V and 650 V limits apply to voltages induced on aerial telephone lines and not directly to transfer voltages at substations. In a different publication [125], it is mentioned that these limits were adopted as a result of a compromise involving economic considerations, and it can be assumed that they were not derived from human tolerable body current assumptions. If the limit was required to avoid danger to a person, it is considered that it should be set to as low as 60 V for the longer fault clearance times. The continued application of the 430 V and 650 V limits over many years may be explained by the extremely low probability of the transfer potential hazard. This may be an example of how, during a period when deterministic approaches were used, a probabilistic consideration entered the specification by artificially raising the safety threshold. The recently published standard IEC 61936-1 states that 'It must be recognised that fault occurrence, fault current magnitude and presence of human beings are probabilistic in nature' [126]. Section 8.5.4 of this chapter considers in more detail how probabilistic considerations affect the risk management of earthing systems.

By contrast, the approach adopted in the American standard IEEE Std.80 is to acknowledge that it is impractical or impossible to restrict the rise of earth potential to prevent dangerous voltages of this kind, and therefore, either bonding or isolating best avoids such hazards [30].

8.5.3 Methods for limiting hazardous potential differences and dimensioning of earthing systems

The area occupied by the substation or installation initially sets the overall dimensions of an earth grid, and normally all the available area is occupied in order to achieve the lowest resistance. The grid mesh density should be sufficient to facilitate connection to the substation plant by short earthing conductors or downloads.

If the basic design of an earth grid is insufficient to satisfy the adopted safety criteria, a number of different measures can be taken to either reduce step and touch potentials or restrict current flow through the accidental human body circuit. These measures can be grouped into four categories.

8.5.3.1 Reducing earth grid current

Impedance-earthed systems, which are described in detail in section 8.2.3 of this chapter, enable the single line to earth fault current to be restricted. On existing solidly-earthed systems, however, it may not be possible to introduce impedance earthing because of limitations in the insulation withstand capabilities of the system. As an alternative, instead of reducing the total earth fault current, the earth return component of the fault current can be reduced by providing an additional above ground metallic return path. For example, on unearthed overhead lines, a continuous earth wire could be installed. On circuits already equipped with an earth wire, a second earth wire could be provided.

8.5.3.2 Reducing earth impedance

If all the available land area has been utilised, making direct connections to nearby existing grids can reduce the overall earth impedance. If extended earthing systems formed by tower lines or cables are available, normally these should also be connected to the main earth grid. Often, such extended earthing systems have much lower earth impedances than the grid to which they are connected. As already noted in section 8.4.2.1, grid earth resistance is directly proportional to the earth resistivity. Therefore, if an earth resistivity survey indicates that the grid is in relatively high resistivity earth and there is a nearby low resistivity earth region, it can be advantageous to connect to a satellite grid installed in the low resistivity earth area. It should be noted that although grid extensions or connections to extended earthing systems are beneficial by lowering grid resistance, the earth potential rise is exported over a much wider and often uncontrolled area.

Instead of extending the earthing system outwards, sometimes it is beneficial to install deep-driven earth rods or piles or drilled earth wells. Such deep-driven earths are particularly useful for penetrating deeper earth layers of low and stable resistivity. Finally, additives or treatments are sometimes employed to lower the earth resistivity in the immediate vicinity of the earth electrodes. However, these may be of limited benefit for large area earth grids because they have little effect on the resistivity of the deeper earth which influences the grid resistance.

8.5.3.3   Limiting potential differences

A solid plate covering the entire surface area would provide the ultimate protection against step and touch voltage hazards within a substation, but such an installation would be impractical and uneconomic [127]. However, the following practical methods for the limitation of potential gradients may be adopted. In the first place, the overall grid mesh density can be increased to provide a better general equipotential area within the installation. It was noted in section 8.5.1.2 that the mesh voltage in an equally spaced grid increases towards the perimeter. Accordingly, the grid mesh density may be made denser at the grid periphery in order to equalise the mesh voltages. A more complete equipotential zone could be established by separating the operating or fault current carrying earth grid from the protective earth grid, which can be achieved by installing a fine mesh counterpoise mat near the ground surface.

At often-frequented places inside the substation, such as switch and isolator positions, small fine mesh wire mats can be buried just beneath the surface chippings. A further area requiring special attention in terms of earthing is the substation fence. In many cases, the fence is metallic and can be touched from the outside by the public. Therefore, suitable measures to control potential gradients must be employed.

At installations which have a limited earthing area and which may carry high fault current including lightning currents, such as communication masts and transmission line towers, there may be significant step voltage hazards as well as touch voltage risk at the base of such installations. The step voltage risk, specifically, may be controlled by the use of potential ramps [127], which are illustrated in Figure 8.23. Ramp rings

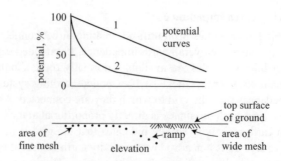

Figure 8.23   Potential distribution above an earth grid with a potential ramp (curve 1) and without ramp (curve 2) (from IEEE Std.81 1983, IEEE Copyright 1983 and IEEE Std.80 2000. All rights reserved [30])

are also recommended in BS6651 for the protection of masts, towers and columns using single and multiple earth electrode systems [128].

8.5.3.4   Increasing resistance/insulation in the accidental path

One of the simplest and most widely used methods for increasing the resistance in the accidental path is to cover the entire surface of the substation with gravel [30]. This type of covering provides a high resistance barrier even under wet conditions but in practice the integrity of the layer is often lost due to lack of maintenance and erosion due to excessive vehicle travel. As an alternative or in addition to the gravel layer, a plastic sheet can be placed under the ground surface [129]. If there are particular routes in frequent use within an installation, wooden passageways can also provide very good insulation.

The previous methods are concerned with providing additional resistance between the ground surface and the body in the accidental path. A further option for touch voltage conditions would be to provide insulation to earthing downleads and to insulate or provide isolation from all aboveground metal work that can be touched from ground level. In this way, the touch voltage hazard is effectively eliminated.

There are various methods for protecting against the danger of transferred potentials, depending on the nature of the circuit involved in transmitting the potential. Such circuits are commonly formed by communications circuits, traction rails, low voltage neutral circuits, potable equipment, piping and metallic fences [30].

8.5.4   Risk management approach to earthing safety

To recall, EA TS 41-24 [2] states that 'if the potential rise of the earth electrode exceeds the appropriate CCITT level (430 V or 650 V),...measures shall be taken to counter the risk of danger'. This standard also specifies limits for touch and step voltages corresponding to probabilities of heart fibrillation based on IEC 60479-1 and certain assumed values of resistance in the accidental circuit. However, experience has shown that, in many situations, it is not reasonably practicable to satisfy these requirements, particularly in respect of third parties. As already mentioned in section 8.5.3, the minimum impedance of a substation earthing system is often limited by the available land area and ground resistivity, and the minimum earth fault current is limited by the electrical system configuration. In many cases, the minimum achievable rise of earth potential may exceed several thousands of volts, and high potential contours may extend considerably beyond the substation. Consequently, transferred potentials affecting third parties in the vicinity of a substation may exceed the levels specified in the standards, and the costs involved in mitigating these potentials could be very substantial.

8.5.4.1   Allowable risk levels

In the UK, legislation that covers the duties of employers to their employees and third parties is set out in 'The Health and Safety at Work etc.' Act 1974 [130]. This act requires that 'it shall be the duty of every employer to conduct his (her) undertaking in such a way as to ensure, so far as is reasonably practicable, that persons. . .are not thereby exposed to risks to their health and safety'. More recently, the Management of Health and Safety at Work Regulations 1999 [131], which were enacted as a result of European Union Framework Directive 89/391/EEC, states that 'Every employer shall make a suitable and sufficient assessment of. . .the risks to the health and safety of persons not in his employment arising out of or in connection with the conduct by him (her) of his (her) undertaking'.

What is reasonably practicable has been established by a number of legal test cases and more comprehensively by the Health & Safety Executive (HSE) [132], and this concept is referred to as the 'as low as reasonably practicable (ALARP)' principle. An individual fatality risk of 1 in 10 million per person per year is considered broadly acceptable for members of the public. Between 1 in 1 million and 1 in 10 000 per person per year, the ALARP principle must be applied; which means that the required expenditure to reduce risk is dependent on the individual risk level, i.e. the higher the risk, the higher the justifiable spend. The HSE has attempted to reflect the price people are prepared to pay to secure a certain averaged risk reduction and advise a benchmark value for preventing a fatality of £1 m (2001 prices). This value corresponds to a reduction in risk of 1 in 10^5 being worth about £10 to the average individual [133].

The Electricity Industry in the UK, through the offices of the Energy Networks Association, is currently investigating applying risk management techniques to hazards resulting from earth potential rise. It is expected that this work will result in the establishment of an industry-wide policy recommending the necessary action to be taken in respect of third parties in the vicinity of electrical installations. This will include a revision to Engineering Recommendation S.36 [134], which covers the provision of telecommunications services near to substations. The risk management approach starts by recognising the probabilistic nature of exposure to earth potential rise. In simple terms, this means that account is taken of the likelihood of the occurrence of an earth fault at the same time as an individual being in a position which bridges a dangerous potential.

8.5.4.2   Case study

The following example will serve to illustrate the main steps involved in such an approach. A member of the public in his or her garden and using the garden tap is exposed to a transferred touch potential of 1500 V for 200 ms, as a result of an earth fault on the transmission system. Assuming standard values of resistance in the accidental circuit (see section 8.5.2.2), a fatality would be likely from this exposure as this voltage is clearly higher than the safe levels specified in the standards. Accordingly, from a deterministic perspective, this hazard would require mitigation. However, let us now consider the probability that the individual will experience this fatal electric shock. This overall probability P can be determined as the product of three separate probabilities:

$$P = P_F P_{FB} P_C$$

where P_F is the probability of an earth fault on the power system, P_{FB} is the probability of heart fibrillation and P_C is the probability of contact i.e., the probability that the individual is in an accidental circuit.

P_F can be estimated from historical fault records. Assuming the typical probability of an earth fault occurring which results in a significant earth potential at a transmission substation is 0.2 per annum; i.e. one significant earth fault every five years on average.

P_{FB} is the likelihood that, if a person were exposed to the earth potential rise, heart fibrillation would occur. Probabilities of heart fibrillation corresponding to current magnitude and duration (in this case 200 ms) are given in IEC 60479-1 [115]. The current magnitude can be estimated from the accidental touch circuit and, in this example, a body resistance R_b of 487 Ω (hand-to-feet current path) will be assumed to cover at least 95 per cent of the population. The additional circuit resistances will depend upon the specific exposure scenario, which depends on what the individual is wearing and doing at the time of exposure. Assumptions may be made to determine the exposure scenario: for example, it may be reasonable to assume that a person would be wearing footwear while outdoors. The ITU standard T K33 [135] specifies a resistance of 3000 Ω for damp elastomer soled shoes on loose soil and so this is adopted in our example. All other circuit resistances and the source impedance are considered to be negligible in this case. In many exposure scenarios there will be an additional insulation in the accidental circuit such as where an individual is using an insulated power tool. In this example, such insulation is neglected. The total resistance of the accidental circuit is calculated to be 3650 Ω, which would result in a current of 0.41 A flowing through the body for a touch voltage of 1500 V. At this current magnitude, IEC 60479-1 predicts a probability of fibrillation of less than five per cent or 0.05.

Finally, it is necessary to estimate P_C, which is the likely time the person is present in the accidental circuit. This will again depend upon the activity undertaken by the exposed individual; in particular the proportion of time in contact with the circuit elements. Let us assume that in this example the person is in contact with the garden tap, on average, for one minute per day. Each day, therefore, the probability of contact is 7×10^{-4}. The estimated annual probability that this person will experience a fatal electric shock as a result of using a garden tap can be estimated as the product of the probabilities P_F, P_{FB} and P_C, viz., $0.2 \times 0.05 \times 7 \times 10^{-4} = 7 \times 10^{-6}$ or 1 in 142 000. This level of individual risk falls within the lower ALARP region, where the cost of mitigation must be balanced against the risk. If mitigation in this case were not prohibitively expensive, it may be considered worthwhile.

It should be noted that the above example covers one exposure scenario only. To cover the total individual risk all reasonably foreseeable scenarios should be considered and all risks added together.

In many cases, it is expected that although the exposure voltages may exceed the limits prescribed in the standards, the level of individual risk does not warrant significant expenditure on mitigation. However, it is worth noting that, in general, it is not prohibitively expensive to design an earthing system to control touch and step voltages inside a substation to within the levels specified in the standards and, consequently, this approach is normally adopted.

Were industry to apply expensive mitigation, the costs would ultimately be borne by society as a whole through increased electricity prices. This is because this issue affects the industry as a whole and, therefore, the effect of competition or regulation would be minimal. But according to the HSE, this price may not be considered by society as worth paying. So, by using risk assessment to determine necessary expenditure, the industry spends responsibly and can also prioritise investment with the end result being improved safety. It should, however, also be remembered that the risks resulting from earth potential rise are borne predominately by people living and working at or close to substations, i.e. they are unevenly distributed within the general population and, consequently, the perceived levels of tolerability of such risks tend to be lower.

8.6   Impulse performance of earthing systems

In the preceding sections, earthing system design and performance under power frequency conditions has been examined in detail and a number of standards which provide very detailed design guidelines were referred to. In this section, to begin with, the standards have been examined in respect of their consideration of earthing requirements under fast transients and a summary of their recommendations is presented.

8.6.1   Standard guidelines for transient earthing

8.6.1.1   IEEE Std.80 and IEEEStd.142

IEEE Std.80 (Guide for safety in substation grounding) does not provide detailed guidance for designing earthing systems subjected to lightning surges but considers that grounding systems designed according to power frequency principles will 'provide a high degree of protection against steep wave front surges. . .'. This is based on the assumption that the human body can withstand higher currents for very short duration. This standard also recommends that surge arresters should 'always be provided with a reliable low resistance ground connection' and have as 'short and direct a path to the grounding system as practical'.

Some utilities provide separate downlead connections for surge arresters and others use the metallic mounting structure because it offers a lower impedance path to the earthing system [30].

8.6.1.2   EA-TS 41-24

The UK earthing design standard EA TS 41-24 [2] (Guidelines for the design, testing and maintenance of main earthing systems in substations), recognises that 'equipment such as surge arresters and CVTs are more likely to pass high frequency current due to the low impedance they present to steep fronted surges' and 'unless a low impedance earth connection is provided. . .the effectiveness of the surge arrester could be impaired'.

This standard also recommends that the connection from the equipment to earth should be 'as short, and as free from changes in direction, as is practicable' and that 'the effectiveness of the arrester can be improved. . .by connecting. . .(it) to a high frequency earth electrode in the immediate vicinity, for example an earth rod'.

8.6.1.3 CENELEC – HD 637 S1

The CENELEC harmonisation document, HD 637 S1 [1], which is likely to form the basis of a new IEC standard, suggests measures to reduce the amount of interference created when surges are dissipated to earth. These include minimising the inductance of the current paths by the significant meshing of earth electrodes and earthing conductors. Where high transient currents are more likely to occur, the density of the earthing mat should be increased.

8.6.1.4 BS 6651

A quantitative value applicable to transient earthing appears in BS 6651 (Protection of structures against lightning) [136]. This standard recommends that earthing systems designed for lightning protection should have an earth resistance of less than 10 Ω. Many electricity companies for power system earthing systems have adopted this value. In Japan, the 10 Ω resistance figure is the target value for transmission line tower footing resistance [137]. Further work is required to establish values based on more detailed models which include inductance.

8.6.2 Soil ionisation

Many factors influence the behaviour of soil under high current magnitudes. First, the characteristics of the medium: the soil type and resistivity, the moisture content, temperature and pressure. Second, the nature of the energisation: the waveform or impulse shape, its magnitude and polarity.

The reduction in the earth impedance of earth electrode systems under high magnitude impulse currents has been long recognised and this phenomenon is now generally accepted to be as a result of soil ionisation. Since the performance of earthing systems under fast transients is vital to overvoltage protection, this area has attracted much interest. Early work concentrated on experimental studies of earth electrode systems experiencing soil ionisation effects. Recent work not only involved earth electrode studies, but also the development of soil ionisation models and experimentation to understand the soil ionisation process.

A review of work investigating soil ionisation effects in earthing systems is presented. Four main areas of investigation are considered: soil ionisation effects on earth electrode systems, estimation of critical electric field intensity, soil ionisation models, and analysis of soil ionisation mechanisms.

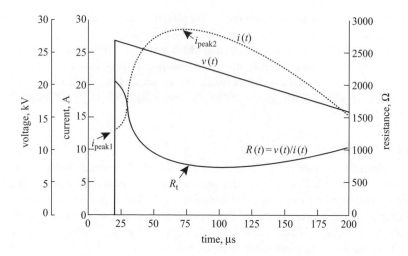

Figure 8.24   Typical voltage and current records in the ionisation region, and the ratio $R(t)$ derived from $v(t)/i(t)$ (water content, $wc = 3\%$, $V_{ch} = 27$ kV [154])

8.6.2.1   Response of earth electrode systems to high impulse currents

(i) Early work

The early experimental work on the characterisation of earth electrodes under transient conditions [138–145] demonstrated that:

a   transient impedance, defined as the ratio between instantaneous voltage and current, was shown to decrease from an initial impedance, during the fast front of the surge, to a minimum corresponding to maximum ionisation which is as little as one third of the power frequency value and, subsequently, recovering to the leakage resistance value at the end of the impulse tail (Figure 8.24)

b   soil ionisation occurs above a certain current threshold, which can be related to the electric field intensity in the soil

c   the instantaneous voltage–current curves were shown to form the shape of a hysteresis loop

d   spiked electrodes could enhance the soil ionisation effect

e   time delays in ionisation and deionisation required a dynamic modelling approach.

(ii) Recent work

Work continued with investigations by Kosztaluk *et al.* [146] on tower footings and further evidence of the complicated dynamics of the soil ionisation process was provided by Oettlé and Geldenhuys [147]. They described the corona ionisation time constant, and found it to be inversely proportional to the current magnitude. They considered that this was due to the formation of individual streamers in the soil.

Tests were carried out [148] on a meshed earthing system of a 77 kV power substation and also a steel-reinforced concrete pole. Lightning impulses with current magnitudes of up to 40 kA were applied, and it was found that the transient impedance during soil ionisation, of both the mesh and the concrete pole, decreased with increasing peak current magnitude. This effect was more pronounced for the smaller concrete pole. Subsequently, tests on single rod electrodes, grids, horizontal electrodes and tower footings [149] showed that no further reduction in resistance occurred at magnitudes of current above 20 kA. It was also established that the reduction in resistance was more marked in high resistivity soils and that the dependency is less, the more extensive the earthing systems. Tests on a 500 kV transmission tower base in medium resistivity soil did not show the marked reduction in resistance found with rods and crowsfeet [137].

8.6.2.2 Estimation of critical electric field intensity

A key parameter in representing the ionisation threshold level in soils is the critical electric field intensity, E_C. This critical electric field, E_C, can be calculated from the product of the measured soil resistivity and the current density at the electrode surface assuming uniform current distribution. The value of current used corresponds to the instant when the impulse resistance decreased below the low voltage resistance value. A wide range of E_C values (between 1.3 and 17.3 kV/cm), depending on soil type, test electrode geometry and homogeneity of soil have been reported [140, 142–145, 150].

In reviewing these experimental-based investigations, some consider that an average E_C value of 3 kV/cm is realistic [151] and others [152] showed that plots of E_C values against the corresponding soil resistivity were largely scattered which indicated that there was no definite relationship between E_C and soil resistivity. Mohamed Nor *et al.* [153, 154] obtained an E_C value of 5.5 kV/cm for sand which was found to be independent of moisture content. Loboda and Scuka [155], based on their own experimental studies, considered that streamers played an influential role in the soil ionisation process and demonstrated that the calculated E_C values could be up to 70 per cent higher than those determined by assuming a uniform and perfectly conducting ionised zone.

8.6.2.3 The mechanisms of soil ionisation

A significant amount of research work to understand the mechanisms of soil ionisation has been carried out in the last few decades. Two contending theories have emerged to explain the initiation of ionisation.

(i) Electric field enhancement

Between 1981 and 1993, many studies [156–159] of electrical breakdown characteristics of small-scale soil samples have been carried out. Emerging from these is the proposition that ionisation occurs in soil air voids due to electric field enhancement. Key points from these investigations are that;

a the ionisation process takes a finite time to initiate from application of the impulse, and delay times range from 10 to 100 μs *b*   delay times exhibited a statistical variation, but average delay times and their deviation decreased as the applied voltage was increased

c   delay times decreased with water content and applied voltage

d   breakdown thresholds with soil samples immersed in SF_6 were 2.5 times higher than with air, which provided strong evidence in support of the theory.

(ii) Thermal process

In 1982, van Lint and Erler [160] investigated breakdown in earth samples under high electric field stress using coaxial test rigs. Their results confirmed that initiation delay times of ionisation decreased with increasing voltage. However, unlike the results obtained by Flanagan *et al.* [157], initiation delay times of up to 1500 μs were reported. It was suggested that the field enhancement model could not explain such long delay times. Instead, it was proposed that the negative temperature coefficient of ionic solutions would produce a concentration of current flow in narrow parallel water filaments. Further heating of these channels would result in vaporisation and eventually breakdown along the filament path. Further investigations into the proposed thermal mechanisms of soil breakdown [161–164] suggested that neither the thermal model nor the field enhancement model alone could account for the wide ranging behaviour observed in soil conduction under high voltage. Srisakot *et al.* [159], based on fast impulse experimental studies on sand samples, proposed that two distinct conduction phases could be identified; the first associated with thermal effects and the second due to soil ionisation.

8.6.3   Models of concentrated earth electrodes exhibiting soil ionisation

An example of a finite-element (FE) modelling approach was used by Nekhoul *et al.* [165] to describe the behaviour of an earth rod exhibiting ionisation effects. Srisakot *et al.* [159] carried out numerical electric field modelling of a simple hemispherical earth electrode system using FE and boundary element BE software. Sekioka *et al.* [166] developed linear ($R–L$) and non-linear ($R–L$) models to describe the earth impedance of an earth grid, and these are compared with the simple single resistance model. It is demonstrated that the inductance effects of a grounding grid should be included to achieve an accurate lightning surge analysis and that the non-linear model should be used for high magnitude, steep fronted impulses. It was also demonstrated experimentally that reinforced concrete poles provide effective conduction paths under impulse conditions [167]. A non-linear resistance can model such concentrated electrodes, and it is suggested that an energy-dependent resistance model is more accurate than a current-dependent resistance model [168].

8.6.4   Models of earthing systems under high frequency and transient conditions

Three different types of model have been applied in the analysis of high frequency and transient performance of earthing systems. These correspond to increasing complexity according to the type of earthing system under study [169].

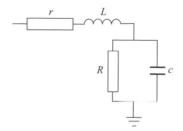

Figure 8.25   Equivalent circuit model for a vertical earth electrode

Table 8.1   Formulae for the earth resistance [29], inductance [170] and capacitance [35] of a vertical earth electrode

$$R = \frac{\rho}{2\pi \ell}\left(\ln\frac{2\ell}{a} - 1\right) \tag{8.9}$$

$$L = \frac{\mu \ell}{2\pi}\left(\ln\frac{2\ell}{a} - 1\right) \tag{8.10}$$

$$C = \frac{2\pi \ell \varepsilon_0}{(\ln(2l/a) - 1)} \tag{8.11}$$

8.6.4.1   Transmission line per unit length series impedance and shunt admittance

This model, either in lumped parameter or distributed parameter form, can be used to describe single horizontal and vertical earth conductors (Figure 8.25). The per unit length impedance and admittance quantities can be computed using expressions for earth resistance, shunt capacitance and conductor self-inductance. Approximate formulae applicable to a vertical electrode of length *l* and radius *a* are shown in Table 8.1.

Even at high frequency, the series resistance *r* of the earth conductor can be neglected. The corresponding per unit length impedance and admittance can be used to obtain the input impedance Z_{oc} of the earth electrode:

$$Z_{oc} = \sqrt{\frac{Z_C}{Y_C}}\coth(l\sqrt{Z_C Y_C}) \tag{8.12}$$

where Z_C is the per unit length series impedance, Y_C the per unit length shunt admittance and *l* the length of the electrode.

8.6.4.2   Network analysis of transmission line segments

For extended earthing systems that do not exhibit complex coupling of the inground electrode, e.g. transmission tower line earthing systems, it is possible to build-up circuit models from the transmission line components in section 8.6.4.1 using circuit analysis techniques. In such models, the effects of soil ionisation have also been included [171, 172].

8.6.4.3   Electromagnetic model

The electromagnetic model, first developed to analyse ground losses from antennas due to the finite conductivity of earth, is based on the method of moments. It was further developed and applied to analyse grounding systems [173, 174]. Such an approach offers a more accurate theoretical treatment than the models in sections 8.6.4.1 and 8.6.4.2, but the model still requires a laterally homogeneous equivalent earth model. Examples of software based on the electromagnetic field theory approach are found in the literature [175–177]. Olsen *et al.* [178] distinguish between the low frequency or quasi-static model and an exact or full wave model valid at high frequency. They point out that even low frequency models can be considered to be field theory models in contrast to equivalent circuit models.

A comparison of methods in sections 8.6.4.1 and 8.6.4.2 indicated that both methods were suitable for the simulation of simple structures such as horizontal electrodes but that significant differences were apparent for large earthing grids [179]. The limits of low frequency quasi-static models have been investigated and it has been shown that if the buried electrode is greater than one-tenth of a wavelength in earth, then an exact full wave model would be required [178]. Heimbach and Grcev [180] proposed an alternative method of combining the frequency-dependent characteristics of an earthing system in EMTP in order to evaluate overvoltages in power systems.

8.6.5   Simulations of earthing system performance under transient and high frequency conditions

The performance of earthing systems under transient conditions is different from power frequency behaviour because of the more significant influence of inductive and capacitive effects. In contrast to the 50 Hz response, at high frequency the inductance of a small earthing system, such as a rod, has a significant effect and the effective length of such systems can be very small [175, 181–185]. Inductive and capacitive effects of rod earths are also evident under fast transients [186]. Figure 8.26 shows the simulated voltage front time resulting from a 1/5 current impulse injected at the top of a 5 m rod. For low values of resistivity, the inductance of the system has a greater effect and a sharpening of the voltage impulse shape is seen. At high resistivity, the magnitude of the displacement current becomes significant compared with the conducted leakage current, and the effect of earth permittivity is influential on the voltage rise time.

Figure 8.27 shows the frequency response of a 100×100 m^2 earth grid for a range of soil resistivities. As can be observed, each curve has a lower frequency range over which the impedance is nearly constant. For each earth resistivity value, the impedance increases rapidly above a threshold frequency, and this behaviour can be attributed to the inductance of the conductor. Transient simulations, of the same grid in 100 Ωm earth were carried out for a 1/5, 1 A current impulse. The results

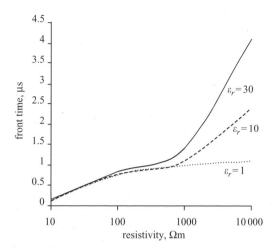

Figure 8.26   *Voltage front time for a 1/5 current impulse on a 5 m rod, $\rho = 100\ \Omega m$ (reproduced from ERA report no. 2002-0113, published 2002)*

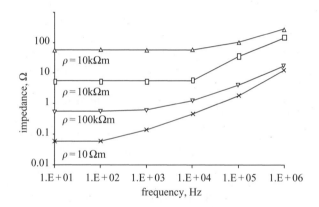

Figure 8.27   *Earth impedance of a four-mesh grid as a function of frequency for different soil resistivities*

(Figure 8.28) indicate significant attenuation of the surge as it propagates across the grid and a slowing of the voltage rise time. The voltage drop across the down lead is very high for the conditions considered [186].

Other investigations based on simulations using the electromagnetic model support the principle of an effective area and confirm that increased grid meshing in the vicinity of the injection point plays an important role in reducing the peak voltage [187].

At low frequencies, the impedance of a grid is independent of the injection point for the same earth conditions. However, above a critical frequency, the impedance

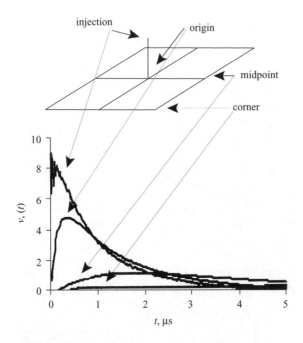

*Figure 8.28   Voltage at different points on a grid for a 1/5 current impulse
(reproduced from ERA report no. 2002-0113, published 2002)*

seen from the grid extremities compared with a central injection point increases
considerably [180].

8.7   References

1   HD637 S1: 'Power installations exceeding 1kV AC'. European Committee for
    Electrotechnical Standardisation (CENELEC), 1999
2   EA-TS 41-24: 'Guidelines for the design, installation, testing and maintenance
    of main earthing systems in substations'. Electricity Association, Technical
    Specification, 1992
3   THE ELECTRICITY COUNCIL: 'Engineering recommendation S.5/1 – earth-
    ing installations in substations'. Thirty-fourth chief engineers' conference,
    July 1966
4   SIMPSON, I.B.K., *et al.*: 'Computer analysis of impedance effects in large
    grounding systems', *IEEE Trans. Ind. Appl.*, 1987, **23**, (3), pp. 490–497
5   NATIONAL GRID: 'The seven year statement'. Market Analysis and Design,
    Commercial Systems and Strategy, Coventry, UK, 2000
6   NATIONAL GRID: 'FADS – faults and defects database on the high volt-
    age transmission system'. Engineering and Technology, Leatherhead, UK,
    2000

7 JONES, P.: 'A guide for evaluating earth return currents at national grid substations'. Internal report, quality of supply, National Grid Company, 2000

8 NATIONAL GRID: 'Technical and operational characteristics of the NGC transmission system'. Issue 1.1, Coventry, UK, 2000

9 HMSO: 'Main prospectus for the sale of the regional electricity companies share offers'. November 1990

10 ACE REPORT NO. 51: 'Report on the application of engineering recommendation P2/5 security of supply'. British Electricity Boards (Electricity Association), 1979

11 WESTINGHOUSE ELECTRIC CORPORATION: 'Electrical transmission and distribution reference book' (PA, 1950, 4th edn.)

12 B.E.I: 'Modern power station practice, vol. K' (Pergamon Press, London, 1990, 3rd edn.)

13 SHREIR, L.L.: 'Corrosion, vol. 12' (G. Newnes, London, 1963)

14 BELL, H.: 'Protection of a.c. systems', *J. IEE*, 1960, **6**, (70), pp. 571–575

15 SAY, M.G.: 'Electrical earthing and accident prevention' (G. Newnes, London, 1954)

16 WILLHEIM, R., and WATERS, M.: 'Neutral grounding in high voltage transmission' (Elesevier, New York, 1956)

17 BEEMAN, D.: 'Industrial power systems handbook' (McGraw-Hill, New York, 1955)

18 VUKELJA, P., *et al.*: 'Experimental investigations of overvoltages in neutral isolated networks, *IEE Proc. C, Gener. Transm. Distrib.*, 1993, **140**, (5), pp. 343–350

19 HANNINEN, S., and LEHTONEN, M.: 'Characteristics of earth faults in electrical distribution networks with high impedance earthing', *Electr. Power Syst. Res.*, 1998, **44**, (3), pp. 155–161

20 SUGIMOTO, S., *et al.*: 'Thyristor controlled ground fault current limiting system for ungrounded power distribution systems', *IEEE Trans. Power Deliv.*, 1996, **11**, (2), pp. 940–945

21 IEEE Std.142-1991: 'IEEE recommended practice for grounding of industrial and commercial power systems'. IEEE Press, New York, 1991

22 THE ELECTRICITY SAFETY, QUALITY AND CONTINUITY REGULATIONS, 2002 No. 2665

23 PETERSON, H.: 'Transients in power systems' (Wiley, New York, 1951)

24 OPENSHAW TAYLOR, E., and BOAL, G.: 'Electric power distribution 415 V–33 kV' (Arnold, London, 1966)

25 KERSHAW, S.S., *et al.*: 'Applying metal-oxide surge arresters on distribution systems', *IEEE Trans. Power Deliv.*, 1989, **4**, (1), pp. 301–307

26 SAY, M.: 'Electrical earthing and accident prevention' (G. Newnes, London, 1954)

27 FRANKLIN, A.C., and FRANKLIN, D.P.: 'The J&P transformer book: a practical technology of the power transformer' (A.C. FRANKLIN & D.P. FRANKLIN, 1983, 11th edn.)

28   SCHÖN, J.H.: 'Physical properties of rocks: fundamentals and principles of petrophysics', *in* HELBIG, K., and TREITEL, S. (Eds): 'Handbook of geophysical exploration. Section 1, seismic exploration: vol. 18' (Elsevier Science Ltd, Oxford, 1998)

29   SUNDE, E.D.: 'Earth conduction effects in transmission systems' (Dover Publications Inc., 1968)

30   IEEE Std.80-2000: 'IEEE guide for safety in AC substation grounding'. The Institute of Electrical and Electronic Engineers, New York, 2000

31   IEEE Std.81-1983: 'IEEE guide for measuring earth resistivity, ground impedance, and earth surface potentials of a ground system'. The Institute of Electrical and Electronic Engineers, New York, 1983

32   BRITISH GEOLOGICAL SURVEY: URL:http://www.bgs.ac.uk [accessed 1 June 2002]

33   GRIFFITHS, D.H., and KING, R.F.: 'Applied geophysics for geologists and engineers – the elements of geophysical prospecting' (Pergamon Press, Oxford, 1981, 2nd edn.)

34   BELL, F.G.: 'Ground engineers reference book' (Butterworth, London, 1987)

35   TAGG, G.F.: 'Earth resistances' (G. Newnes Ltd., England, 1964, 1st edn.)

36   THE ELECTRICITY ASSOCIATION: 'Engineering recommendation S.34 – a guide for assessing the rise of earth potential at substation sites'. The Electricity Association, 1986, p. 23

37   WENNER, F.: 'A method for measuring earth resistivity'. Bureau of Standards scientific paper, 1915, no. 258 (Frank Wenner was a physicist at the American Bureau of Standards)

38   VAN NOSTRAND, R.G., and COOK, K.L.: 'Interpretation of resistivity data'. Geological Survey professional paper 499, US Dept. of the Interior, Washington, 1966

39   AVO International Ground/Earth resistance products [www] URL:http://www.avo.co.uk/products/groundtster/ground.html [accessed 24 April 2002]

40   MA, J., and DAWALIBI, F.P.: 'Study of influence of buried metallic structures on soil resistivity measurements', *IEEE Trans. Power Deliv.*, 1998, **13**, (2), pp. 356–365

41   BHATTACHARYA, P.K., and PATRA, H.P.: 'Direct current geoelectric sounding – part 9, principles and interpretation' (Elsevier, London, 1968)

42   McNeill, J.D.: 'Applications of transient electromagnetic techniques'. Technical note TN-7, Geonics Limited, 1980

43   THE BRITISH ELECTRICAL AND ALLIED INDUSTRIES RESEARCH ASSOCIATION: 'Section M: communication interference – notes on electrical resistivity maps (Refs. M/T32 and M/T33)'. Reference M/T35, November 1934

44   THE BRITISH ELECTRICAL AND ALLIED INDUSTRIES RESEARCH ASSOCIATION: 'Electrical resistivity of England and Wales – apparent resistivity at a depth of 500 feet'. Ref. D.263, February 1963 (Reproduced from ERA Report M/T32, 1934)

45   INSTITUTE OF GEOLOGICAL SCIENCES: 'Geological map of the UK (South)'. 3rd edition solid, 1979

46   THE BRITISH ELECTRICAL AND ALLIED INDUSTRIES RESEARCH ASSOCIATION: 'Electrical resistivity map of Southern Scotland'. (Reproduced from ERA report M/T33, 1934)

47   CLAXTON, J.: 'Investigation into the performance of electrical earthing systems'. PhD thesis, Cardiff University, UK, 1997

48   HABBERJAM, G.M.: 'Apparent resistivity observations and the use of square array techniques' (Gebrüder Borntraeger, Berlin, 1979)

49   BRITISH ELECTRICITY INTERNATIONAL: 'Modern power station practice' (Pergamon, London, 1992, 3rd edn.)

50   THAPAR, B., and GROSS, E.T.B.: 'Grounding grids for high voltage systems', *IEEE Trans. PAS*, 1963, **82**, pp. 782–788

51   ENDRENYI, J.: 'Evaluation of resistivity tests for design of station grounds in non-uniform soil', *AIEE Trans.*, 1963, **82**, pp. 782–788

52   BS7354: 1990: 'Code of practice for the design of high-voltage open terminal stations'. British Standards Institution, 1990

53   MA, J., *et al.*: 'On the equivalence of uniform and two-layer soils to multilayer soils in the analysis of grounding systems', *IEE Proc., Gener. Transm. Distrib.*, 1996, **143**, (1)

54   TAYLOR, M.J.: 'Soil resistivity assessment for electrical earthing applications'. PhD thesis, Cardiff University, UK, 2001

55   DAWALIBI, F., and BARBEITO, N.: 'Measurements and computations of the performance of grounding systems buried in multilayer soils', *IEEE Trans. Power Deliv.*, 1991, **6**, (4), pp. 1483–1487

56   LOKE, M.H.: 'Electrical imaging surveys of environmental and engineering studies'. 1999, [www] URL:http://www.abem.com [accessed 27 April 2002]

57   DWIGHT, H.D.: 'Calculation of resistances to ground', *Electr. Eng.*, 1936, **55**, pp. 1319–1328

58   SCHWARZ, S.J.: 'Analytical expressions for resistance of grounding systems', *IEEE Trans. Power Appar. Syst.*, 1954, **73**, (13), pp. 1011–1016

59   NAHMAN, J., and SALAMON, D.: 'Analytical expressions for the resistance of grounding grids in nonuniform soils', *IEEE Trans. Power Appar. Syst.*, 1984, **103**, (4), pp. 880–885

60   SULLIVAN, J.A.: 'Alternative earthing calculations for grids and rods', *IEE Proc., Gener. Transm. Distrib.*, 1998, **145**, (3), pp. 270–280

61   CHOW, Y.L., ELSHERBINY, M.M., and SALAMA, M.M.A.: 'Resistance formulas of grounding systems in two-layer earth', *IEEE Trans. Power Deliv.*, 1996, **11**, (3), pp. 1330–1336

62   EPRI TR-100622: 'Substation grounding programs'. Vol. 1–5, Electric Power Research Institute, Palo Alto, US, 1992

63   XISHAN, W., *et al.*: 'Computation and experimental results of the grounding model of Three Gorges power plant'. 11th international symposium on *High voltage engineering*, London, UK, 1999, p. 4

64   MA, J., and DAWALIBI, F.P.: 'Analysis of grounding systems in soils with finite volumes of different resistivities', *IEEE Trans. Power Deliv.*, 2002, **17**, (2), pp. 596–602

65   COLOMINAS, I., *et al.*: 'A numerical formulation for grounding analysis in stratified soils', *IEEE Trans. Power Deliv.*, 2002, **17**, (2), pp. 587–595

66   ENDRENY, J.: 'Analysis of transmission tower potential during ground fault', *IEEE Trans. PAS*, 1967, **86**, (10), pp. 1274–1283

67   MGHAIRBI, A., GRIFFITHS, H., and HADDAD, A.: 'The chain impedance of shielded transmission lines'. 36th universities *Power engineering* conference, Swansea University, 2001, session 6C1, p. 5

68   CCITT(ITU), THE INTERNATIONAL TELEGRAPH AND TELEPHONE CONSULTATIVE COMMITTEE: 'Directives concerning the protection of telecommunications lines against harmful effects from electric power and electrified railway lines'. Volume I–IX, The International Telecommunications Union, Geneva, 1989

69   ROGERS, E.J., and WHITE, J.F.: 'Mutual coupling between finite lengths of parallel or horizontal earth return conductors', *IEEE Trans. Power Deliv.*, 1989, **4**, (1), pp. 103–113

70   NAHMAN, J.: 'Earthing effects of coated underground cables with metallic shields bonded to earth electrodes', *IEE Proc., Gener. Transm. Distrib.*, 1997, **144**, (1), pp. 22–30

71   POPOVIC, L.M.: 'Practical methods for the analysis of earthing systems with long external electrodes', *IEE Proc. C, Gener. Transm. Distrib.*, 1993, **140**, (3), pp. 213–220

72   NAHMAN J., and SALAMON, D.: 'Effects of the metal sheathed cables upon the performances of the distribution substations grounding systems', *IEEE Trans. Power Deliv.*, 1992, **7**, (3), pp. 1179–1187

73   MELIOPOULOS, A., and MASSON, J.: 'Modelling and analysis of URD cable systems', *IEEE Trans. Power Deliv.*, 1990, **5**, (2), pp. 806–815

74   IEEE/ANSI C37: 'IEEE standard rating structure for AC high-voltage circuit breakers rated on a symmetrical current basis'. The IEEE Inc., November 1985, 1986 edn.

75   IEEE Std.242-1975: 'IEEE recommended practice for protection and co-ordination of industrial and commercial power systems, buff book'. IEEE Press, New York, 1980

76   IEEE/ANSI, C37.12-1981: 'Guide to specifications for AC high-voltage circuit breakers rated on a symmetrical current basis and a total current basis'. Approved on August 1981, The IEEE Inc., November 1985

77   IEEE Std.399-1980: 'IEEE recommended practice for power system analysis, brown book'. IEEE Press, New York, 1980

78   IEC 909: 'Short circuit calculation in three phase ac systems'. IEC International Electrotechnical Commission publication, 1988

79   BRIDGER, B.: 'All amperes are not created equal: a comparison of current of high voltage circuit breakers rated according to ANSI and IEC standards', *IEEE Trans. Ind. Appl.*, 1993, **29**, (1), pp. 195–201

80 KNIGHT, G., and SIELING, H.: 'Comparison of ANSI and IEC 909 short-circuit current calculation procedures', *IEEE Trans. Ind. Appl.*, 1993, **29**, (3), pp. 625–630

81 BERIZZI, A., MASSUCCO, S., SILVESTRI, A., and ZANINELLI, D.: 'ANSI/IEEE and IEC standards for short-circuit current evaluation: methodologies, computed values and results'. 6th international symposium on *Short-circuit currents in power systems*, September 1994, Liege, Belgium, pp. 1.1.1–1.1.8

82 DUNKI-JACOBS, J.R., LAM, B.P., and STRATFORD, R.P.: 'A comparison of ANSI-based and dynamically rigorous short-circuit current calculation procedures', *IEEE Trans. Ind. Appl.*, 1988, **24**, (6), pp. 1180–1194

83 ROENNSPIESS, O.E., and EFTHYMIADIS, A.E.: 'A comparison of static and dynamic short-circuit analysis procedures', *IEEE Trans. Ind. Appl.*, 1990, **26**, (3), pp. 463–475

84 BERIZZI, A., MASSUCCO, S., SILVESTRI, A., and ZANINELLI, D.: 'Short-circuit current calculation: a comparison between methods of IEC and ANSI standards using dynamic simulation as reference', *IEEE Trans. Ind. Appl.*, 1994, **30**, (4), pp. 1099–1106

85 PROFESSIONAL GROUP P9: 'IEE colloquium on fault level assessment – guessing with greater precision?'. The IEE, London, 30 January 1996, digest no.: 1996/016

86 ARRIILAGA, J., and ARNOLD, C.P.: 'Computer analysis of power systems' (J. Wiley and Sons, London, 1990)

87 VIEL, E., and GRIFFITHS, H.: 'Fault current distribution in HV cable systems', *IEE Proc., Gener. Transm. Distrib.*, 2000, **147**, (4), pp. 231–238

88 VERMA, R., and MUKHEDKAR, D.: 'Ground fault current distribution in substations, towers and ground wires', *IEEE Trans. PAS*, 1979, **98**, pp. 724–730

89 NAHMAN, J. M.: 'Proximity effects on the ground fault current distribution within the earthing system formed by a substation and the associated transmission lines', *IEE Proc. C, Gener. Transm. Distrib.*, 1988, **135**, (6), pp. 497–502

90 GUVEN, A.N., and SEBO, S.A.: 'Analysis of ground fault current distribution along underground cables', *IEEE Trans. PD*, 1986, **PWRD-1**, (4), pp. 9–15

91 POPOVIC, L.M.: 'Practical method for evaluating ground fault current distribution in station supplied by an inhomogeneous line', *IEEE Trans. PD*, 1997, **12**, (2), pp. 722–727

92 PETERS, O.S.: Bureau of Standards technologic paper 108, 1918

93 MORGAN, A., and TAYLOR, H.G.: 'Measurement of the resistance of earth electrodes', *World Power*, 1934, **XXI**, (CXXI), pp. 22–27, 76–81, 130–135

94 JONES, P.: 'Electrical measurement of large area substation earth grids'. PhD thesis, Cardiff University, UK, 2002

95 CURDTS, E.B.: 'Some of the fundamental aspects of ground resistance measurements', *AIEE Trans.*, 1958, **77**, (1), pp. 760–767

96   TAGG, G.F.: 'Measurement of earth-electrode resistance with particular refer-
     ence to earth-electrode systems covering a large area', *IEE Proc.*, 1964, **111**,
     (12), pp. 2118–2130
97   TAGG, G.F.: 'Measurement of the resistance of an earth-electrode system
     covering a large area', *IEE Proc.*, 1969, **116**, (3), pp. 475–479
98   TAGG, G.F.: 'Measurement of the resistance of physically large earth-electrode
     systems', *IEE Proc.*, 1970, **117**, (11), pp. 2185–2189
99   DAWALIBI, F., and MUKHEDKAR, D.: 'Ground electrode resistance mea-
     surements in non uniform soils', *IEEE Trans. Power Appar. Syst.*, 1974, **93**,
     (1), pp. 109–116
100 MA, J., *et al.*: 'Ground impedance measurement and interpretation in various
     soil structures'. IEEE Power Engineering Society winter meeting, 2000, **3**,
     pp. 2029–2034
101 MA, J., and DAWALIBI, F.P.: 'Influence of inductive coupling between leads
     on ground impedance measurements using the fall-of-potential method', *IEEE
     Trans. Power Deliv.*, 2002, **16**, (4), pp. 739–743
102 HARID, N., MGHAIRBI, A., and GRIFFITHS, H., *et al.*: 'The effect of
     AC mutual coupling on earth impedance measurement of shielded transmis-
     sion lines'. 37th universities *Power engineering* conference, Staffordshire
     University, 2002, pp. 822–826
103 CARSON, J.R.: 'Wave propagation in overhead wires with ground return', *Bell
     Syst. Tech. J.*, 1926, **5**, (10), pp. 539–554
104 VELAZQUEZ, R., REYNOLDS, P.H., and MUKHEDKAR, D.: 'Earth return
     mutual coupling effects in ground resistance measurements', *IEEE Trans.
     Power Appar. Syst.*, 1983, **102**, (6), pp. 1850–1857
105 IEEE Std.81.2-1991: 'IEEE guide for measurement of impedance and safety
     characteristics of large, extended or interconnected grounding systems'. The
     Institute of Electrical and Electronic Engineers, New York, 1991, appendices A
     and B, pp. 81–94
106 FOSTER, R.M.: 'Mutual impedance of grounded wires lying on the surface of
     the earth', *Bell Syst. Tech. J.*, 1931, **X**, (3), pp. 408–419
107 BOAVENTURS, W.C., *et al.*: 'Alternative approaches for testing and eval-
     uating grounding systems of high voltage energised substations'. 10th
     international symposium on *High voltage engineering*, Montreal, Canada,
     1997, p. 4
108 WANG, G., and ZHOU, W.: 'A frequency conversion measurement sys-
     tem of grounding resistance'. 11th international symposium on *High voltage
     engineering*, London, UK, 1999, p. 4
109 GILL, A.S.: 'High-current method of testing ground grid integrity', *NETA
     WORLD*, International Testing Association, **10**, (2), 1988–89
110 THE ELECTRICITY ASSOCIATION: 'Engineering recommendation P.34 –
     AC traction supplies to British Rail'. The Electricity Association, 1984,
     Addendum 1990, pp. 29–32
111 NATARAJAN, R., *et al.*: 'Analysis of grounding systems for electric traction',
     IEEE Trans. Power Deliv., 2001, **16**, (3), pp. 389–393

112 International Telecommunication Union (ITU): 'Limits for people safety related to coupling into telecommunications system from a.c. electric power and a.c. electrified railway installations in fault conditions'. Series K: Protection against interference, 1996

113 THE ELECTRICITY ASSOCIATION: 'Engineering recommendation G78 – recommendations for low voltage connections to mobile telephone base stations with antennae on high voltage structures'. London, 2003, Issue 1

114 IEC/TR 61400-24: 'Wind turbine generator systems – lightning protection'. International Electrotechnical Commission IEC publication, 2002

115 IEC 479-1: 'Guide to effects of current on human beings and livestock, part 1 general aspects'. International Electrotechnical Commission IEC publication, 1994

116 DAWALIBI, F., and MUKHEDKAR, D.: 'Optimum design of substation grounding in two-layer earth structure: part I – analytical study, part II – comparison between theoretical and experimental results and part III – study of grounding grids performance and new electrodes configurations', *IEEE Trans. PAS*, 1975, **PAS-9**, (2), pp. 252–261, 262–266, 267–272

117 HEPPE, R.J.: 'Computation of potential at surface above an energised grid or other electrode allowing for non-uniform current distribution', *IEEE Trans. PAS*, 1979, **98**, (6), pp. 1978–1989

118 BIEGELMEIER, G., and LEE, W.R.: 'New considerations on the threshold of ventricular fibrillation for a.c. shocks at 50–60 Hz', *IEE Proc. A, Phys. Sci. Meas. Instrum. Manage. Educ. Rev.*, 1980, **127**, (2), pp. 103–110

119 DIN VDE 0141:1989: 'Earthing systems for power installations with rated voltages above 1 kV'. Technical Help to Exporters Translations, British Standard Institution, Milton Keynes, UK

120 DALZIEL, C.F.: 'Dangerous electric currents', *AIEE Trans.*, 1946, **65**, pp. 579–585, also discussion on pp. 1123–1124

121 DALZIEL, C.F., and LEE, W.R.: 'Re-evaluation of lethal electric currents', *IEEE Trans. Ind. Gen. Appl.*, 1968, **4**, (5), pp. 467–476

122 INTERNATIONAL TELECOMMUNICATIONS UNION RECOMMENDATION K.33: 'Limits for people safety related to coupling into telecommunications system from a.c. electric power and a.c. electrified railway installations in fault conditions', ITU 1997

123 MELIOPOULOS, A.P.S.: 'Power system grounding and transients – an introduction' (Marcel and Dekker, 1988)

124 DAWALIBI, F.P., SOUTHEY, R.D., and BAISHIKI, R.S.: 'Validity of conventional approaches for calculating body currents resulting from electric shocks', *IEEE Trans. Power Deliv.*, 1990, **5**, (2), pp. 613–626

125 KLEWE, H.R.J.: 'Interference between power systems and telecommunication lines'. Report M/T 126, Electrical Research Association (E.R.A.), London, 1958

126 IEC 61936-1: 'Power installations exceeding 1 kV a.c. – part 1: common rules'. International Electrotechnical Commission IEC publication, 2002

127   KOCH, W.: 'Grounding methods for high-voltage stations with grounded neutrals', *Elektrotechnische Z.*, 1950, **71**, (4), pp. 89–91 (appendix J, IEEE Std.80 1986)

128   BS6651:1999: 'Code of practice for protection of structures against lightning'

129   THAPAR, B., and ZIAD, A.A.: 'Increasing of ground resistance of human foot in substation yards', *IEEE Trans. Power Deliv.*, 1989, **4**, (3), pp. 1695–1699

130   HMSO: 'The Health and Safety at Work etc. Act 1974'. ISBN 0105437743, UK

131   HSC, Management of health and safety at work regulations 1999, ACOPS, HMSO 2000, ISBN 0 7176 2488 9

132   HSE: 'The tolerability of risk from nuclear power stations'. HMSO 1992

133   HSE: 'Reducing risks, protecting people, HSE's decision-making process'. HMSO 2001

134   ENGINEERING RECOMMENDATION S.36: 'Procedure to identify and record 'hot' substations'. The Electricity Association, 1988

135   INTERNATIONAL TELECOMMUNICATIONS UNION RECOMMENDATION K.33: 'Limits for people safety related to coupling into telecommunications system from a.c. electric power and a.c. electrified railway installations in fault conditions'. ITU 1997

136   BS6651: 1999: 'Code of practice for protection of structures against lightning'. British Standard 1999

137   TAKEUCHI, M., *et al.*: 'Impulse characteristics of a 500 kV transmission tower footing base with various grounding electrodes'. 24th international conference on *Lightning protection*, ICLP-98, Birmingham, UK, 1998, pp. 513–517

138   TOWNE, H.M.: 'Impulse characteristics of driven grounds', *General Electric Review*, 1928, **31**, (11), pp. 605–609

139   BEWLEY, L.V.: 'Theory and tests of the counterpoise', *Electr. Eng.*, 1934, **53**, pp. 1163–1172

140   DAVIS, R., and JOHNSTON, J.E.M.: 'The surge characteristics of tower and tower-footing impedances', *Journal of the IEEE*, 1941, **88**, pp. 453–465

141   BELLASHI, P.L.: 'Impulse characteristics of driven grounds', *AIEE Trans.*, 1941, **60**, pp. 123–128

142   BELLASHI, P.L., *et al.*: 'Impulse and 60-cycle characteristics of driven grounds, part II', *AIEE Trans.*, 1942, **61**, pp. 349–363

143   BERGER, K.: 'The behaviour of earth connections under high intensity impulse currents'. CIGRE paper 215, 1946

144   PETROPOULOS, G.M.: 'The high-voltage characteristics of earth resistances', *Journal of the IEE*, Part II, 1948, **95**, pp. 59–70

145   LIEW, A.C., and DARVENIZA, M.: 'Dynamic model of impulse characteristics of concentrated earths', *Proc. IEE*, 1974, **121**, (2), pp. 123–135

146   KOSZTALUK, R., LOBODA, M., and MUKHEDKAR, D.: 'Experimental study of transient ground impedances', *IEEE Trans. Power Appar. Syst.*, 1981, **PAS-100**, (11), pp. 4653–4656

147 OETTLÉ, E.E., and GELDENHUYS, H.J.: 'Results of impulse tests on practical electrodes at the high-voltage laboratory of the national electrical engineering research institute', *Trans. South Afr. Inst. Electr. Eng.*, 1988, pp. 71–78

148 MORIMOTO, A., *et al.*: 'Development of weatherproof mobile impulse voltage generator and its application to experiments on non-linearity of grounding resistance', *Electr. Eng. Jpn.*, 1996, **117**, (5), pp. 22–33

149 SEKIOKA, S., *et al.*: 'Measurements of grounding resistances for high impulse currents', *IEE Proc., Gener. Transm. Distrib.*, 1998, **145**, (6), pp. 693–699

150 ARMSTRONG, H.R.: 'Grounding electrode characteristics from model tests', *IEEE Trans. Power Appar. Syst.*, 1953, **73**, (3), pp. 1301–1306

151 MOUSA, A.M.: 'The soil ionisation gradient associated with discharge of high currents into concentrated electrodes', *IEEE Trans. Power Deliv.*, 1994, **9**, (3), pp. 1669–1677

152 OETTLÉ, E.E.: 'A new general estimation curve for predicting the impulse impedance of concentrated earth electrodes', *IEEE Trans. Power Deliv.*, 1988, **3**, (4), pp. 2020–2029

153 MOHAMED NOR, N., HADDAD, A., and GRIFFITHS, H.: 'Impulse characteristics of low resistivity soils'. 36th universities *Power engineering* conference, Swansea University, 2001, Session 3C, p. 5

154 MOHAMED NOR, N., SRISAKOT, S., GRIFFITHS, H., and HADDAD, A.: 'Characterisation of soil ionisation under fast impulses'. Proceedings of the 25th international conference on *Lightning protection*, Rhodes, Greece, 18–22 September 2000, pp. 417–422

155 LOBODA, M., and SCUKA, V.: 'On the transient characteristics of electrical discharges and ionisation processes in soil', Proceedings of the 23rd international conference on *Lightning protection*, Firenze, Italy, 1996, pp. 539–544

156 FLANAGAN, T.M., *et al.*: 'Electrical breakdown properties of soil', *IEEE Trans. Nucl. Sci.*, 1981, **NS-28**, (6), pp. 4432–4439

157 FLANAGAN, T.M., *et al.*: 'Electrical breakdown characteristics of soil', *IEEE Trans. Nucl. Sci.*, 1982, **NS-29**, (6), pp. 1887–1890

158 LEADON, R.E., *et al.*: 'Effect of ambient gas on arc initiation characteristics in soil', *IEEE Trans. Nucl. Sci.*, 1983, **NS-30**, (6), pp. 4572–4576

159 SRISAKOT, S., GRIFFITHS, H., and HADDAD, A.: 'Soil ionisation modelling under fast impulse'. Proceedings of the 36th universities *Power engineering* conference (UPEC), Swansea (UK), 12–14 September 2001

160 VAN LINT, V.A.J., and ERLER, J.W.: 'Electric breakdown of earth in coaxial geometry', *IEEE Trans. Nucl. Sci.*, 1982, **NS-29**, (6), pp. 1891–1896

161 ERLER, J.W., and SNOWDEN, D.P.: 'High resolution studies of the electrical breakdown of soil', *IEEE Trans. Nucl. Sci.*, 1983, **NS-30**, (6), pp. 4564–4567

162 SNOWDEN, D.P., and ERLER, J.W.,: 'Initiation of electrical breakdown of soil by water vaporisation', *IEEE Trans. Nucl. Sci.*, 1983, **NS-30**, (6), pp. 4568–4571

163   SNOWDEN, D.P., and VAN LINT, V.A.J.: 'Vaporisation and breakdown of thin columns of water', *IEEE Trans. Nucl. Sci.*, 1986, **NS-33**, (6), pp. 1675–1679

164   SNOWDEN, D.P., BEALE, E.S., and VAN LINT, V.A.J.: 'The effect of gaseous ambient on the initiation of breakdown in soil', *IEEE Trans. Nucl. Sci.*, 1986, **NS-33**, (6), pp. 1669–1674

165   NEKHOUL, B., *et al.*: 'Calculating the impedance of a grounding system'. Collection de notes internes de la Direction des Etudes et Recherches – Matérial électrique, transport et distribution d'énergie, Electricité de France, Report 96NR00048, 1996.

166   SEKIOKA, S., *et al.*: 'A time and current dependent grounding impedance model of a grounding net'. International conference on *Electrical engineering* (ICEE'98), **1**, pp. 832–835

167   SEKIOKA, S., *et al.*: 'Surge characteristics of grounding system(s) of power distribution lines in Japan with reference to a reinforced concrete pole'. International conference on *Grounding and earthing*, Belo Horizonte Brazil, 18–21 June 2000, pp. 73–77

168   SEKIOKA, S., *et al.*: 'Development of a non-linear model of a concrete pole grounding resistance'. International conference on *Power system transients*, Lisbon, 3–7 September 1995, pp. 463–468

169   GRCEV, L., and ARNAUTOVSKI, V.: 'Comparison between simulation and measurement of frequency dependent and transient characteristics of power transmission line grounding'. 24th international conference on *Lightning protection* ICLP-98, Birmingham, UK, 1998, pp. 524–529

170   ROSA, E.B.: 'The self and mutual inductances of linear conductors', *Bulletin of the Bureau of Standards*, 1907, **4**, (2), p. 301

171   MENTER, F., and GRCEV, L.: 'EMTP-based model for grounding system analysis', *IEEE Trans. Power Deliv.*, 1994, **9**, pp. 1838–1849

172   CIDRAS, J.: 'Nodal frequency analysis of grounding systems considering the soil ionisation effect', *IEEE Trans. Power Deliv.*, 2000, **15**, (1), pp. 103–107

173   GRCEV, L.: 'Computer analysis of transient voltages in large grounding systems', *IEEE Trans. Deliv.*, 1996, **11**, (2), pp. 815–823

174   GRCEV, L., and DAWALIBI, F.: 'An electromagnetic model for transients in grounding systems', *IEEE Trans. Power Deliv.*, 1990, **5**, (4), pp. 1773–1781

175   GRCEV, L., and GRCEVSKI, N.: 'Software techniques for interactive optimisation of complex grounding arrangements for protection against effects of lightning'. 24th international conference on *Lightning protection* ICLP-98, Birmingham, UK, 1998, pp. 518–523

176   DAWALIBI, F., and SELBY, A.: 'An electromagnetic model of energised conductors', *IEEE Trans. Power Deliv.*, **8**, (3), pp. 1275–1284

177   ANDOLFATO, R., *et al.*: 'Ariel and grounding system analysis by the shifting complex images method', *IEEE Trans. power Deliv.*, 2000, **15**, (3), pp. 1001–1009

178   OLSEN, R.G., *et al.*: 'A comparison of exact and quasi-static methods for evaluating grounding systems at high frequencies', *IEEE Trans. Power Deliv.*, 1996, **11**, (2), pp. 1071–1081

179 HEIMBACK, M., and GRCEV, L.D.: 'Simulation of grounding structures within EMTP'. 10th international symposium on *High voltage engineering*, Montreal, Canada, 1997, p. 5

180 HEIMBACH, M., and GRCEV, L.D.: 'Grounding system analysis in transients programs applying electromagnetic field approach', *IEEE Trans. Power Deliv.*, 1997, **12**, (1), pp. 186–193

181 MAZZETTI, C., and VECA, G.M.: 'Impulse behaviour of ground electrodes', *IEEE Trans. Power Appar. Syst.*, 1983, **PAS-102**, (9), p. 3148

182 VISACRO, S., *et al.*: 'Evaluation of communication tower grounding behaviour on the incidence of lightning'. 24th international conference on *Lightning protection* ICLP-98, Birmingham, UK, 1998, pp. 556–561

183 GUPTA, B.R., and THAPAR, B.: 'Impulse impedance of grounding grids', *IEEE Trans. Power Appar. Syst.*, 1980, **PAS-99**, (6), p. 2357

184 DAVIES, A.M., GRIFFITHS, H., and CHARLTON, T.: 'High frequency performance of a vertical earth rod'. 24th international conference on *Lightning protection* (ICLP), Birmingham, UK, 1998, pp. 536–540

185 GRIFFITHS, H., and DAVIES, A.M.: 'Effective length of earth electrodes under high frequency and transient conditions'. 25th international conference on *Lightning protection*, Rhodes, Greece, 2000, pp. 469–473

186 GRIFFITHS, H., and HADDAD, A.: 'Earthing systems under lightning impulse conditions', *in* 'ERA Technology report no. 2002-0113, Lightning protection 2002: standards and practices', pp. 12.1–12.11

187 GRCEV, L.D., and HEIMBACH, M.: 'Computer simulation of transient ground potential rise in large earthing systems'. 23rd international conference on *Lightning protection* ICLP-96, Firenza, Italy, 1996, pp. 585–590

Chapter 9

Circuit breakers and interruption

H.M. Ryan

9.1 Introduction

Much literature exists relating to the historical developments worldwide associated with switchgear related technology [1–29]. A circuit breaker (CB) can be described [3] as a device used in an electrical network to ensure the uninterrupted flow of current in that network under normal operating conditions, and to interrupt the flow of excessive current in a faulty network. Under some circumstances, it may also be required to interrupt load current and to perform open–close–open sequences (autoreclosing) on a fault on others. The successful achievement of these, and indeed all, duties relies on extremely careful design, development and proving tests on any circuit breaker. Good and reliable mechanical design is required [1] to meet the demands of opening and closing the circuit breaker contacts, and an effective electrical design is also essential to ensure that the CB can deal with any of the electrical stresses encountered in service [4, chap. 7].

During the opening and closing sequences, an electric arc occurs between the contacts of a circuit breaker and advantage is taken of this electrical discharge to assist in the circuit interruption process, as has been discussed in detail [4, chap. 7]. In an AC electrical network, the arc is tolerated in a controlled manner until a natural current zero of the waveform occurs when the discharge is rapidly quenched to limit the reaction of the system to the interruption. With asymmetrical waveforms and for DC interruption, advantage is taken of the arc resistance for damping purposes or to generate a controlled circuit instability to produce an artificial current zero.

As discussed elsewhere, the arc control demanded by such procedures may require gas pressurisation and flow, which in turn, make additional demands upon the circuit breaker mechanism [1, 3] (see also chapters 7, 8, 10, 11 of Reference 4). The reader is directed specifically to G.R. Jones [4, chap. 7] for a fuller treatment of important system-based effects together with the principles of current interruption in HV systems. However, Figures 9.1–9.3, and their respective subcaptions,

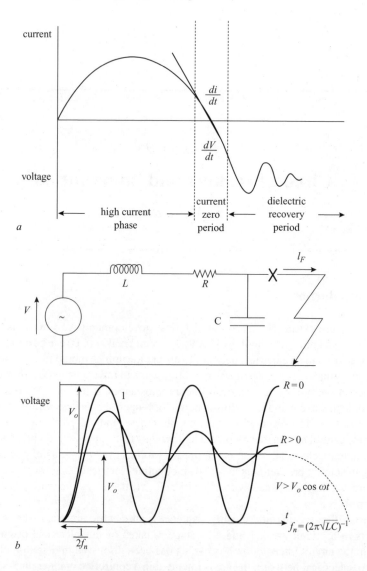

Figure 9.1   *Typical circuit breaker current and voltage waveforms (after Jones [4, chap. 7]). Voltage transients produced by short circuit faults and short line faults are of particular practical interest. Short circuit faults occur close to CB b, producing the most onerous fault conditions. Short line faults occur on transmission lines a few km from the CB c, and constitute the most onerous transient recovery voltages, typically at current zero (dV/dt) 10–20 kV/μs. Note: as pointed out by Jones, the situation is made more complicated in three-phase systems because current zero occurs at different times in each phase, implying that the fault is interrupted at different times leading to different voltage stresses across the interrupter units in each phase*

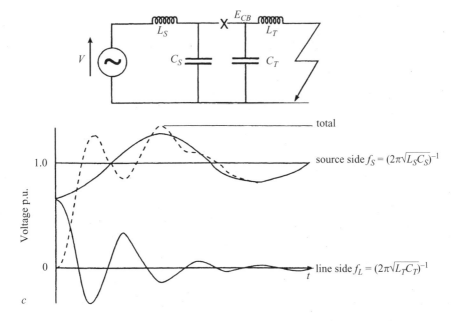

Figure 9.1   Continued

    a   voltage and current waveforms during the current interruption process for a simple symmetrical power–frequency (50 Hz) current wave; here, current reduces naturally to zero once every half cycle, at which point interruption is sought (i.e. minimum natural rate of di/dt); for conventional power systems with inherently inductive circuits, the contact gap is less severely stressed transiently during both the thermal and the dielectric recovery phases

    b   short line fault – onerous fault current

    c   short line fault – onerous transient recovery voltages

provide a brief indication of some of the strategic considerations covered by Jones. The above simplified description of circuit interruption processes [3], still serves to illustrate the complexity of the interactions involved, which are determined, on the one hand, by the nature of the arcing and arc quenching medium and, on the other, by the electrical network demands. It is of strategic interest to note that since Garrard's excellent review in 1976 [7], there has been a major swing to SF_6 as the arc quenching medium for modern transmission and distribution GIS (gas-insulated switchgear) [3, 4, 8], combining the excellent insulating and arc interruption characteristics of SF_6 gas.

    In passing it should be noted that there is also significant current interest in the development of GIL (gas-insulated lines) for long distance transmission circuits, i.e. longer than 500 m. (To date experience has been mainly limited to transmission

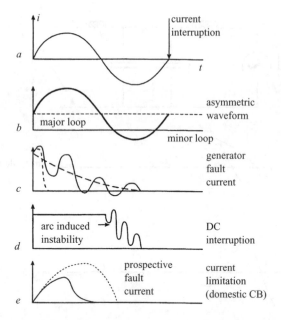

Figure 9.2   Circuit breaker current waveforms (after Jones [4, chap. 7])

    a shows sinusoidal current waveform with its natural current zero

    b shows sinusoidal current waveform superimposed on a steady current to form an asymmetric wave with major and minor loops which cause different CB stresses

    c related condition to *b* which occurs in generator faults; corresponds to the power frequency wave superimposed on an exponentially decaying component (consequently, the current zero crossing may be delayed for several half-cycles)

    d shows a further situation relating to the interruption of DC faults which is achieved by introducing an oscillatory current via arc instability in the CB so forcing the current to pass through zero eventually

    e finally, at the lower domestic voltages, current limitation can be conveniently induced, leading to an earlier and slower approach to zero current than occurs naturally; with the additional benefit of reducing the energy absorption demands made on the interrupter module

over short distances, typically less than 500 m.) Two recent CIGRE technical papers provide valuable technical overviews of typical GIL circuits [29].

At distribution voltage ratings, vacuum circuit breakers are now finding increased application [9, 10] and are challenging the new range of evolutionary commercial SF_6 designs [4], utilising puffer, suction, self-pressurising and rotary arc interrupters. Hybride CB designs are sometimes preferred at distribution ratings as illustrated in

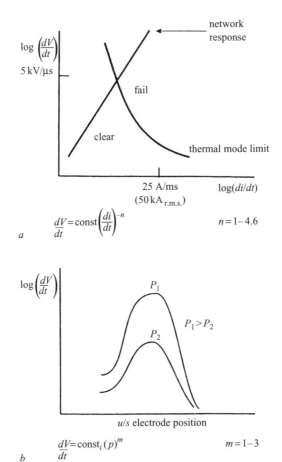

$$\frac{dV}{dt} = \text{const} \left(\frac{di}{dt}\right)^{-n}$$

$n = 1\text{--}4.6$

a

$$\frac{dV}{dt} = \text{const}_i\,(p)^m$$

$m = 1\text{--}3$

b

Figure 9.3   *Typical circuit breaker characteristics (after Jones [4, chap. 7])*

    a thermal recovery characteristics, network response and interrupter characteristic; this characteristic is in the form of a critical boundary separating fail and clear conditions on a rate of rise of recovery voltage (*dV/dt*) and rate of decay of current (*di/dt*) diagram, typically, the boundary condition obeys the relationship $dV/dt = \text{constant}\,(di/dt)^n$ with $n = 1\text{--}4.6$

    b thermal response characteristics; effect of pressure and geometry on performance, thermal response characteristic may also be improved by increasing the CB gas pressure, the nature of the gas or the geometry of the interrupter

Figure 9.4. Here, vacuum interrupters are used, but the designers have opted for SF_6 gas to provide insulation surrounding the interrupter. Two further designs of vacuum switchgear are illustrated in Figures 9.5*a–b* and Figure 9.5*c* shows a sectional view of a vacuum interrupter.

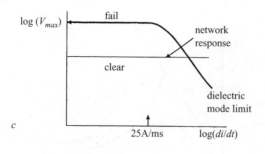

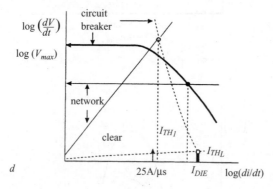

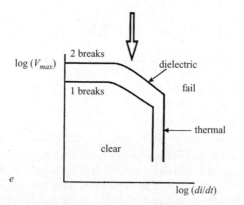

Figure 9.3   Continued

    c   dielectric recovery characteristic, here, for the dielectric regime, the characteristic is represented by the critical boundary separating successful clearance and fail on a maximum restrike voltage (V_{MAX}) and rate of decay of current (di/dt) diagram, the dielectric recovery performance may be improved by increasing the number of contact gaps (interrupter units) connected in series

    d,e overall CB performance – superposition of thermal and dielectric, note that the overall limiting curves for the CB performance are obtained by combining the thermal and dielectric recovery characteristics

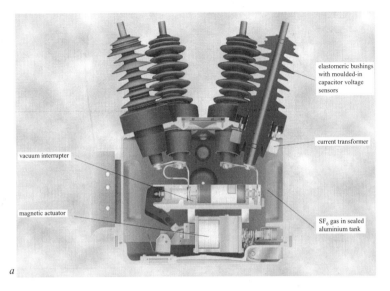

elastomeric bushings
with moulded-in
capacitor voltage
sensors

current transformer

vacuum interrupter

magnetic actuator

SF$_6$ gas in sealed
aluminium tank

a

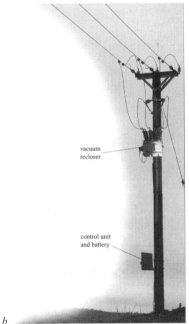

vacuum
recloser

control unit
and battery

b

Figure 9.4　*GVR vacuum recloser (courtesy Whip and Bourne, after Blower [9])*
　　　a　sectional view of recloser
　　　b　recloser connected to an overhead line

Figure 9.5   Examples of vacuum switchgear

    a   type Hadrian 12 kV vacuum switchgear (after Stewart [10], courtesy VA TECH Reyrolle)

    b   type WS 33 kV vacuum switchgear (after Stewart [10], courtesy ALSTOM T&D Distribution Switchgear Ltd)

9.2   Circuit interruption characteristics, arc control and extinction

The ability to interrupt electrical power circuits represents an essential function, especially in situations of overloads or short circuits when, as a protective measure, immediate interruption of the current flow must take place. In the early days of current interruption, circuits were broken merely by the separation of contacts in air followed by drawing the resulting electric arc out to such a length that it could no

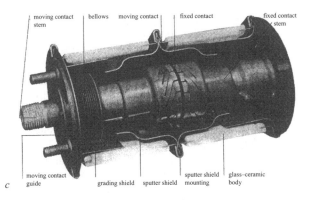

Figure 9.5   *Continued*

    c   sectional view of vacuum interrupter (courtesy Vacuum Inter-
         rupters Ltd) and Reference 4, chapter 3, also published earlier in
         chapter 6, Bradwell (Ed.) 'Electrical insulation' (Peter Peregrinus
         Ltd, 1983)

longer be maintained [1, 8]. Such a means of interruption soon became inadequate as
the voltages and current capacity of power systems grew and it became necessary to
develop special switching devices called circuit breakers.

At the outset, it should be appreciated that within the power industry the general
term switchgear is widely used and this covers the combination of (i) switching
devices and (ii) measuring devices (which monitor the system, detect faults and, as
appropriate, initiate circuit breaker operation). High reliability is essential with any
circuit breaker design as the device has to protect costly power system networks.
In service, they can remain closed for long periods until called upon to open under
even the most onerous short circuit situations.

An essential requirement in any CB design concerns its in-service operating performance reliability, the effectiveness of arc control and efficient arc quenching,
removal of power dissipated in the arc (via convection, conduction and radiation) which is essential to ensure rapid voltage withstand capability after current
interruption [4, chap. 7]. Jones provides a useful description of the principles of
current interruption in HV systems when considering the various phases of current
interruption.

As highlighted in an earlier comprehensive book in this series, dealing solely with
power circuit breaker theory and design [1], circuit breakers have strategic roles to
perform. They are required to control electrical power networks by:

(i)   switching circuits on
(ii)   carrying load
(iii) switching circuits off under manual or automatic supervision.

Flurscheim [1], observed (appropriately) that the character of their duty is unusual
and they will normally be in a closed position carrying load or in the open position

providing electrical isolation. Typically, in a service life of up to 30 years, they are called upon to change quickly from one condition to the other (possibly only occasionally), and to perform their special function of closing onto a faulty circuit, or interrupting short circuit current (sometimes only on very rare occasions). Hence, the earlier comment, 'reliability is essential'! The circuit breaker must be reliable in static situations and also be effective instantaneously when called upon to perform any switching operation even after long periods without movement [1]. All circuit breakers have designated ratings, and they must satisfy rigorous standard IEC tests, which are set to guarantee compliance in service (see IEC 62271-100 High voltage alternating current circuit breakers [30]), covering a wide range of conditions.

Ali [4, chap. 8] informs his readers that a circuit breaker has to be capable of successfully:

- interrupting (i) any level of current passing through its contacts from a few amperes to its full short circuit currents, both symmetrical and asymmetrical, at voltages specified in IEC 62271-100, and (ii) up to 25 per cent of full short circuit making currents at twice the phase voltage
- closing up to full short circuit making current (i.e. $2.5 \times I_{sym}$) at phase voltage and 25 per cent of full making currents at twice the phase voltage
- switching (making or breaking) inductive, capacitive (line, cable or capacitor bank) and reactor currents without producing excessive overvoltages to avoid overstressing the dielectric withstand capabilities of a system
- performing opening and closing operations whenever required
- carrying the normal current assigned to it without overheating any joints or contacts.

As pointed out recently by Ali [4, chap. 8], interrupting devices become more complex as the short circuit currents and voltages are increased and, at the same time, the fault clearance times are reduced to maintain maximum stability of the system.

The major components of a circuit-breaker are:

- interrupting medium, e.g. sulphur hexafluoride (SF_6) gas
- interrupter device (see Figure 9.19)
- insulators
- mechanism (importance of opening energy is shown in Figure 9.18b)

(Note: the writer found that regular participation in IEC/BSI standardisation committees, together with corresponding CIGRE study committee/colloquia activities, provided him with first class opportunities for ongoing professional empowerment and career development.)

The interruption of short circuits represents a severe duty imposed on circuit breakers, and short circuit ratings have progressively risen during the past 80 years to meet the increasing needs of power networks, as voltages have increased up to 420 kV in the UK and up to 1000 kV worldwide. Modern transmission circuit breakers are capable of interrupting faults of up to 60 kA, and for distribution systems (at voltages in the range 1 to about 150 kV) CB ratings up to 25 kA are typical. Over the years, the total break time of CBs, required to interrupt short circuits, has been significantly reduced in the interests of system stability:

- 10–20 cycles, for early plain break oil circuit breaker (required because of long arc durations)
- 6–8 cycles, achievable by introduction of arc controlled interrupters
- <2 cycles, now readily achievable with modern SF_6 CB designs.

A circuit breaker is an electromechanical device which is used to control and protect distribution and transmission networks (see chapters by Ali, Jones, Fletcher and Pryor in Reference 4). It can be called on to operate and to interrupt currents as low as 10 A up to its full short circuit rating (i.e. up to 40/50 or even 60 kA) – changing the circuit breaker from a perfect conductor to a perfect insulator within only a few milliseconds (see chapters by Jones and Ali in Reference 4). Great care is required to ensure that appropriate dynamic operating mechanism characteristics (i.e. speed/travel) are designed into any circuit breaker to ensure effective and safe opening and closing of the interrupter throughout its life and for all operating conditions [1–4, 8]. As Ali [4, chap. 8], has pointed out, the number of operating sequences and the consistency of closing and opening characteristics generally determines the performance of the mechanism (see Figure 9.18b). He further states that:

(i) although only 2000 satisfactory type-test operations are required to achieve compliance with IEC62271-100, the current trend (or tendency) is to carry out 5000 extended trouble-free operations tests to demonstrate compatibility of mechanism/CB interface with the current second generation of SF_6 interrupters

(ii) the task with third-generation SF_6 circuit breakers becomes much easier as these devices are fitted with low energy mechanisms and lightweight moving parts, and can satisfactorily perform 10 000 trouble-free operations without any stresses and excessive wear and tear on the moving and fixed parts of the circuit breaker; detailed coverage of interrupter mechanisms is outside the scope of this brief overview chapter but the reader is directed to a major body of available resource material [e.g. see CIGRE publications http://www.cigre.org].

It is sufficient to comment here that a circuit breaker may have either a single-phase or a three-phase mechanism [1, 4, 8], depending on network operational requirements or client preference. Mechanisms used with modern interrupters, are generally hydraulic, pneumatic, or spring or a combination of these [4, chap. 8]):

- pneumatic close and pneumatic open
- hydraulic close and hydraulic open
- spring (motor charged) close and spring open
- hydraulic close and spring open
- pneumatic close and spring open.

Some examples of mechanisms used with distribution switchgear are illustrated by Ali, in chapter 11 of Reference 4, and Figure 9.4 illustrates a vacuum reclosure with magnetic actuator [9].

Much incremental research and development of improved interrupter devices has taken place during the past 80 years and continues to the present time. This strategic effort has led to a greatly improved understanding of the basic principles of interruption and the development of advanced modelling and testing techniques culminating in the design of a wide range of advanced circuit-breaker devices [1–10]. These developments, sometimes involving increasing levels of sophistication, have kept pace with the requirements of expanding power systems, deregulation and frequently special operating situations [11].

To illustrate this fact, consider the following example. On the basis of current technology [12], the controlled switching systems is one of a number of potential solutions to overcome transient problems. However, the performance of modern CBs is extremely complex and the development of effective testing methodologies, which can be influenced by a number of considerations, is crucial. A thorough understanding of circuit breaker characteristics, together with effective testing strategies for testing of controlled switching systems, is complex (see section 9.6). Briefly, it must be appreciated that a complete modern transmission substation can involve many components, and the entire installation may be produced, delivered and installed by one manufacturer or the components may be delivered by different producers and erected and commissioned by the user or one of the manufacturers. Because of such diversity, a modular testing procedure has recently been suggested and this will be considered later (see section 9.6).

Extensive literature exists relating to the historical developments [1–4, 8] associated with switchgear technology, detailing numerous and varied techniques adopted for improving the interrupting ability of modern circuit breakers (see also www.cigre.org).

Whatever technique is used, the basic problem is one of controlling and quenching (or extinguishing) the high power arc which occurs at the separating contacts of the interrupter when opening high current circuits. The reliability of a circuit breaker is interdependent on several factors: insulation security, circuit breaking capability, mechanical design and current carrying capacity. The range of arc quenching media used in commercial circuit breakers has included air, compressed air, oil, small oil volume, strong magnetic field devices, SF_6 and vacuum environments [1–10].

It must be appreciated that circuit breakers (i.e. circuit interrupters) are essentially mechanical devices for closing or opening electrical circuits. Historically, circuit breakers have been classified:

(i)   according to the insulating media used for arc extinction
(ii) sometimes based on special interruption features (e.g. simple autoexpansion and magnetic rotation techniques of arc interruption in SF_6 designs).

Fuller consideration of these specialist design concepts is outside the scope of this chapter and these aspects are discussed elsewhere [1–10].

Detailed information relating to the dielectric and arc interruption characterisation of atmospheric air, compressed air, oil, vacuum, SF_6, various other gases and gas mixtures, and solid insulating materials, exists and is now readily available to the switchgear designer who also has recourse to a wide range of advanced simulation packages [4].

At transmission voltages, sulphur hexafluoride gas (SF_6) has been used for arc interruption and part insulation since the 1960s, and has now fully replaced compressed air or small oil volume devices in new substation installations in most of the developed countries [1–4, 8]. Indeed, as has already been stated, SF_6 interrupter technology is very mature, as evidenced by the third generation devices now available commercially [4]. Despite this, much of the switchgear in service on transmission systems in the UK still uses compressed air or small oil volume interruption techniques. At distribution voltage levels, sales of new CBs, in most developed countries, are now exclusively centred on devices where interruption is based on vacuum and SF_6 techniques which has led to more efficient and cost effective designs [9]. Although these modern types of CB have progressively replaced older types of interrupter, using air or oil-based interruption methodologies, here again, many of the older designs are still in service and continue to perform effectively [1, 4, 9].

9.2.1 Principles of current interruption in HV systems (after Jones [4, chap. 7])

All methods of interruption current, to date, rely on introducing a non-conducting gap into a metallic conductor, achieved by mechanically separating two metallic contacts so that the gap formed is either automatically filled by a liquid, a gas, or even vacuum. In practice, such inherently insulating media may sustain a variety of different electrical discharges which then prevent electrical isolation from being achieved.

Jones considers three major facets to such electrical discharges which then prevent electrical isolation being achieved:

(i) As contacts are separated, an arc discharge is inevitably formed across the contact gap. The problem of current interruption then transforms into one of quenching the discharge against the capability of the high system voltage of sustaining a current flow through the discharge. Since this physical situation is governed by a competition between the electric power input due to the high voltage and the thermal losses from the electric arc, this phase of the interruption process is known as the thermal recovery phase and is typically of a few microseconds duration.

(ii) The second facet of current interruption relates to the complete removal of the effects of arcing which only occurs many milliseconds after arc formation even under the most favourable conditions. The problem then is one of ensuring that the contact geometry and materials are capable of withstanding the highest voltage which can be generated by the system without electrical breakdown occurring in the interrupter.

(iii) The third facet bridges the gap between the thermal recovery phase and the breakdown withstand phase. The problem in this case is that the remnant effect of the arcing has cleared sufficiently to ensure thermal recovery but insufficiently to avoid a reduction in dielectric strength. This is known as the dielectric recovery phase.

Based on this understanding, circuit interruption technology is concerned on the one hand with the control and extinction of the various discharges which may occur, and on the other it relates to the connected system and the manner in which it produces post-current interruption voltage waveforms and magnitudes.

9.3 Distribution switchgear systems

Historically, distribution switchgear technology was based on oil or air break circuit breaker devices. Although good reliability was achieved with CBs based on both these interrupting media for several decades, evolving strategic health and safety guidelines and lifetime ownership considerations have resulted in a strategic swing in the 1970s towards new medium voltage switchgear, i.e. 1–70 kV, using either vacuum or SF_6 as the interrupting media. Oilless circuit breaker designs have been developed since the late 1970s [1, 4, 8–10]. Initially:

- SF_6 devices offered a proven replacement technology, with excellent dielectric and interruption characteristics
- early vacuum interrupters (bottles) were difficult to manufacture and were prone to produce voltage transients due to severe current chopping [1, 4, 8].

Consequently, SF_6 circuit breakers were deemed by users to offer a more popular and reliable replacement interrupting media. In recent years, vacuum interrupters have found increased acceptance and application as a result of advances in material technology and improved manufacturing techniques. Modern vacuum interrupters have a good performance record [4, 8–10] especially with respect to frequent high fault clearance operation (e.g. see design shown in Figures 9.4*a* and *b*). SF_6 is used almost exclusively at voltages above 36 kV. Low voltage systems (as originally stated) (>1 kV) have in the past used air and oil CBs, but advances in moulded case air circuit breakers have made air CBs more cost effective and this is now the most widely used technology in this range (see Reference 9 and companion articles).

Recent progress in distribution switchgear has been discussed extensively in Reference 4 and was also considered in a special IEE feature covering four valuable articles. In his introduction to these articles, Blower chose to group distribution switchgear into three areas of use as summarised below and detailed in his original paper [9], namely:

(i)   primary distribution
(ii) secondary distribution in cable connected networks
(iii) secondary distribution in overhead line connected networks.

Some useful technical observations relating to companion articles by other authors are also listed in [9]. Blower illustrated a new generation of autoreclosures which use vacuum interrupters for arc interruption, driven by magnetic actuators (e.g. Figure 9.4). He considers that distribution switchgear (often referred to as medium voltage switchgear), which covers the voltage range above 1 kV up to about 150 kV, can be divided into the three groups identified above. Blower also points out certain special aspects/considerations relating to this interrupter. These include:

- the use of permanent magnets in the actuator has arisen with the introduction of new magnetic materials which are both powerful and not affected by additional magnetic fields
- the magnets provide positive on and off position for the vacuum interrupter, and the addition of a magnetic field from a surrounding coil to change the flux distribution is used to change the state
- the vacuum interrupter has a very long life on this type of duty and the actuator has a very low energy consumption and can operate thousands of times from small primary cells (e.g. lithium-type cells); a life of 10+ years, before replacement is required, has been claimed
- the vacuum device can be contained in a sealed vessel filled with SF_6 to ensure a high reliability of insulation and independence from climatic conditions (see Figure 9.4*a*).

9.4   Substation layouts and control aspects

9.4.1   Substation layouts

Typical substation switching layouts have been discussed recently by Fletcher [4, chap. 9], Pryor [4, chap. 10] and also Ali [4, chaps 8 and 11] for transmission networks and primary distribution substations for single and double busbar switching arrangements and also for urban and rural distribution systems. Figures 9.6–9.11 present a summary of some strategic switching arrangements considered, with supplementary information provided in subcaptions to these figures and in the résumé reproduced below:

- Pryor has recently reported that primary distribution substations, within the UK, typically operate at 11 kV and are utilised to supply a relatively large number of consumers within the local area (Figure 9.6*a*).
- The double busbar arrangement, shown in Figure 9.6*b*. This system is widely used for large important supplies (i.e. as at bulk supply points), which in the UK typically operate at 132/33 kV, or perhaps for large industrial customers, higher security may be built into the substation design. Pryor comments that:

   (i)   modern busbar protection is very reliable such that the same degree of availability can normally be achieved without the need to resort to a second busbar system

430   *Advances in high voltage engineering*

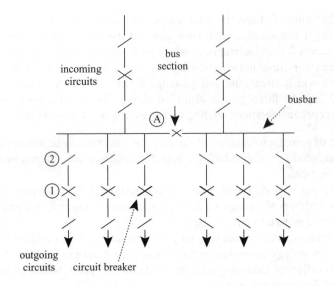

Figure 9.6a   Typical distribution primary substation layout: single busbar switching arrangement (after Pryor [4, chap. 10])

- a primary substation feeding a rural network is commonly of the open terminal air-insulated design whereas a primary substation feeding urban networks is more commonly of the indoor metalclad design
- a common switching arrangement for either substation type is of the single busbar design but with the busbar being split into two sections and interconnected via a bus section CB, A
- there are usually two incoming circuits – one feeding each section of the busbar; there may be typically five outgoing circuits feeding either multiradial networks for overhead rural systems or ring circuits for urban connected networks
- for maintenance purposes disconnectors, 2, are fitted on either side of the CB, 1
- facilities are also provided for earthing outgoing or incoming circuits and for earthing each section of the busbar; current transformers, CTs, may be fitted within the outgoing circuit for either protection or, less commonly, for tariff metering purposes (e.g. for large single customers)

(ii)   it is also considered necessary to ensure that each section of busbar is normally kept energised because, if a section remains deenergised for a significant period of time, then deterioration may occur such that possible busbar failure may result on reenergisation.

- Urban distribution system switching arrangement (see Figure 9.7). This system is fed from a primary distribution switchboard, each outgoing feeder on one section

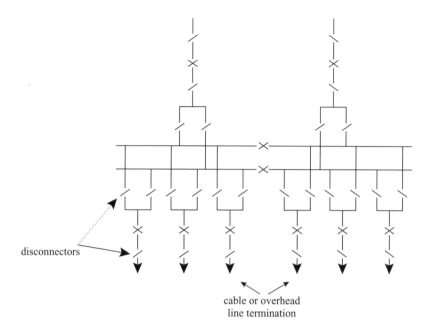

disconnectors

cable or overhead
line termination

Figure 9.6b   *Typical distribution primary substation layout: double busbar switch-*
               ing arrangement (after Pryor [4, chap. 10])

- arrangement allows the option of four separate sections of busbar to
  be utilised such that if any one busbar fails, supplies could generally
  be maintained via the other three busbar sections; all circuits can
  be selected to either of the two main busbar sections
- busbar protection is usually provided to ensure very rapid clearance
  of any faulted section of busbar (such arrangements are, however,
  very costly)

of the busbar feeds via an 11 kV cable network to typically 10 to 12 secondary
distribution substations which are electrically connected within the ring circuit.
This ring circuit is connected back to a feeder on the adjacent section of busbar
at the same substation [4]. Pryor goes on to describe general network principles
which operate in most industrialised countries, although some may employ radial,
as well as ring, circuits. Typical distribution voltages used elsewhere may range
from 10 kV to 20 kV. 6.6 kV systems were common at one time within the UK,
but have now been largely phased out.

- A rural distribution system switching arrangement is illustrated in Figure 9.8
  (see Pryor [4, chap. 10]). Here, the primary switchboard will typically comprise
  two sections of busbar connected via a bus section CB, A. The number of outgoing feeders tends to be less than with an urban network with typically six
  panels. Surge arrestors, commonly fitted at the cable sealing end and mounted
  physically at the top of the pole at the junction between the cable and overhead

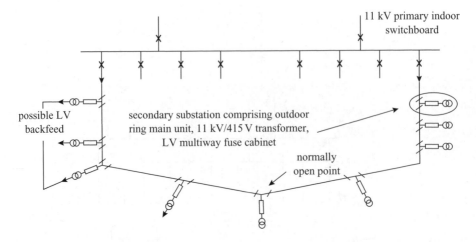

Figure 9.7   Urban   distribution   system   switching   arrangement   (after   Pryor [4, chap. 10])

- this ring circuit usually has a normally open point, the purpose of which is to minimise the number of customers affected by the faulted section of the ring circuit
- ring main switches are manually operated; hence fault location and customer reenergisation is time consuming; a modern tendency is for the ring main switches to be fitted with remotely operable mechanisms such that switching times can be reduced by allowing remote substation facilities or, with modern intelligence systems, to allow automatic deenergisation and isolation of a faulted section and reenergisation up to the point of the fault
- the tee-off point of the ring circuit feeds via an 11 kV/415 V, three-phase transformer to an LV fuse board, typically having up to five outgoing circuits which feed directly to large customers or groups of customers; in most of these LV circuits, it is possible to achieve an LV backfeed from an adjacent ring main unit, thus maximising the number of customers on supply while fault repairs are in progress

line, reduce the probability of transient induced overvoltage (caused by lightning strikes to the overhead line) affecting either the cable or associated switchgear. (See Pryor [4, chap. 10], for fuller details of urban and rural distribution networks.)

- Turning now to larger substations at both transmission and distribution voltages (see Fletcher [4, chap. 9]), where security of supply is critical, the double busbar arrangement shown in Figure 9.9 has been extensively used in the UK. A variation of this is shown in Figure 9.10 which illustrates a multisection double busbar arrangement. A third layout, reproduced in Figure 9.11, is known as a one and

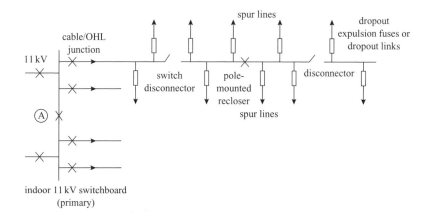

Figure 9.8 *Rural distribution system switching arrangement (after Pryor [4, chap. 10])*

- the primary switchboard may be of open-terminal form or in the outdoor weatherproofed form; it is now more common, however, for these primary switchboards to be enclosed within brick-type enclosures, in which case, conventional indoor metalclad type switchgear is more commonly used (most 11 kV rural primary switchboards are of the metalclad-type, whereas at 33 kV, open-type busbar connected switchgear is more common)
- most UK rural networks feed overhead line distribution circuits and there is usually a short connection of cable connecting from the switchboard to the terminal pole of the overhead line

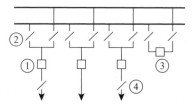

Figure 9.9 *Double busbar switching arrangement (after Fletcher [4, chap. 9])*

Here:

- each circuit is provided with a dedicated circuit breaker CB (1) and can be selected to connect to either of the two independent busbars by appropriate switching of busbar selector disconnectors (2)
- the busbars can be coupled by means of a bus coupler circuit breaker CB (3), the line disconnectors (4) being normally included at transmission voltages to facilitate maintenance, but this feature may be omitted in many cases, without loss of operational functionality
- service continuity can be maintained with a busbar out of service; however, Fletcher points out that a single circuit outage is necessary for fault repair or maintenance in the CB area

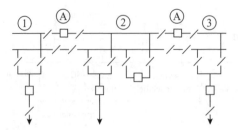

Figure 9.10   Multisection, double busbar switching arrangement (after Fletcher [4, chap. 9])

Here:

- bus section CBs (A) are added to the standard double busbar config-uration, providing the ability to segregate circuits across a number of sections of busbar separated by CBs and this is used to limit system disruption under worst case fault conditions
- Fletcher provides an example in the above figure of the use of this configuration for the connection of large generating units to a system
- if each unit is arranged so as to feed a separate section of busbar (1, 2, 3 in the diagram above), then normally only one unit will be tripped for a busbar fault, leaving the remainder in service
- under worst case fault conditions (a bus section or bus coupler CB fault), a maximum of two generating units will trip
- the system can thus be operated with reserve generation capacity sufficient only to cater for the loss of these two units

a half switch mesh–double busbar switching arrangement and this arrangement is widely used abroad at transmission voltages.

For fuller information on substation layouts, the reader should follow the activities and readily available literature of appropriate CIGRE study committees (e.g. SC. 23, see also References 7–12, 15, 16, 19, 26, 27, 29) which provide strategic infor-mation of global activities, and trends in this sector. This includes consideration of such important issues as environment, short circuit forces on busbars, pollution, and minimum safety clearances for personnel working within substations.

9.4.2   Intelligent networks

Before considering intelligent networks, it is pertinent to comment on certain strategic effects of deregulation within the electricity supply sector (see also section 9.9.2). Several workers [4, 22–24], the most recent being Ziegler [26] whose personal views are briefly reproduced below, have reported that increased competition since deregulation has forced utilities to go into cost saving asset management with new

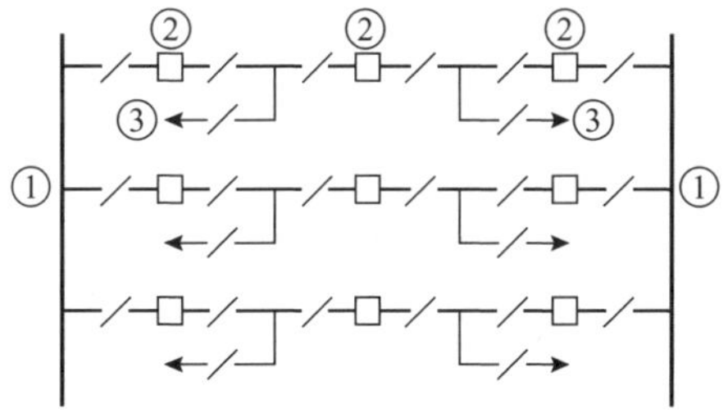

Figure 9.11   *One and a half switch, double busbar switching arrangement (after Fletcher [4, chap. 9])*

Here:
- this arrangement has two busbars (1) interconnected by chains of three CBs (2)
- outgoing circuits are taken off from between pairs of CBs (3)
- this switching arrangement is widely used internationally at transmission voltages, but is not common in the UK
- when fully operational, this arrangement offers a high level of service continuity of the outgoing circuits and is beneficial where line availability is critical
- however, this arrangement can be vulnerable to single fault outages, during periods of busbar or CB maintenance outages

risk strategy:

- plants and lines are higher loaded up to thermal and stability limits
- existing plants are operated to the end of their lifetime and not replaced earlier by higher rated types
- redundancy and backup for system security are provided only with critical industrial load
- corrective event-based repair has replaced preventative maintenance
- considering this changed environment, Ziegler feels that power system protection and control face new technical and economical challenges [26].

A summary of Ziegler's opinions, recognising his authority as a former chairman of CIGRE, SC34, is reproduced below:

(i) protection and substation control have undergone dramatic changes since the advent of powerful microprocessing and digital communication

(ii) smart multifunctional and communicative feeder units, so called IEDs (intelligent electronic devices) have replaced traditional conglomerations of mechanical and static panel instrumentation

(iii) combined protection, monitoring and control devices and LAN-based integrated substation automation systems are now state of the art

(iv) modern communication technologies including the internet are used for remote monitoring, setting and retrieval of load and fault data

(v)   higher performance at lower cost has resulted in a fast acceptance of the new technology

(vi)   the trend of system integration will continue, driven by the cost pressure of competition and technological progress

(vii)   the ongoing development towards totally integrated substations is expected to pickup speed with the approval of the open communications standard IEC 61850 in the next few years.

W.J. Laycock [27] has recently written an interesting and thought provoking article on intelligent networks, covering:

(i)   the development of existing networks

(ii)   the need for more intelligence

(iii)   future developments and opportunities.

He discusses how engineers and managers in diverse locations are being challenged to find solutions that will simultaneously drive down cost and encourage better utilisation of distribution sector assets by the development of intelligent networks. Some strategic issues raised by Laycock will now be briefly touched on and are reproduced below, including the application of intelligence, centres of advanced intelligence, network services, protection intelligence and drivers for the future (see also Tables 9.1 and 9.2 and Figures 9.12–9.15). (The reader is directed to the original article for fuller treatment.)

Figure 9.12 illustrates a classical MV distribution network which may include SCADA and FPIs but the presence onsite of trained staff is still normally required. Intelligence is defined by Laycock as 'whatever independent action can be taken on receipt of information', and is apparent at the following locations:

- network control centre
- protection and autoreclose equipment
- voltage control equipment

Table 9.1   Centres of advanced intelligence (after Laycock)

- distributed network control
- automation
- adaptive protection
- plant condition assessment
- voltage and waveform control
- generation scheduling
- transport and supply of electricity
- network services
- ownership boundaries

Table 9.2   Network services (after Laycock)

On-network	Off-network
• frequency control	• asset management
• voltage control	• performance measurement
• waveform control	• plant condition monitoring
• reactive power supply	• refurbishment
• dynamic stability	• meter reading
• load following	• billing
• fault level control	• connection management

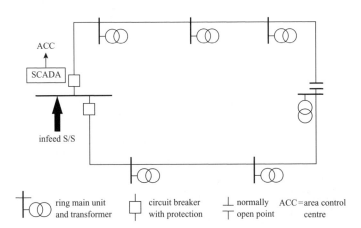

Figure 9.12   Classical open ring MV distribution network configuration (after Laycock [27])

- network arrangement is in common use worldwide, employing a single infeed S/S and ring main units (RMUs) at load tee-off points
- it is passive (i.e. has no active generation), its sole purpose being to distribute energy to customers
- may have some form of remote control (SCADA) at the infeed point and may be equipped with fault passage indicators (FPIs), but normally requires trained people on site to respond to any kind of incident

- network owners office for:

    (i)   network design
    (ii) asset management
    (iii) purchase and sales of electricity
    (vi) performance assessment.

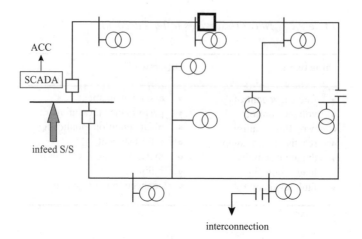

Figure 9.13   Open ring with tees and interconnection (after Laycock [27])

- networks commonly have multibranch tees, additional CBs and interconnection with other networks for emerging backfeeds etc
- it must be realised that all the functions relating to application of intelligence (control/management) become more complex, often requiring compromises between conflicting requirements

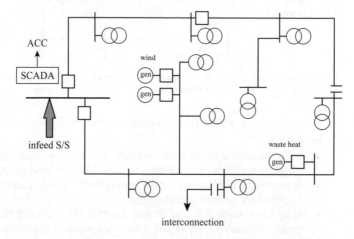

Figure 9.14   Open ring with embedded generation (after Laycock [27])

Laycock informs the reader that intelligence may be supplied directly by human intelligence, by relatively simple control/relaying equipment or, on more modern networks, by digital equipment and computers, the mix varying widely between networks and over time as a network develops. He goes on to indicate the functions that need the application of intelligence: (i) in the hard sense of network control and (ii) in the soft sense of network management, pointing out that there are many links

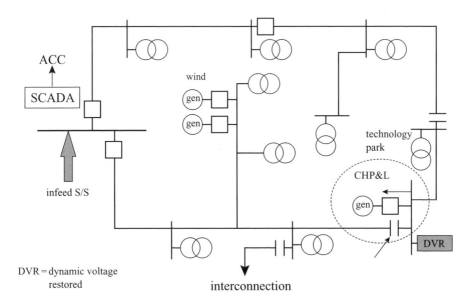

Figure 9.15   *Open ring with an IPP, CHP scheme (after Laycock [27])*

- Laycock identifies the rapid changes post-liberalisation as more players are attracted to this sector, creating the need for significantly more intelligence than was necessary with the earlier passive nature of network design

between these functions:

Application of intelligence

Control	Management
loading	access for work
match supply to demand	asset management
security	billing
short circuits	new connections
supply restoration	supply contracts
voltage control	service contracts
dips, sags, spikes and harmonics	performance measurement

Laycock also recognised that, in practice, networks are rarely as simple as the open ring shown in Figure 9.12, since they commonly have multibranch tees, additional CBs and interconnection with other networks for emergency feedbacks etc. Consequently, Laycock considers that Figure 9.13 presents a more typical network, where all the functions, as detailed above, become more complex, often necessitating compromises between conflicting requirements. He also points out that although Figure 9.13 is still a passive network, many previously passive networks are now becoming active with the addition of embedded or dispersed generation (EG). Laycock comments that some generation might come from renewable energy sources to reduce global warming, others to provide improved security or quality of supply or will simply be an independent power producer (IPP) who is exploiting a business opportunity. Thus, according to Laycock, modern networks may now take on the appearance of Figure 9.14 and, moreover, the characteristics of these modern networks require the application of considerably more intelligence than previously.

This raises some strategic issues, which according to Laycock, include:

- small stochastic generators (output dependent on primary source, e.g. wind)
- operation with alternative infeeds
- concept of active distribution (do minimum to enable connection in the first place!)
- ownership and money flows
- action on loss-of-grid
- generator scheduling (at transmission level).

It is recognised that, as a consequence of the development of privatised energy markets and the connection of more sophisticated plant and changes in perception as to ownership, operation and outsourcing of services, networks are being further stretched. These concepts have been illustrated by Laycock in Figure 9.15. This figure shows the imposition of a combined heat and power scheme (CHP) which has been designed to operate islanded with its own load under certain conditions, but not necessarily optimised with network operation.

As a consequence, Laycock considers that the following may be added to the issues raised by Figure 9.14:

- complex generation, supply and connection contracts
- island operation and power wheeling
- quality and security of supply
- provision of network services
- electricity supply regulation.

Laycock also points out that this opening-up or liberalisation of the electricity market creates opportunities for increased entrepreneurial input as individuals and business groups identify opportunities for new business ventures, which could increase utilisation of networks (e.g. see Figure 9.15) while reducing the cost of ownership. However, it must be recognised that although here is undoubtedly a strategic need for a significantly increased level of intelligence with modern active distribution networks, it is not yet clear to Laycock and others at this time the precise form(s) that such new intelligence systems will take. In sections 9.4.3 and 9.4.4, we will briefly review Laycock's perspective relating to future needs/developments in this sector. (The reader should also become familiar with recent reviews relating to intelligent condition monitoring strategies relating to switchgear [4] and also current strategic developments in the CIGRE sector.)

9.4.3 Need for more intelligence (after Laycock)

Laycock considers that intelligence will be applied in three primary areas:

(i)   innovative solutions to network design and operation to (enable) facilitate new business streams

(ii) use of intelligent devices on the network to improve responses and collect information

(iii) resolution of problems as they occur including disputes over responsibilities.

Table 9.1 details anticipated centres for advanced intelligence which, according to Laycock, will appear on future networks. In the past, responsibility for provision was generally centred on a single entity, normally a Regional Electricity Company (REC). However, responsibility for provision of future intelligent systems will involve several interest groups operating within the still evolving liberated energy market sector, including possibly several utilities, manufacturers, traders etc., all with access to data via sophisticated software and computer systems.

As Laycock and others have pointed out:

- the degree of automation using intelligent devices to control the network and to collect and analyse information will vary according to the owner's philosophy, its legacy systems and the pace of network development;
- provision of outsourced network services is receiving increased attention and is a prime area for innovative thinking
- the kinds of services that may be needed are listed in Table 9.2, but Laycock considers that this cannot be exclusive at this stage in their development.

9.4.4 Network protection and future developments (after Laycock)

Laycock provides a valuable perspective relating to the critical provision of protection as networks move from passive to active operation, pointing out that what was previously acceptable on radial distribution circuits will fail to satisfy requirements of dynamic operation with active generation and with loads that are susceptible to prolonged voltage dips. Major strategic considerations raised by Laycock relating to network protection in the future are set out in his original article [27].

Finally, looking towards possible future developments/trends, Laycock points out that one must have a good understanding of the drivers for the future development of intelligent networks in order to apply human intelligence to produce effective advanced solutions. He also sets out his current thinking relating to a range of strategic factors. Further consideration of the topic of substation control, as identified by other CIGRE experts, is given in section 9.5.

9.5 Substation control in the system control (CIGRE. WG.39.01 [11])

Brief reference will now be made to the strategically important issue of substation control (SCS) which now represents an integral part of the electrical power system's

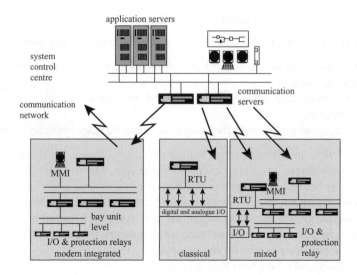

Figure 9.16   Types of substation control system [11]

control hierarchy. A recent CIGRE working group paper [11] provides the reader with a useful positioning overview of this theme. The paper is focused on the placement, functionality and coordination of a substation control in a hierarchical system control scheme. It presents various scenarios as to where a substation control centre may be placed and what types of function are required from the substation control to improve the overall control performance. Feedback from a small survey carried out by WG.39.01 also forms part of this paper. Certain major strategic aspects discussed in this publication, including the outcomes from the survey, will be extensively reproduced and highlighted in this section as set out in the following paragraphs, figures and tables (see Figure 9.16). (The original source [11] is gratefully acknowledged.)

This CIGRE WG.39.01 paper [11] comments that the SCS may have different roles to play in an electrical utility operation and it fits in different roles in the hierarchical control centres schemes, as shown in Figure 9.16:

(i)   classical monitoring, control and protection using remote terminal units (RTUs) and local applications
(ii)   the SCS existing in parallel to the classical RTU configuration
(iii) a modern SCS combining protection, monitoring and control and without use of a RTU.

The paper reports that if intelligent electronic devices (IEDs) are used to provide integrated monitoring, protection and control functionality, these have to be connected over a coupler or gateway to the station LAN. The I/O units may be distributed throughout the substation on the process bus, and the interface from the process bus to the station bus depends on the bay control units. Redundant configuration may be implemented to suit the availability requirements of the substation control. Optical fibre is the preferred communication media rather than the hardware connections of the earlier generation. The optical-fibre-based communication is claimed to have lower interference and allow reconfiguration at a later stage with the benefit of lower engineering costs.

Further strategic facts detailed in [11], are:

- Data engineering and management: important aspects of data engineering and maintenance play an important role for the SCS and, as reported [11], must be planned and coordinated to fit in the system control hierarchy.
- Functional requirements which influence SCS and higher level centres.
- Interfaces from substation control to high level centres have been considered as follows (see [11]):

    (i)   multimaster configuration
    (ii) protocols in use
    (iii) multimedia possibilities.

It is possible to review experience with the SCS in system control from feedback from a limited CIGRE survey carried out by WG.39.01 [11], based on returns from mainly large utilities with more than 100 substations and for voltage levels of 150 kV and above:

(i)   Implementation ratio of SCS to the number of substations varied from 0 to more than 50 per cent. The majority of the substations were unmanned and allowed remote switching. Almost all utilities owned the higher level centre and substations, which communicated with them. The higher level centres may have supervisory control and data acquisition (SCADA) or energy management systems (EMS).

(ii)   In the emerging unbundled and competitive market which now exists, a distinction needs to be made between the owner of the substation and the system operator. Very often, with an independent system operator (ISO) or even a transmission system operator (TSO), the ownership of the substation differs and a contractual agreement between the two parties must be negotiated.

(iii) The SCS must ensure that only those commands which come from authenticated higher level centres be executed. Networking and remote logins are weak spots in the security of such a system. Firewalls and other security schemes must be considered to inhibit unauthorised access to the SCS or SCC database.

(iv) The data must be refreshed with the resumption of the link, and system supervision events from SCS should be logged or alarmed in SCC. However, most of the utilities do not see any technical problems in achieving this. The experience from the survey [11] shows that the remote switching is a regular feature for the substations, which have been implemented with SCS. In some cases, the remote switching was done only when a maintenance crew was present in the substation for testing and maintenance purposes.

(v)   The performance, in terms of control actions, information updates, communication link speeds and total data transfer is of keen interest to most utilities.
Most typical response times required for reporting of a breaker tripping from
the substation to the control centre varied in the range of one to three seconds.
However, responses for less than one second and higher than three seconds
were also reported for such an event.

(vi)   In comparison with the tripping events, measurements were refreshed at
a range between five and ten seconds. The SCS performed some filtering
and data selection on the measurements and indications before transferring
them to the higher level centre. More than 80 per cent of the measurements and about 50 per cent of the indications, which were measured and
monitored at the SCS level, were transferred to the higher level centre.
The transfer of disturbance data is normally requested in the case of disturbances, but some utilities apparently prefer to transfer these data cyclically
once a day.

(vii) Communications speeds varied between 300 b.p.s. and 9600 b.p.s. with some
links being as fast as 64 kb.p.s. Most of the stations have power line carrier
and optical fibre links but radio and leased lines are also in use. Redundant
communication links are one way to increase the total availability of the SCS
from the SCC. As the cost of communications is going down, almost all
utilities are opting for redundant links.

(viii) The availability of SCS is of crucial importance for total system control
performance. At first level, the availability of the control system at the
SCS should be high, which means that the critical control components are
redundant. In addition to the SCS availability, the availability of the communication link influences the total availability of the SCS as seen from
the SCS. Most of the utilities which have been asked in the survey [11]
apparently require an availability level higher than or equal to that of the
classical RTU controlled units. Quantitatively, most of the availability figures were given as 99.5 per cent, with one value positioned between 98 and
99.5 per cent.

(ix)   The protocols used for communication between SCS and SCC have in practice
been driven towards standard non-proprietary protocols. Utility communication architecture-driven ICCP and IEC870-5-101 telecontrol protocols were
the most favoured ones for the implementation. Exceptions were some substations where proprietary protocols were used in order to match the implemented
protocol in the SCC.

To summarise, this valuable WG.39.01 paper [11], representing just one of many
strategic published outputs from CIGRE/IEC organisations in recent years, has
provided valuable evidence 'that the present state of the substation control system
from the hierarchical control system viewpoint is slowly gaining acceptance and the
technology has been shown to be maturing. The results from the small survey indicated that most of the utilities have acquired some experience with SCS systems
and are in the process of upgrading the classical RTU links to the SCS and
SCC link'.

These authors [11] state that the factors that must be considered while designing or planning a substation control system are:

- the operational hierarchy of the system
- the level of acceptable integration of protection and monitoring devices
- whether the substations are manned or unmanned
- the complexity of the substation
- the data flow and communication performance.

From this evidence [11], it would seem that practically all new substations are being considered for SCS. Apparently, some utilities prefer a mixed solution of SCS with RTU-type links to the SCS, others install an RTU and plan to upgrade to SCS in the near future and certain other utilities go purely for the SCS-type solution.

There is a need to differentiate between real-time and non-real-time data to the SCC in order to leave enough capacity for fast real-time activities.

9.6 Planning specification and testing of controlled HVAC switching systems [12]

9.6.1 Background

The uncontrolled switching of inductive and capacitive elements can create electrical transients which may cause equipment damage and system disturbances. This problem has been well known for many years. CIGRE Study Committee 13 recognised that controlled switching offered a potential solution to this problem, as restated below and also in Sections 9.6.2 and 9.6.3:

- In 1992, they appointed a task force TF13.00.1 which prepared a document summarising the state of the art [1]; subsequently, Working Group WG.13.07 was created in 1995 to study, in greater detail, the strategic issues associated with controlled switching from both supplier and user perspectives [12].
- The first publication of WG.13.07 addressed fundamental technical issues such as the potentially achievable transient reductions for various switching conditions. It is claimed [12] that the focus was on the most common applications of switching of shunt reactors, shunt capacitor banks and lines. The fundamental theoretical and scientific aspects were presented and the dominant role of the circuit breaker characteristics was clearly identified.
- On the basis of the information presented in the first document, a second document was prepared which addresses the more practical aspects of controlled switching. Extensive guidance is given on how to approach controlled switching projects and, in particular, how to study, specify and test the circuit breaker and control system. (Note: The full text of this report can be obtained via the CIGRE website under the title 'Controlled switching of HVAC circuit breakers – planning, specification and testing of controlled switching systems' [12].)

Initial planning questions are introduced, in order to assist the decision making process for any application, such as, 'is controlled switching a technically suitable and cost effective solution?'. There is considerable interaction between the component parts of a controlled switching system (CSS), and the hierarchy of responsibility to ensure the overall performance is discussed in some detail [12]. Subsidiary issues are also introduced, such as the interfaces between the controller and the auxiliary systems of a substation.

Recommendations for the type testing of CSS components and also the integrated system are made, including detailed guidance on difficult issues such as the determination of the circuit breaker characteristics. The document concludes with guidance on complete system performance checks and commissioning tests.

The key aspects of this CIGRE report will now be considered and restated.

A case study prior to decision taking [12], with a schematic decision process for capacitor bank switching, is also set out.

9.6.2   Specification of controlled switching installations [12]

This recent abridged CIGRE report [12] comments that the specification of a controlled switching system must consider the interactive combination of a switching device, a controller and a variety of associated items of auxiliary equipment:

- The system may be assembled from components from a single source or disparate sources, and it is important to identify which parameters or requirements are of particular relevance to each component of the system and which to the system as a whole.
- Any application of controlled switching must be undertaken with due regard to the characteristics, interactions and compatibilities of its component parts. The authors [12] go on to point out that the performance of the individual components must be well defined and understood but the performance of individual components of the system is of limited interest in isolation since it is the overall accuracy requirement which should be met. The principle of interaction between the power system and the CSS is clearly illustrated in the original paper [12].
- There are likely to be numerous parties involved in a CSS application and clear identification of responsibilities is vital. The requirement may be to construct a total system on the basis of a comprehensive specification, but may equally require the application of controlled switching into an existing uncontrolled scenario. In this latter case, although the overall specification remains valid, certain parameters will be predetermined (or may not be available in an accurate form) resulting in a need to ensure that any newly installed equipment is sufficiently robust in its capabilities to overcome such potential difficulties.
- In order to take these basic conditions into account, an approach is adopted which considers the individual components of the CSS to be discretely identifiable units. However, considering ongoing developments in integration and digitisation of substation protection and control facilities, the authors [12] felt that the principle of this document should be equally applicable in all cases whether of analogue, digital or hybrid type.

Three levels of specification requirements were advanced [12]:

(i)   *Performance*: defines the basic functional compatibility requirements to which the chosen CSS will be subjected.
(ii) *System*: defines specific technical and material requirements for the CSS itself.
(iii) *Component*: addresses the internal interactions of the CSS taking into account the external requirements of the system specification.

The requirements set out and discussed in [12] are incorporated at the highest possible level and cascade to lower levels as appropriate.

Finally, it should be emphasised that this important paper [12], which summarises and explains strategic aspects relating to the testing of controlled switching systems and circuit breaker testing, represents an invaluable reference source for any reader wishing to achieve a good understanding of this subject area.

9.6.3   Concluding remarks

In their closing remarks the authors [12] state that:

- controlled switching has already been widely applied by many utilities
- controllers and CBs available from a variety of manufacturers have suitable characteristics for controlled switching
- they consider that the information and studies undertaken in this programme of work [12] go some considerable way towards consolidating available knowledge on issues of preparing studies, specifications and testing
- the contents of the full report are intended to provide guidance on the subject of controlled switching in a broad context and, as far as possible, on a scientific level as well
- the potential variety and complexity of the applications of controlled switching is such that many issues can only be covered in general terms – but the authors feel that the guidance presented in the extended version of their paper should form the building blocks for more specific cases to be developed
- moreover, they suggest that this work would be well suited as the basis for future standardisation activities.

The authors [12] also point out that the idle time behaviour of a CB is another issue of major importance. Supported by experimental evidence, they point out that depending on the period of quiescence, there may be a change in the operating time of the first operation.

Note: This further example of recent international collaborative CIGRE work, as restated and described above, illustrates to the reader the real benefits of participating in such a forum (e.g. CIGRE/IEC committee, working group, task force activities, or even merely obtaining such information directly as a CIGRE member!) as a means of empowerment by keeping oneself abreast of current relevant strategic international collaborative technical work in the sector.

9.7   Dielectric and global warming considerations

At this point, very brief mention can be made to the strategic role of [4,13,14]:

- undertaking appropriate characterisation development studies on potential solid/gaseous insulation systems, for a wide range of conditions
- analytical techniques to enable the switchgear designer to develop efficient designs with the robust expectation that the switchgear insulation systems for GIS and GIL plant will achieve a long and trouble-free life in service.

For many years, the writer was extensively involved in developing numerical field techniques which have found widespread application in the insulation design of GIS, GIL and other switchgear [4, chap. 3]. Simple empirical and semiempirical breakdown estimation methods by Ryan *et al.* [31], which are extensions of the work by Pederson [32], together with an available experimental database obtained from extensive Paschen's Law/similarity-type high voltage studies in gaseous insulants – covering a wide range of electrode systems, gas pressures and temperatures, were thoroughly developed to such a degree that minimum breakdown voltages of practical GIS/GIL design layouts (as well as a host of SF_6 gas-gap arrangements [4, chap. 3, 13, 14]) can be estimated to within a few per cent, at the design stage, often without recourse to expensive development testing. Such derived voltages are generally the minimal withstand levels attainable under practical conditions. Undoubtedly, there is still considerable scope in the future to predict dielectric performance by utilising advanced planning simulation tools incorporating genetic algorithms and AI techniques, linked to various extensive databases relating to breakdown characteristics of gaseous insulation, including equipment service performance data (e.g. from surveys similar to Reference 15). Further discussion relating to dielectric issues is presented in section 9.9.1.

Finally, it is anticipated that the continuing, and growing, environmental concerns relating to SF_6, which was put on the list of greenhouse gases of the Kyoto Protocol, as discussed in a valuable CIGRE paper [16] will influence the development of the next generation of gas-insulated switchgear. The writer recommends that the reader(s) keep abreast of this developing theme. (This positioning paper can be downloaded from the CIGRE website.) However, he is of the opinion that a commercial replacement for 100 per cent SF_6 gas, for interruption purposes at the higher ratings, still seems remote!

9.8   Some examples of modern switchgear

9.8.1   *SF_6 live-tank and dead-tank switchgear*

Live-tank and dead-tank types of SF_6 circuit breaker (see Figures 9.17–9.19) have been developed for ratings up to 63 kA at 525 kV. They have given satisfactory service worldwide for more than 20 years. The particular choice of CB type for any substation depends on many factors [4, chap. 8] which include cost of switchgear, level of atmospheric pollution, potential environmental restrictions, price and availability of the land, individual preferences, security against third party damage etc.

Figure 9.17 Outdoor installation of 145 kV dead tank switchgear in the USA (courtesy ALSTOM T&D Ltd)

(i) *Live-tank designs*: the interrupters in live-tank circuit breakers are housed in porcelains. The interrupter heads are live and mounted on support insulators on top of a steel structure to conform with safety clearances. The live-tank circuit breakers are normally installed in open terminal outdoor substations.

(ii) *Dead-tank designs*: the interrupters in dead-tank circuit breakers are housed in an earthed metal tank, usually aluminium, mild or stainless steel, depending on the current rating [4, chap. 8]. GIS dead tank circuit breakers can be of either horizontal or vertical configuration. They can be used in GIS indoor and outdoor substations (see Figure 9.17).

Finally, it can be seen that Figure 9.18a, after Jones [4, chap. 7], provides some useful interrupter performance characteristics, showing the effect of various parameters on the thermal recovery characteristics of gas blast interrupters, and Figure 9.18b shows the strategic incremental improvements of opening energy as a function of the interrupting technique used and the interrupting capacity expressed in kA.

Turning now to modern distribution switchgear, Figure 9.4 presents an example of a new generation of reclosure, recently discussed at greater length by Blower [9], which uses long life vacuum interrupters for arc interruption. In this figure, particular attention should be directed towards three design aspects:

(i) the disposition of
  • the vacuum interrupter
  • the magnetic actuator
  • the toroidal current transformer, mounted within an SF_6 insulated gas-sealed aluminium enclosure

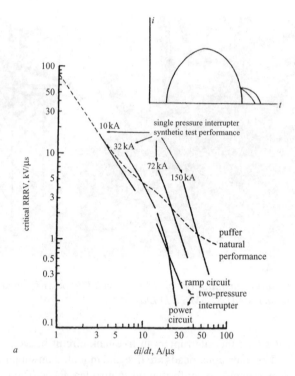

Figure 9.18   *Typical interrupter performance characteristics (after Jones [4, chap. 7])*

        a   effect of various parameters on the thermal recovery characteristics of gas blast interrupters

(ii) the use made of elastomeric bushings with moulded-in capacitive voltage sensors (which provide good environmental performance and are vandalism and mishandling-damage resistant)
(iii) signals from the current and voltage sensors can be fed to intelligent digital relays which can be programmed for a wide range of protection requirements e.g. autoreclosing functions based on overcurrent, earth-fault and sensitive-earth-fault protection.

Figure 9.4b illustrates this type of reclosure pole-mounted and connected to an overhead distribution system. This device apparently uses a low energy consumption magnetic actuator, capable of thousands of operations. Permanent magnets using new magnetic materials are used in the actuator energised from small primary type cells (e.g. lithium) with a claimed life expectancy of up to ten years providing positive on and off positions for the vacuum interrupter [9].

    In passing, the reader should be aware of the exciting recent developments and improvements in transformer technology (including improved ratings) achieved by the application of advanced magnetic materials and highly sophisticated design tools (e.g. [28]).

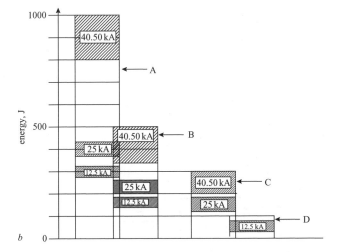

Figure 9.18   Continued

   b   evolution of opening energy as a function of the interrupting tech-
         niques used and the interrupting capacity expressed in kA; the
         increasingly simple design and the utilisation of the arc energy
         result in a reduction in switching energy
   (A) autopneumatic technique
   (B) autopneumatic + thermal expansion technique
   (C) thermal expansion + rotating arc technique
   (D) thermal expansion or rotating arc technique

9.9   Equipment life expectancy: condition monitoring strategies

Before considering modern condition monitoring strategies it is appropriate to look
first at:

(i)   the knowledge available to switchgear designers/power utilities in the early
      1980s, prior to the development and widespread implementation of CM
      methodologies
(ii) the strategic changes in network structure/operation since the advent of the open
      market.

9.9.1   Evaluation of solid/gaseous dielectric systems for use
            in HV switchgear

As reported by Milne and Ryan [33], the application of synthetic resins in the design
and construction of switchgear for power equipment, originated over 100 years ago
with the introduction of phenolic resins. Since this period, a very large and ever
widening number of synthetic materials and their compounds have been developed
but the one dominating in the heavy electrical industries for the past 50 years is the
epoxy-resin-based system. The aim of good electrical insulation design is to provide
the required electrical characteristic and service life at minimum cost. In making the

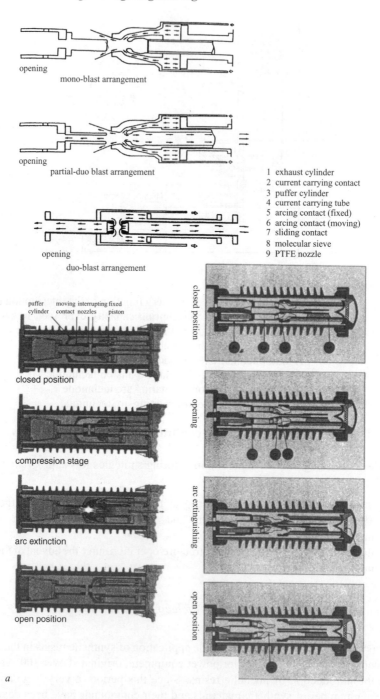

opening

mono-blast arrangement

opening

partial-duo blast arrangement

1 exhaust cylinder
2 current carrying contact
3 puffer cylinder
4 current carrying tube
5 arcing contact (fixed)
6 arcing contact (moving)
7 sliding contact
8 molecular sieve
9 PTFE nozzle

opening

duo-blast arrangement

puffer   moving interrupting fixed
cylinder contact nozzles piston

closed position

compression stage

arc extinction

open position

closed position

opening

arc extinguishing

open position

a

Figure 9.19   SF$_6$ puffer-type interrupters (after Ali [4, chap. 8])
       a principles

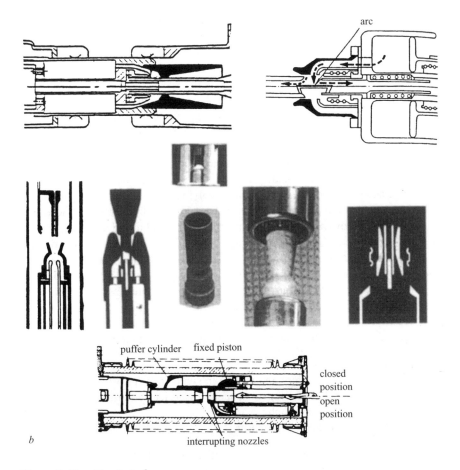

Figure 9.19   *Continued*

   b   examples of PTFE nozzles

Notes: puffer type interrupters can be classified according to the flow of compressed SF_6 gas. They are often referred to as mono, partial duo, and duo blast interrupters [4]. The nozzle is a vitally important component of a puffer type interrupter. The interruption characteristic of an interrupter is governed by nozzle geometry, shape, size and nozzle material (see Ali [4, chap. 8])

final selection (see examples in Figure 9.19) and deciding on thickness and shape, the designer is seldom fortunate to have adequate technical data on new alternative materials applicable to his/her particular design. Because of this, insulation designs in switchgear change very slowly both in type of material used, component size, shape and electrical operating stresses. Evaluation of a new insulating material is a costly and time consuming process, which should include testing under conditions closely representative of those that are experienced in service.

Solid insulation in switchgear is expensive and a failure at high voltage is often catastrophic because it is not usually self-restoring. Operating stresses are, therefore, usually a few orders of magnitude lower than the established intrinsic electrical strength determined under controlled laboratory conditions. The requirement of an insulating material is governed by the specific application and can be as listed principally as electrical, mechanical, physical and chemical properties. Frequently, properties other than dielectric are equally as important, such as compatibility, mechanical, thermal and processability [20]; consideration of these may cause an otherwise attractive material to be rejected.

Milne and Ryan [33] reported that medium and high voltage insulation material research in their company at that period was oriented around the evaluation of basic material characteristics, followed by studies involving thermal ageing, long term electrical stress ageing (both at power frequency and at higher frequencies, to accelerate ageing by electrical discharges), environmental testing and multifactor stress ageing involving in some cases full scale test rigs. This 1984 contribution was primarily concerned with solid insulation operating in an atmosphere of air or SF_6, as used in modern switchgear in the voltage range 11–565 kV which was (at that time and still is) a technical subject which does not achieve the same prominence in the literature as other insulating media such as gases or liquids.

Table 9.3 provides a convenient summary of insulation materials, characteristics of interest and evaluation procedures for switchgear application, listing:

- material characteristics of interest
- evaluation procedures before application in switchgear (prior to the application of approved CM techniques for subsequent in-service condition monitoring and asset management of the switchgear)
- certain electrical/mechanical properties.

Clearly, it is possible to build on such earlier physical characterisation studies [4, 13, 14, 21] to create extensive databases which can be integrated into modern in-service condition monitoring techniques which will be briefly considered in section 9.9.3.

9.9.2 Open market: revised optimal network structure

During the past 12 years, there have been massive strides in the liberalisation of the electricity supply industry in the UK and worldwide [22–24]. The main driving forces for network development before liberalisation were increase of demand, connection of new power stations built to meet the demand and high reliability of the network. The cost incurred to develop the network was borne by the customers [24, 4, chap. 1]. Some consequences of this practice [24] were that:

- reliability and voltage quality increased continuously
- the technical complexity of the networks gradually increased by the use of the latest technology components
- the economic point of view was of less importance.

Table 9.3   Insulation materials: characteristics of interest and evaluation procedures (after Ryan and Milne)

Characteristics of interest

Electrical

electrical strength
surface flashover
surface resistivity
volume resistivity
permeability
loss angle (tan δ)
tracking resistance
dry arc resistance
voltage endurance

Thermal

heat distortion
expansion
ageing
flammability
toxic gases

Mechanical

tensile
flexural
impact
hardness

Chemical

compatibility
resistance
environmental
stability

Physical

density
moisture absorption
weathering
fungi resistance

Evaluation procedures for switchgear applications

basic material characteristics
short term testing
life testing
thermal ageing
environmental ageing
voltage stress ageing
full scale duration testing of switchgear assemblies

Note: references to further strategic information relating to the dielectric performance of gaseous insulation and solid support insulation, or barriers, as used in high voltage SF_6 insulated switchgear systems are given in chapter 4 of Reference 4

After liberalisation (deregulation), the electricity industry worldwide has gone through several changes in organisation and structure, attitude to cost and relationship with customer [24, 4, chap. 1]. Some examples of the change of roles have recently been given [24]:

> after roles are separated with different ownership or control, the power producers are organised for competition, the transmission companies have become natural monopolies, the consumers can shop around for cheap power and good service, and investments in new technology must result in increased profit or reduced cost. There is a tendency to consider

the short term rather than take a longer term view in decision making. This new environment for the network owners has resulted in changed planning procedures with emphasis put on economic performance.

It can be appreciated from the foregoing, and from a thorough evaluation of recent excellent CIGRE task force reports, that:

- one of the effects of market liberalisation is the unbundling of roles within the electricity supply industry
- new principal role categories are those who manage the transmission networks, those who use the transmission networks to trade, those who facilitate energy trading and statutory authorities
- major stakeholders, identified as the demand customers, power producers system operator, network owners and regulator, exert new and different driving forces, and the long standing driving forces have reduced in importance
- the authors also studied the effects of the new driving forces upon the planning processes and propose a methodology for accounting for them in a pseudo-optimisation.

From the foregoing paragraphs, it can now be appreciated that the effective management of assets to ensure that the user obtains the optimum life for the plant is becoming more vital as electricity transmission and distribution systems are operating to different criteria and are being worked harder [4, chap. 23, 22–24]. Lowen [24] provides a high level view of the electricity industry and considers the effects of reform in the industry since privatisation from the perspective of the England and Wales transmission system operator (National Grid). He considers that the single biggest driver for the future is the world growth in electricity generation and consumption. Based on the levels in 1995, forecasts show that there will be a 54 per cent and 33 per cent increase in energy and electricity consumption, respectively, in 2015 worldwide.

9.9.3   Condition monitoring strategies

Recently, in a chapter devoted to Condition Monitoring (CM) of high voltage equipment, White [4, chap. 22] has:

(i)   equated the asset value of electrical plant owned by the electricity supply generation and distribution industry in the UK as billions of pounds sterling

(ii)   estimated the replacement value of the transformers alone on the National Grid to be about 1 billion pounds sterling.

White goes on to observe that:

- to be able to buy, fit and forget such valuable assets is not an option in today's financial environment
- unless a transformer is operating or is operational, it is worth only the value of its recyclable components
- an asset can very quickly lose its value if it is not correctly managed to give optimum life performance

- management of an asset such as a transformer will almost certainly involve CM in some form or other.

Similarly, CM techniques are now in use worldwide to monitor a variety of network equipment, including overhead lines, GIL, cables, motors, generators and switchgear, and are the focus of this section. Clearly, the safety of personnel is essential for all CM studies whatever the network equipment being monitored or the particular CM technique adopted.

To be able to effectively manage the assets of a network system, it is important to know whether an item of HV equipment continues to remain in a serviceable condition or not. Fortunately, this can be assessed by monitoring many different quantities using a wide variety of techniques – hence the adoption of the term condition monitoring or in-service condition monitoring. As the term CM suggests, some form of ongoing measurement process is performed periodically and comparisons made with earlier measurements – on the same item of equipment – to determine whether this item of equipment is still in a serviceable condition. As Coventry *et al.* [4, chap. 22] have recently pointed out in a valuable overview on CM of HV equipment, this topic now attracts hundreds of papers per year and adopts methodologies ranging in complexity from simple manual checks, at one extreme, to state of the art computer-based systems at the other.

It must be recognised at the outset that a vast amount of design data have been accrued for many years by equipment designers and network users relating to strategic information concerning the design, construction, performance characteristics of equipment and in-service performance of similar items of equipment [13–17]. In addition, a recent report [19], briefly summarising the findings of a second CIGRE survey concerned with GIS service reliability, considers that GIS technology has contributed very effectively to increasing the reliability of new substations and to improving the asset lifecycle of existing ones. The report provides a valuable database of GIS service experience survey on SF_6-insulated equipment relating to more than 13 500 circuit breaker bays and 118 500 bay years service. Much valuable information is presented in the report which provides an analysis of results on installations and GIS major failure report data, including general data about GIS installations, data concerning GIS failure frequencies, major failure characteristics and consideration of life expectancy, maintenance and environmental issues [19]. This work, typical of CIGRE surveys in other sectors of electrical plant, provides a valuable resource and benchmark for those involved in condition monitoring methodologies including users and manufacturers operating within the area of GIS substation planning, design, construction and in-service performance.

Three broad aims of CM were recently identified by Coventry [4, chap. 22]:

(i) To prevent component/network failures which can have strategic consequences on network safety and costs. Electrical failure of an item of equipment can result in a short circuit fault which can result in significant damage (and sometimes even catastrophic damage). Consequently, there is an obvious benefit of adopting CM methodologies which can provide advanced warning of potential problems relating to an item of equipment.

(ii) To enable maintenance programmes to be planned/performed within an appropriate timescale, according to the latest measured condition of the equipment.

(iii) To estimate the remaining life of the item of equipment, based on its present condition, together with knowledge on the rate of degradation. This approach allows equipment to be used closer to the end of their life and facilitates the implementation of an effectively planned asset replacement strategy (e.g. see Reference 4, chaps 21–23).

Some examples, originally reported by Coventry and Jones [4] are reproduced in Figures 9.20–9.24 and in associated text below describe the application of a variety of strategic condition monitoring methods based on collaborative studies in the UK relating to intelligent CM strategies. Aspects illustrated include:

(i) Figure 9.20, which illustrates a UHV coupler, defect recognition and decision support approach used in GIS equipment [4, chap. 22]. Figure 9.20*a* shows a schematic of a UHF coupler, perhaps the most commonly used sensor for the detection of UHV signals in GIS:

- the coupler consists of a disc-shaped electrode mounted on a hatch cover plate of a GIS chamber and connected to the outside world by means of a gas-tight feed-through
- as a precaution, the coupler may be shunted by a resistance of a few kilo ohms to reduce the level of the voltage induced by the incident electric field at power frequency
- in the UK, it is current practice for all new GIS at 420 kV to be equipped with UHV couplers as standard equipment; satisfactory sensitivity is achieved with the couplers spaced at intervals of up to 20 m.

(ii) Figure 9.20*b* and *c* provide typical discharge characteristics. Curve *b* shows equivalent circuit and discharge pattern with poor contact between conducting components (e.g. where a stress shield has become loose and this poorly bonded component will tend to charge capacitively so that the potential difference across the contact gap rises until the gap sparks over). This spark reduces the potential difference across the contact gap to zero [4, chap. 22].

(iii) Figure 9.20*c*, shows typical discharge pattern with free conducting particles (e.g. this pattern is accumulated over a number of cycles). A free particle moving under the influence of the electric field may initiate breakdown if it enters a region of critical electric field. Coventry [4, chap. 22] points out that the behaviour of different types of defect has been studied extensively in a library environment to assist with defect recognition and risk analysis. Information on defects detected in service is normally maintained in a database to assist further decisions. (Note: the reader should also be aware that such characterisation of discharge patterns has also been known (and accrued) for more than 30 years from first generation SF_6-insulated GIS, and since the 1960s for 300/420 kV SF_6-insulated instrument transformers which were first developed by UK manufacturers in conjunction with the then UK transmission utilities. GIS and free standing SF_6 instrument transformers have been very

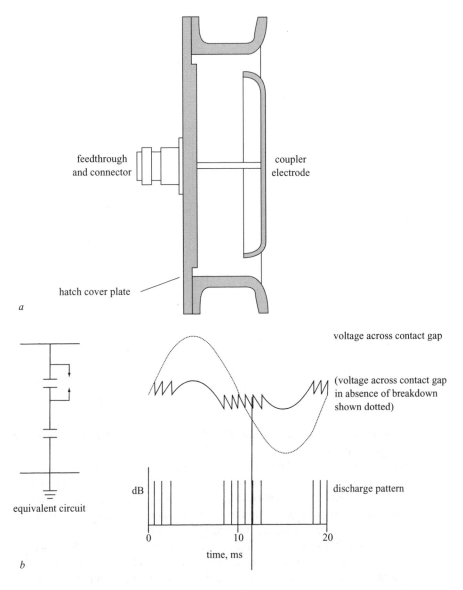

Figure 9.20   UHV coupler defect recognition and decision support (after Coventry [4, chap. 22])

  a   schematic of UHV coupler
  b, c discharge patterns
  b   poor contact between conducting components

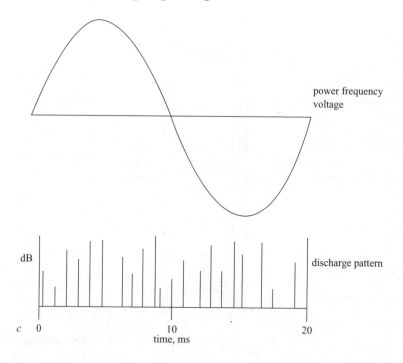

Figure 9.20   Continued

    c free conducting particles

        also included: defect recognition and decision support

reliable and have seen more than 30 and 40 years service, respectively, in the UK grid and elsewhere.)

(iv) When UHF signals are detected, decisions must be made on whether corrective action is necessary and in what timescale. Information may be obtained from the UHF signals on which to base these decisions. Frequently, it is possible to recognise the type of defect and the way in which it is discharging from the point on the power frequency wave at which discharges occur.

Figure 9.20*b* shows an example of poor contact between conducting components, such as where a stress shield has become loose. The poorly bonded component will tend to charge capacitively so that the potential difference across the contact gap rises until the gap sparks over. The spark reduces the potential difference across the contact gap to zero.

A free conducting particle, such as a piece of swarf, lying on the floor of the GIS enclosure becomes charged by induction under the influence of the power–frequency electric field. The induced charge may be sufficient to overcome the particle's weight, so that it moves under the combined influence of the electric field and gravity. Coventry comments that the particle may return to the enclosure at any point on the power–frequency wave. In general, this will be a different point on the wave to that at which it lifted

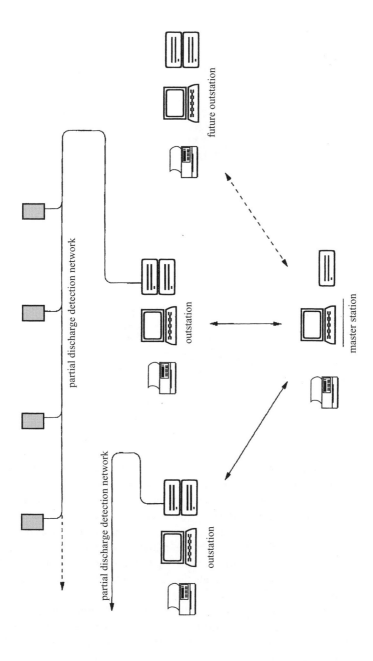

Figure 9.21   GIS continuous monitoring system and NGC partial discharge monitoring strategy (after Coventry [4, chap. 22)

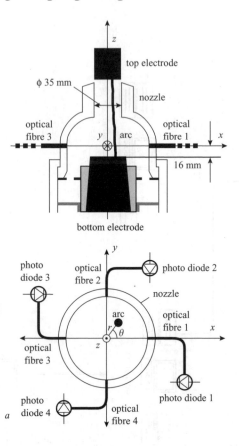

Figure 9.22   Arc optical emission probes deployment on various types of circuit breaker (after Coventry [4, chap. 22])

    a   SF$_6$ puffer

off, and a discharge occurs as charge is exchanged between the particle and the enclosure. The particle bounces and the sequence continues. The particle moves comparatively slowly, typically returning to the enclosure once every few cycles. The resulting pattern is characterised by discharges occurring at any point on the power–frequency wave and a repetition rate of one discharge every few cycles. A typical pattern, accumulated over a number of cycles, is shown in Figures 9.20c. A free conducting particle moving under the influence of the electric field may initiate breakdown if it enters a region of critical electric field.

    Several researchers have studied the behaviour of different types of defect has been studied extensively in the laboratory to assist with defect recognition and risk analysis. Information on defects detected in service is maintained in a database to assist future decisions. Artificial intelligence techniques are being introduced for defect recognition and decision making in GIS.

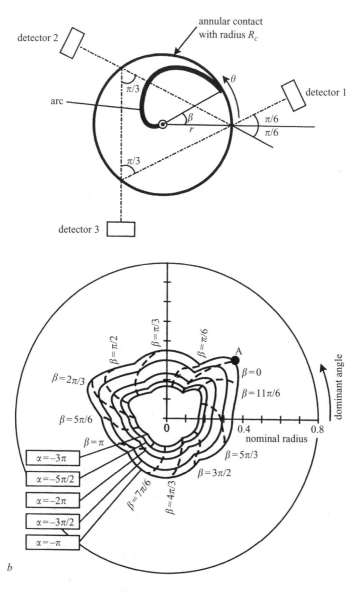

Figure 9.22 *Continued*

    b   SF$_6$ spinner

(v)   Figure 9.21 presents a schematic of a continuous monitoring system for partial discharge monitoring in GIS.

    •   Coventry [4, chap. 22] reports that this system consists of a partial discharge detection network which collects and transmits data from the UHV couplers to an outstation

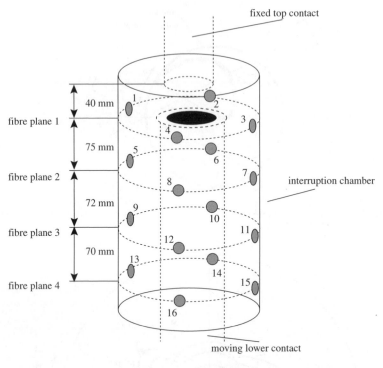

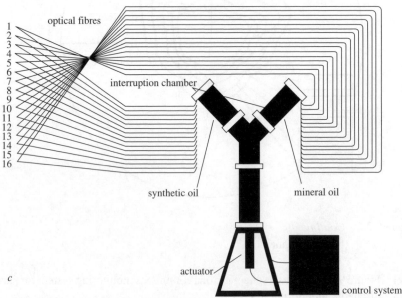

Figure 9.22   Continued
    c   oil

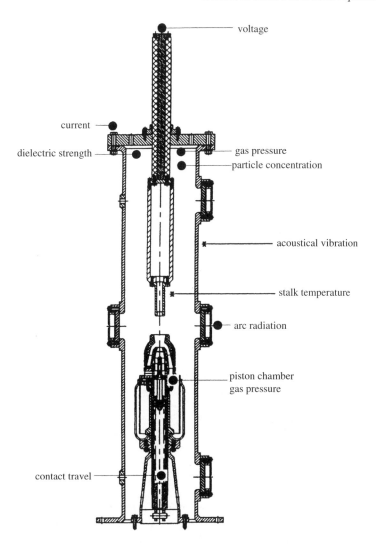

Figure 9.23   Location of various optical fibre sensors on a model high voltage SF₆ circuit breaker (after Jones et al. [4, chap. 22])

- the outstation consists of PC peripherals and software to control the partial discharge detection network and store the collected data
- outstations at a number of substations with GIS connected by modem link to a master station at the headquarters location
- data are transferred from the outstations automatically and at least once per week [4, chap. 22].

(vi) Coventry comments that this continuous monitoring system (Figure 9.21) is intended to produce more effective cover than that provided by routine manual

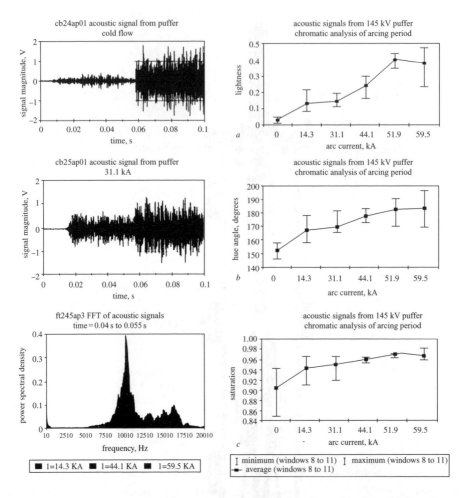

Figure 9.24　Optoacoustic vibration monitoring (after Jones [4, chap. 21])

a　typical time variation of acoustic signals from an SF_6 puffer circuit breaker without a fault current and with a 31.1 kA fault

b　power spectral density curves for acoustical signals from an SF_6 puffer circuit breaker for part of the fault current arcing period

c　signal discrimination in terms of time window compressed coordinates L.H.S

surveys and to reduce its demands on engineers' time. The system is able to detect, store and display information on all defects that occur. It operates unattended, and an engineer need only intervene when the system indicates that captured data require their attention.

The normal mode of operation is for the system to signal when captured data require the attention of an engineer. Such an indication is generated where

UHF signals are captured for the first time or where a change occurs in the behaviour of UHF signals as indicated by preset thresholds based on amplitude or count rate. On being notified by the system of the above condition, the engineer may retrieve and display partial discharge data on which to base decisions on action to be taken.

Coventry comments that the system must be designed to provide the engineer with sufficient information on which to base decisions, without handling and storing excessive quantities of data. Typically, displays showing the point on the power–frequency wave at which discharge occurs (phase-resolved display) and displays showing changes in discharge behaviour over a period of time (trend analysis) are provided.

(vii)   Coventry states that all GIS on the NGC system are monitored for partial discharge using the UHF technique. The decision to use continuous rather than periodic monitoring is based on a cost–benefit analysis. Continuous monitoring systems are installed at the most strategic GIS where the costs of unplanned outages are higher.

(viii) Figure 9.22 illustrates some arc optical emission probes (and indicates typical dispositions of sensor/detectors) deployed on various types of circuit breakers. Parts *a*, *b* and *c* relate to SF_6 – puffer, SF_6 – spinner and oil-interrupters, respectively (for further information, see Coventry [4, chap. 22]).

Since the early 1980s, the fundamentals of optical fibre systems (OFS) have been well covered by G.R. Jones [4, chap. 21] and his research group at Liverpool University. In numerous quality research papers, these researchers have demonstrated that optical fibre technology has the scope of being used for a variety of purposes within the electrical power systems sector as:

- it has the advantage of inherent bandwidth availability and relative immunity to lightning strokes and EMI and EM interference
- it is becoming more attractive and increasingly economically viable year on year as data transfer rates of local area networks (LANs) increase
- research into optical fibre based parameter sensing has progressed well and Jones considers that it is at the stage where properly engineered systems are becoming available utilising purely optical monitoring.

Jones [4, chap. 21] describes examples of optical fibre sensors, such as for measuring electrical, mechanical, aerodynamic/thermodynamic and chemical monitoring which his research group has investigated for monitoring parameters such as in high voltage equipment and systems,

As examples, consider:

(i)   Figure 9.23, which illustrates the deployment and monitoring locations of various optical fibre sensors on a model high voltage SF_6 circuit breaker. Specific parameters monitored by Jones *et al.* include current, voltage, gas pressure, contact stalk temperature, contact travel, particulate concentration and arc radiation. Jones has discussed a range of different forms of optical fibre based current transducers investigated by his research group at Liverpool University [4, chap. 21]. Jones and his colleagues found that by winding an optical fibre around part of the circuit breaker structure acoustical and mechanical vibrations produced by the arcing and other mechanical effects associated with the operation of the CB can be conveniently monitored [4, chap. 21]. In a similar manner Jones states that analysis of other features from acoustical records taken from tests on CBS, using chromatic approaches, yields 'deeper levels' of information regarding the mechanical operation of the circuit breaker e.g. hydraulic drive pressure [4, chap. 22].

(ii)   Finally, Figure 9.24 summarises certain strategic information attained by Jones *et al.* [4, chap. 21], using optoacoustic vibration monitoring. The reader is directed to the original sources for fuller explanation(s) of the complex phenomena represented herein, but brief consideration of subcaptions *a*, *b* and *c* to this figure provides the reader with an appreciation of how interpretation of time-dependent acoustical signals, which are representative of records obtained during the operation of an SF_6 puffer circuit breaker under various fault arcing current conditions, has led to effective vibrational characterisation of practical circuit breakers. Moreover, the writer of this chapter wishes to emphasise that the ongoing high quality fundamental research by G.R. Jones *et al.* at Liverpool for over 20 years, working in association with manufacturers and utilities, has resulted in:

- incremental improvements in the knowledge of fundamental/physical characterisation of arc interruption performance phenomena; this has resulted in improved computer modelling and computer prediction of CB interruption performance capabilities – which can now result in the attainment of near optimal performance levels
- the meaningful and effective development of a variety of intelligent in-service condition monitoring, CM, methodologies – now finding increasing application in practical equipment installed in modern transmission networks.

Recent CIGRE working group reports [19, 20, 25] considers the excellent reliability and availability of modern switching equipment which continues to get better with the introduction of modern improved designs of CM and DT devices. The role of condition monitoring systems (CMS) has assumed even greater importance with the recent thrust to work network systems harder [22–24]. In the future, maintenance of switchgear will increasingly only be carried out when the condition of the equipment warrants intervention [25]. This approach provides the user with benefits of reduced cycle costs, improved availability due to fault prevention and the ability to plan for any outages necessary for maintenance. Moreover, the application of CMS together with other new technologies offers the user other advantages such as increased functionality and performance enhancement [20, 25]. (C.J. Jones also points out that this development creates a need for guidelines regarding the efficient utilisation of this increased functionality, see also earlier comments in section 9.6.)

The application of CMS and diagnostic techniques to substation equipment (when adopting approved operational guidelines) achieves a variety of benefits for the equipment operator:

- maintenance prediction
- failure prevention
- improved commissioning tests
- more accurate end of life assessments.

C.J. Jones, convener of W.G.13.09 CIGRE brochure 167 [20], has recently distilled the contents of this report into a later conference paper [25]. The following aspects are covered in this guideline brochure [20], by section:

1 definitions given for the commonly used terms in the area of monitoring and diagnostics (necessary given the diversity of understanding that exist for the same terms)
2 the need for monitoring is discussed, based on the objectives of the equipment technology being considered
3 justification of the application of monitoring and diagnostics is analysed and examples given of possible approaches (note that a single approach cannot be prescribed for this process)
4 state of the art review of the diagnostic techniques and sensors suitable for the application to switching equipment
5 requirements for the design and testing of the switching equipment and the monitoring systems considered
6 the issue of dependability, monitoring and diagnostics is covered
7 management of information; strategy for how data from sensors are handled to provide useful information
8 consideration of future of monitoring and diagnostics
9 main conclusions and report recommendations are given.

Item 9 in the above report includes the following information:

- There is now clear evidence that the application of diagnostic techniques DT and CM can bring benefits to users and suppliers of switching equipment. (These benefits can generally be given some financial value.)
- The application of diagnostics and condition monitoring will become more commonplace in the future.
- There is a trend towards the use of continuous online CM to supplement offline or periodic monitoring.
- Users require simple information in a timely manner rather than large quantities of data which require analysis. Monitoring systems that incorporate expert systems to provide such information will be of greater benefit to users.
- A variety of sensors, intelligent electronic devices (IEDs) and monitoring systems already exists in substations. Unfortunately, at present, there is little standardisation on the communication of data. The work that is currently ongoing in this area will allow significant benefits to be achieved.

- It is not possible to provide a single approach that will cover the justification of the implementation of CM to switching equipment.
- From the utilities point of view, in some cases it is much more important to monitor the old, unreliable equipment than the new equipment. Unfortunately, this task is not easy to accomplish.
- An essential matter in the context of diagnostic testing and monitoring is the interpretation of the measurements. In some cases, it is straightforward to convert measurements to useful information about the condition of the equipment, in other cases it is not.
- It is suggested that the supplier of sensors and monitoring equipment should be responsible for ensuring the continued availability of spare parts or interchangeable units for at least ten years from the date of final manufacture of the monitoring equipment. This includes also the tools that are necessary for maintenance and commissioning of these spare parts.
- The CIGRE Working Group.13.09 (now disbanded), identified the need for future work in the area of lifecycle cost evaluation not just for circuit breakers but all primary equipment.

C.J. Jones has recently reported that his company has had online condition monitoring experience since 1991 and offered the following brief perspectives [25a]:

- some benefits of CM are directly relevant to the manufacturer such that sensors can really be fitted at no cost to the user
- as we apply substation information systems it becomes obligatory to utilise extensive CM and DT on switching equipment (and other items of plant)
- try to keep the diagnostic technique as simple as possible (i.e. coil, current?)
- a hierarchy of diagnostic techniques can be applied
  - high level on an online basis
  - low level on an as needs basis
- monitoring must be matched to the equipment technology
- don't fall into the trap of monitoring something because it can be monitored
- processing/communications technology has developed dramatically, so think about what is useful not what is possible, as things will become possible
- changing as fast as technology are utilities, so consider training issues; systems will need to give not data nor information but decisions.

Finally, further important aspects of CM strategies, considered and discussed by WG.13.09 and C.J. Jones [20, 25], are reproduced in Tables 9.4 and 9.5.

9.9.4   General discussion

As mentioned earlier, the use of diagnostic techniques is not new having been widely used and developed for more than 30 years. In the UK, a (DTI) government initiative in the early 1980s encouraged research into intelligent switchgear etc. and, after a slow start, great progress has been made on this theme. The current intense interest in the application of diagnostics and condition monitoring, CM, techniques to HV

Table 9.4 SF₆ circuit breaker and circuit switches (after C.J. Jones) (reproduced from Table 2.1, CIGRE W.G. 13.09 [20])

Function/parameter continuously monitored (C) or periodically diagnosed (P)	Equipment operation/status		Failure prevention		Maintenance support/ life assessment		Optimum operation		Commissioning tests	
	need	rank	need	rank	need	rank	need	rank	need	rank
Switching:										
• operation times			C – to detect any abnormal condition	H	C – to evaluate CB condition	H	C – controlled switching	E	P – to establish reference data	H
• pole discrepancy			C – to detect any abnormal condition	H	C – to evaluate CB condition	H	C – controlled switching	H	P – to establish reference data	H
• arcing time			C – to detect any abnormal condition	L			C – controlled switching	H		
• contacts velocity			C – to detect any abnormal condition	H			C – controlled switching	H		
• main contacts position			C – to detect any abnormal condition	E			C – controlled switching	H		
• contact wear (I^2t)					C – to support overhaul planning	H				
Mechanical drive:										
• number of operations					C or P – to support overhaul planning	H			P – to establish reference data	H
• stored energy (spring position, pressure)	C – to assess CB availability	E	C – to assess CB availability	E			C – controlled switching	E		
• latch position			C – to detect any abnormal condition	H						
• velocity			C – to detect any abnormal condition	H	P – to evaluate CB condition	H			P – to establish reference data	H
• vibration finger print					P – to evaluate CB condition	L			P – to establish reference data	L
• state of mechanism										
– number of motor pump starts/motor current/running time	C – to detect any abnormal condition	H	C – to detect any abnormal condition	H	P – to evaluate CB condition	H				
– charging current/time (motor protection)	C – to detect any abnormal condition	H	C – to detect any abnormal condition	H	P – to evaluate CB condition	H				

rank: E – essential; H – high; L – low

Table 9.5   Diagnostic techniques and sensors for testing the switching function of circuit breakers and other types of switching equipment (reproduced from Table 4.3, CIGRE WG.13.09 [20]) (after C.J. Jones)

Parameter	Application(s)	Method/sensor	
position of primary contacts	all	position transducer, e.g.:	
		• auxiliary switch, contactor	C
		• electronic proximity sensor	C
		• optical sensor	C
contact travel characteristics (position, velocity, acceleration)	all	dynamic position sensor with analogue output:	
		• resistance potentiometer	C, P
		• magneto-resistive sensor	C, P
		• linear variable differential transformer	C, P
		dynamic position sensor with digital output:	
		• optical with incremental coding	C, P
		• optical with non-incremental (absolute) coding	C, P
operating time	all	electrical recording of time to close/open of primary circuit	P
	GS, D	motor running time	C, P
pole discrepancy in operating times	all	electrical recording of time to close/open of primary circuit	C, P
	GS, D	motor running time	C, P
arcing time	all	combined recording of load current profile and travel characteristics	P
arcing contact wear	CB	accumulated $I^2 t_{arc}$ by:	
		• current and time measurement	C
		• statistical estimates	P
	all	dynamic contact resistance	P

Abbreviations:
CB: SF_6, air blast, minimum oil and oil tank circuit breakers
GS: grounding switches
D: disconnectors
C: used for continuous monitoring
P: used for periodic diagnostic testing

switchgear and switching equipment and also a wide range of other HV equipment, including transformers, cables, overhead lines, surge arrestors, motors etc. has gained further impetus with the increased emphasis on working network systems harder and the strong focus on asset management (see section 9.9.2 and References 21 to 24).

As C.J. Jones [25] and others have pointed out, the primary need for CM systems in the switchgear sector has been driven mainly by equipment manufacturers to support a strong move from predetermined interval based maintenance to predictive (or condition-based) maintenance.

Without doubt, we are witnessing a transitional period with regard to diagnostics and CM strategies, because of increased utility interest in this subject area [20], as a holistic network-wide approach is developed. Subsequently, the present writer anticipates that in the future one will look at not only the benefits of increased equipment availability, through a range of diagnostic and condition monitoring strategies, but the increased functionality will facilitate monitoring asset management, life extension. It is also likely that significant improvements in software support will result in the widespread adoption of evolving data warehousing techniques which will facilitate greatly expanded data evaluation capability to provide even better, i.e. a more comprehensive, diagnostic monitoring service and possible integration with SCADA and other systems.

The topic of condition monitoring has assumed increasing strategic importance as the electrical power sector drives relentlessly towards cost saving, asset management, long life and improved plant performance [4, 21–24]. In this chapter, it has not been possible to critically review strategic issues associated with CM methodologies but rather to indicate, and provide a brief flavour of the wide range of published work in this area supported by generous referencing of strategic publications [4, 21–25].

A recent IEE event (within the HVET2002 international school) described a wide range of monitoring systems used in the various sectors of the power industry, with contributions from experts covering developments from the perspectives of electricity supply industry, ESI, users, manufacturers, equipment designers, system planners, monitoring equipment designers, consultants and academics.

A broad range of contributions were presented, with reference to earlier strategic studies by these authors (including coverage of intelligent condition monitoring systems as applied to GIS, transformers, arresters, bushings, cables etc.), who discussed many strategic techniques such as RF emissions, acoustic emissions, dissolved gas analysis, optical fibre sensing, remote E-field monitoring etc. This event clearly reflected how CM of electrical equipment is assuming an increased importance as the commercial drive to improve efficiency is pursued.

9.10 Summary

This chapter has provided a general introduction to the vast subject of circuit breakers, interruption strategies, characterisation and performance evaluation of gaseous/solid insulation systems, the identification of strategic online condition monitoring (CM) methods and diagnostic techniques (DT) as applied to modern switchgear systems, before and post deregulation. Although it has only been possible to touch very briefly on these aspects together with certain related strategic issues including substation control systems (SCS), also working networks harder and environmental matters, important strategic issues have been considered briefly and a generous list of references, including CIGRE electronic sources is included, which will enable the reader to explore this developing topic in greater depth.

9.11　Acknowledgements

The author wishes to thank the directors of NEI Reyrolle Ltd (now VA TECH REYROLLE) and several other organisations for permission to publish earlier papers which have been briefly referred to in this chapter. He has called heavily on materials originally published by CIGRE, Professor G.R. Jones, Mr C.J. Jones, Mr S.M. Ghufran Ali, Mr W.J. Laycock and Dr P. Coventry to whom he extends grateful thanks. He also gratefully acknowledges the assistance given and contributions made by many of his former colleagues and students at Reyrolle, the University of Sunderland, Cigrean colleagues and those within the industrial, IEC, IEE Professional Groups and academic communities worldwide and for their help and generous support over the years.

9.12　References

1　FLURSCHEIM, C.H. (Ed.): 'Power circuit-breaker theory and design' (Peter Perigrinus Ltd, London, 1982)

2　RAGALLER, K.: 'Current interruption in high voltage networks' (Plenum Press, 1978)

3　RYAN, H.M., and JONES, G.R.: 'SF$_6$ switchgear' (Peter Perigrinus Ltd, London, 1989)

4　RYAN, H.M. (Ed.): 'High voltage engineering and testing' (IEE Publishing, Power series no. 32, 2001, 2nd edn.)

5　WRIGHT, A. (Ed.): 'Arcs sparks and engineers' (Reyrolle Heritage Trust Press, Hebburn, 2001)

6　CLOTHIER, H.W.: 'Switchgear stages'. A collection of articles written by Henry Clothier, bound and published by G.F. Laybourne, 1933

7　GARRARD, C.J.O.: 'High voltage switchgear', *IEE Proc.*, 1976, **123**, pp. 1053–1080

8　BROWNE, THOMAS E. Jr. (Ed.): 'Circuit interruption: theory and techniques' (Marcel Dekker Inc, 1984)

9　BLOWER, R.: 'Progress in distribution switchgear', *Power Eng. J.*, 2000, **14**, (6), pp. 260–263

10　STEWART, S.: 'Primary switchgear', *Power Eng. J.*, 2000, **14**, (6), pp. 264–269

11　CIGRE Working Group 39.01 report: 'Substation control in the system control', *Electra*, 2002, (200), pp. 40–53 (www.cigre.org)

12　CIGRE Working Group 13.07 report: 'Controlled switching of HVAC circuit breakers: planning, specification and testing of controlled switching systems', *Electra*, 2001, (197), pp. 23–33 (www.cigre.org)

13　RYAN, H.M., and WHISKARD, J.: 'Design and operation perspective of a British UHV laboratory', *IEE Proc. A, Phys. Sci. Meas. Instrum. Manage. Educ. Rev.*, 1986, **133**, (8), pp. 501–521

14 RYAN, H.M., LIGHTLE, D., and MILNE, D.: 'Factors influencing dielectric performance of SF$_6$ insulated GIS', *IEEE Trans.*, 1985, **PAS-104**, (6), pp. 1527–1535

15 CIGRE Working Group 13.09: 'Monitoring and diagnostic techniques for switching equipment' (convenor JONES, C. J., United Kingdom) *Electra*, 1999, (184), p. 27

16 O'CONNELL, P. *et al.*: CIGRE, WG.23.02 Paper 'SF$_6$ in the electric industry, status 2000', *Electra*, 2002, (200) pp. 16–25

17 JONES, C.J., HALL, W.B., JONES, G.R., FANG, M.T.C., and WISEALL, S.S.: 'Recent development in theoretical modelling and monitoring techniques for high voltage circuit-breakers'. CIGRE 1994, paper 13-109

18 OLKEN, M.I. *et al.*: 'Restructuring and reregulation of the electric industry in North America: challenges and opportunities', *Electra*, 2000, (45), pp. 44–45 (special issue)

19 CIGRE Working Group 23.02, Task Force 02 report: 'Report on the second international survey on high voltage gas insulated substations (GIS) service experience', *Electra*, 2000, (188), p. 127

20 CIGRE Working Group 13.09: 'User guide for the application of monitoring and diagnostic techniques for switching equipment for rated voltages of 72.5 kV and above'. Brochure 167

21 'Online condition monitoring of substation power equipment – utility needs'. CEA report 485T1049

22 JEFFERIES, D.: 'Transmission today: lessons from a decade of change', *Electra*, 1999, (187), pp. 9–18. Reproduced opening speech, London CIGRE symposium *Working plant and systems harder: enhancing the management and performance of plant and power systems*, 7–9 June 1999

23 URWIN, R.J.: 'Engineering challenges in a competitive electricity market'. Keynote address, ISH *High voltage* symposium, paper 5.366.SO, 22–27 Augst 1999, London

24 LOWEN, J.: 'AC/DC power transmission: keynote address', *Power Eng. J.*, June 2002, **16**, (3), pp. 97–101

25 JONES, C.J.: 'CIGRE Working Group 13.09 – monitoring and diagnostic techniques for switching equipment'. Based on paper first published at IEE/PES TandD conference, Atlanta, November 2001 (*a* also presented at IEE, HVET 02, international *High voltage* Summer School, Newcastle upon Tyne, 10 July 2002)

26 ZIEGLER, R.G.: 'Protection and substation automation; state of art and development trends', *Electra*, 2003, (206), pp. 14–23

27 LAYCOCK, W.J.: 'Intelligent networks', *Power Eng. J.*, 2002, pp. 25–29

28 BAEHR, R.: 'Transformer technology: state of art and trends for future developments', *Electra*, 2001, **198**, (13), pp. 13–19

29 CIGRE Working Group 23/21/33 report: 'Gas insulated lines (GIL)', *Electra*, 2003, (206) pp. 55–57 also KOCH, H., and SCHOEFFNER, G.: 'Gas insulated transmission line (GIL) an overview', *Electra*, 2003, (211), pp. 8–17

30   IEC 62271-100: 'High voltage alternating current circuit breakers.

31   RYAN, H.M.: 'Prediction of alternating sparking voltages for a few simple electrode systems by means of a general discharge – law concept', *IEE Proc.*, 1967, **114**, (11), pp. 1815–1821

32   PEDERSON, A.: 'Calculation of spark breakdown or corona starting voltages in non-uniform fields', *IEEE Trans.*, 1967, **PAS-86**, pp. 200–206

33   MILNE, D., and RYAN, H.M.: 'The evaluation of solid dielectric systems for use in high voltage switchgear', IEE Conf. Publ. 239, 1984, pp. 76–79

Chapter 10

Polymer insulated power cable

J.C. Fothergill and R.N. Hampton

10.1 Introduction

10.1.1 Structure

The name cables is given to long current-carrying devices that carry their own insulation and present an earthed outer surface. In this context, overhead lines for example, are not considered as cables. Power cables have a coaxial structure: essentially, they comprise a central current-carrying conductor at line voltage, an insulation surrounding the conductor and an outer conductor at earth potential. AC cables are generally installed as a three-phase system and hence the outer conductor should only carry fault and loss currents. In practice, a more sophisticated construction is adopted. The interfaces between the metal conductors and the polymeric insulation would tend to include protrusions and voids; features that would lead to electrical stress enhancement and premature failure [1]. To overcome this, a polymer *semicon*, a conductive polymeric composite, is placed at both interfaces. The inner semicon, the insulation and the outer semicon are co-extruded to ensure the interfaces are smooth and contaminant free. Surrounding this cable are layers to protect the cable during installation/operation and carry the loss/fault currents. These layers also serve to keep out water, which may lead to water treeing (section 10.4.2). A schematic diagram of a power cable is shown in Figure 10.1.

10.1.2 Voltage ratings

10.1.2.1 MV, HV and EHV

High voltage cables that are used for distribution and transmission purposes are generally categorised according to the voltage rating:

- medium voltage (MV) 6–36 kV
- high voltage (HV) 36–161 kV
- extra high voltage (EHV) 161–500 kV (or more)

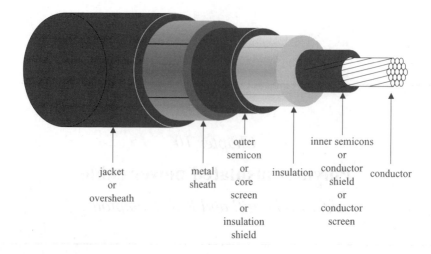

jacket or oversheath

metal sheath

outer semicon or core screen or insulation shield

insulation

inner semicons or conductor shield or conductor screen

conductor

Figure 10.1   Cut-away section of a power cable

There is no international consistency on the distinction between distribution and transmission. In the UK, transmission is at 275 kV and 400 kV, i.e. using EHV lines and cables, whereas in Italy, for example, transmission starts at 150 kV. The distributors then transfer power from the transmission system across their local distribution systems to the domestic customer.

10.1.2.2   Electrical stresses

The electrical stress within an AC cable is given by:

$$E = \frac{V}{x \ln(R/r)} \tag{10.1}$$

where V is the applied voltage, r is the radius over the inner semiconductive screen, R is the diameter over the insulation and x is the intermediate radius (between r and R) at which the electric stress is to be determined.

The probability of failure depends upon the electrical stress and increases with stress. The effect of an increased stress can be estimated using the Weibull probability function [2]:

$$P_f = 1 - \exp\left\{-\left(\frac{E}{\alpha}\right)^{\beta}\right\} \tag{10.2}$$

where P_f is the cumulative probability of failure, i.e. the probability that the cable will have failed if the stress is increased to a value E. The two parameters, α and β are known, respectively, as the characteristic stress and the shape parameter.

Inspection of Equation (10.1) shows that the electric stress varies with the position within the cable. There are three potentially useful stresses that can be considered:

- maximum stress at conductor screen
- mean geometric average stress for the whole insulation
- minimum stress at the core screen.

The decision as to which of these to consider is an important one and is guided by the potential modes of failure:

- *Maximum*: the highest level of stress also corresponds to the highest probability of instantaneous failure or, equivalently, the highest rate of electrical ageing. This is most important if the most serious cable defects are located on or near the conductor screen.
- *Mean*: this is most important if the most serious defects are uniformly located throughout the bulk of the insulation.
- *Minimum*: this is most important if cable system reliability is determined by the performance of accessories or if the electrical design or installation method of accessories degrades cable performance. It is also important if the most serious cable defects are located on or near the core screen.

10.1.3 Uses of cables

Power cables are commonly used in underground or underwater (submarine) connections. Cables are placed at strategic points of the transmission grid to supplement overhead lines or, in some cases, they can form the whole backbone. Interconnection between networks is particularly well suited to cable solutions [1].

They may also be used in other applications. For example, overhead covered conductors allow smaller phase clearance between the conductors on medium voltage overhead lines. Objects, particularly tree branches, may touch the lines without tripping or customer outage. This has led to substantial improvements in service reliability (e.g. Reference 4). In most cases, the aluminium alloy conductor is covered with black UV-resistant crosslinked polyethylene and filled with grease, to provide corrosion protection and longitudinal water-tightness. Arcing guides are applied at insulator tops, to protect the line from arcing damage. Another application is the Powerformer™ [5] and the related Windformer™ and Motorformer™. These new generators are able to supply electricity directly to the high voltage grid without the need for a step-up transformer. They are suitable for power generation at output voltages of several 100 kV. The new concept is based on circular conductors for the stator winding, and it is implemented by using proven high voltage cable technology. Thus, the upper limit for the output voltage from the generator is only set by that of the cable.

10.1.4 AC and DC

Cables are used in both alternating current (AC) and direct current (DC) schemes. The cable designs used in each case are outwardly very similar and have many identical design elements. However, the detailed engineering and the materials used are very different.

AC is the globally preferred means of transferring electric power. This form of transfer makes it straightforward to generate electricity and to transform voltages up and down. This means of transfer accounts for more than 98 per cent of the global power infrastructure.

In long distance transmission schemes, there are advantages in using DC over AC. System instabilities caused by connecting regions with slightly different AC phases and frequencies are obviated. Capacitive charging current implies that there is a maximum useful length of AC cable without the use of shunt reactors. DC cables are therefore particularly useful for long distance submarine connections. Furthermore, the lack of electromagnetic effects under DC conditions eliminates the skin effect in which the conductor resistance can rise by up to 20 per cent at 50 Hz. The drawbacks for a DC solution are that the terminal equipment for AC–DC conversion is more costly and less efficient than an AC transformer. (Reviews on AC/DC power transmission are contained in Reference 7.) Thus, the system must be sufficiently long to be economically viable. In addition, the control of the electrical stress in a DC cable insulation is much more difficult.

10.1.5   Cable types

There are, in general, four types of underground cable used characterised by the type of insulation:

(i)   *Polymeric:* low density polyethylene (LDPE), high density polyethy-lene (HDPE), crosslinked polyethylene (XLPE) or ethylene propylene rubber (EPR).

(ii)   *Self contained fluid filled (FF or LPOF):* paper or paper polypropylene lami-nated (PPL) insulated with individual metal sheaths and impregnated with low pressure biodegradable fluid – common on land >1 kV.

(iii)   *Mass impregnated non draining (MIND or solid):* paper-insulated with individ-ual metal sheaths and impregnated with an extremely low viscosity polybutene compound that does not flow at working temperatures – common at MV and submarine DC.

(iv)   *High pressure fluid filled (pipe type or HPOF):* paper-insulated and installed in trefoil in steel pressure pipes and impregnated with high pressure non-degradable fluid which is maintained at high pressures by pumping plants – common in the USA.

The typical electrical stresses employed in cables are shown in Table 10.1.

Up until the mid 1980s, paper-insulated cables were the system of choice at high voltages. However, improvements in polymeric cables and accessories plus environ-mental concerns has led to a significant reduction in the use of paper cables for land applications. In particular, there has been a strong preference for XLPE cables over EPR (except in Italy and the USA).

XLPE and EPR have emerged as the favoured polymeric insulations through the 90°C continuous operating temperature that can be achieved when they are used. This temperature matches that which can be attained when fluid-filled lapped insulations (paper and PPL) are used. In contrast, LDPE and HDPE are limited to

Table 10.1   Average stress levels for selected insulations

Type	Stress, kV/mm	EHV	HV	MV
fluid	average at core screen	10	8	8
filled	average at conductor screen	17	14	10
EPR	average at core screen	4	3	2
	average at conductor screen	8	5	3
XLPE	average at core screen	5	3	2
	average at conductor screen	11	6	3

operating temperatures of 70°C and 80°C, respectively. Table 10.2 identifies some of the main advantages of the respective technologies.

The situation today is:

- almost all new HV systems that are being installed are new build/expansion
- the majority of HV cables, already installed within the existing system, are insulated with paper (83 per cent paper and 17 per cent polymeric)
- there is very little replacement of existing paper cables: most of installed capacity is based on paper
- most HV transmission is by overhead lines (OHLs)
- HV cables are replacing OHLs in environmentally sensitive areas but there is limited impact (globally) on new OHLs
- XLPE cables make up two to five per cent of the total installed HV cable capacity in the range 115–161 kV
- XLPE cables make up 50–70 per cent of the HV cable capacity presently being installed.

The reasons that there has been little replacement of paper cables are:

- Paper cables continue to operate reliably.
- The previous designs of XLPE cables were too large to fit existing rights of way or pipes that have been designed for paper cables. This is of special importance in the USA where there are many paper cables. In this area, the key task is to develop polymeric cable designs that are flexible, small (i.e. working at high stresses) and easy to joint, as it is essential to use the existing pipes/ducts. Present designs address these issues, and now cables exist that match the size and performance of paper cables [6].

10.2   The components of the polymeric cable

10.2.1   Conductor

Conductors in the USA tend to be based on the American wire gauge (AWG), but in the rest of the world are based on IEC228 and are therefore metric. Stranded conductors

Table 10.2   Advantages (+) and disadvantages (−) of HV cable insulations (quantitative comparisons given in italics)

XLPE	EPR	Paper
+ no risk of oil leakage	+ no risk of oil leakage	− oil leakage
+ simple design	+ simple design	− complicated design high maintenance burden
+ very low dielectric losses *losses for a 132 kv 1000 mm²cable are 5.1 mA/m*	− high dielectric losses	+ low dielectric losses *losses for a 132 kV 1000 mm² cable are 14.7 mA/m*
+ acceptable fire performance	+ acceptable fire performance	− fire performance in tunnel and substation applications is a concern due to the copious supply of flammable oil
− complicated accessories (possible long repair times)	− complicated accessories	+ simple accessories
− limited but growing track record *15 years of experience*	− limited track record	+ extensive track record *>60 years of experience*
− most of existing system is paper, transition joints difficult	− most of existing system is paper, transition joints difficult	+ most of existing system is paper
− may suffer from water trees if no metal sheath	+ longer life in wet conditions, metal sheath not generally required	+ metal sheath required to contain fluid thus water ingress is not possible
− high conductor losses *132 kV 1000 mm² can transmit 250 MVA of power*	− high conductor losses	+ low conductor losses *132 kV 1000 mm² can transmit 275 MVA of power*
− large size *132 kV 1000 mm² cable has a 98 mm diameter and a weight of 25 kg/m*	− large size	+ small size *132 kV 1000 mm² cable has a 70 mm diameter and a weight of 18 kg/m*

have generally comprised concentric layers (Figure 10.2a) but, in order to make them smoother and more compact, they are often specially shaped in rolling mills nowadays (Figure 10.2b). When operating at high voltages and currents, the AC current is preferentially carried more in the outer than in the inner conductors (the skin effect). In addition, the electromagnetic fields induce eddy currents (proximity effect). These effects tend to increase the conductor resistance under AC above that which is seen under DC. The AC/DC ratio ($R_{AC/DC}$) can be as large as 1.15. This increased resistance serves to increase the joule heating losses within the conductor and increases

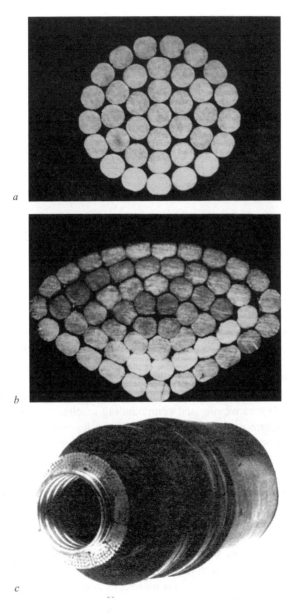

Figure 10.2   Methods of laying conductors
    a   concentric-lay
    b   shaped compacted
    c   Milliken (an oil-filled cable) (from [3])

the temperature of the cable. Thus, special Milliken conductors, Figure 10.2c, may be used for large conductor designs to reduce the AC/DC ratio.

Conductors are virtually all made from either copper or aluminium. Copper has the advantage of being more conductive, and, therefore, requires less material to carry a given current. Copper conductors, therefore, have the advantage of being small. However, the cost of aluminium is lower than copper, and even though it has a lower conductivity it is often used as the conductor. An additional advantage of aluminium over copper is that the conductors are lower in weight even though more volume of material is used. At MV, aluminium is preferred whereas, at HV and EHV, the smaller size of copper provides the greatest advantage.

10.2.2   Semicon

Semiconductive screening materials are based on carbon black (manufactured by the complete and controlled combustion of hydrocarbons) dispersed within a polymer matrix. The concentration of carbon black needs to be sufficiently high to ensure an adequate and consistent conductivity. The incorporation must be optimised to provide a smooth interface between the conducting and insulating portions of the cable. The smooth surface is important as it decreases the occurrence of regions of high electrical stress [8]. To provide the correct balance of these properties, it is essential that both the carbon black and polymer matrix be well engineered.

The same care needs to be paid to the manufacture of the matrix polymer for semicons as for the XLPE insulation. However, the chemical nature of the polymers is subtly different because of the need to incorporate the carbon black. The carbon black and other essential additives (excluding crosslinking package) are compounded into the matrix. The conveying and compounding machinery used are designed to maintain the structure of the carbon black within a homogeneous mix. Before the addition of the crosslinking package, filtration may be applied to further assure the smoothness of the material to a higher standard than that provided by the compounding process.

The smoothness of the extruded cable screens is assured by extruding a sample of the complete material in the form of a tape. The tape is optically examined for the presence of pips or protrusions. Once detected, the height and width of these features are estimated, thereby enabling width-segregated concentrations to be determined. When using such a system, care needs to be exercised when examining the present generation of extremely smooth (low feature concentration) screens, as the area of tape examined needs to match the likely number of detected features: smooth screens require larger areas of examination.

10.2.3   Insulation

It is clear from Table 10.2 that XLPE is the most common cable insulation currently being installed. Insulating XLPE compounds need to fulfil a number of requirements. They should act as thermoplastic materials within the extruder and crosshead. They should crosslink efficiently with the application of high temperatures and pressures within the vulcanisation tube. They should be immune to thermal degradation throughout the cable manufacture process and operation at the maximum cable temperature for the life of the system. They must display an extremely low occurrence of the features that can enhance the applied electrical stress and thereby lead to premature failure. To deliver these requirements, it is essential that the greatest care is paid to the design and manufacture of the polymer and the engineering of the appropriate crosslinking and stabilising packages.

The manufacturing technology employed for XLPE compounds to be used for power applications needs to ensure the highest level of cleanliness at all points of the production chain. The sequence comprises three main parts: base polymer manufacture, addition of a stabilising package and addition of the crosslinking package.

There are two types of crosslinking process that are commonly used for power cables:

(i) Peroxide cure – thermal degradation of an organic peroxide after extrusion causes the formation of crosslinks between the molten polymer chains.
(ii) Moisture cure – chemical (silane) species are inserted onto the polymer chain. These species form crosslinks when exposed to water; this process occurs in the solid phase after extrusion.

The peroxide cure method is the most widely used crosslinking technology. It is used for MV, HV and EHV. The moisture cure approach is limited to MV where it has proven itself as a flexible and economic solution.

10.2.3.1 Peroxide cure

This first part of the production sequence starts with the polymer reactor, which must be designed to deliver polymers with consistent properties that are free from particulate contaminants (free from moving metal parts) and chemical species that degrade the dielectric properties. In practice, this means that the sequence of polymers manufactured in the reactor and the materials employed in its construction must be controlled. For example, the materials used to seal the closed conveying system must be made from thermoplastic polyethylene. Within the manufacture, it is essential that materials be conveyed in the safest way possible to ensure that extraneous contaminants are not introduced. Experience has proved that the best solution is to use the shortest possible distance between the reactor and the compounding equipment. The transport from the polymerisation reactor should be affected by stainless steel pipelines and vessels with dense phase or gravity conveying, as this significantly reduces the formation of polymer strings and dust.

The addition of the crosslinking and stabilising packages is the next step in the chain. These processes incorporate the active chemical species into the polymer matrix. Before the addition of the crosslinking package, the material passes through extremely fine filters and screens that operate in cascade. The purpose of these processes is to ensure that the infrequently occurring contaminants do not proceed to the packaging process. These contaminants may be within the matrix or adhering to the surface.

At all interprocess points samples are taken for the essential quality assurance processes. These processes are in two parts:

(i)   assurance of the physical and chemical properties
(ii) measurement of the cleanliness.

The physical and chemical analyses serve to establish that the extrudability and crosslinking characteristics are both within specification. These checks ensure the consistency of both the polymer flow, within the extruder and the crosshead, and the formation of the chemical crosslinks within the vulcanisation tube.

10.2.3.2   Moisture cure

The key part of the moisture cure approach is the insertion of the chemically active species onto the polymer backbone. There are three methods for achieving this:

(i)   *Sioplas:* a suitable silane is melt compounded with a peroxide and a polymer. During this process, the silane becomes chemically grafted to the polymer chain. The material is then pelletised for use in the cable extruder. The catalyst for the formation of the crosslinks, and other additives, are added at the extrusion step by way of a masterbatch.

(ii) *Monosil:* this process is similar to Sioplas except that all of the components (including the catalyst) are added at the cable extruder. The silane grafting reaction therefore takes place at the same time as the cable is being extruded.

(iii) *Ethylene-vinyl silane copolymers (EVS):* this process does not graft silanes onto existing polymer chains, but inserts the silane groups into the polymer chains while they are being constructed within the reactor. The polymer exits the reactors ready for extrusion. The catalyst for the formation of the crosslinks, and other additives, are added at the extrusion step by way of a masterbatch.

After the core exits the extrusion line, it is solid and in a thermoplastic state. It must be crosslinked offline. The amount of time taken for this process will depend upon the technology used, the water or steam temperature and thickness of insulation.

10.2.3.3   Cleanliness

The cleanliness of insulation materials (peroxide curable, Sioplas and EVS) may be assessed by converting a representative sample of the polymer into a transparent tape and then establishing the concentration of any inhomogeneities. This is not possible for monosil as the chemistry occurs immediately prior to the cable extrusion. The inhomogeneities are detected by identifying variations in the transmission of light through the tape. To gain the required level of consistency and sensitivity, the tape is inspected by an automated optical system (Figure 10.3). This approach has an excellent signal to noise level and good rejection of spurious data. The data

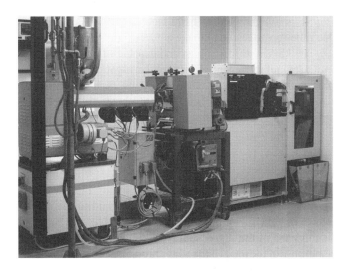

Figure 10.3   *Automated equipment for determining the contamination concentration of XLPE (courtesy of Borealis AB)*

processing is carried out by a microcomputer, which is able to produce size segregated concentration data for a number of selected levels of obscuration.

10.2.4   Metal sheath

For many years, lead, or lead alloys, were the main materials used for the metal sheath layer. This is principally because the low melting temperature allows the lead to be extruded at a temperature of approximately 200°C over the polymeric cable. The main disadvantages of lead are its high density (11 400 kg/m^{-3}) leading to a heavy product and the tendency to creep, flow or embrittle under cyclic temperature loadings. This latter effect has led to a number of cases where the sheath has ruptured. There are also environmental concerns relating to the use of lead, and its use may become restricted by EU directives. In 2000, the EU Commission officially adopted the waste electrical and electronic equipment (WEEE) and Reduction of Hazardous Substances (ROHS) proposals. The ROHS proposals required replacement of lead and various other heavy metals from 2008. To address these problems a number of materials have been used:

- *Extruded aluminium:* this has excellent mechanical performance but requires a large bending radius and can be difficult to manufacture since it requires corrugations. It can be heavy.
- *Aluminium foil:* this is light and easy to manufacture but small thicknesses (0.2–0.5 mm) do not give mechanical protection; the strength comes from the polymer oversheath. It relies on adhesive to make a water tight seal, and it can suffer from corrosion.

- *Copper foil:* this is light, easy to manufacture but not as flexible as aluminium foil. It also relies on adhesive to make a water tight seal.
- *Welded copper:* this is strong, robust and capable of carrying significant current but difficult to manufacture since it requires corrugation and it is difficult to ensure perfect longitudinal welds in practical situations.
- *Welded stainless steel:* this is strong, robust and capable of carrying significant current but has similar manufacturing difficulties to welded copper.

10.2.5   Oversheath (jacket)

In addition to the meticulous attention that must be paid to the insulation system, care also needs to be taken with the oversheath layer. The vast majority of HV and EHV XLPE cables are of the dry design type, which means that a metal barrier is included. The purpose of the metal barrier is to protect the core within from mechanical damage, carry fault and loss currents and to exclude water from the construction (the electrical ageing rate is significantly higher in the presence of moisture, see section 10.4.2). The metal barrier is a key part of the cable design and much care needs to be taken as this significantly affects how a cable system may be installed in practice. The metal layer is itself protected by a polymeric oversheath. Due to the critical performance needed from the oversheath, there are a number of properties that are required: good abrasion resistance, good processing, good barrier properties and good stress crack resistance. Experience has shown that the material with the best composite performance is an oversheath that is based on polyethylene.

10.3   Cable manufacture

10.3.1   Stages of cable manufacture

The stages in cable manufacture may be summarised as:

(i) *Conductor manufacture:* this involves:
- wire drawing to reduce the diameter to that required
- stranding in which many wire strands and tapes are assembled
- laying up; the assembly of non-circular (Milliken) segments into a quasi-circular construction

(ii) *Core manufacture:* this involves:
- triple extrusion in which the core of the cable is formed comprising the inner semicon, insulation and outer semicon
- crosslinking which can be carried out directly after extrusion (peroxide cure) or offline (moisture cure)
- degassing in which peroxide crosslinking byproducts are removed by heating offline; the diffusion time depends upon temperature and insulation thickness.

(iii) *Cable manufacture:* this involves:
- core taping during which cushioning, protection and water exclusion layers, inner semicon, insulation and outer semicon are applied over the extruded core
- metal sheathing; the application of a metal moisture and protection layer
- armouring; the application of high strength metal components (steel) to protect the cables; essential for submarine cables.

10.3.2 Methods of core manufacture

All of the production processes will be common to all methods of manufacture, with the exception of the extrusion process where there are three types of peroxide crosslinking method (crosslinking is often referred to as vulcanisation or curing):

(i) VCV: vertical continuous vulcanisation
(ii) CCV: catenary continuous vulcanisation
(iii) MDCV: Mitsubishi Dainichi continuous vulcanisation, often called long land die.

In all of these processes (moisture cure and peroxide cure), the three layers of the cable core are extruded around the conductor. This uncrosslinked core then passes directly into the curing tube; this is where differences in the processes become apparent. In the moisture cure approach, which takes place offline after extrusion, the manufacturing process is considerably simplified as the length of the tube following extrusion only has to be enough for the thermoplastic core to cool sufficiently to prevent distortion.

10.3.2.1 Vertical continuous vulcanisation

The curing tube is arranged vertically, and the cable is maintained within the centre of the tube by control of the cable tension. A schematic diagram showing VCV is given in Figure 10.4. The conductor is fed from a payoff through an accumulator. This allows a new reel of conductor to be loaded onto the payoff when the old reel runs out without the continuous extrusion process being stopped; the accumulator allows time for the two conductors to be welded together. The conductor is pulled to the top of an extrusion tower, which may be 100 m tall, over a large capstan and is fed via a preheater through a triple extrusion head supplied by three extruders. Curing is effected by heating the cable with hot nitrogen gas. The gas is pressurised to ensure that gas-filled voids do not form from decomposition of the peroxide. The VCV technique ensures concentricity of the conductor within the cable core because of the vertical alignment of the cure tube. VCV lines are particularly effective for producing cables with large conductors >1600 mm^2 as there is not the same difficulty of maintaining the tension compared with the CCV technique. Cables with insulation thickness up to approximately 35 mm can be produced with VCV lines.

Following production, the cable is checked for concentricity and possibly for gross imperfections, such as detachment of the inner semicon, using x-rays or ultrasonic techniques before being laid up with the outer protection layers. In land cable, the

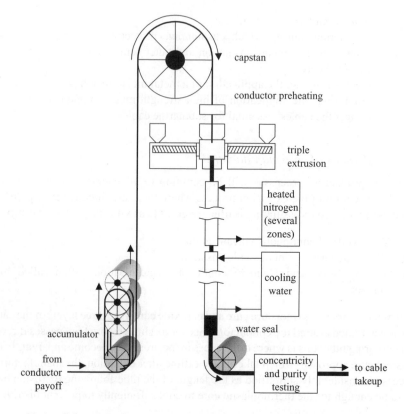

Figure 10.4   Schematic diagram showing the principal components of a VCV line

cable is taken up on drums, but for subsea cables it is preferable to avoid jointing by laying up in long lengths and loading directly onto a waiting ship.

10.3.2.2   Catenary continuous vulcanisation

This technique is similar to VCV except that the curing tube is arranged in a catenary and the cable is maintained within the centre of the tube by very careful control of the cable tension. This has become much easier with sophisticated automatic control systems. Care needs to be taken to ensure that the polymer, which is molten in the stages prior to crosslinking, does not drip or droop off the conductor under the influence of gravity. This effect becomes stronger as the thickness of insulation increases or the ratio of the insulation thickness to conductor size increases. Techniques including special polymers, rotation of the cable and quenching of the insulation are all effective at reducing the drooping effect. An additional problem can arise with large conductors (heavy cables), when it becomes difficult to both apply a large tension (needed to keep the cable in the centre of the tube) and control the tension. This practically limits the conductors sizes to below 1400–1600 mm². Cables with insulation thicknesses of up to 25 mm can be produced with CCV lines. The VCV technique does not suffer from the effects of drip or droop off of the conductor under the influence of gravity to the same extent as for the CCV technique. However, due to construction costs of large vertical structures VCV lines are shorter than CCV lines: 80–100 m for VCV and 200–250 m for CCV. This difference enables CCV lines to be faster as identical cables require the same residence time.

10.3.2.3 Mitsubishi Dainichi continuous vulcanisation

In this technique, the cure tube is arranged horizontally after the extruder but, unlike VCV and CCV, it does not employ nitrogen gas to maintain the pressure. The MDVC technique arranges that the diameter of the die is the same size as the final cable and thus fills the tube/die. The thermal expansion resulting from heating the polymer to the molten state to initiate the crosslinking exerts a pressure that suppresses the formation of voids. The MDCV process does not experience the same droop issues as those faced by the CCV process due to being fully enclosed by the die. However, it is important to ensure that the conductor remains centralised within the tube while the polymer is molten before crosslinking. Centralisation may be achieved by ensuring a very high tension (thereby essentially eliminating the catenary) or by using special polymers with high viscosities. These special methods are normally required for conductors above 1000 mm^2.

10.4 Failure processes

Polymeric insulated power cables are designed to be high reliability products. The failure of a major power cable is likely to have a considerable effect on the power transmission grid and may take several days/weeks to repair. If it is under the sea, it may take months to repair and cost well in excess of £1 million ($1.5 million). Cables have a good service history. The majority of cable failures are caused by external influences such as road diggers or ship anchors. Cable system failures may occur at joints and terminations. A study of MV cable systems in France [9] has shown that the failure rate for paper cables is 3.5 failures per 100 cable km per year which compares with a rate of two failures per 100 cable km per year for XLPE cables. The failure rate for XLPE has been classified in Table 10.3. This analysis shows that the

Table 10.3   20 kV XLPE system performance (after [9])

Class	Fault rate (#/yr/100 cct km)
Total system	2.0
Third party damage	1.0
Accessories	0.9
Cable	0.1

XLPE cables form the most reliable component of the cable system. Further inspection would suggest that the XLPE cable system is 2.5 times more reliable than the paper system when third party damage is excluded (we assume that a paper cable is as likely to get dug up as an XLPE one!). This estimate may well overstate the case and illustrates one of the major problems of evaluating field failure statistics; namely the differing ages of the populations. XLPE data are based on a population which is 15 years old whereas the paper data are based on a population that is 30–40 years old; thus we would expect higher failure rates of the paper system since it contains older devices; this would be quite independent of the intrinsic reliabilities of the systems.

Causes of insulation breakdown include:

- Extrinsic defects (contaminants, protrusions or voids) caused during manufacture or installation. These would normally lead to electrical treeing or direct breakdown soon after production of the void and cable energisation.
- Water treeing (wet ageing) caused by water leakage through the sheath or inappropriate design or deployment of a cable without a water barrier. Water trees lead to a weakening of the insulation and electrical treeing or direct breakdown.
- Thermoelectric ageing: the combination of the electric field, acting synergistically with a raised temperature, causes the insulation to weaken over time and for breakdown to occur eventually. This process may not always be significant within the lifetime of a well designed cable.

10.4.1   Extrinsic defects

We have already described the efforts made by cable manufacturers to exclude and detect contaminants, protrusions and voids (CPVs) in their products. Contaminants within the bulk of the insulation and protrusions into the insulation from the semicon cause field intensifications that lead to premature failure of the polymer.

10.4.1.1   Contaminants and protrusions

Many studies [10–12] have shown the degradation caused by large metallic contaminants. Figure 10.5 shows the influence of size on the electrical breakdown strength. The effect of a five-fold increase in concentration at the 100 micron level is to further reduce the strength by 17 per cent and 14 per cent for AC and impulse, respectively. As well as showing the reduction in the characteristic strengths, the increasing size of contaminants changes the statistical nature of the failures, making them less scattered or more certain.

The effects displayed in Figure 10.5 may be explained by the fact that metallic contaminants increase the electric stress within their immediate locality such that the local electric stress is higher than the breakdown strength. This effect is best described in terms of a stress enhancement factor that acts as a multiplier for the Laplacian stress (i.e. the geometrically calculated stress). The data from Figure 10.5 are in line with the trends predicted by the theoretical estimates for the increase in local electrical stress found at the tip of sharp metallic contaminants. It is interesting to note that it is possible to get large stress enhancements but that the magnitude of the enhancement

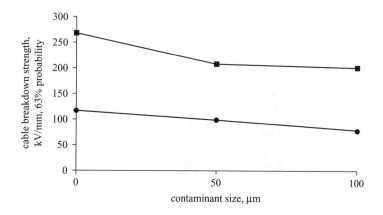

Figure 10.5   *Electrical breakdown strength (AC lower circles and impulse upper squares) of model cables with artificially added contaminants (concentration of 5 per kg) [12]*

falls dramatically with distance from the tip; for a 5 micron radius, the field falls by 50 per cent within 1.5 radii of the tip [8]. Thus, the calculated stress enhancements should be viewed as providing the upper limits of any assessment.

The electrical stress enhancements are not only based on the size and concentration but they have a significant influence from the nature (conducting, insulating, high permittivity), the shape (sharp or blunt) and the way that they are incorporated into the matrix. The effect of the shape of contaminants has been assessed [13] using XLPE cups with a Rogowski profile. It was shown that contaminants with irregular surfaces reduced the AC ramp breakdown strength by a greater degree than those with smooth surfaces (Figure 10.6).

The local stress enhancement experienced within an insulator will have contributions from the size of the contaminants, their concentration and the nature (conducting or high permittivity) of the contaminants. This is shown in Equation (10.3).

$$\eta = 1 - \frac{1}{\alpha}\left(0.5\ln\frac{\lambda+1}{\lambda-1} - \frac{\lambda}{\lambda^2-1}\right) \tag{10.3}$$

where:

$$\alpha = 0.5\ln\frac{\lambda+1}{\lambda-1} - \frac{1}{\lambda} + \frac{1}{(k-1)\lambda(\lambda^2-1)}$$

$$k = \frac{\varepsilon_2}{\varepsilon_1} \quad \lambda = \frac{1}{\sqrt{1-(r/a)}}$$

η = stress enhancement factor

r = radius of the ellipse

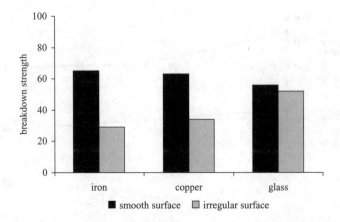

Figure 10.6   The effect of shape of added contaminants on the AC ramp breakdown strength (relative to base) of XLPE cups [13]

$2a$ = length of the ellipse

ε_1 = permittivity of the matrix

ε_2 = permittivity of the defect

10.4.1.2   Voids

Voids are likely to lead to breakdown if discharges occur inside them. These discharges are known as partial discharges since they are not, in themselves, a complete breakdown or full discharge. The least problematic (and perhaps the least likely) shaped void is the sphere. The field inside an air-filled sphere-shaped void is higher than the field in the insulation by a factor equal to the relative permittivity of the solid. For other shaped voids, the field in the void will be higher than this. For example, a void inside XLPE (relative permittivity = 2.3) will have a field inside it of at least 2.3 times that of the field in the XLPE itself. The criterion for a discharge in a void is that the void field must exceed the threshold described by the Paschen curve (see Reference 14, for example); this is dependent on the size of the void. The Paschen field has a minimum for air at atmospheric pressure for a void diameter of 7.6 μm. The breakdown voltage at this diameter is 327 V. This equates to a void field (ignoring the non-linear effects) of 43 kV/mm or an applied field of 19 kV/mm within the XLPE. Below this void size, the Paschen field increases rapidly; for example, a 2 μm void would require an applied field of approximately 150 kV/mm to cause discharging. It is clear then that voids of diameter exceeding a few microns are likely to allow severe electrical damage to occur.

10.4.1.3   Partial discharges and electrical treeing

In an electrical discharge, electrons are accelerated by the electric field such that their kinetic energy may exceed several electron-volts. With such energies, collisions with gas molecules may cause further electrons to be released, thus strengthening the discharge, or they may cause electroluminescence and the release of energetic photons. The surface of the void is therefore likely to be bombarded by particles (photons or electrons) with sufficient energy to break chemical bonds and weaken the material. In the case of sharp protrusions or contaminants, which give rise to high local electric fields, electrons may be emitted and very quickly acquire the kinetic energy required to cause permanent damage to the insulation. Electrons may accumulate – i.e. they may be trapped – around such defects and cause a further increase in local electric fields. Mechanical stress, which may already be increased due to the modulus and thermal expansion coefficient differences between the host materials and the CPVs, may be further enhanced by electromechanically induced stress. These effects, catalysed by CPVs in an electric field, may lead directly to a breakdown path, but are more likely to lead first to the formation of an electrical tree.

An electrical tree is shown in Figure 10.7. For the sake of clarity, this has been grown in a translucent epoxy resin from a needle acting as a protrusion. A breakdown path can be seen to be growing back through the electrical tree from the plane counter-electrode. Electrical trees have a branched channel structure roughly oriented along the field lines. Typically, the diameters of the channels are 1–20 μm, and typically each channel is 5–25 μm long before it branches. There is considerable evidence that the branching is determined by the local electric field, which is grossly distorted by trapped electrons emanating from the discharges within the tree. There is a considerable body of work on this subject (e.g. Reference 14). Trees tend to grow more directly across the insulation if they are spindly, so-called branch trees. The other type of tree, the bush tree, uses the energy to produce a lot more dense local treeing,

Figure 10.7   An electrical tree grown in epoxy resin

and therefore tends to take longer to bridge the insulation. Because branch trees occur at lower voltages, there is a non-monotonic region between branch and bush growth in which failure occurs more quickly at lower voltages. Trees grown from larger voids can be clearly distinguished from those due to protrusions; it is noticeable that the fields must cause the voids to discharge before the trees initiate. After an initial period in which the tree grows quickly, the rate of growth decreases. Finally, as the tree approaches the counter electrode, a rapid runaway process ensues.

Although the electric tree processes are now quite well understood, apparently little can be done to prevent electrical tree propagation in polymeric insulation once it has started. It is thought that electrical trees take relatively little time to grow and cause breakdown in cables, in perhaps a few minutes to a few months. Once an electrical tree has been initiated, the cable can be considered to be terminally ill. It is therefore vital to prevent the inclusion of CPVs during cable manufacture and installation. The triple extrusion continuous vulcanisation techniques and appropriate cable protection and deployment techniques appear to have been successful in this regard, and cables rarely suffer from these problems. There are more likely to be problems at joints or terminations where high field stresses may inadvertently be introduced or if water trees, described in the following section, are allowed to grow.

10.4.2   Wet ageing – water trees

In the early days of polymer-insulated cables, it was assumed that the polymers would be essentially immune to the deleterious effects of water that were well known in paper cables. Consequently, the first designs of cable were installed with little or no water precautions. Within a few years, a large number of cables started to fail in service. Upon examination, tree-like structures were seen to have grown through the insulation. It was assumed that they continued to grow and failure occurred when the whole insulation was breached. This is the phenomenon of water treeing [15].

Many studies have been carried out into the phenomenon and its solution. Looking back, it is clear that a number of improvements in cable design, manufacture and materials have reduced the incidence of cable failures by water treeing. These improvements have included:

- water barriers (metal or polymeric) to exclude the water
- triple extrusion (all polymer layers extruded at the same time)
- semiconductive polymer screens to replace carbon paint or paper tapes
- cleaner insulations
- smoother semicons
- internationally recognised approval methods
- special long life insulations based either on additives or polymer structure.

The laboratory studies have concluded that the growth of water trees is affected by:

- test voltage
- test frequency
- mean temperature

- temperature gradient
- type of material
- presence of water (external and within the conductor).

There are essentially two types of water trees: vented trees that originate from the surface of the insulation and are potentially the most dangerous and bow tie trees that grow from contaminants or voids within the insulation and tend to grow to a limiting size without breaching the insulation. These trees do not comprise tubules containing water as might be surmised from the earlier description of electrical trees. The branches of a water tree actually appear to comprise a high density of water-filled voids of typical diameter 1–10 µm. Such branches are therefore similar to a string of pearls, but in practice even branches of water trees are not usually discernible. They are simply diffuse regions of water-filled voids. If dried up, reimmersion in water reopens the voids. Boiling stabilises the structure but probably also produces extra small voids. There is limited evidence that a percolation network does interconnect the voids, but the size scale of the interconnecting features is around 10 nm. Electrolyte material accompanies the water into the voids and the ability of cationic dyes, such as rhodamine B, to stain the trees permanently indicates that some oxidation must have taken place. Chemical modification has also been shown using IR and FTIR spectroscopy and by fluorescence techniques.

Water trees grow much more slowly than electrical trees. Typically, they may not be observed at all for several years, even if the prevailing conditions for their growth are in place. They will then grow fast initially and then very slowly. Indeed, in the case of bow-tie water trees, there is much evidence that they stop growing completely after a given length (dependent on prevailing conditions) and that they might not precipitate breakdown. Vented water trees may cross the insulation completely without breakdown occurring, but they do greatly weaken the insulation. Generally, an electrical tree or a breakdown path may grow back through a water tree. Figure 10.8 shows a badly degraded EPR cable containing water trees. The water trees are not as clear as have been seen in XLPE cables, but this may be because EPR is not transparent or even translucent as is XLPE. Also seen in this figure are breakdown paths starting to grow back through the water-treed areas. The cable had broken down within 1 m of this area. The general trend is for the ultimate failure of the cable to be due to the conversion of a vented tree to an electrical tree.

There are many proposed mechanisms of water treeing and these have been critically reviewed in Reference 14. Essentially, it is likely that solvated ions are injected at partially oxidised sites. These catalyse further oxidation by maintaining the ion concentration. A sequence of metal–ion catalysed reactions is proposed in which bonds break and cause microvoids to develop. Alternating electromechanical stresses open up pathways for solvated ions; these initiate new microvoids. Many tree-retardant polymers contain ion catchers to prevent the metal–ion catalysis, and these have been found to successfully delay the onset and growth rate of water trees.

The tree inception time, i.e. the time between the conditions being right for water tree growth and the first observation of water trees, is highly dependent upon the electrical stress. Typically, the inception time is inversely proportional to a high

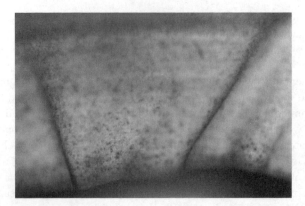

Figure 10.8   Water trees and breakdown paths in an EPR cable. The black streaks are breakdown paths starting to cross the cable. The black dots, which are intense around these breakdown paths, are stained water trees. EPR is more resistant to water trees than XLPE, but this cable had been taken from service where it had been saturated in water with no metal sheath

power ($\approx$4–10) of electric field. For this reason, low voltage cables, which tend to run at lower electric fields, may not have a water barrier to prevent water ingress and hence electrical treeing. In such cases, with fields typically <4 kV/mm (see Table 10.1), the probability of failure through electrical treeing is low and a water barrier would make little difference. HV and EHV polymer-insulated cables generally use water barriers, and these become mandatory above 66 kV. Furthermore, the conductor is often water blocked (a water-swellable compound or an extruded mastic) to prevent the transport of water along the conductor. The water may enter the conductor either after a cable breakdown or during installation or through an incorrectly installed accessory.

10.4.3   Dry ageing – thermoelectric ageing

The requirement for extra high voltage (EHV) underground power cables is increasing [16–22]. There is commercial pressure to push the mean electric field in the insulation of such cables towards 16 kV/mm, and the most common insulation used is crosslinked polyethylene (XLPE). Long term experience of XLPE, however, is limited to moderately stressed cables with mean fields of 5 to 7 kV/mm. Furthermore, the introduction of crosslinking processes has permitted the continuous operating temperature of polymeric cables (XLPE and EPR) to be increased to 90°C, equalling that of oil-filled (LPOF and HPOF) paper and polypropylene paper laminate (PPLP) cables. The use of XLPE as the insulation for transmission cables has grown steadily since the early 1990s. Many extruded power cables have been operating for 20 years and are approaching the end of their 30-year design life. If robust methodologies could be found for improving or/and evaluating the reliability of AC power cables, it may be possible to continue to use them without compromising the reliability of the system.

Such methodologies require considerable improvements in the understanding of any ageing or degradation mechanisms of cable insulation. They would enable XLPE cables to be more competitive at EHV levels.

Recently, there has been considerable discussion in the literature about whether semicrystalline polymer insulation, such as XLPE, will age naturally under combined thermal and electric fields, without water or extrinsic damage (e.g. CPVs). Such considerations suggest that the polymer is gradually weakened and that this weakening may eventually lead to failure. Some of the theories of ageing suggest that a minimum field (perhaps around 20 kV/mm) is required for such ageing to occur, others assume just that the ageing slows up considerably at lower fields. The three main current theories are: the Dissado–Montanari–Mazzanti (DMM) theory [23], that of Lewis *et al.* [24] and the Crine theory [25]. There is not space to contrast these models here (see, e.g. Griffiths *et al.* [26]) but the essential elements are described. It is assumed that this ageing is macroscopic whereas the extrinsic and treeing degradation is localised and leads directly to a breakdown channel. This ageing leads ultimately to partial discharging or/and electrical treeing. The ageing process is likely to start at nanometre size scales, and for this reason is difficult to observe directly. It is likely to lead to the production of nano-voids which gradually increase in both size and number. Once these approach the micron size scale, they may be capable of precipitating breakdown through energetic mechanisms, such as hot electron damage, which can take place close to the design field in the presence of large defects. The ageing mechanisms proposed therefore involve the production of macrodefects ($>10 \ \mu m$) from microdefects (<10 nm).

The ageing models lead to equations describing the life of the insulating material in terms of the field and temperature. There has been considerable work (e.g. European ARTEMIS[1] programme) to understand these mechanisms better in order to be able to diagnose the state of a cable, develop a replacement policy based on such diagnosis and develop life models to be used in the resource management of cable systems. Such an improved understanding is also useful in considering whether it is worthwhile investing further effort in improving manufacturing processes.

The three models referred to earlier differ slightly in physical details and some-what more in their mathematical development but their commonality can be found in considering the different size scales of the various processes involved. The Lewis model considers the breaking of chemical bonds as a starting point for the ageing. The problem is stated more generally in thermodynamic terms in the other two ageing models. They consider that moieties – i.e. small regions – of the polymer may exist in either of two states. In an unaged polymer, most of the moieties are in state 1 whereas, as ageing progresses, more and more moieties switch to state 2. An energy barrier exists between the states and moieties may switch between states by thermal activation. In the Lewis model, these states would comprise unbroken and broken bonds; the other two models are more general and may include molecular chain reconfigurations, for example. The presence of a local electric field changes the dynamic equilibrium between the states, and, according to the DMM theory, causes irreversible changes when a threshold field is exceeded. It is assumed that changes from state 1 to state 2 will eventually strain, and, possibly, lead to nanometre and submicron sized voids

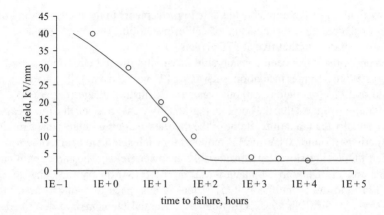

Figure 10.9   Electrical life line with model (line) and experimental data points (from [27]) for AC polyimide data. Experimental points at 63.2% failure probability are displayed

when a sufficient concentration of moieties have switched to state 2. Areas that are mechanically weakened in this way will increase in size leading to super-micron-sized voids and the more rapid degradation associated with hot electron injection, partial discharging and electrical treeing, for example.

All three models assume that a section of the polymer can transfer between alternative local states by surmounting an energy barrier. The presence of an electric field alters the energies of the two states, thereby altering the relative proportions of the polymer in each of the alternative states at equilibrium. It also accelerates the rate of approach to the equilibrium distribution from any arbitrary starting distribution. It is important to note that these processes are temperature dependent. A typical result from such modelling is shown in Figure 10.9.

10.5   Mathematical design models for cables

Cables are designed to operate at electrical stresses in the range of 3–11 kV/mm (Table 10.1). However, the breakdown strengths of virgin cables are often in the range of 80–100 kV/mm. It is interesting to examine why the practical stresses are so much lower than the strength of the material. Many studies have shown that the electrical strength of practical cables is reduced with:

- increased operating temperatures – cable tests are often carried out at ambient temperature whereas the cables are required to operate at 90°C
- length of operation – cables age (use up some of their performance) during operation, such that the performance capability after ten years' operation is less than at the start
- length of cable and thickness of insulation – the more insulation materials employed, the larger the probability of occurrence of the harmful particles.

These effects need to be accommodated when designing the cable. The largest of these effects and the one that is most difficult to address is the length of operation. This is difficult because the lifetime of cables is generally very long, typically 30–40 years. This time span is longer than can be practically tested. Thus, engineers adopt the approach of testing for 0.5–2 years and then extrapolating to the end of life. To enable this extrapolation, an ageing model is required. Section 10.4.3 describes a number of ageing models that have been developed. However, these ageing models have yet to gain any significant acceptance amongst cable manufacturers and users, probably because of their high complexity and the lack of good data to support them. Currently, the inverse power law model is still widely used together with Weibull statistics to indicate the reliability. In this section, we do not propose to describe these models in detail, they are described elsewhere (e.g. Reference 14), or to describe the mathematics of the new ageing models referenced above, we merely want to comment on how important this area is for development.

The main inputs into mathematical design models are:

- electrical ageing models which enable the extrapolation of test data out to service lives
- the experimental ageing data that serve as input to the ageing models
- the breakdown characteristics of the cable systems.

However, as can be seen from Figure 10.10, even the global experience of cable design data in this area is somewhat limited.

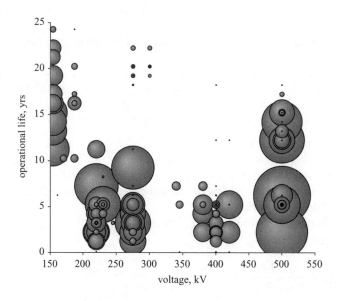

Figure 10.10   Global operational experience of XLPE EHV cables systems as a function of system voltage (test loops excluded) – the size of the bubble indicates the length of the cable circuit (courtesy of Borealis AB)

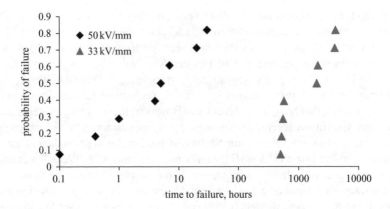

Figure 10.11   *Time to failure of XLPE cables installed in air when tested at two different stresses*

Furthermore, it can be seen from Figure 10.11 that the likelihood of cable failure increases with time under electrical stress and that the ageing rate is very sensitive to the applied electrical stress. The typical time to failure can be described by a number of statistics (Gaussian, Weibull, Parameter Free, etc., see, e.g., Reference 28); the relationship of these times to the stress and temperature is the ageing models. Once the correct model has been established and parameterised, it may then be used to determine the typical failure time for any desired stress and temperature. It should be recognised that the ageing models deal with the relationship of the typical values but do not, in general, deal with the statistical scatter. This scatter can be quite large (an order of magnitude in the case for 33 kV/mm) and can have a profound influence on the engineering predictions that emerge.

The importance of ageing considerations can be seen by considering one of the many cable design approaches presently used for HV and EHV XLPE cables. It is known that the AC voltage breakdown stress reduces with time and temperature. Thus, the approach is based on four steps: [30]

(i)   determining the breakdown performance of the virgin cables
(ii) determining how much of the initial performance will be lost due to elevated temperature operation
(iii) determining how much of this performance will be consumed during the service life of the cable system
(iv) determining an appropriate safety factor.

The following equations are then used:

$$E_{design} = \frac{E_{fail}}{F_{safety} F_{age} F_{temp}} \qquad (10.4)$$

$$\text{thickness} = \frac{V/\sqrt{3}}{E_{design}} \qquad (10.5)$$

where:

E_{design} = insulation design stress [kV/mm]
E_{fail} = reference stress for cable failure; taken from breakdown curves [kV/mm]
F_{safety} = safety factor
F_{age} = ageing factor relating test duration to the required life of the cable
F_{temp} = temperature factor relating test temperature to service temperature
V = cable voltage
thickness = thickness of insulation.

This approach has been used for the design of 275–500 kV XLPE cables. By way of an example, the thickness of a 400 kV cable may be determined for the AC condition by:

$$E_{design} = \frac{40}{1.1 \times 2.3 \times 1.1} = 14\,\text{kV/mm} \qquad (10.6)$$

$$\text{thickness} = \frac{400/\sqrt{3}}{14} = 17\,\text{mm} \qquad (10.7)$$

In this case, the level of ageing was assessed using the inverse power law for a design life of 30 years. Inspection clearly shows that the largest factor that influences the design stress and thus the thickness of a cable is the allowance made for ageing; the 2.3 factor. If the allowance is too much, the cable will be too large and, hence, there may be issues associated with weight, cost, length on a drum and number of accessories. If the allowance is too little, then the cable will be too small and the reliability will be compromised.

Thus, there can be no doubt that the correct understanding of the ageing phenomena and the ability to accurately model them are absolutely critical to the development, approval and successful operation of XLPE cables.

10.6   Direct current transmission

10.6.1   Economics

The incorporation of DC transmission into the pan-European network is not likely to make massive savings in efficiency. However, even small savings may have very large environmental benefits. For example, KEMA has estimated [31] that the savings from DC distribution in the EC would be approximately 9.5 Mtonnes of CO_2, approximately one per cent of the total produced by electrical sources in the EC at the moment.

Long distance HV cable interconnects are possible with reactive compensation [29]. A simple calculation shows that a typical 100 MW capacity AC cable uses approximately two per cent of its power in losses over 250 km, approximately 13 per cent over 500 km and almost 90 per cent over 1000 km. These large losses do not occur under DC. Since high voltage DC cables are expected to be more compact and recyclable, it may be possible to use them as longer links. This would also make their installation less controversial as visual impact and noise would be reduced.

10.6.1.1 Economic drivers

In 1999, overall European electricity generation capacity was 575 GW with 322 GW being conventional thermal plant. The corresponding actual energy generated for all plant was 2533 TWh of which 1281 TWh was conventional thermal which is worth $150 billion per annum.

KEMA, in their analysis of the potential benefits of the DC society [31], consider that the following advantages exist in incorporating DC transmission into the current AC grid:

- decrease of transmission and distribution losses by 40 per cent
- simple embedding of renewable energy sources such as photo-voltaics and fuel cells
- simple connection of energy storage systems
- better use of existing transmission infrastructure due to higher energy densities
- improved control of energy transport by fast power electronic devices, no reactive power and limited short circuit currents
- space saving with associated economic benefits.

Andersen *et al.* [33], in a 2002 review article, state, 'High voltage direct current (HVDC) is often the economic means for delivering electric power over long distance and/or for interconnecting two unsynchronised AC networks, which may be at different frequencies. HVDC schemes totalling 60 GW with individual scheme ratings between 50 MW and 6300 MW have been installed worldwide, and many more installations are being considered'. The controllability of HVDC means that the power delivered can be modulated to give improved damping to the AC transmission, sometimes allowing additional power to be transmitted safely through the AC interconnection.

Such benefits are expected to generate a European internal market and a global market for high voltage DC links. For polymer-insulated HVDC cable systems operating with new converter technology, the opportunity is illustrated well by experience to date with ABB's HVDC Light cable system which, in three years, has achieved the penetration shown in Table 10.4.

Table 10.4   HVDC light cable projects

Project	Power (MVA)	Voltage rating (kV)	Cable length (km)	In service (year)
Gotland/SE	50	80	140	1999
Directlink/AU	180	84	354	2000
Tjæreborg/DK	8	10	8	2000
Murraylink/AV	200	150	355	2002
Cross Sound/US	330	150	82	2002*

* successfully commissioned, commercial operation delayed

Although these systems are still unproven in long-life installations and with new voltage source converter technology (VSC) which is currently limited to 150 kV, it is expected that over the long term they will be capable of operating in excess of 150 kV and could achieve or probably exceed power ratings equivalent to current HVAC power cable systems i.e. 500–1000 MVA. The operational benefits of being able to transmit higher powers over equivalent HVDC cables without the need for static VAR compensation as required with AC systems, and the reducing cost of new VSC technologies, are expected to produce both capital investment and operational savings that will lead to continual improvements in the cost benefit case. When the environmental benefits are added, it is expected that HVDC interconnections will present a compelling case for a variety of interconnection projects worldwide.

10.7 Testing

Polymeric power cable testing recommendations are contained in IEC 60840 [34]. These are classified as:

- type tests
- routine tests
- sample tests
- electrical tests after installation.

Prequalification tests are included for EHV and MV cables, but not HV (IEC 60840). The status of HV prequalification tests are the subject of much discussion.

10.7.1 Prequalification and development tests

Prequalification tests are usually carried out by manufacturers to satisfy themselves before formal development testing. As such, predevelopment tests vary widely although CIGRE Working Group 21.03 has made some suggestions [34] mainly regarding checking for contaminants, protrusions and voids.

Prequalification tests are intended to be representative of onsite conditions. Accelerated ageing is usually provided by increased electrical stresses ($<1.7 \ U_0$) and by heating and cooling cycles[2]. In addition, a prequalification test may include lightning and switching impulse testing and wet testing to ensure that water trees do not grow. Typical protocols are described in References 35 and 36. Prequalification tests require approximately 1 year at EHV and between 1 and 2 years at MV.

10.7.2 Type approval testing

There are industry standards for type, sample and routine tests [34–40]. Type tests are project based; they qualify a particular design of cable and accessories and therefore, in principle, only need to be carried out once. Type approval tests require approximately 30–40 days to complete.

A typical type test includes both electrical and non-electrical testing including:

- partial discharge testing
- dissipation factor
- switching and lightning impulse testing
- withstand voltage during bending and crossbonding leads.
- various mechanical properties before and after thermal ageing
- ozone resistance
- density measurement
- carbon black content
- cold elongation and cold impact testing.

10.7.3 Sample testing

To check that the cable is within specified limits, a reel of cable from each manufacturing run is tested. Typically, this includes [41]:

- conductor examination and measurement of its electrical resistance
- checks of dimensional accuracy
- hot set test for XLPE and EPR insulation
- measurement of capacitance
- measurement of density of HDPE
- lightning impulse test followed by a power frequency voltage test.

10.7.4 Routine testing

Routine tests are carried out on each manufactured component to check that it meets specifications. These take the form of voltage withstand and partial discharge tests, Table 10.5 [41]; all delivered cables must successfully complete this test before dispatch.

10.7.5 Future trends in testing

At present, there is no agreement about whether tests are necessary, or what form such tests should take, to monitor dry (thermoelectric) ageing (section 10.4.3). Ageing, in this context, is considered to be the change within the polymer, brought about by electrical and thermal stresses, that leads to macroscopic degradation, thus loss of serviceability, ultimately leading to partial discharging and electrical treeing. Experimental techniques are therefore required that can be used in a complementary manner and that can be expected to provide chemical–physical, microstructural and electrical characterisation. It is also important that these experimental techniques provide ageing markers that can be used for the prognosis of the cable and the diagnosis of ageing leading to both bulk and localised degradation, which can take place close to the design field.

The European ARTEMIS programme, whose partners include cable manufacturers, material suppliers, electricity distributors and a number of universities throughout Europe, has been tackling this problem. In the programme, a variety of

Table 10.5   Typical requirements for routine tests [41]

Rated voltage(kV)	Withstand test voltage (kV)	Withstand test duration (Min)	Partial discharge test voltage (kV)
45–47	65	30	39
60–69	90	30	54
110–115	160	30	96
132–138	190	30	114
150–161	218	30	131
220–230	318	30	190
285–287	400	30	240
330–345	420	60	285
380–400	420	120	330
500	550	120	435

investigative techniques have been applied to lengths of cables and to samples peeled from full-sized XLPE-insulated cables. The techniques used range from those with micron and submicron resolution, which are intended to determine submicron features, through to 'bulk' measurements such as differential scanning calorimetry (DSC) that give information on larger characteristics, e.g. crystallinity, in addition to lamella thickness information. Spectroscopic techniques have been used to identify the chemical composition (e.g. in terms of crosslinking byproducts such as acetophenone and cumyl alcohol) and aid the interpretation of the electroluminescence processes. The electrical properties of the material are characterised by dielectric measurements, conduction currents, electroluminescence, space charge accumulation and packet formation. Relationships between the various electrical properties are attempted taking into account the transit of space charge across the sample and the onset field at which electroluminescence and space charge can be detected.

It is expected that tests based on such work will be developed by cable manufacturers as the basis for the prognosis of the health of the cable systems [42].

10.8   Notes

[1] Ageing and reliability testing and monitoring of power cables: diagnosis for insulating systems

[2] U_0 is the rated r.m.s. power frequency voltage between each conductor and screen or sheath for which cables and accessories are designed

10.9   References

1   DENSLEY, J.: 'Ageing and diagnostics in extruded insulations for power cables'. 1995 IEEE 5th international conference on *Conduction and breakdown in solid dielectrics*, IEEE pub. 95CH3476-9, 1995, pp. 1–15

2   ABERNETHY, ROBERT B.: 'The new Weibull handbook' (Robert B. Abernethy, North Palm Beach, Fla., 1996, 2nd edn.)

3   MOORE, G.F. (Ed.): 'Electrical cables handbook' (Blackwell Science Ltd, 1998)

4   VOLDHAUG, L., and ROBERTSON, C.: 'MV overhead lines using XLPE covered conductors Scandinavian experience and NORWEB developments'. IEE Conf. Publ., 406, 1995, proceedings of the 2nd international conference on the *Reliability of transmission and distribution equipment*, March 29–31 1995, Coventry, UK, pp. 52–60

5   LEIJON, MATS: 'Novel concept in high voltage generation: Powerformer^{TM}'. IEE Conf. Publ., 467, 1999, proceedings of the 11th international symposium on *High voltage engineering* August 23–27 1999, London, UK, pp. 5.379.S5–5.382.P5

6   KARLSTRAND, J., SUNNEGARDH, P., ZENGER, W., GHAFURIAN, R., and BOGGIA, R.: 'Water-cooled 345 kV solid-dielectric cable system'. CIGRÉ paper 21-111, 2002

7   *Power Eng. J.*, June 2002, **16**, (3) special issue on AC/DC power transmission

8   BARTNIKAS, R., and EICHHORN, R.M.: 'Engineering dielectrics' *in* 'Electrical properties of solid insulating materials', *American society for testing and maetrials (ASTM)*, 1983, **IIA**, (STP 783), pp. 1–721

9   BRINCOURT, T., and REGAUDIE, V.: 'Evaluation of different diagnostic methods for the French underground MV network'. JICABLE99 451

10 OHATA, K., TSUCHIYA, S., SHINAGAWA, N., FUKUNAGA, S., OSOZAWA, K., and YAMANOUCHI, H.: 'Construction of long distance 500 kV XLPE cable line'. JICABLE99, pp. 31–36

11 COPPARD, R.W., BOWMAN, J., DISSADO, L.A., ROLAND, S.M., and RAKOWSKI, R.T.: 'The effect of aluminium inclusions on the dielectric breakdown strength of polyethylene', *J. Appl. Phys.*, 1990, **23**, pp. 1554–1561

12 ANTONISCKI, J.R., NILSSON, U., and GUBANSKI, S.M.: 'Influence of metal inclusions on the breakdown strength of model cables'. Conference on *Electrical insulation and dielectric phenomena (CEIDP)*, Annual Report, **1**, 1997, pp. 283–286

13 HAGEN, S.T., and ILDSTAD, E.: 'Reduction of AC breakdown strength due to particle inclusions in XLPE cable insulation'. Conference on *Power cables and accessories 10 KV–500 kV*, 1993, **382**, pp. 165–168

14 DISSADO, L.A., and FOTHERGILL, J.C.: 'Electrical degradation and breakdown in polymers' (Peter Peregrinus, London, 1992)

15 STEENNIS, E.F.: 'Water treeing – the behaviour of water trees in extruded cable insulation'. ISBN 90-353-1022-5, pp. 132–133

16 PESCHKE, E., SCHROTH, R., and OLSHAUSEN, R.: 'Extension of XLPE cables to 500 kV based on progress in technology'. 1995, JICABLE 6–10

17 ANDERSEN, P., DAM-ANDERSEN, M., and LORENSEN, L., *et al.*: 'Development of a 420 kV XLPE cable system for the Metropolitan Power Project in Copenhagen'. CIGRE, 1996, pp. 21–201

18 OGAWA, K., KOSUGI, T., KATO, N., and KAWAWATA, Y.: 'The worlds first use of 500 kV XLPE insulated aluminium sheathed power cables at the Shimogo and Imaichi power stations', *IEEE Trans. Power Deliv.*, 1990, **5**, (1), pp. 26–32

19 POHLER, S., BISLERI, C., NORMAN, S., PARMIGIANI, B., and SCHROTH, R.G.: 'EHV XLPE cables, experience, improvements and future aspects'. CIGRE2000, pp. 21–104

20 HENNINGSEN, C.D., MULLER, K.B., POLSTER, K., and SCHROTH, R.G.: 'New 400 kV XLPE long distance cable systems, their first application for the power supply of Berlin'. CIGRE1998, pp. 21–109

21 'First 525 kV XLPE extra high voltage cables destined for Dachaoshan', *Mod. Power Syst.*, 2000, pp. 39–41

22 '400 kV XLPE cable going underground', *Mod. Power Syst.*, 1999, pp. 35–37

23 DISSADO, L.A., MAZZANTI, G., and MONTANARI, G.C.: 'The role of trapped space charges in the electrical aging of insulating materials', *IEEE Trans. DEI*, 1997, **4**, (5), pp. 496–506

24 LEWIS, T.J., LLEWELLYN, P., VAN DER SLUIJS, M.J., FREESTONE, J., and HAMPTON, R.N.: 'A new model for electrical ageing and breakdown in dielectrics'. 7th DMMA, 1996, pp. 220–224

25 CRINE, J.P.: 'A molecular model to evaluate the impact of ageing on space charges in polymer dielectrics', *IEEE Trans. DEI*, 1997, **4**, (5), pp. 487–495

26 GRIFFITHS, C.L., BETTERIDGE, S., and HAMPTON, R.N.: 'Thermoelectric ageing of cable grade XLPE in dry conditions'. IEEE international conference on *Conduction and breakdown in solid dielectrics*, 1998, pp. 279–282

27 HIROSE, H.: 'A method to estimate the lifetime of solid electrical insulation', *IEEE Trans. EI*, 1987, **22**, pp. 745–753

28 IEEE P930: 'Guide for the statistical analysis of electrical insulation breakdown data'. IEEE, 2002 (to be published by IEEE Standard Activities Department)

29 HALVARSSON, P., KARLSTRAND, J., LARSSON, D., LARSSON, M., REINHOLDSDÓTTIR, K., and SIGURÕSSON, E.: 'A novel approach to long buried AC transmission system'. CIGRÉ paper 21-201, 2002

30 GREGORY, B., GRIFFITHS, C.L., HAMPTON, R.N., and MAINWARING, S.P.: 'A probabilistic design method for polymeric structures to operate at high electrical stress'. *Dielectric materials, measurements and applications* 2000, IEE Conf. Publ. 473, pp. 419–424

31 KEMA Electricity Technology Roadmap: 'Technology for a sustainable society'. April 4, 2002

32 CIGRÉ: 'Recommendations for electrical tests (prequalification and development) on extruded cables and accessories at voltages >150 (170) kV and <400 (420) kV', *Electra*, 1993, **151**, pp. 15–19

33 ANDERSEN, B.R., XU, L., HORTON, P.J., and CARTWRIGHT, P.: 'Topologies for VSC transmission', *Power Eng. J.*, 2002, **16**, (3), pp. 142–150

34 IEC Committee Draft 20A/407/CD: 'Tests for power cable systems. Cables with extruded insulation and their accessories for rated voltage above 150 kV ($U_m = 170$ kV) up to 500 kV ($U_m = 525$ kV)'. 'International Electrotechnical Commission, Geneva, Switzerland, February 1999

35   PARPAL, J.-L.: 'Prequalification testing Alcatel 290/500 (525) kV extruded cable system at IREQ'. IEEE/PES Insulated Conductors Committee, minutes of the 102nd meeting, appendix 13-A, St Pete Beach, FL, November 1997

36   PARPAL, J.L., AWAD, R., and BELEC, M. *et al.*: 'Prequalification testing of 345 kV extruded insulation cable system'. CIGRE1998, pp. 21–101

37   IEC Publication 60840: 'Tests for power cables with extruded insulation for rated voltage above 30 (36) kV up to 150 (170) kV'. IEC, Switzerland, 1993

38   AEIC Specification CS7-93: 'Specifications for crosslinked polyethylene insulated shielded power cables rated 69 kV through 138 kV'. AEIC, Birmingham, AL, June 1993

39   IEEE Standard 48-1996: 'IEEE standard test procedures and requirements for alternating-current cable terminations 2.5 kV through 765 kV'. IEEE, New York, June 1996

40   IEEE Standard 404-1993: 'IEEE standard for cable joints for use with extruded dielectric cable rated 5000–138000 V and cable joints for use with laminated dielectric cable rated 2500–500000 V'. IEEE, New York, July 1994

41   HIVALA, L.J.: 'Extruded solid-dielectric power transmission cables', *in* BARTNIKAS, R., and SRIVASTAVA, K.D. (Eds): 'Power and communication cables' (IEEE Press, 2000), chap. 4, p. 231

42   FOTHERGILL, J.C., MONTANARI, G.C., and STEVENS, G.C. *et al.*: 'Electrical, microstructural, physical and chemical characterisation of HV XLPE cable peelings for an electrical ageing diagnostic data base', *IEEE Trans. Dielectr. Electr. Insul.*, 2003, **10**, (3), pp. 514–527

Chapter 11

Numerical analysis of electrical fields in high voltage equipment

A.E. Baker, T.W. Preston and J.P. Sturgess

11.1   Introduction

The accurate determination of the electrical field distribution is vitally important in the production of safe and economic designs of high voltage electrical equipment.

With transmission voltages typically 400 kV (r.m.s.), line and test voltages of 630 kV (r.m.s.) at power frequencies and 1425 kV (peak) lightning impulse, the problems of configuring the high voltage (HV) winding, leads, stress control rings and bushings of an HV transformer become apparent. On the rare occasions when something goes wrong, the consequences can be very serious. Figure 11.1 indicates the severe damage that can be caused by an HV flashover, in this case between the corona shield at the coil end of the HV bushing and a neutral lead. The insulation on the lead had been severely ruptured, and as a result a costly repair was needed.

The HV transformer is only one of many examples that show the importance of designers having analysis techniques capable of assessing or revalidating their designs accurately under all anticipated steady-state and transient operating conditions.

Besides the quasi-electrostatic problems associated with HV equipment discussed above, there is much interest in, and concern over, the large voltage gradients that occur along the end windings of machines fed by inverters. The inverter voltage waveforms, seen by the end windings, contain high harmonic components that can lead to sparking and erosion at the insulation surface unless special measures are taken. To alleviate the stresses, special insulating tapes have been developed whose conductivity increases with the applied electrical stress [1, 2]. These tapes reduce the stress concentration where the stator winding emerges from the stator slot.

As shown by the examples given, physical characteristics such as the geometric arrangement, the material properties of the insulation and stress grading materials

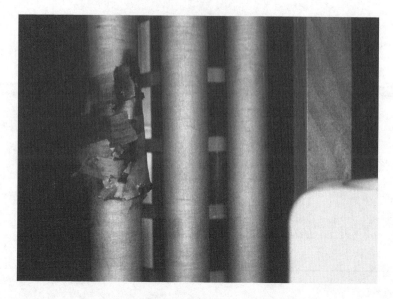

Figure 11.1   Insulation rupture of the neutral lead (courtesy of ALSTOM T&D – Transformers, Stafford (TST))

and the applied voltage must be taken into consideration. Thus, as new materials are developed and new configurations conceived, it is essential to be able to model them at the design stage rather than to rely solely on intuition and experience.

Modern design analysis is based on numerical methods, several of which are available to the designer. The question is, which one offers the best facilities in terms of speed of analysis, flexibility of modelling and range of investigative tools?

11.2   Which numerical method?

11.2.1   The finite-difference method

The finite-difference method was originally developed by Southwell [3] in 1946 but was not extensively used until the early 1960s when digital computers could be used to improve solution speed.

The method relies on the problem area being divided into a grid structure, as shown in Figure 11.2, from which a numerical equation for each grid intersection can be obtained through a Taylor expansion of the differential equations describing the solution field. From the resulting matrix the formulation potential can be evaluated at each node. Normally the matrix of equations is solved using an iterative method.

The method has been mainly used for two-dimensional geometries but can also be applied to three-dimensional arrangements. However, the representation of curved or irregular volumes is poor and, in electrostatic calculations, the electrical stresses can be misleading in these regions.

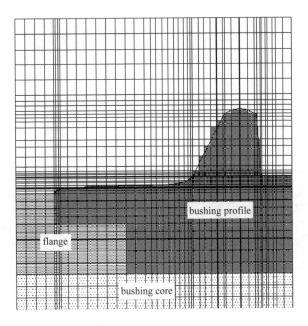

Figure 11.2   Example of a finite-difference mesh

11.2.2   The finite-element method

The finite-element method was originally used in the aeronautical and civil engineering industries and was promoted extensively by Zienkiewicz and Taylor [4] in the late 1950s and early 1960s. It was the late sixties before the method entered the electrical engineering industry, but it has been continually developed since that time.

The method relies on transforming the problem area of complicated form over a continuous region into one in which the region is divided into small elements (discretised) and an approximate solution is used for the field quantity in each element (Figure 11.3). By suitable choice of elements, complicated geometries can be represented and each element can have different physical properties, that may be non-linear (i.e. field dependent). Transforming the governing partial differential equations into an energy form, an algebraic expression may be created for the potential at each node through the extremisation of the energy functional. The resulting set of equations is sparse and there are many techniques that can be used to invert the matrix efficiently.

Although a two-dimensional problem is shown in Figure 11.3, the method is equally adaptable to three-dimensional situations. However, the generation of the mesh is more complex and time consuming. In excess of 1 million unknowns is not uncommon.

11.2.3   The boundary-element method

This is another analysis technique that has been available to designers for many years. It models the problem under consideration by representation of the surfaces of the

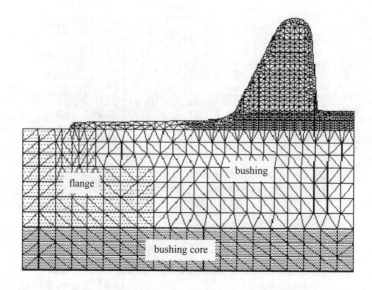

Figure 11.3   Example of a finite-element mesh

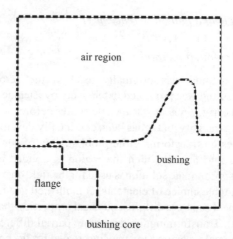

Figure 11.4   Example of a boundary-element mesh

regions concerned rather than their interiors, in contrast with the last two methods. This results in the problem being reduced by a dimension, i.e. a two-dimensional problem is modelled with one-dimensional (line) elements, a three-dimensional one with two-dimensional elements. A significant advantage of the method is the fact that the air regions between solid objects do not need to be discretised. This can be of major benefit when it comes to three-dimensional mesh generation. An example of a boundary-element discretisation is shown in Figure 11.4.

The numerical equations are formed from the application of Green's theorem to the partial differential equations that describe the field distribution within the region concerned. An equation is formed at the nodes on the outer surfaces of all regions and results in an unsymmetrical, densely packed matrix. An efficient method for inverting the matrix is difficult to find and leads to large solution times.

The method is applicable to both two-dimensional and three-dimension problems. However, the latter results in a very large unsymmetrical matrix to be inverted, so restricting the complexity of the problem that can be solved.

11.2.4   Comparative summary

All the techniques described have their advantages and disadvantages, so complicating choice. For instance, although it is easy to set up the problem, the finite-difference technique is not easily applied to modelling irregular geometries, such as slanted and curved surfaces. Nodal distribution can also be very inefficient. This is not so with finite elements. Equally, the boundary-element method can model efficiently regions in which the material properties are linear, but it is not efficient at handling non-linear materials. Furthermore, the type of matrix formed is not amenable to efficient solution. The finite-element method is well suited to the modelling of non-linear materials and the sparse form of the resulting matrix makes it readily adaptable to efficient solution, but it is necessary to discretise the free space surrounding the object to be modelled.

It is impossible to be definitive as to which method is best – different establishments will arrive at different conclusions according to their specific range of applications and problems. However, history has shown that as a robust and flexible design method, the finite-element method is generally best suited to the needs of designers, although in some cases there is little to choose between it and the boundary-element method.

11.3   Formulation of the finite-element equations in two and three dimensions

11.3.1   General

The finite-element method transforms a problem of complicated form in a continuous region into one in which the region is subdivided into a large number of well defined regions, known as elements. An approximate (though good) solution is obtained for the potential at the nodes that lie at the vertices of the element. It will be evident that complicated geometries can be represented well and that, in principle, each element can have different properties.

In order to obtain the field distribution in a defined space, a formulation is obtained either through an approach such as the Galerkin method [5] or directly via a functional derived from the stored electrical energy of the region under consideration. This functional is extremised with respect to the potential at the nodes defining the elements and then integrated over all elements associated with the node concerned, given an assumed polynomial variation of potential over the element (the shape function). An equation is thus formed at each node, resulting in a large set of simultaneous equations that replaces the governing partial differential equation.

To illustrate the process, the method will be applied to the modelling of an electrostatic field. The partial differential equation that describes the voltage potential distribution within any given region is derived as follows:

$$\nabla \cdot \boldsymbol{D} = \rho_v \tag{11.1}$$

where $\boldsymbol{D}$ is electrical flux density, Cm^{-2}, ρ_v is free charge density, Cm^{-3}, and since $\boldsymbol{D} = \varepsilon_0 \varepsilon_r \boldsymbol{E}$ then Equation (11.1) becomes:

$$\nabla \cdot (\varepsilon_0 \varepsilon_r \boldsymbol{E}) = \rho_v \tag{11.2}$$

where $\boldsymbol{E}$ is electrical stress, Vm^{-1}, ε_0 is 8.85419×10^{-12}, Fm^{-1} and ε_r is relative permittivity.

To formulate in terms of the voltage potential, V can be written in terms of $\boldsymbol{E}$:

$$-\nabla V = \boldsymbol{E} \tag{11.3}$$

Substituting this in Equation (11.2) results in the following equation:

$$\nabla \cdot \{\varepsilon_0 \varepsilon_r (-\nabla V)\} = \rho_v \tag{11.4}$$

Equation (11.4) can be rewritten for a homogeneous region as:

$$\nabla \cdot \nabla V = -\frac{\rho_v}{\varepsilon_0 \varepsilon_r} \tag{11.5}$$

Equation (11.5) is called Poisson's equation and applies to a homogeneous medium. If ρ_v is zero the equation reduces to Laplace's equation for homogeneous media. This is shown in full by Equation (11.6) (an expansion of Equation (11.5) in Cartesian coordinates):

$$\frac{\partial^2 V}{\partial x^2} + \frac{\partial^2 V}{\partial y^2} + \frac{\partial^2 V}{\partial z^2} = 0 \tag{11.6}$$

Equation (11.6) is the partial differential equation that describes the voltage potential distribution in high voltage situations were the medium is homogeneous and the charge density is zero. If the medium is non-homogeneous then Equation (11.6) is written as:

$$\varepsilon_0 \varepsilon_r \left(\frac{\partial^2 V}{\partial x^2} + \frac{\partial^2 V}{\partial y^2} + \frac{\partial^2 V}{\partial z^2} \right) = 0 \tag{11.7}$$

for each region. This is to allow for the different permittivities in the different regions.

To use Equation (11.7) in a finite-element formulation requires the equation to be transformed into an energy functional form that relates directly to the electrical energy of the system. This equation can be derived by several methods as described in the following section.

11.3.2   Forming the functional equation

Several approaches are possible for formulating the functional equation. One is to use the variational method, which is based on Euler's theorem [6]. Unfortunately, it is not always possible to deduce the functional from it, and it is not regularly used. However, Euler's equation can always be used to check the validity of the functional since the resulting equation reverts to the original partial differential form.

An alternative method, the energy-related functional, can often be written directly from electrical stored energy considerations. This approach is, perhaps, more intuitive for the engineer who can more easily relate to the energy of a system rather than a strictly mathematical approach.

A third and increasingly popular technique is a particular implementation of the weighted residual method known as the Galerkin method. The Galerkin method starts by supposing that a trial solution potential exists at each node. Insertion of this set of potentials into the governing equation will lead to a residual at each node, i.e. the equation will not be perfectly satisfied since, in general, the trial values will not be correct (there is a similarity to the finite-difference formulation here). It is clear that an approximate solution can be obtained by adjusting the potentials to minimise the sum of the residuals at all the nodes (this method is sometimes called the collocation method). However, a better solution can be obtained by introducing weighting functions at each node to try to force the sum of the local residual errors to zero over the whole domain. This method is described as the weighted residual method. Where the weighting functions used are the chosen shape functions for the discretised region, the method becomes the Galerkin method.

11.3.3   The energy functional illustrated

The total electrical energy in a system of volume Ω may be written as:

$$F = \frac{1}{2} \int_{\Omega} D \cdot E \, d\Omega \tag{11.8}$$

i.e.

$$F = \int_{\Omega} \frac{1}{2} \varepsilon_0 \varepsilon_r E^2 d\Omega \tag{11.9}$$

If it is assumed that the permittivity is constant within the region concerned, then Equation (11.9) may be used to write the energy (in a Cartesian coordinate system) as:

$$F = \int_{\Omega} \frac{\varepsilon_0 \varepsilon_r}{2} \left[\left(\frac{\partial V}{\partial x} \right)^2 + \left(\frac{\partial V}{\partial y} \right)^2 + \left(\frac{\partial V}{\partial z} \right)^2 \right] dx \, dy \, dz \tag{11.10}$$

It can be shown that this is the functional that, when differentiated with respect to V and equated to zero, gives a distribution of V that satisfies the governing partial differential equation. Physically, the process corresponds to minimising the stored electrical energy i.e. the potential energy of the system for the imposed boundary conditions. The differentiation is most conveniently carried out on the discretised system, and this will be shown in relation to the two-dimensional case.

For element e:

$$F_e = \frac{\varepsilon_0 \varepsilon_r}{2} \int_e \left[\left(\frac{\partial V}{\partial x} \right)^2 + \left(\frac{\partial V}{\partial y} \right)^2 \right] dx\, dy \tag{11.11}$$

where suffix e indicates integration over an element.

Hence, the contribution to the rate of change of F with V from the variation of potential of node i in element e only, χ_e, is:

$$\chi_e = \frac{\partial F_e}{\partial V_i} = \frac{\varepsilon_0 \varepsilon_r}{2} \int_e \frac{\partial}{\partial V_i} \left[\left(\frac{\partial V}{\partial x} \right)^2 + \left(\frac{\partial V}{\partial y} \right)^2 \right] dx\, dy \tag{11.12}$$

$$\chi_e = \frac{\varepsilon_0 \varepsilon_r}{2} \int_e \left[2 \frac{\partial V}{\partial x} \cdot \frac{\partial}{\partial V_i} \left(\frac{\partial V}{\partial x} \right) + 2 \frac{\partial V}{\partial y} \cdot \frac{\partial}{\partial V_i} \left(\frac{\partial V}{\partial y} \right) \right] dx\, dy \tag{11.13}$$

11.3.4   Numerical representation

To solve high voltage problems using the finite-element method requires Equation (11.13) to be equated to zero. To represent the problem numerically, the problem region is divided into elements and Equation (11.13) is applied at the nodes forming the element vertices. The variation of the potential over the elemental shape has then to be approximated by a polynomial distribution (known as the shape function). The order of the chosen polynomial will dictate the type of element used, for example a linear distribution would only require a simple triangular element. For higher order shape functions, the number of nodes describing the element must be capable of defining the order used, e.g. a quadratic shape function over a triangular element requires nodes at the middle of each element side. Many books on the finite-element method, e.g. Reece and Preston [7], cover the numerical derivation of the functional, so it will not be discussed further here.

There will be contributions to the rate of change of the region functional χ with respect to the potential V_i at node i from all the elements connected to i. In the case shown in Figure 11.5, there will be contributions from elements 1 to 6.

Hence, generally, the contribution to $\partial F / \partial V$ from a change in V_i is:

$$\frac{\partial F}{\partial V_i} = \sum_e \frac{\partial F_e}{\partial V_i} \tag{11.14}$$

where $\sum_e$ represents the summation of contributions from all elements associated with V_i, i.e., all the elements connected to node i.

When these derivatives are equated to zero, a group of simultaneous equations is formed. These can be written in matrix form as:

$$[S_t]\{V_i\} = 0 \tag{11.15}$$

where $[S_t]$ is a square matrix known as the stiffness matrix and is formed from the geometric coordinates of the nodes defining the elements and the material properties. V_i is a column matrix containing all the nodal potentials. The coefficient stiffness matrix will be sparse, i.e. it will contain many zeros since, from the discussion above,

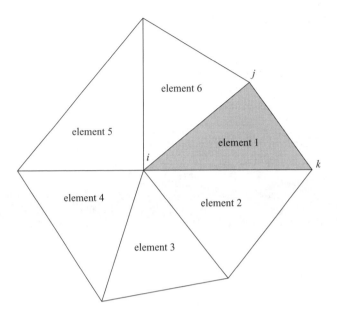

Figure 11.5   Elements associated with node i

it will be evident that any node will only be coupled to nodes directly connected to it by an element edge. Some of the nodes included in the equations will fall on boundaries, and their potentials may be known, or some other boundary condition may apply.

11.4   Variations on the basic formulation

11.4.1   General

When solving engineering problems, certain features arise where it is necessary to modify the original formulation either to improve the accuracy of modelling or speed of solution. Such features could be the representation of very thin metal foils that are used in bushings to grade the potential, so avoiding high stress regions, or the representation of partially conducting layers such as may occur on overhead lines due to contamination from the environment. To model these types of region requires the use of very thin elements (which can lead to numerical inaccuracy) or a modification to the formulation to allow elements of zero thickness to be used. The section below details several engineering implementations which have enabled a wider range of problems to be solved.

11.4.2   Representation of foils

In the design of high voltage systems, there is a need, in particular regions, to grade the voltage distribution or to shield regions of potentially high stress. Normally, this stress

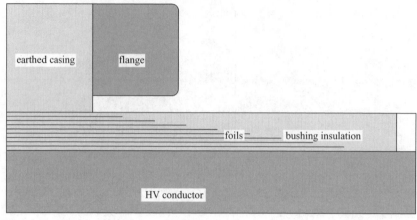

Figure 11.6   Generator terminal bushing

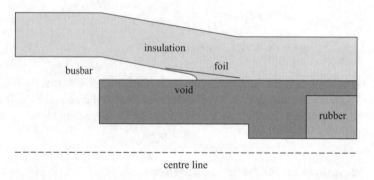

Figure 11.7   Conical bushing

grading is achieved through the use of foils which are thin metal strips. Figures 11.6 and 11.7 illustrate two examples where foils are used.

The electrical characteristic of a foil is a surface of constant voltage, whose value is dependent upon the capacitance of the system in which it is used.

To model foils using finite elements the foil is either represented as a region of high permittivity or as a constant voltage surface. The first approach has the disadvantage that foils are extremely thin (~0.1 mm) and to model them would result in skinny elements having a large aspect ratio. This, combined with the very large permittivity, makes the stiffness matrix ill conditioned with the result that it cannot be inverted accurately. The second approach is by far the most convenient and accurate way of modelling foils. Figure 11.8 illustrates, schematically, how the foil can be represented by line elements.

All the nodes connected by the line elements (marked 'F' in the figure) are constrained to be at the same (unknown) potential, whose value floats to a level that is

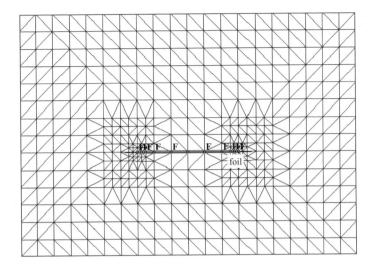

Figure 11.8   Foil representation

consistent with the capacitive distribution of the problem. A fine mesh is required at the end of the foils to ensure an accurate value of stress.

11.4.3   Directional permittivity

In high voltage equipment, there is a considerable amount of insulating material used for support purposes and, in some cases, these structures can have permittivities that depend upon the method of construction. For instance, a material used regularly in high voltage transformers is made from a wood/resin mixture. The electrical grade consists of laminates of wood bonded together with resin to form a layered board having different permittivities in different directions. However, the mechanical grade is formed from a homogeneous wood/resin mixture, so the bulk permittivities are the same in all directions.

To model directional permittivities the functional equation is modified simply by relating the directional permittivity to the appropriate coordinate as indicated in the following equation:

$$F = \int_\Omega \frac{\varepsilon_0}{2} \left[\varepsilon_x \left(\frac{\partial V}{\partial x} \right)^2 + \varepsilon_y \left(\frac{\partial V}{\partial y} \right)^2 + \varepsilon_z \left(\frac{\partial V}{\partial z} \right)^2 \right] dx\, dy\, dz \qquad (11.16)$$

11.4.4   Modelling resistive/capacitive systems

There are numerous occurrences in industrial design where the material used has both capacitive and resistive properties. Examples are:

(i)   stress grading paints or tapes used to redistribute the stresses on the end windings when they emerge from the stator core of machines

(ii)   contamination on bushings derived from environmental effects
(iii)   the coating of optical fibres, intended for use in high voltage environments, with a semiconductor to prevent high stress due to surface defects.

The current density in a partially conducting material has two components: J_{cond}, the conduction current and J_{cap}, the capacitive current:

$$J = J_{cond} + J_{cap} \tag{11.17}$$

These can be related to the permittivity, ε, the conductivity, σ, and the time derivative of the electric field, E:

$$J = \sigma E + \frac{\partial \varepsilon E}{\partial t} \tag{11.18}$$

If the electric field is assumed to vary sinusoidally with time at a radian frequency ω, then:

$$J = \sigma E + j\omega\varepsilon E \tag{11.19}$$

or, a complex permittivity may be defined to relate D and E:

$$D = \left(\varepsilon - j\frac{\sigma}{\omega}\right) E \tag{11.20}$$

The electric energy over a volume Ω for this system is given, from Equation (11.8), by:

$$F = \left(\varepsilon - j\frac{\sigma}{\omega}\right) \int_{\Omega} \frac{E \cdot E}{2} \, d\Omega \tag{11.21}$$

Replacing E by the derivative of the voltage gives the functional:

$$\chi = \frac{1}{2}\left(\varepsilon - j\frac{\sigma}{\omega}\right) \int_{\Omega} \left(\frac{\partial V}{\partial x}\right)^2 + \left(\frac{\partial V}{\partial y}\right)^2 + \left(\frac{\partial V}{\partial z}\right)^2 d\Omega \tag{11.22}$$

The functional can be discretised using the finite-element method and then minimised to give the voltage distribution. The problem can be solved for a three-dimensional region which is divided into tetrahedral elements or reduced to two-dimensional and solved over triangles, for example. The voltage over each tetrahedron can be described by a first-order linear shape function or by higher order elements. Once the voltage distribution has been found, it is possible to derive further field quantities such as electric stress, current density and loss intensity.

11.4.5   Modelling partially conducting tapes and paints

Partially conducting regions such as stress grading tapes or contaminated surfaces are typically only tenths of a millimetre thick, orders of magnitude smaller than typical dimensions associated with other components of the problem concerned. In a normal finite-element formulation, this leads to elements in the mesh with poor aspect ratios which can, in turn, lead to numerical instability, particularly in a non-linear regime. For this reason, a surface element formulation can be introduced to represent those types of partially conducting region whose relative thickness is small. The surface elements are assigned the material properties of the partially conducting layer but the elements occupy no physical thickness in the model. This allows the energy associated with the stress grading layer to be included without the numerical errors that skinny elements would introduce. The formulation assumes that there is a voltage drop along the length and width of the tape but not across its thickness.

The energy functional at the surface is similar to that for the volume:

$$\chi = \frac{1}{2}\left(\varepsilon - j\frac{\sigma}{\omega}\right)\int_{\Gamma}\left(\frac{\partial V}{\partial p}\right)^2 + \left(\frac{\partial V}{\partial q}\right)^2 d\Gamma \tag{11.23}$$

where p and q are two mutually orthogonal directions tangential to the plane of the surface element.

11.4.6 Space charge modelling

The effects of space charge can be included via the Poisson equation:

$$\nabla^2 V = -\frac{\rho}{\varepsilon_0} \tag{11.24}$$

where ρ is the charge density.

In general, the space charge distribution will not be known *a priori* and must be calculated iteratively during the finite-element solution. This is achieved by including an additional relationship between the voltage and the charge density. There are many possibilities for this relationship, depending on the physics of the problem. For example, in the case of a multielectrode valve device under space charge limited conditions, the voltage and current at a point are related by the three-halves power law:

$$V^{3/2} = \frac{9}{4}\frac{J}{\varepsilon_0}\left(\frac{m}{2e}\right)^{1/2}x^2 \tag{11.25}$$

where J is the current density at the given point, Am^{-2}, m is the mass of an electron, kg, e is the charge on the electron, C, x is the distance of the point from the cathode, m and V is the voltage at the given point, volts.

The current density is related to the charge density by:

$$J = \rho u \tag{11.26}$$

where u is the electron velocity, ms^{-1}.

In this case, the electron velocity can be obtained by equating the electron kinetic energy and the electrical potential energy.

This gives a relation between V and ρ which can be used in the finite-element iterations until a consistent voltage and charge distribution is found:

$$\rho = -J\left(\frac{m}{2eV}\right)^{1/2} \tag{11.27}$$

For other space charge problems, the physical relationship between voltage and charge density may be different and alternative formulations may be required such as ray tracing simulation or surface charge modelling – Hanke *et al.* [8].

11.4.7   Time variation

When the applied voltage cannot be represented by a sinusoid, for example when transient or containing harmonics, it is necessary to use a time domain solution to time step through the waveform. At each time step, an iterative method is used to reach the correct conductivity for the electric stress distribution based on the non-linear characteristics of the material.

The finite-element equation for the voltage can be written in terms of the stiffness matrix $[S]$, the nodal voltages $\{V\}$, and the material properties as:

$$\sigma[S]\{V\} + \varepsilon[S]\frac{d\{V\}}{dt} = 0 \tag{11.28}$$

The time derivative can be approximated by:

$$\frac{\partial V}{\partial t} = \frac{V_{new} - V_{old}}{\Delta t} \tag{11.29}$$

Then, evaluating the voltage at the mid-point of the time step gives:

$$\sigma[S]^{1/2}(\{V_{new}\} + \{V_{old}\}) + \frac{\varepsilon[S](\{V_{new}\} - \{V_{old}\})}{\Delta t} = 0 \tag{11.30}$$

At a given time step, V_{new} is unknown while V_{old} is known from the previous step. Thus, the finite-element equation to be solved becomes:

$$[S]\left(\frac{\sigma}{2} + \frac{\varepsilon}{\Delta t}\right)\{V_{new}\} + [S]\left(\frac{\sigma}{2} - \frac{\varepsilon}{\Delta t}\right)\{V_{old}\} = 0 \tag{11.31}$$

11.4.8   Open boundary problems

Of significant concern in most electrical field analysis is the representation of the infinity of free space in which many devices are assumed to sit. If free space is discretised along with the rest of the model, there is a question of where the mesh should be truncated. The simple, but unhelpful, answer is that the region should be truncated at a point where placing it any further away from the sources has no effect on the solution. Clearly, this involves at least two solutions to the problem.

To avoid this, there have been quite a number of methods proposed that may be characterised as:

(i)   using elements with special properties
(ii)   applying special boundary conditions.

In the former category come so-called ballooning elements and infinite elements. These methods often require boundaries of predetermined shapes or discretisation, which may be restrictive. In the latter come hybrid finite-element/boundary-element methods (requiring special post-processing techniques) and, more recently, the charge iteration method of Aiello *et al.* [9]. This last method is showing great promise, as the boundaries of the mesh can be of arbitrary shape and placed quite close to the sources of excitation. It is even possible to have some parts of the mesh physically disconnected from other parts of the mesh, linked only through the special boundary conditions.

11.5 Applications

11.5.1 General

Numerous problems arise in electrical engineering due to the presence of high voltages. These range from insulation failure in large transformers and high voltage generators through contamination effects on bushings and overhead lines to potential health hazards from power lines and cables. It is not possible to illustrate all such problems but the following examples attempt to show the versatility of the finite-element method and how the variants of the functional are used to good effect in solving unusual problems.

11.5.2 High voltage transmission line

There has been controversy recently concerning the effect that electric and magnetic fields, emitted by high voltage overhead lines, might have on human beings both psychologically and physically. Thus, there is a need to determine the magnitude and distribution of these fields and if necessary to reduce them. Previously, this type of investigation would be done experimentally, requiring a long length (500 m to 1000 m) of overhead line to be constructed and powered, and electromagnetic field measurements to be made. To obtain results from such a system is time consuming and expensive. The study below shows how this type of investigation can be done using the finite-element method.

The problem to be solved is to determine the electric field distribution due to a set of three-phase high voltage cables as shown in Figure 11.9. The position of the cables is taken midway between the pylons allowing for the catenary effect due to the cable weights. The arrangement considered is a 440 kV, twin three-phase overhead system. (The complete cable arrangement between pylons can be modelled but would require a three-dimensional solution that was considered unnecessary for the problem under investigation.)

Figures 11.10 and 11.11 show the horizontal and vertical components, respectively, of the electrical field distributions. The distribution shows clearly how the cables act like aerials in emitting the field. By changing the phase sequence on one side of the cable layout, as shown in Figure 11.12, a different distribution can be obtained. Thus, the phase sequence can be selected to minimise the field at ground level as shown in Figure 11.13.

Figure 11.13 indicates that a person or dwelling 45 m from the centre of the tower (50 m along axis) will be exposed to a lower electric field with phase sequence *b* than phase sequence *a*.

11.5.3 Foils in high voltage bushings

Figure 11.7 illustrated a conical bushing in which it was required to reduce the stress at the junction of the insulation and conductor. Due to manufacturing tolerances, it is almost inevitable that a void is created at such junctions between the busbar and the insulation resulting in a high stress region with subsequent breakdown. To reduce

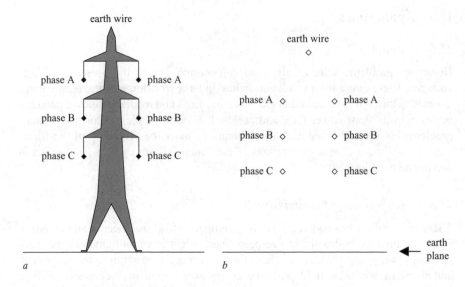

Figure 11.9   High voltage overhead cable arrangement
a   outline of pylon
b   position midway between pylons

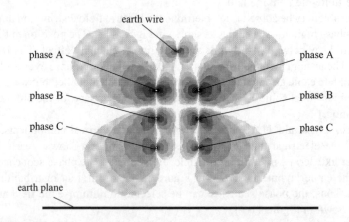

Figure 11.10   Variation of the horizontal component of electric stress

the high stresses in the void region a foil can be strategically placed as shown in Figure 11.14 which also compares the electrical stresses for the cases without and with foils.

However, care is needed in the placement of the foil since higher stresses can, as shown also in Figure 11.14, occur at the ends of the foil. Normally, these higher stresses are situated in a region that can withstand the higher values than in an air void.

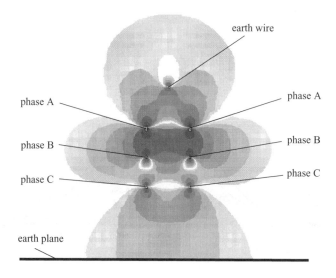

Figure 11.11   Variation of the vertical component of electric stress

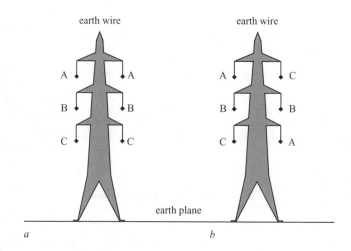

Figure 11.12   Phase sequence layouts

11.5.4   Modelling the effect of contamination on an insulating system

Many insulating materials are used in a polluted environment; the sheds of high voltage bushings, aerial optical fibres strung from power pylons and the end winding conductors of some motors are just three examples. Atmospheric pollution settles on the surface of the insulator in a thin layer. It contains a proportion of debris that, although of very high resistivity when dry, can become partially conducting when damp.

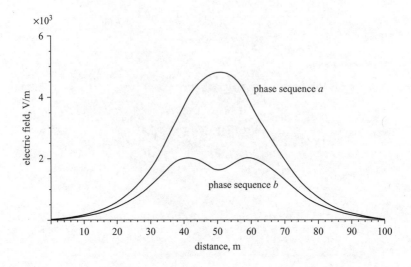

Figure 11.13   Variation of the total electrical stress at ground level

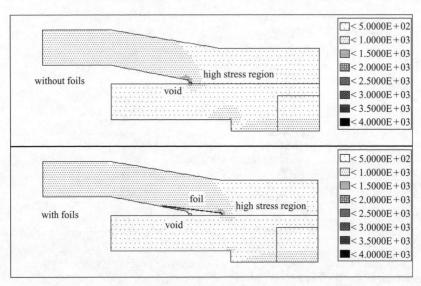

Figure 11.14   Comparison of the electric stress distributions

This thin, partially conducting layer at the surface of an insulator causes the voltage and stress distribution at the surface to change substantially from the dry (and as-designed) state. This is particularly true if the partially conducting layer is absent in one location for some reason, for instance:

(i)   the pollution layer has dried out locally due to a local source of heat (external or arcing)

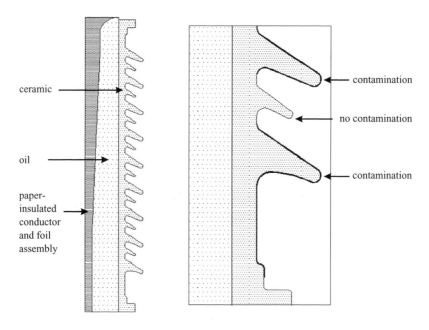

ceramic

oil

paper-
insulated
conductor
and foil
assembly

contamination

no contamination

contamination

Figure 11.15   HV bushing with partial contamination on the outer sheds

(ii)   the pollution layer has been removed – examples include an incomplete manual cleaning of the sheds of a bushing or birds landing on aerial optical fibres, wiping them clean with their claws

(iii)   variable exposure of the insulating system to the source of pollution.

A major difficulty for the modeller is finding an appropriate value to use for the electrical conductivity, although many problems are not strongly sensitive to the value chosen. If critical, studies will have to be made covering the range of possible values.

Figure 11.15 shows a section through a high voltage bushing in which foils are used to create a uniform stress distribution in the insulation.

Three studies were performed:

(i)   with all the sheds free of pollution – the as-designed state

(ii)   with the surface of the sheds covered in a 1 mm thick layer of partially conducting pollution

(iii) as (ii), but with one of the sheds free from pollution (having been cleaned, perhaps).

Modelling partially conducting layers requires the use of an electrodynamic (or even time-stepping) solver that can handle variations in both permittivity and electrical conductivity. For this problem, a frequency domain (i.e. electrodynamic) formulation was used.

The partially conducting layer of pollution was assigned an electrical resistivity of 500 Ωm. Of principal interest is the maximum stress suffered by the oil, occurring at the interface between the oil and the paper insulation.

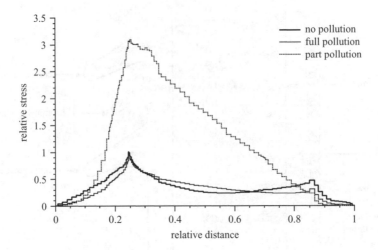

Figure 11.16   Variation of stress along the paper/oil interface

Figure 11.16 shows the stress variation along this surface, from bottom to top of the bushing. The scale has been normalised so as to have a maximum stress value of 1 for the as-designed case. The graph demonstrates that the effect of the overall pollution layer is small, but that cleaning one shed near the bottom of the bushing can increase the maximum stress in the oil by a factor of three.

The maximum stress will be dependent upon the electrical resistivity chosen for the pollution layer. This is rarely known accurately and, in any case, will vary over time with changes in atmospheric and environmental conditions. In these circumstances, it is appropriate to perform a sensitivity analysis, running through a range of possible values of resistivity. With modern computing power, this is a trivial task and gives increased confidence in the results obtained and the conclusions drawn.

Figure 11.17 shows the results of such an analysis. The maximum stress at the oil/air interface (at about 0.25 of the device height, Figure 11.16) has been plotted against the resistivity. For convenience, the stress has been normalised by division by the maximum stress for the non-polluted, as-designed case. As can be seen, the solution is relatively independent of the resistivity for values less than 5000 Ωm. Above this value, the maximum relative stress falls rapidly to a point where, beyond 100 000 Ωm, it is having negligible effect on the stress.

11.5.5   Stress grading of high voltage windings

11.5.5.1   General

As machine end windings leave the stator core their surface can be subject to high voltage gradients. The high electric stresses arising here can lead to sparking and erosion. To alleviate the stresses, special stress grading materials are applied whose resistivity varies with the applied stress. Since the tape is very thin in comparison with other dimensions, a surface representation can be implemented. These materials have

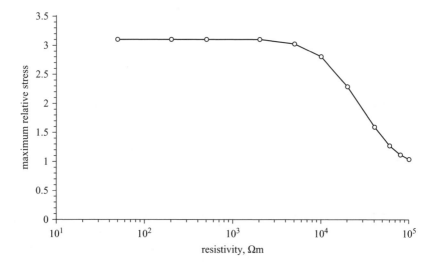

Figure 11.17   Variation of maximum stress with resistivity

been modelled both in the frequency and time domains as an aid to stress grading system design.

11.5.5.2 Non-linearity

The stress grading materials have resistivity that varies with the applied electric stress in a highly non-linear manner. Experimental measurements (Figure 11.18) have shown that the resistivity varies very rapidly with the stress so that a simple linear interpolation representation of the curve is not appropriate.

Examination of the experimental data has indicated an exponential variation of the resistivity with the $\frac{2}{3}$ power of the electric stress, Gully and Wheeler [10]:

$$\rho_s = K \exp\{-nE^{2/3}\} \tag{11.32}$$

where ρ_s is the surface resistivity, ohm/□ (ohm per square unit of material surface area[1]), E is the surface electrical stress, kV/cm, and K and n are constants derived from measurements. K is the resistivity constant in ohms and n is the exponential constant. K may be thought of as the theoretical resistivity at zero stress and n represents the degree of non-linearity the material possesses. A high conductivity tape is denoted by a relatively low value of the resistivity constant, K, and a high value of the exponential constant, n. As tape conductivity decreases, K increases and n decreases

In the finite-element solution, an under-relaxation process is used to reach the appropriate resistivity for the electric stress. At each iteration, the peak resultant electric stress is calculated in each element. This value is used to derive a new resistivity using Equation (11.32) and the provided parameters K and n.

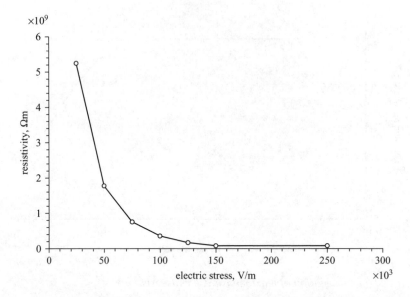

Figure 11.18   Variation of resistivity with electric stress in a stress grading material (based on the paper 'Finite element modelling of non-linear stress grading materials for machine end windings', by A.E. Baker, A.M. Gully and J.C.G. Wheeler; as published on pp. 265–268 of the IEE Conference Publication (no. 487) on the International Conference of Power Electronics, Machines and Drives in Bath, 16–18th April 2002)

The values of K and n must be measured for each stress grading tape under investigation. In addition, the properties of the tape are known to vary with ageing and temperature. At high temperatures, the n values of stress grading tape tend to decrease and the tape's effectiveness reduces. An n value of zero means that the tape has lost its non-linear properties and is only resistive.

11.5.5.3   Experimental comparison

An experimental test bar has been used as in Figure 11.19. The copper bar has a rectangular cross section, and is wrapped with a 3 mm thick layer of insulation and covered with an outer, 0.2 mm thick, layer of stress grading tape.

A voltage was applied to the bar, varying sinusoidally at 50 Hz. The peak voltage was measured at points along the tape surface. The external surface of the insulation at the centre of the bar is earthed to simulate the earthing of the stator core in a machine. The test measurements for a typical bar are compared with the finite-element prediction for the bar with and without stress grading tape in Figure 11.20. Note how the voltage gradient is much reduced by the presence of the tape. For this test, the parameters for the tape were $K = 2.8 \times 10^{10}\,\Omega$ and $n = 4.3$.

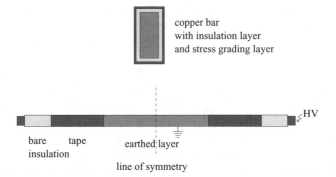

Figure 11.19 *Experimental bar (based on the paper 'Finite element modelling of non-linear stress grading materials for machine end windings', by A.E. Baker, A.M. Gully and J.C.G. Wheeler; as published on pp. 265–268 of the IEE Conference Publication (no. 487) on the International Conference of Power Electronics, Machines and Drives in Bath, 16–18th April 2002)*

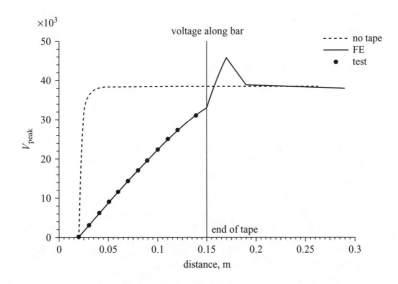

Figure 11.20 *Comparison with test (based on the paper 'Finite element modelling of non-linear stress grading materials for machine end windings', by A.E. Baker, A.M. Gully and J.C.G. Wheeler; as published on pp. 265–268 of the IEE Conference Publication (no. 487) on the International Conference of Power Electronics, Machines and Drives in Bath, 16–18th April 2002)*

A frequency domain solution was used, where a single value of conductivity applies through the time cycle (section 11.4.4).

For the case shown in Figure 11.19, there is also a layer of stress grading paint applied after the partially conducting tape. The graph shows that for this particular bar the taped portion was too short, as there is a sharp rise in the stress at the interface of the tape and paint. The high impedance probe used to measure the voltage along the bar does not give reliable results on the sample bar in the paint region because it is not of sufficiently high impedance. This is a common problem, particularly as higher voltages are applied to the bar to simulate overvoltage tests rather than normal service conditions. The impedance of the probe dominates, resulting in inaccurate measurement. The finite-element method clearly has no such difficulties and the stress can be predicted along the whole length of the bar. The inaccuracy of measurements is exacerbated further when non-sinusoidal voltage waveforms are applied to the bar.

11.5.5.4   Effect of parameters for stress grading materials

In selecting the best tape for an application, it is useful to look at the variation of stress with the parameters K and n. Figure 11.21 shows the variation with voltage along a bar for different values of K and with n fixed. Figure 11.22 shows similar curves for n varying and with K fixed.

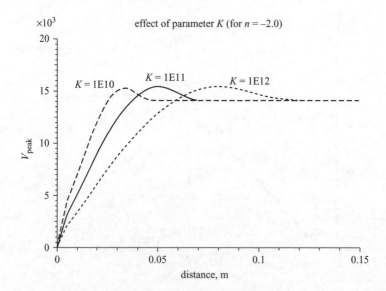

Figure 11.21   Variation of voltage with K (based on the paper 'Finite element modelling of non-linear stress grading materials for machine end windings', by A.E. Baker, A.M. Gully and J.C.G. Wheeler; as published on pp. 265–268 of the IEE Conference Publication (no. 487) on the International Conference of Power Electronics, Machines and Drives in Bath, 16–18th April 2002)

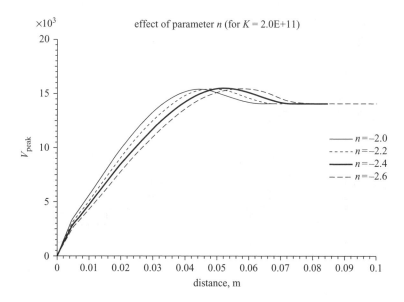

Figure 11.22   *Variation of voltage with n (based on the paper 'Finite element modelling of non-linear stress grading materials for machine end windings', by A.E. Baker, A.M. Gully and J.C.G. Wheeler; as published on pp. 265–268 of the IEE Conference Publication (no. 487) on the International Conference of Power Electronics, Machines and Drives in Bath, 16–18th April 2002)*

In some cases, a particular stress grading system may give a good stress distribution but at the expense of increased heating in the tape. The finite-element model can provide the current density in the tape by applying Equation (11.19) as a post-processing function. The loss distribution can then be calculated using:

$$\text{loss density} = \frac{J^2}{\sigma} \ \text{W/m}^3 \tag{11.33}$$

The loss density values could be used as input to a thermal finite-element model to calculate temperature rises. However, the loss density distribution alone gives a good indication of hot spots on the bar and can be used to compare different stress grading tape behaviour. Figure 11.23 shows the loss density for two tapes with different K and n values. Note that the lower value of K denotes higher conductivity and hence more loss.

11.5.5.5   Example of time-stepping solution

Using the time-stepping technique, a sinusoidal voltage was applied to a copper bar as in Figure 11.19. Figure 11.24 shows the variation of voltage with time for a point in the middle of the stress-grading layer. The applied voltage was a pure sine wave but the voltage at the tape clearly contains other harmonic components. Figure 11.25 shows

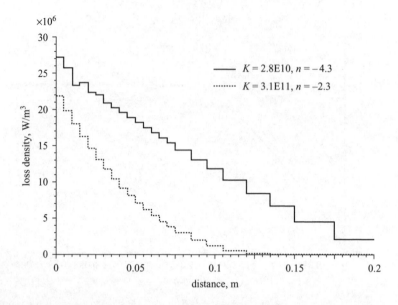

Figure 11.23   *Effect of K on loss density (based on the paper 'Finite element modelling of non-linear stress grading materials for machine end windings', by A.E. Baker, A.M. Gully and J.C.G. Wheeler; as published on pp. 265–268 of the IEE Conference Publication (no. 487) on the International Conference of Power Electronics, Machines and Drives in Bath, 16–18th April 2002)*

the time derivative of voltage at the same point. The shape of this curve compares well with measured results such as those given in Wood and Hogg [11].

These curves illustrate the effects of the tape non-linearity on the distribution of the voltage and its derivative (which is related to current). The frequency domain model would have forced both of these curves to be pure sine waves like the applied voltage. When the applied voltage is non-sinusoidal as in the PWM waveforms of inverter fed machines, the need for a time domain solver is increased.

11.6   The choice of the order of the finite-element approximation

11.6.1   General

In the simplest form of the finite-element approximation to the electric field distribution, the potential varies linearly over an element. This implies that the electric stress is constant within the element volume. Such elements are known as linear, or first-order, finite elements. It is perfectly possible, however, to assume a higher-order variation (quadratic, cubic, etc.) of potential and electric stress[2].

Higher-order elements are usually implemented by the introduction to the element definition of additional nodes, located on the sides, faces or within the

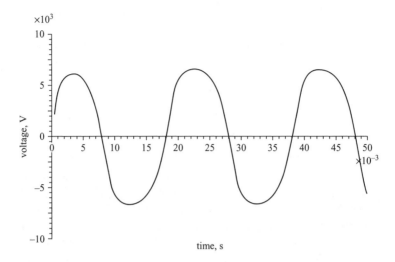

Figure 11.24   *Variation of tape voltage with time (based on the paper 'Finite element modelling of non-linear stress grading materials for machine end windings', by A.E. Baker, A.M. Gully and J.C.G. Wheeler; as published on pp. 265–268 of the IEE Conference Publication (no. 487) on the International Conference of Power Electronics, Machines and Drives in Bath, 16–18th April 2002)*

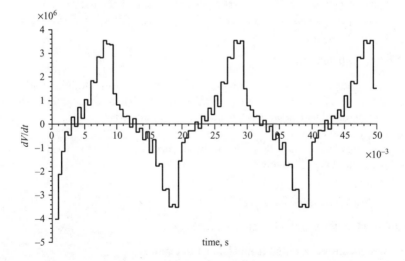

Figure 11.25   *Variation of voltage time derivative (based on the paper 'Finite element modelling of non-linear stress grading materials for machine end windings', by A.E. Baker, A.M. Gully and J.C.G. Wheeler; as published on pp. 265–268 of the IEE Conference Publication (no. 487) on the International Conference of Power Electronics, Machines and Drives in Bath, 16–18th April 2002)*

element volume. With higher-order elements, the locations of the additional nodes can be chosen to allow curved element edges. These are called isoparametric elements and permit the modelling of curved surfaces without introducing the abrupt changes in shape that arise when the surface is modelled by a series of straight lines (flat patches in three dimensions).

So, which should be chosen?

11.6.2   First-order elements

The main advantages of first-order elements are:

(i)   Simplicity in mesh generation: nodes only occur at element vertices; conversely, elements are formed by a simple interconnection of nodes.

(ii)   Simplicity in solution formulation: this is particularly true where non-linear material properties are concerned (i.e. permittivities or electrical conductivities that depend upon the value of the electric field). With first-order elements, the electric field is constant in an element, so the permittivity and conductivity are single-valued in each element.

(iii)   Speed of construction of the finite-element equations: since the element's material properties are constant within the element, simple, analytical expressions can be written for the element's contribution to the set of finite-element equations.

(iv)   Simplicity in post-processing: displaying a single value of electric stress in each element is a simple and rapid process. Plotting two- and three-dimensional graphs of the stress variation in space is similarly a simple process of identifying the elements involved.

And their disadvantages are:

(i)   The boundaries between regions are formed from straight lines (two-dimensional) or flat patches (three-dimensional). There are geometric discontinuities distributed all across the model of a curved surface, each one introducing an artificial stress concentration.

(ii)   The stress varies in a step-wise fashion from one element to the next, i.e. the stress field is not smooth.

11.6.3   Higher-order elements

The main advantages of higher elements are

(i)   smooth stress variation from element to element
(ii)   smooth stress variation at curved surfaces
(iii)   smaller number of elements needed to model the field variation.

And the disadvantages are:

(i)   more complex mesh construction and post-processing
(ii)   slower construction of the finite-element equations

(iii) a numerical integration involving the variation of the material properties over the element is required for each element

(iv) slower solution – there are more nodes per element.

In simple theory, the higher-order elements will give a solution, to any given level of accuracy, using fewer elements than a linear solution. The complexity of practical problems, however, leads to many nodes being required simply to describe the geometry of the objects. The result is that a higher-order solution often takes longer (mesh generation + solution + post-process) than the linear one.

For general field evaluation, the memory capacity and processor speed of modern computers allows a good solution to be obtained using a large number of first-order elements. Higher-order elements rarely give a significant improvement in solution for the same total computing time. However, where the fields are changing rapidly in space and need to be calculated to a high degree of accuracy and smoothness, higher-order elements may be appropriate. It should be remembered, however, that knowledge of the fields *per se* is often not the ultimate purpose of the investigation, as discussed in the next section.

11.7 Assessment of electrical stress distribution

11.7.1 General

It is rare, in power engineering, that the magnitude of the voltage or the electrical stress distribution is the ultimate aim of the analysis. What is usually required is an assessment of whether or not there will be electrical breakdown of the insulation system. Unfortunately, relating computed electrical stress distribution to breakdown is far from simple, for two reasons:

(i) mathematical singularities in the numerical solution can result in very high, localised stresses being calculated

(ii) there is no simple relationship between maximum stress and breakdown.

11.7.2 Mathematical singularities

Where there is an abrupt change in the orientation of a surface there will be a change in the electrical stress, even if the surface is an equipotential one. As a general rule of thumb, the finite-element mesh should be concentrated in the regions where the fields are changing most rapidly. It would be natural, therefore, to place smaller elements in the vicinity of a sharp corner.

It is often found, however, that as the element size gets smaller and smaller, so the computed peak stress gets higher and higher although the stored electrical energy for the problem shows no such trend. This reflects an attempt to capture a form of mathematical singularity and the maximum value of the electrical stress appears to be determined by the discretisation, rather than by the excitation.

One solution to this problem (if the value of the maximum stress shows no sign of reaching a limit as the discretisation is increased in the vicinity) is to assess the average stress in the local region. This average is formed using a volume weighting, so reducing the effect of small, high stress regions. A more sophisticated average is formed using the volume-weighted root mean squared stress. This is related to the mean electrical stored energy in the region.

11.7.3   Relationship between stress and breakdown

The physics of electrical breakdown is complex, and most numerical models do not consider such phenomena as corona discharge – it would be an unwarranted complication in the face of other modelling approximations. Nevertheless, the designer must make a judgement about whether a given stress is going to lead to insulation failure.

The first, and most useful, tool for this is experience. Given a restricted class of electrical equipment, construction methods and insulation systems and given experience of a certain number of failures(!), the engineer can build a base of experience relating calculated stress to observed breakdown (or lack of it). From this, design rules-of-thumb may be derived.

Other, more technically-based, assessments may also be made. One such concept is the stressed volume – the volume of the insulation system that is experiencing electrical stress above a nominated level. This volume has been related to breakdown and is associated with the mean electrical stored energy, as discussed above.

11.8   Pre and post processor developments

11.8.1   General

The most sophisticated field solver is of limited use if it is difficult and time consuming to enter data and to examine the results. Considerable effort has been expended in improving the pre processing and the post processing stages and in integrating these with the solver to permit automatic design optimisation.

11.8.2   Description of the problem geometry

Often, there will be a description of the problem geometry already in existence in a CAD package. Most numerical analysis packages will interface directly with the most popular CAD file formats, but the process is not totally straightforward.

The drawing is not only a description of the geometry: it also contains dimensions, annotation and other parts not relevant to the analysis. These all need to be removed.

A drawing does not necessarily describe a series of closed geometric shapes. Lines that should meet exactly may, on the drawing, fall short of each other (and leave a gap) or cross each other. Code is needed to cope with these cases.

A recent development is to make a parametric description of the object, i.e. in terms of algebraic variables such as radius, width and angle rather than 32.1 mm or 20°. The engineer can then study several designs by changing a few, characteristic design variables rather than redrawing the whole object.

Once a parametric description of the problem is available, it is a short step to varying the parameters in a systematic search for an optimal solution and this area is a major, recent development.

For two-dimensional analysis, two-dimensional drawings are a good starting point. This is less true for three-dimensional problems. For these, it may be most convenient to describe objects in terms of basic, primitive shapes such as spheres, cylinders and rectangular bricks and the intersections between them. Such solid modellers are widely available for mechanical analysis, although not all will cope with the problem of describing the air surrounding the solids.

Extrusion methods for three-dimensional mesh generation from a two-dimensional starting mesh are still widely used. When coupled with sophisticated mesh transformation and joining techniques, extrusion methods can handle most mesh generation problems.

11.8.3   Creation of a discretisation from the problem geometry

Automatic discretisation from the geometry description is routine in two dimensions and possibly in three. Following this automatic process, the engineer should be able to change the discretisation by hand through some form of graphical tool. This is useful for the cases where first analysis shows the discretisation to be inadequate in some local region of the problem, or where a small design change is required (such as the introduction of a small radius between two objects).

An extension of automatic mesh generation is to link it to an error evaluator that makes an assessment of the solution's accuracy in relation to the discretisation used. Areas showing a large solution error can then be rediscretised. By this means, the solution can be performed to a specified degree of accuracy. It should be emphasised, however, that although such a solution is accurate, it is not necessarily correct (the material properties may be wrong, for instance).

11.8.4   Assigning material properties

An important point concerns the existence of accurate and appropriate material data. With more powerful and sophisticated analysis techniques being used to create designs stressed nearer and nearer to material limits, the accuracy of the overall solution can become limited by the accuracy of the material properties used. As a first step, a sensitivity analysis should be performed to find how the solution might be influenced by uncertainty in material properties.

11.8.5   Post processor developments

Post processing of a field solution is the means by which the engineer gets to the object of the analysis. Normally, this involves inspecting the field solution in relation to the problem geometry, so some form of visualisation is required.

Field plots in two- and three-dimensions are now routine. Three-dimensional viewers can rotate the image easily, rapidly and at will to enable all facets of the solution to be examined. Sections may be sliced from the three-dimensional volume and examined either separately or in the context of the full three-dimensional model.

There is still a conceptual difficulty in viewing a field in a volume of space, rather than on a surface, but this would still be the case even if a stereo view or a virtual reality image were used (both are technically possible). One approach is to attempt to draw equivalued surfaces as a series of semitransparent shells.

For two-dimensional slices, a whole range of analysis is on offer and easy export of the results to other formats (particularly popular PC-based packages such as MatLab, Mathcad and Microsoft Word and Excel) is required.

11.8.6   *Design optimisation*

The process of design optimisation starts with the analysis of an initial design. This solution is post processed to see how the initial design performs in terms of the target for improvement and to check whether any constraints on the performance or dimensions have been violated. Typical values that might be used to characterise a design are the capacitance between an electrode and earth or the maximum electrical stress in a region or at a surface.

The results of this post processing operation are fed into one of many algorithms to attempt to generate a better design and establish an iterative loop, leading to an optimal design.

11.9   Notes

[1] Surface resistivity is defined as the resistance between opposite edges of a square area of the surface of the material. If the area of the square is increased, then the change in the measured resistance due to the increased distance between the electrodes will be cancelled exactly by the reduction in resistance due to the increased width of the electrodes. Surface resistivity is, thus, independent of the area used for its measurement [12]

[2] The order of the stress variation is always one less than the order of the potential variation, e.g. quadratic (second-order) potentials give linear (first-order) stresses

11.10   References

1   RIVENC, J.P., and LEBEY. T.: 'An overview of electrical properties for stress grading optimization', *IEEE Trans. Dielectr. Electr. Insul.*, 1999, **6**, (3), pp. 309–318

2   ROBERTS, A.: 'Stress grading for high voltage motor and generator coils', *IEEE Electr. Insul. Mag.*, 1995, **11**, (4), pp. 26–31

3   SOUTHWELL, R.V.: 'Relaxation methods in theoretical physics' (OUP, Oxford, 1946)

4   ZIENKIEWICZ, O.C., and TAYLOR, R.I.: 'The finite element method, vols 1 & 2' (McGraw-Hill, Maidenhead, 1991, 4th edn.)

5   CHARI, M.V.K., and SALON, S.J.: 'Numerical methods in electromagnetism' (Academic Press, 1999)

6   FRANKLIN, P.: 'Methods of advanced calculus' (McGraw-Hill Book Company Inc., New York and London, 1944, 1st edn.)

7   REECE, A.B.J., and PRESTON, T.W.: 'Finite element methods in electrical power engineering' (Oxford University Press, 2000)

8   HANKE, K., HEISING, S., PROBERT, G., and SCRIVENS, R.: 'Comparison of simulation codes for the beam dynamics of low-energy ions'. Presented at ICIS2001, Oakland, California, USA, 3–7 September 2001

9   AIELLO, G., ALFONZETTI, S., COCO, S., and SALERNO, N.: 'Placement of the fictitious boundary in the charge iteration procedure for unbounded electrical field problems', *IEEE Trans. Magn.*, 1995, **31**, (3), pp. 1392–1395

10 GULLY, A.M., and WHEELER, J.C.G.: 'The performance of aged stress grading materials for use in electrical machines'. Proceedings of IEE Conference on *Dielectric materials, measurements and applications*, No. 473, 2000, pp. 392–396

11 WOOD, J.W., and HOGG, W.K.: 'Endwinding stress in large generator stators'. International Seminar on *High voltage*, 1989

12 GOLDING E.W., and WIDDIS, F.C.: 'Electrical measurements and measuring instruments' (Pitman Press, Bath, 1968, 5th edn.)

Chapter 12

Optical measurements and monitoring in high voltage environments

G.R. Jones, J.W. Spencer and J. Yan

12.1   Introduction

Optical techniques are attractive for measurements and monitoring under high voltage conditions for several reasons. They enable measurements to be made remotely via free space so providing a high degree of geometric isolation of the measuring equipment from the high voltage environment. They also provide a means via optical fibre transmission for penetrating into high voltage enclosures for monitoring purposes with a high degree of inherent electrical insulation.

Phenomena associated with high voltage conditions such as electrical discharges (corona, sparks, arcs etc.) are conveniently addressed since the latter produce both optical emissions themselves, generate changes in the optical properties of the surrounding media (e.g. pressure and temperature-induced refractive index changes etc.) and often produce optically identifiable byproducts such as microparticulates. On the other hand, high voltage equipment needs monitoring not only via the measurement of electrical parameters such as voltage and current but also thermal (e.g. temperature etc.), mechanical (e.g. vibration, linkage movement, catenary swings etc.) and chemical (e.g. degradation of insulating media etc.) parameters, all of which in principle can be optically addressed. The optical techniques are also immune to electromagnetic interference (EMI) (which can be copiously produced by and via high voltage systems) so leading to more reliable and accurate monitoring means.

By way of an example Figure 12.1 illustrates these two major facets of high voltage related optical measurements and monitoring. The free space aspect corresponds to photographic recording of arcing flashover on the overhead high voltage power lines of electrified railways. Such discharges are erratic (Figure 12.1*a*) and potentially destructive in nature, requiring if they occur proper control and extinction.

Figure 12.1

    a Free space monitoring; photographic record of an arc flashover on an overhead high voltage railway power line as an example of free space monitoring

    b Optical-fibre-based-monitoring; hybrid optical fibre current transformer mounted on the high voltage end of a line side transformer for monitoring transient currents produced by discharges of the forms shown in *b*

Their properties and behaviour need to be properly understood and free space optical techniques are well suited for investigating these.

    The use of optical fibre sensing is illustrated by a hybrid optical current transformer (HOCT) which was mounted on the high voltage side of the line side transformer of the Channel Tunnel Rail Link shown on Figure 12.1*b*. This enabled transient currents flowing during the occurrence of flashover of the type shown on the free space photograph of Figure 12.1*a* to be monitored at locations where conventional electric current measurement techniques could not be used because of high voltage insulation limitations and electromagnetic interference.

Consideration is given in this chapter to some fundamental principles of optics directly relevant to measurement and monitoring under high voltage conditions. There is a description of various optical measurement techniques both in free space and via optical fibre systems. Finally, some examples are given of typical experimental and test results obtained with various optical systems.

12.2   Fundamental optical principles

12.2.1   Introduction

Traditionally, radiation from high voltage discharges has been investigated remotely via free space transmission at various levels of sophistication ranging from simply the spatial extent of the emitting plasma (using high speed photography), to detailed spectral analysis (using conventional spectroscopy). The interaction of the discharge plasma and its surroundings with probing light beams via scattering processes has been used to provide localised property information such as gas density variations.

When such discharges are enclosed within high voltage equipment (e.g. circuit breakers, gas insulated busbars) access for such optical measurements may be gained with optical fibres. Such optical fibres may also be used in conjunction with appropriate optical sensing elements for monitoring the condition of high voltage equipment via the measurement of electrical, mechanical and thermal parameters of the device itself.

A brief overview is therefore given of the fundamental principles which govern free space and fibre transmission optics as well as relevant facets of spectroscopy, light scattering and optical signal modulation.

12.2.2   Optical intensity

Two-dimensional optical intensity images provide quantitative information about the location, size and shape of various electrical discharges. Such information is important for tracking the formation time variation and decay of such discharges in relation to the operation of high voltage equipment e.g. insulator flashovers (Figure 12.1*a*), arcing in circuit breakers etc. (Figure 12.2). Alternatively, the emission from a narrow section across a discharge may be observed continuously in time as a streak photograph.

12.2.3   Spectroscopy

Spectroscopy involves resolving an optical signal into its different wavelength components so that the wavelength-dependent structure becomes known. A knowledge of such wavelength detailed structure of light from electrical discharges can provide

parasitic arcing   optical fibre arc monitor

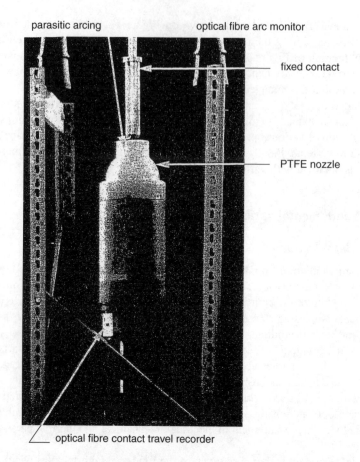

fixed contact

PTFE nozzle

optical fibre contact travel recorder

*Figure 12.2   High speed framing photograph of electrical breakdown outside the
PTFE nozzle of a high voltage interrupter unit*

information concerning the spatial and time variation in these discharges of the concentration and temperature of various ionic species and the density of electrons [1–3].
These are the fundamental properties which govern the nature and behaviour of the
discharge.

The use of such techniques is facilitated if the discharge plasma is in local
thermodynamic equilibrium (temperature and composition of species in equilibrium),
a condition which is not always satisfied in constricted discharges (e.g. lamps and wall
stabilised arcs etc.) or possibly in rapid time varying discharges. Spectral analysis
is also complicated by the presence of turbulence and discharge instabilities, which
may well exist in discharges occurring naturally and in high voltage equipment.

Consideration also needs to be made regarding the validity of other assumptions
[1] such as the path length over which the emitted radiation is transmitted before
being reabsorbed (optical depth). Other constraints are indicated in sections 12.2.3.1
to 12.2.3.3.

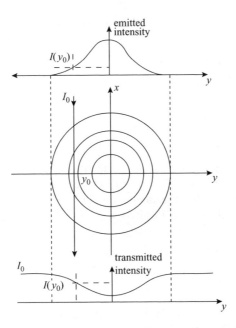

Figure 12.3   *Abel inversion of measured radial intensity variations across a cylindrical electrical discharge (emitted and transmitted intensities shown) (courtesy of G.R. Jones, 'High pressure arcs in industrial devices: diagnostic and monitoring techniques', published by Cambridge University Press, 1988)*

Spectral measurements involve monitoring the intensity of radiation at a given wavelength when viewed along a chord ($y = y_0$) through a cylindrical electrical discharge (Figure 12.3). This radiation may be either emitted by the discharge, as in emission spectroscopy, or from an external beam passed through and absorbed to some degree by the discharge, as in absorption spectroscopy. In both cases, the change in intensity along the chord is governed by [1].

$$\frac{dI}{dx} = e(x) - k(x)I(x) \tag{12.1}$$

where I is the intensity of the radiation propagating in the x direction, $e(x)$ is the spontaneous emission coefficient and $k(x)$ the absorption coefficient, all parameters being referred to the coordinate (x, y_0). With emission spectroscopy, the emission coefficient $e(x)$ is to be determined, but with absorption spectroscopy the absorption coefficient $k(x)$ is to be measured.

Since the intensity measured along any chord y under conditions of cylindrical symmetry (Figure 12.3) has contribution from discharge plasma elements at different radii, it is necessary to deconvolve the intensity data in order to obtain true radial profiles of emission or absorption coefficients. A number of deconvolution techniques based upon a technique called Abel inversion are available [4]. These yield an expression for the emission coefficient for an optically thin plasma at a radius r of the form:

$$e(r) = -\frac{1}{\pi} \int_{r}^{r_0} \frac{\{d[I(y)]/dy\}}{(y^2 - r^2)^{1/2}} dy \qquad (12.2)$$

and a similar expression for $k(r)$.

12.2.3.1   Temperature

The intensities of spectral lines are strong functions of temperature, some examples of which are given for an SF_6 and copper discharge on Figure 12.4 [1].

Of the many methods for determining discharge temperatures from the intensity of such spectral lines, the relative line intensity method offers several advantages. The ratio of the intensities of two lines (e_{Lmn1}, $/e_{Lmn2}$) from the same species depends upon temperature (T) according to [1]:

$$\frac{e_{Lmn1}}{e_{Lmn2}} = \frac{A_{mn1} g_{m1} \lambda_{mn2}}{A_{mn2} g_{m2} \lambda_{mn1}} \exp\left[-\frac{E_{m1} - E_{m2}}{KT}\right] \qquad (12.3)$$

where the suffices 1 and 2 refer to the two spectral lines, g_m is the probability that the atom will be in state m with energy E_m, A_{mn} is the transition probability of spontaneous relaxation from level m to n, λ is the wavelength of the emitted radiation, k is Boltzmann's constant.

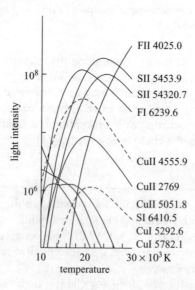

Figure 12.4   Variation of selected S, F and Cu line intensities with temperature for an SF_6–Cu plasma (courtesy of G.R. Jones, 'High pressure arcs in industrial devices: diagnostic and monitoring techniques', published by Cambridge University Press, 1988)

An alternative scheme for determining temperature is by using the intensity ratio of ion to neutral species for a given element or the ion of one element and the neutral of another element.

The presence of continuum radiation complicates the measurement of line intensity and must be subtracted from the total measured intensity. The use of a narrow spectral line is therefore to be preferred since the recorded continuum radiation is a less significant fraction of the total measured intensity.

12.2.3.2 Electron density

Electron densities in the plasma columns of electrical discharges such as high pressure arcs have been mainly determined from measurements of the broadening of particular spectral lines [5–7]. However, there are other line broadening mechanisms which may need to be taken into account under particular conditions[1] (see Note 12.7). If these can be successfully taken into account then typically electron densities as low as 2×10^{22} m^{-3} can be measured [1].

12.2.3.3 Species concentration

The concentration of impurities in electrical discharges (e.g. entrained electrode material) may be determined from a comparison of the intensities of the host and impurity spectral lines according to [1]:

$$\frac{N_{0g}(C_g)}{N_{0m}(C_m)} \propto \frac{e_{Lg}(c_g)}{e_{Lm}(c_m)} \tag{12.4}$$

where C_g, C_m are the concentrations of the host and impurity, respectively, N_{0g}, N_{0m} are the number densities of host and impurity, respectively, and e_{Lg}, e_{Lm} are the line emission coefficients of the host and impurity, respectively. An example of the relationship between the percentage copper vapour in an SF$_6$ arc discharge and the intensity ratio of copper (5293 CuI, 5153 CuI) and sulphur lines (5321 SII) is shown in Figure 12.5.

The method requires that the host and impurity species be in local thermal equilibrium and that the temperature be accurately known.

12.2.4 Light scattering

Electrically charged particles are capable of scattering electromagnetic radiation as a result of their interaction with the electric vector of the radiation. For instance, a bound electron executing simple harmonic motion and driven by the oscillating electric vector of an electromagnetic wave, reradiates the electromagnetic energy. This constitutes the phenomenon of light scattering. Two major categories of such scattering may be defined on the basis of the degree of coherence of the scattering process.

12.2.4.1 Incoherent scattering

Incoherent scattering refers to the electromagnetic radiation that may be scattered in a random direction relative to the incident radiation (e.g. by atmospheric particles).

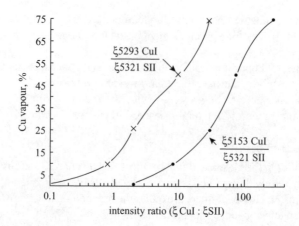

Figure 12.5   Variation of CuI: SII line intensity ratios with percentage of copper vapour impurity in SF_6 (T = 17000 k, P = 2 bar) (courtesy of D.R. Airey, PhD thesis, University of Bath, 1977)

The intensity of the incoherently scattered light is considerably less than that of the incident or coherent light scattered radiation ($\sim 10^{-12}$). Consequently, incoherent light scattering experiments necessitate the use of high powered light sources such as lasers. Incoherent light scattering forms the basis of powerful and versatile diagnostic techniques. For instance, these techniques are capable of yielding highly localised values of temperature and particle density without resorting to complex unfolding procedures [4], which are necessary with emission and absorption spectroscopy.

Provided that a number of conditions are satisfied, the light scattered from a discharge plasma [1] is composed of the sum of two components:

(i)   One component is governed by scattering from free electrons (Thomson scattering). When the light wavelength (λ) is small compared with the Debye length (λ_D), the scattered radiation consists of a spectral line centred on the incident radiation wavelength but broadened by the Doppler motion of the electrons. When (λ/λ_D) is large, the line shape changes into two narrow lines separated by twice the electron plasma frequency symmetrically with respect to the radiation frequency. The temperature and density values encountered in high pressure electric arcs correspond to conditions between these extreme cases. Also, for high density, low temperature arc plasmas the number of particles in a Debye sphere may become small (<5), in which case the scattered spectrum is modified [8–10].

(ii) The second discharge scattered component is governed by the bound electrons of the ions or atoms (Rayleigh scattering), the intensity of the scattered light decreasing as λ^{-4}. The classical Rayleigh scattering from neutral atoms or molecules has been measured for a range of common gases by George *et al.* [11].

When the incident radiation wavelength is of the order or greater than the particle size, the scattering becomes a weaker function of wavelength and is known as Mie scattering [12]. Such scattering may be important from solid particles of size a few microns produced within and around electrical discharges (Figure 12.6). Laser Doppler techniques rely upon detecting light scattered from such microparticles to determine their velocities.

The velocity with which a discharge plasma or surrounding gas flows may be monitored by measuring the Doppler shift of the frequency (wavelength) of light scattered by micron-sized particles moving with the flow. There are advantages in using a differential Doppler system [13], in which the scattered light from two equal intensity coherent beams is mixed in the test section to produce the Doppler signal (Figure 12.7). In this case, the frequency shift is given by:

$$f_s - f_0 = \left(\frac{2v}{\lambda_0}\right) \sin\left(\frac{\theta}{2}\right)$$ (12.5)

where λ_0 is the wavelength of the probing beams, v is the particle velocity and θ is the angle between the two incident beams. Since the frequency shift is therefore independent of the scattering angle (ϕ in Figure 12.7), a large collection solid angle may be used without the occurrence of aperture broadening.

In practice, small micron-sized particles are utilised in the flow to provide a scattering signal of sufficient amplitude for detection. It is necessary to ensure that these particles are in equilibrium with the local flow and adjust their motion sufficiently in rapid flow accelerations or transients.

12.2.4.2 Coherent scattering

Coherent scattering is responsible for the transmission of light through a medium being manifest in the refraction of light. The effect can be described in terms of the polarisability of the medium, which is related to the refractive index according to [14]:

$$n^2 - 1 = \frac{P}{\varepsilon_0 E} = \frac{\sum N_i \alpha_i}{\varepsilon_0} \cong 2\delta$$ (12.6)

where P is the electric polarisation per unit volume, E is the electric field strength, ε_0 is the permittivity of free space, N_i is the concentration of electric dipole constituent i, α_i is its electrical polarisability and δ is the refractivity.

For a fluid of known composition, the mass density and hence N_i varies with gas pressure and temperature leading to corresponding changes in refractive index n. As an example, the variation of refractivity ($n - 1$) with temperature for air and copper vapour is given on Figure 12.8. Consequently, the measurement of refractive index provides a method for mapping changes of mass density in discharge heated and pressurised fluids.

If electronic effects are important then the electron polarisability contribution, α_e, may dominate in Equation (12.6) [14]. If in addition the wavelength of the light (λ) is sufficiently short [15] and with small refractive index differences from unity the

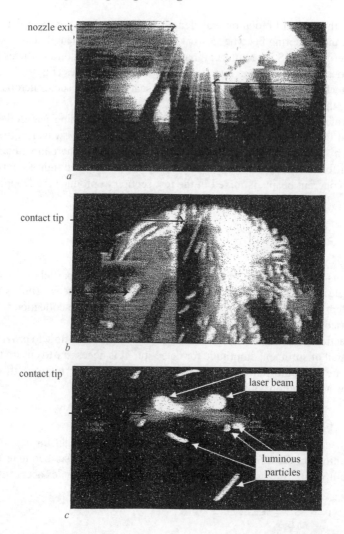

Figure 12.6   High speed video camera frames of particulates formation following high fault current arcing in a circuit breaker

    a N_2 30 kA 30 ms after arc extinction (4 ms exposure)
    b SF_6 30 kA 22 ms after arc extinction
    c SF_6 10 kA 84 ms after arc extinction (luminous particles and laser scattered light from non-luminous particles)

refractive index is given by:

$$n \cong 1 - 4.46 + 10^{-14} N_e \lambda^2 \tag{12.7}$$

where N_e is the electron concentration in cm^{-3}. Equations (12.6) and (12.7) indicate that the electronic refractivity (δ) is of opposite sign to the atomic and molecular

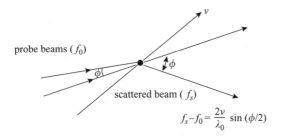

Figure 12.7   *Schematic of light scattering geometries for laser Doppler velocimetry (courtesy of G.R. Jones, 'High pressure arcs in industrial devices: diagnostic and monitoring techniques', published by Cambridge University Press, 1988)*

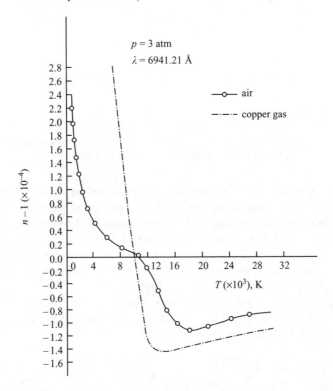

Figure 12.8   *Refractivities of air and copper vapour as functions of temperature (courtesy of F.A. Rodriguez, PhD thesis, University of Liverpool, 1974)*

refractivities. For sufficiently long wavelengths (λ), the refractive index becomes imaginary, implying that longer wavelength radiation is unable to penetrate the electrical discharge. This provides a basis for remotely monitoring the radial extent of an electrical discharge [16] in a confined environment.

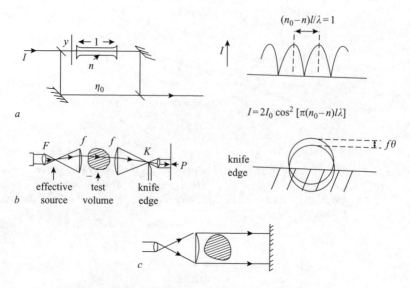

Figure 12.9   Light scattering due to refractive index changes
        a   interferometry
        b   Schlieren
        c   shadowgraph (courtesy of G.R. Jones, 'High pressure arcs
            in industrial devices: diagnostic and monitoring techniques',
            published by Cambridge University Press, 1988)

A light ray passing through a refractive medium may be affected in one of three ways. It may suffer a phase lag governed by a change in wavelength due to the refractive index n; it may be deflected through an angle θ determined by a refractive index gradient (dn/dy) (e.g. mirage formation); it may be displaced laterally through a distance dz caused by the second derivative of the refractive index (dn^2/dy^2). These three cases are the basis of three measurement techniques known respectively as interferometry, Schlieren and shadowgraphy.

In interferometry, the phase difference produced by the refractive index n is measured by recombining probing and reference beams (Figure 12.9a), yielding a resultant intensity I [1].

$$I = 2I_0 \cos^2 \left[\frac{\pi(n_0 - n)l}{\lambda} \right] \tag{12.8}$$

where I_0 is the peak intensity, l the path length, and n_0 a reference refractive index. Thus, a series of intensity fringes are formed which are separated by $(n_0 - n)l/\lambda$. The light source needs to be coherent to ensure a constant phase of the reference beam over a distance as great as the optical path difference (a laser light source with bandwidth $\Delta\lambda = 0.01$ Å has a coherence length of 50 cm).

With a Schlieren system, the displacement θ of light by a refractive index gradient (dn/dy) may be detected as an intensity variation. By focusing a light beam on a sharp edge, the angular deflection θ (Figure 12.9*b*) is given by [1]:

$$\theta = \frac{1}{n_0} \int_0^1 \left(\frac{dn}{dy}\right) dz \tag{12.9}$$

where n_0 is a reference refractive index, l is the optical path length. The brightness pattern produced at a detector array therefore maps the refractive index gradients in the test volume perpendicular (y) to the direction of the propagation (z). Optimum sensitivity is obtained when $D/f \cong \theta$, where D is the beam diameter and f the focal length of the focusing lens.

In shadowgraphy parallel rays passing through a refractive location where $(d^2n/dy^2) \neq 0$ are displaced inhomogeneously whereby some rays converge and others diverge. The medium, therefore, behaves as a lens increasing the intensity at the focal point on axis while decreasing the intensity of the parallel beam elsewhere to form a shadow (Figure 12.9*c*).

An example of a Schlieren photograph is given in Figure 12.10 which is a streak photograph of a shock wave passing through an electrical arc (time vertical, radius horizontal). The different light intensity profiles correspond to refractive index gradients enabling the discharge and shock front to be identified.

12.2.5 Optical fibre propagation

The propagation of light through an optical fibre relies upon coherent scattering as manifested in refractive index differences. On the other hand, incoherent scattering can contribute to performance limitations, e.g. due to reduction in the optical power guided. An optical fibre consists of a cylindrical core of a transparent medium of refractive index n_1 surrounded by an outer annulus of a second transparent medium of refractive index n_2 [17]. Provided the diameter of the core, d_1, is greater than the wavelength of light, light propagates through the fibre via a series of total internal reflections (Figure 12.11). This has a number of implications regarding light transmission:

(i) Only light rays falling on the end face of the optical fibre core within an angle θ_1 will be propagated by total internal refection where [18]:

$$NA = (n_1^2 - n_2^2)^{1/2} = \sin\theta_1 \tag{12.10}$$

(ii) Patterns formed by light rays launched at different angles θ form propagation modes within the fibre core. The number of such modes (patterns) propagating in a fibre is given by

$$N = \frac{1}{2}\left(\frac{\pi d}{\lambda} NA\right) \tag{12.11}$$

where λ is the wavelength of the light. Typical multimode fibres, with core diameters of 50–200 μm, propagate 100–1000 modes. The significance of multimode propagation is that different modes propagate at different velocities so limiting the data transmission rate.

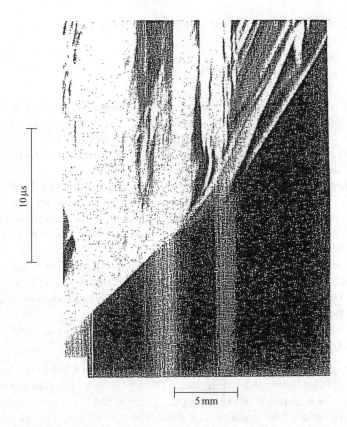

10 μs

5 mm

*Figure 12.10   Streak Schlieren photography of shockwave propagating through a
cylindrical discharge (courtesy of G.R. Jones, 'High pressure arcs in
industrial devices: diagnostic and monitoring techniques', published
by Cambridge University Press, 1988)*

(iii)   Since the refractive index of materials in general varies with wavelength and
since the refractive index is inversely proportional to velocity, different wavelengths propagate with different velocities leading to another limitation to data
transfer rates i.e. due to pulse broadening [17]. For instance, a light emitting
diode (LED) source operating at 850 nm and with a line width of 40 nm will
give pulse spreading of some 4 ns/km.

Incoherent scattering of the Rayleigh type leads to attenuation in optical fibres
which varies as λ^{-4} so that shorter wavelengths ($\leq$800 nm) are significantly suppressed. At longer wavelengths ($\geq$1.6 μm), electron absorption effects dominate
leading to an attenuation which increases with wavelength. There is therefore an
optimum range for low attenuation propagation between these limits.

Optical fibre sensing differs from optical fibre telecommunications in that optical
elements, which are sensitive to external parameters, form component parts of the

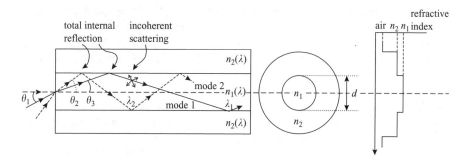

Figure 12.11   Propagation of polychromatic light in a step index optical fibre
showing:
   • *total internal reflection*
   • *wavelength dependent refraction* (λ_1, λ_2)
   • *different propagation modes (1, 2)*
   • *incoherent scattering*

system, whereas in telecommunications systems such external influences are delib-
erately excluded. In sensing applications, the optical signal is modulated to carry
information about the magnitude of a perturbing parameter. The fundamental light
wave properties which can be modulated to carry sensing signals are:

- intensity
- phase (via interferometer measurements, section 12.2.4.2)
- polarisation (via rotation of the plane of polarisation)
- wavelength (e.g. laser Doppler velocity measurement, section 12.2.4.1).

In addition, it is possible to encode the spectrum of polychromatic light carried by an
optical fibre (chromatic modulation [17]).

The general relationship between the output voltage (V) from a photodetector and
the optical modulation $M_1(\lambda)$ of the optical fibre sensor is given by [17]:

$$V = q\left[\int_\lambda (P(\lambda)\left[\int_l T(\lambda)dl\right]R(\lambda)M_1(\lambda)d\lambda\right]^p \qquad (12.12)$$

where $P(\lambda)$ is the source power, $T(\lambda)$ the fibre transmission loss, $R(\lambda)$ is the wavelength dependent responsivity of the photodetector, λ is the optical wavelength. l is
the fibre length and p, q are numerical constants.

For interferometric sensors, the modulation factor $M_1(\lambda)$ is

$$M_1(\lambda) \sim \cos^2 \delta(\lambda) \qquad (12.13)$$

where $\delta(\lambda)$ is the phase difference between the interfering light waves of wavelength
λ. For a Fabry–Perot cavity of thickness $(\nabla x/2)$ and refractive index n:

$$\delta(\lambda) = \frac{(\pi n \nabla x)}{\lambda} \qquad (12.14)$$

For chromatic sensors the spectral modulation $M_1(\lambda)$ is sampled with two or three photodectors having non-orthogonal responsivities $R_n(\lambda)$, $n = 1, 2, 3$. The outputs from the three detectors are then processed to yield as the output optical signal, a dominant wavelength (H), an effective bandwidth (S) and a nominal signal strength (L) [19].

Shifts in dominant wavelength (∇H_d) as small as 0.01 nm are detectable with conventional photodiode detectors using such chromatic systems [19]. Changes in measurand are then detectable in terms of variations in H, L and S leading to the possible simultaneous measurement of several parameters.

12.3 Optical equipment and systems

12.3.1 High speed imaging

High speed imaging is concerned with the optical recording of a transient or fluctuating discharge in order to determine the spatial and temporal variation of the discharge size, movement or light intensity. Both framing and streak images are obtainable (Figure 12.2). The former involves recording a succession of two-dimensional images each taken with a given exposure time, and the latter involves continuously streaking a one-dimensional section of the image along the recording medium (Figure 12.10).

The acquisition of high speed images may be either with a conventional optomechanical or an optoelectronic camera, both requiring a radiation energy density of approximately 10^{-8} J cm^{-2} for legible recording.

Optomechanical systems require the optical image to be swept along a recording film by one of the following means [1]:

- rotating prism or mirror (writing speeds of μs per mm for streak photography)
- movement of recording film in synchronisation with the rotating prism (6×10^3 frames per second, limited by film tearing)
- rotation of the film on a drum in synchronisation with the rotating mirror ($\geq 3.3 \times 10^4$ frames per second).

Optoelectronic systems convert an optical image into electronic signals, which are subsequently amplified and displayed photo-electrically. Additionally, amplification may be achieved using an electronic image intensifier before finally displaying the image. Such image converter cameras are capable of providing $> 10^6$ f.p.s. with a few nanosecond exposures but the recording duration is limited (~ 12 frames).

Hybrid cameras are available whereby the primary recording is via conventional high speed cameras with excellent spatial resolution. Thence the film images are digitised optoelectronically. This provides the best compromise between length of recording time and time resolution.

High speed cameras are used in conjunction with precision shuttering to prevent overwriting on the film. Coarse shuttering is achieved with conventional electromechanical shutters, and finer shuttering is achieved with electronic shutters. Such shuttering can lead to problems of synchronisation with the event to be recorded and image breakthrough which could lead to damage to the photo cathode of the electronic cameras.

Substantial electronic amplification makes image converter cameras more sensitive than mechanical cameras but only at the expense of modest resolution and no spectral information (which with mechanical cameras may be retrieved with colour film).

A lens system reduces the light flux density in the ratio $\Gamma[(1+P)^2 4F^2]^{-1}$, where Γ is the transmission, P the magnification and F the aperture number. The aperture number F for mechanical cameras may be a function of the exposure time and limited by the moving parts rather than by the objective lens. This may prove particularly troublesome when the camera is used as part of a larger optical system such as those used for shadowgraphy or interferometry (section 12.2.4.2).

A detailed quantitative evaluation of photographic records requires a careful consideration of the optical system, a knowledge of the recording characteristics of the film and, in the case of image converter systems, a knowledge of the electrooptical response of the camera. Operation needs to be on the sensitive part of the light intensity: response characteristic of the film emulsion (Figure 12.12*a*) to avoid halation at one extreme and lack of sensitivity at the other. Calibrated neutral density filters may be used to ensure operation in the correct range. For image converter cameras, the response characteristics of the photocathode (Figure 12.12*b*) are needed. The image intensity is also affected by the gain of the camera tube, which may be controlled electronically.

12.3.2 Spectrometer systems

Spectrometers for investigating electrical discharge spectra may produce spectral dispersion via a diffraction grating or prism [3]. The wavelength resolution $(\lambda/\nabla\lambda)$ for a prism depends upon the prism base length, b, and the wavelength dependence of the refractive index, n:

$$\frac{\lambda}{\nabla\lambda} = b\left(\frac{dn}{d\lambda}\right)$$
(12.15)

The wavelength resolution for a grating depends upon the number of lines (N_L) on the grating and on the order (m) of the principal maximum:

$$\frac{\lambda}{\nabla\lambda} = mN_L$$
(12.16)

Gratings with large numbers of lines ($\sim 10^4$ cm^{-1}) produce higher resolutions than prisms, but prisms avoid problems of overlapping of orders.

The profile of a spectral line obtained with a spectrometer is influenced by the width of the spectrometer slit, the choice of which is governed by a compromise between having a sufficiently narrow slit to yield a narrow spectral line width but wide enough to provide sufficient light intensity for short exposure recording (μs). The sensitivity of a spectrometer varies with the optical wavelength.

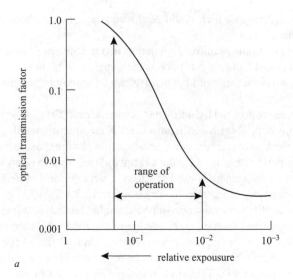

a

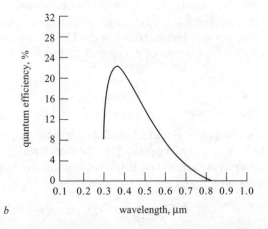

b

Figure 12.12   Some parameters affecting photographic image intensities

a   typical calibration curve for a polaroid film
*b   typical spectral response of the photocathode of an image con-
verter camera (courtesy of G.R. Jones, 'High pressure arcs
in industrial devices: diagnostic and monitoring techniques',
published by Cambridge University Press, 1988)*

A typical spectroscopic system used for arc discharge measurements is presented
on Figure 12.13*a*. This shows an optical system [3] utilising a grating spectrometer
with mirror optics for obtaining survey spectra along with a rapid scanning spectrometer for short exposure spectra. Whereas survey spectra are recorded photographically,
the scanning spectra are detected with a photo multiplier and the output recorded
oscillographically. In the latter case, each spectral line is swept across the focal plane

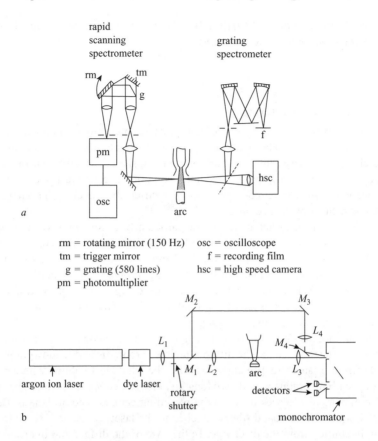

rapid scanning spectrometer

grating spectrometer

rm = rotating mirror (150 Hz)   osc = oscilloscope
tm = trigger mirror   f = recording film
 g = grating (580 lines)   hsc = high speed camera
pm = photomultiplier

Figure 12.13   *Spectrometric systems for arc studies [3, 39]*

    a framing and scanning systems (courtesy of D.R. Airey, PhD thesis, University of Bath, 1977)

    b absorption spectrometric system (courtesy of K. Ibuki, PhD thesis, University of Liverpool, 1979)

of an exit slit by a rotating mirror. This slit samples each line in turn. The use of a half silvered mirror allows the simultaneous recording of high speed photographs.

Streak spectra may be used to monitor continuous time changes in discharge behaviour and properties [20].

Optoelectronic techniques are used for recording short exposure spectra. These range from photomultiplier detection (Figure 12.13*a*) to vidicon image converter cameras (whereby the spectrum from the exit of a polychromator is focused upon the target of a vidicom camera tube). Such systems enable the electronic output to be transferred directly to a microcomputer for displaying the spectra and for additional processing (e.g. deconvolution along a line sight (Figure 12.3)).

Absorption spectroscopy involves the use of an auxiliary spectral source (which may, for instance, be an argon–ion laser tuned with a dye) to provide both a probing and a reference beam (Figures 12.13*b*) [39]. Both beams are passed onto a monochromator before being monitored photo electronically.

12.3.3   Light scattering systems

12.3.3.1   Coherent scattering

Refractive index responsive techniques require a test volume to be illuminated by light which is then analysed. The illuminating radiation should not be strongly absorbed by the electric discharge medium and its intensity should be sufficient to outshine the light emitted by the discharge. The latter condition is easily fulfilled with a laser light source, and the former condition usually restricts the wavelength domain to the visible and near infra-red regions.

Details of optical systems which are particularly useful for investigating discharges have been given by a number of authors [21–24]. Some typical optical systems which have proved useful for electric discharge investigations are shown in Figure 12.14. Such systems normally utilise a laser light source, light conditioning optics and a detector. The laser source may be either pulsed (e.g. ruby) to provide extremely short exposure (a few nanoseconds) or continuously operated ($\sim$4 W argon–ion) to allow long duration ($\sim$10 ms) recording.

The conditioning optics are determined by the type of diagnostic (shadowgraphy, Schlieren or interferometry). The differential interferometric and shadowgraphic systems shown on Figures 12.14*a* and *c* examine the whole test volume with a parallel beam of laser light produced with two long focal length, concave mirrors. The second of these mirrors also serves to focus the recording camera onto a plane in front of the test volume. Narrowband filters matched to the laser wavelength serve to eliminate light emitted from the discharge. In the case of the differential interferometer, two interfering light beams are produced by dividing equally the laser beam with a parallel-sided glass plate (front and back surface reflections) close to the focal length of the first concave mirror.

The distortion suffered by the laser beam in traversing the test volume may be recorded using high speed framing or streak cameras (section 12.3.1). Care is required to ensure that the internal shuttering and optics of the camera do not create any additional distortion.

Analysis of the photographic records from interferometric experiments involves measuring shifts in interference fringes at various locations followed by Abel inversion (section 12.2.3) to determine the radial variation of the refractive index.

12.3.3.2   Incoherent scattering

Plasma light scattering

Since the intensity of incoherently scattered light is extremely small (10^{-12} times smaller than the incident light intensity (section 12.2.4.1)), stringent precautions are needed to eliminate stray scattering from optical windows, and microscopic particles. A powerful laser is required to produce scattered light to outshine the discharge

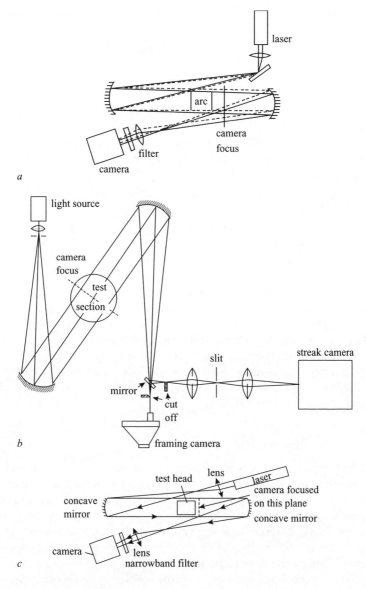

Figure 12.14   *Systems for coherent light scattering measurements*
  a   interferometer
  b   Schlieren
  c   shadowgraph (courtesy of G.R. Jones, 'High pressure arcs
        in industrial devices: diagnostic and monitoring techniques',
        published by Cambridge University Press, 1988)

light within the spectral domain of interest. This requirement suggests that the ion scattered component I_i may be easier to detect than the free electron component I_e (section 12.2.4.1) on account of its narrower spectral width and despite its lower intensity.

An optical system for measuring incoherent light scattered from electrical discharges (Figure 12.15a) uses a ruby laser source whose beam is collimated and focused onto a locality within the electrical discharge by two lenses, a pinhole and a number of irises. The incident beam is dumped after passing through the discharge, the Brewster angle filters avoiding back reflection. Light scattered through 90° to the incident beam passes through a 10 Å optical filter before being detected via a Fabry–Perot etalon and photomultiplier.

Micron particles scattering

The Doppler shift of light scattered from moving micro particles (section 12.2.4.1) may be measured with the system shown on Figure 12.15b. This uses two beams from an argon–ion laser 1.5 W, 4880 Å superposed within an electric discharge being monitored. A double parallel plate compensator is used to reduce the beam separation to 0.3 mm. The scattered light is collected around a disk stop which absorbs the high intensity main beam and determines the axial spatial resolution. A large aperture lens behind the stop collects the maximum possible amount of forward scattered light. The polarised (collimated) scattered light is detected with a photomultiplier together with a 10 Å interference filter and associated polariser. A Fabry–Perot filter of 0.5 Å bandwidth further reduces spurious optical noise. Signal acquisition needs to respect the short duration ($\sim 0.2 \mu s$) and small numbers of scattering events over a prolonged observation period (e.g. ~ 10 ms) [25].

12.3.4   Optical fibre sensing systems

The simplest form of an optical fibre sensing system consists of a light source coupled to an optical fibre which addresses a measurand sensitive element, and which returns a measurand modulated optical signal to a photodetector (Figure 12.16). If the sensing element is distinct from the optical fibre the system is known as extrinsic; if the fibre serves to both transmit and modulate the optical signal in response to a measurand the system is 'intrinsic'. A third form of optical fibre sensing exists whereby the sensing is non-optical but signal transmission is by optical fibre following optoelectronic or optomechanical conversion. Such systems are known as hybrid sensing systems.

Although optical fibres offer many advantages, they need to be carefully deployed. The advantages and difficulties may be summarised as follows:

(i)   high sensitivity (but often respond to more than one parameter)
(ii) inherently electrically insulating (but are surface breakdown sensitive at extra high voltages)
(iii) immune to EMI (but are susceptible to mechanical vibration)
(iv) geometrically flexible (but even gradual bending over of extended lengths causes attenuation)

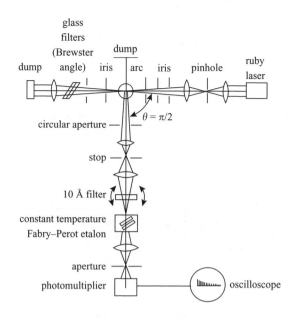

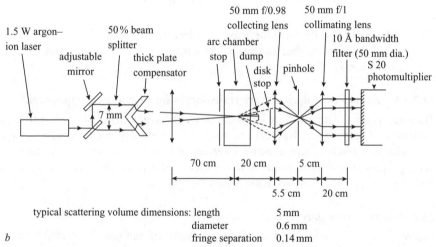

typical scattering volume dimensions: length   5 mm
                                     diameter   0.6 mm
                                fringe separation   0.14 mm

Figure 12.15   *Systems for incoherent light scattering measurement*
        a   scattering from a discharge plasma (courtesy of F.A. Rodriguez, PhD thesis, University of Liverpool, 1974)
        b   scattering from moving micro particles (laser Doppler) (courtesy of G.R. Jones, 'High pressure arcs in industrial devices: diagnostic and monitoring techniques', published by Cambridge University Press, 1988)

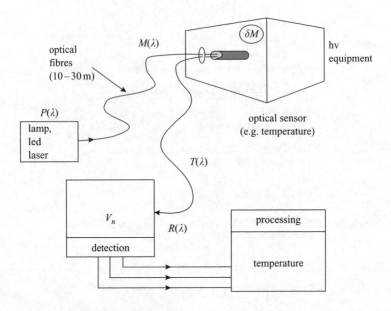

Figure 12.16   Structure of an optical fibre sensing system

(v)   lightweight (but the need for heavier protective sheathing)
(vi) corrosion resistant (but susceptible to ionising radiation, microwave heating and solvents).

12.3.4.1   Examples of optical fibre transducers for high voltage equipment

There are a range of optical fibre transducers which have been tested on high voltage equipment. For example, transducers which have been deployed on high voltage circuit breakers include transducers for current, voltage, arc radiation, solid particulates, gas pressure, contact temperature and mechanical vibration. Some examples of such transducers are discussed in the following section.

Electric current transducers

Figure 12.17 shows a range of different types of optical-fibre-based current transducers. These are [26]:

- an extrinsic magnetooptic element based upon magnetic field polarisation modulation (f)
- intrinsic sensor based on replacing the extrinsic magnetooptical element by a polarisation preserving fibre wrapped around the current carrying busbar (e)
- bulk extrinsic sensor representing an intermediate case between the purely intrinsic and extrinsic magnetooptic sensor (d)
- magnetic field concentrator based on a magnetic yoke around the current carrying busbar concentrating the B-field into a magnetooptic element in a gap on the concentrator (c)

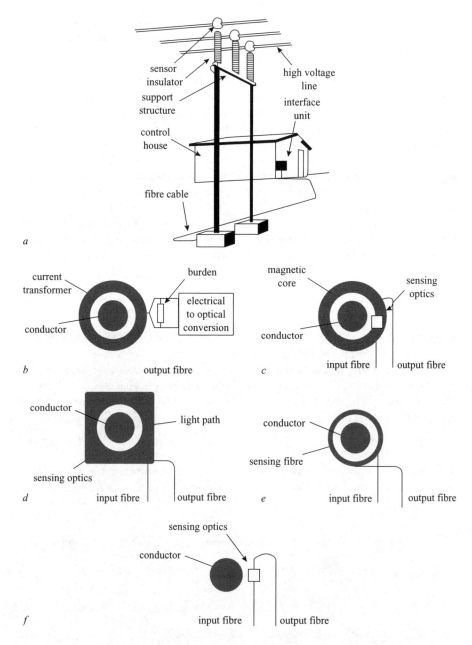

Figure 12.17 *Various forms of optical fibre current transducer (IEEE working group (1994)). a* essential elements of OCT system; *b* conventional CT with optical readout; *c* magnetic concentrator with optical measurement; *d* OCT using bulk optics (e.g. Jackson); *e* fibre optics based current measurement (e.g. Rogers); *f* witness sensor

- hybrid optical current transformer in which optical fibre transmission is used to energise and address an electronic current transformer mounted on the high voltage busbar (*b*).

The hybrid current transducer can draw its operating electrical power either by electromagnetic induction from the current busbar or from a control room laser via a fibre link. A paramount need is to minimise the power consumption of the telecommunications module, which in turn dictates the form of telecommunications encoding used, e.g. pulse frequency modulation, pulse code modulation etc. [27]. Hybrid current transducers are attractive because they rely on established telecommunications techniques, can be adapted to transmit other measurand (e.g. temperature, voltage) and could be coupled into existing telecommunication fibre systems on EHV transmission networks.

The magnetooptic current transducers are based upon the Faraday rotation effect [28], whereby the plane of polarisation of a light wave propagating through a magnetooptic element in a magnetic field of flux density B is rotated through an angle:

$$\theta = \int V(\lambda)Bdl \qquad (12.17)$$

where $V(\lambda)$ is the wavelength-dependent Verdet constant of the magnetooptic element, and dl the optical path traversed in the element.

Since generically it is only possible to measure light intensity, a change in θ is measured as the change in intensity of the light passing through a second polarising filter so that only a resolved component of the polarised light wave is transmitted and detected (Figure 12.18), i.e.:

$$P_0 = P_i \cos^2 \delta \qquad (12.18)$$

Thus, the variation of θ with B is cyclical, so that the range of operation is governed by the range within which $\theta(B)$ has linearity within the required specifications. Practical

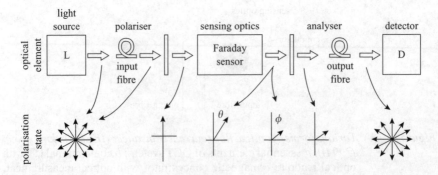

Figure 12.18   *Typical arrangement of optical components and polarisation components in Faraday sensor*

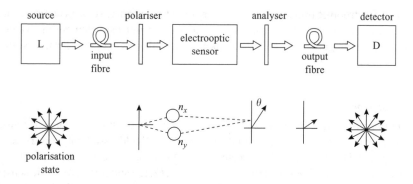

Figure 12.19   Electrooptic sensing system

designs need to accommodate the temperature dependence of $V(\lambda)$ and the strong dependence of θ on stress for high $V(\lambda)$ materials.

Various chromatic modulation forms of the concentrator OCT have been described [28] which seek to overcome the practical limitations described above.

Electrooptic voltage transducers

Electrooptic voltage transducers sense changes in voltage via the effect of the associated electric field on the polarisation of light propagating through an electrooptic element [17]. In such elements, light is propagated via two contrary circularly polarised waves which when recombined constitute a linearly polarised wave (Figure 12.19). The phase difference between the two circularly polarised waves varies with the magnitude of the electric field, as therefore does the orientation of the output linearly polarised wave. Thus, the output optical power varies with the phase difference between the two circularly polarised waves, according to:

$$P_0 = P_i \sin^2 \left(\frac{\delta}{2} \right) \tag{12.19}$$

where

$$\delta = \frac{2\pi}{\lambda_0} n_0 3 T_{ij} V \tag{12.20}$$

λ_0 = optical wavelength, n_0 = refractive index under zero E-field conditions, T_{ij} is the linear electrooptic coefficient of the sensing element (Pockels effect) and V the voltage. Thus, the optical modulation is described by the interference-type modulation factor $M_1(\lambda)$ from Equation (12.13).

Similar problems to magnetooptic elements exist with electrooptic elements in having high sensitivities to temperature and to vibration. In addition, there are difficulties in exposing miniature electrooptic elements to E-fields which are representative of the voltage levels (145, 420 kV) found at distribution and transmission levels without exposure to electrical breakdown across the electrooptic element itself.

Such electrooptic transducers may be deployed as extrinsic sensors across the low voltage end of a high voltage capacitor divider or an intrinsic element of extended length to provide good high voltage breakdown avoidance. A hybrid form of voltage transformer is also possible.

Electrode temperature

The temperature of a high voltage electrode (e.g. circuit breaker contact) can be measured via a Fabry–Perot cavity interferometer [29] addressed polychromatically (section 12.2.5) via optical fibre transmission. As such, the technique is illustrative of how chromatic modulation defined by Equation (2.12) and interferometric sensing defined by Equations (12.12) and (12.13) may be used combinatorially to provide an effective sensing system for such a difficult operating environment.

A fibre addressed Fabry–Perot cavity consists of a silicon wafer into which a cavity is etched and with a semireflective glass plate bonded to cover the cavity (Figure 12.20*a*) [30]. Two multimode step-index fibres for delivering and receiving the polychromatic light are butted to the glass plate. The depth of the cavity ($\sim$0.3–1 μm) is monitored via the optical interference between light reflected from the glass plate surface and that reflected from the silicon surface on the other side of the cavity leading to the phase difference defined by Equation (12.14). The other outer surface of the silicon wafer is coated with aluminium to prevent infra-red radiation from an electrical discharge being transmitted into the cavity via the silicon. The silicon wafer is made sufficiently thick so as not to flex under pressure and the glass plate is arranged to have a thermal expansion coefficient which is very different from that of the silicon. Temperature variations produce internal stresses which lead to changes in the cavity depth and which in turn produce polychromatic optical interference. A typical dominant wavelength:temperature calibration curve is given in Figure 12.20*b* showing a low level of hysteresis between heating and cooling cycles. A resolution of 0.1 per cent of full scale has been shown to be achievable.

Discharge radiation and particulates transducer

The chromatic modulation approach [19, 40] enables the optical emission from an electrical discharge to be monitored along with the concentration of particulate material formed during complex chemical reactions in the discharge plasma using a single optical fibre probe. Particle concentration may be obtained by monitoring the change in the spectral signature of polychromatic light [30] due to incoherent scattering. Monitoring the output optical signal from the probe signal chromatically leads to the arc signature being quantified by the measured dominant wavelength shift. An optical fibre probe for achieving such monitoring is shown on Figure 12.21*a*. It involves transmitting a broad collimated polychromatic beam across a monitoring gap between two optical fibres. Light from the discharge plasma is collected by the receiving fibre superimposed on a collimated beam from a known optical source. The signature of the emission is then obtained as a shift in the dominant wavelength of the superimposed optical signals.

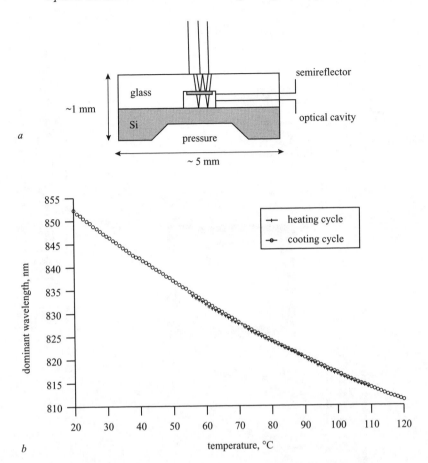

Figure 12.20   *Chromatic Fabry–Perot temperature pressure transducer [29]*
   a   sensor structure
   b   dominant wavelength:temperature calibration curve

Particle concentration monitoring may be achieved with the same optical fibre sensor arrangement as shown in Figure 12.21*a*. The particle concentration is obtained by monitoring the change in the spectral signature of polychromatic light produced by scattering from micron-sized particles once the electrical discharge is quenched. The wavelength of the forward scattered light is a function of the fractional volume of micron-sized particles NR_P^2/R_A^2, where R_A is the radius of the cylindrical volume being optically addressed, N, R_P are the particle concentration and radius, respectively.

Calibration of such a probe with known quantities of well dispersed micron-sized particles yields the dominant wavelength:particle concentration curve of Figure 12.21*b* which shows a high degree of linear correlation.

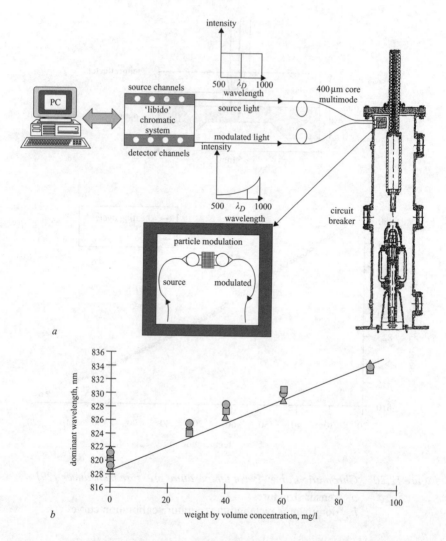

Figure 12.21   Plasma and particulate chromatic monitoring [30]
   a   sensor structure and location in circuit breaker
   b   dominant wavelength:microparticles concentration calibration
        curve

Optical fibre transducers of mechanical parameters

It is also possible to monitor mechanical parameters on high voltage equipment using optical fibre sensing. Examples of high voltage environments in which such optomechanical transduction is important are mechanical operation of high voltage switches (circuit breakers, tap changers etc.), gas pressure in gas-insulated equipment (circuit breakers, transformers, GIS etc.) and vibration on overhead lines, busbars switchgear and transformers.

Pressure measurements may be made using an optical fibre addressed Fabry–Perot cavity (Figure 12.20*a*) with a pressure flexed membrane [30].

Mechanical vibrations may be measured with an optical fibre homodyne interferometer [18]. This is an intrinsic sensor (section 12.3.4) in which changes in the interference between various propagation modes are produced by vibration-induced variations in refractive index and manifest as different light intensity patterns at the fibre output.

Mechanical movements of operating mechanisms etc., are measurable with a fibre-addressed chromatic scale which moves past the fibre end [31] (Figure 12.2).

12.4   Examples of test results

The principles outlined in section 12.2 and the systems described in section 12.3 may be used to provide experimental and condition monitoring results relevant to high voltage systems. Examples of such results are presented in this section and their relevance to high voltage engineering indicated.

12.4.1   High speed photography

Examples of high speed photographs relating to high voltage electrical discharges are given in Figures 12.2 and 12.6. Figure 12.2 is a photograph of an electrical breakdown occurring outside the PTFE nozzle of a high voltage circuit breaker obtained with a rotating mirror and drum type of framing camera (section 12.3.1). Figure 12.6 is a photograph of luminous particles formed during arcing in an SF_6 HV circuit breaker obtained with a high speed video camera.

Sequences of such images enable changes in the structure of electrical discharges with time to be mapped. For example, Figure 12.22 shows how the luminous radii of an arc discharge contracts during an 800 μs period as the discharge current passes through zero following a 41 kA peak current of a 100 Hz current waveform. After the zero current, the arc column contraction is accompanied by an axial displacement of the plasma due to an imposed flow of gas. The axial displacement reflects the timescale of the axial convection losses, and the radial contraction indicates the timescale associated with the combined influence of radial diffusion and flow entrainment effects. Information from such test results is important in understanding the role of the arc discharge in switching fault currents in high voltage networks.

12.4.2   Spectroscopic results

The information that may be derived from spectroscopic measurements of electrical discharges using spectrometer systems of the type discussed in section 12.3.2 includes estimates of electron temperature, impurity concentration etc. (sections 12.2.3.1, 12.2.3.3).

Figure 12.23*a* shows a typical electron temperature map (derived from relative line intensity measurements) for a 10 kA, 10 cm long electric arc in air [20]. The temperature reduces from a maximum value of about 29 000 K on the axis in

576   *Advances in high voltage engineering*

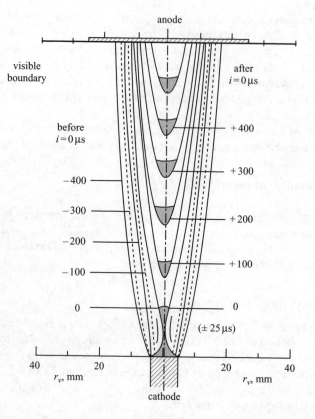

*Figure 12.22   Variation with time of the luminous boundary of a decaying 41 kA
electric arc discharge measured from high speed framing photographs (courtesy of University of Liverpool, Arc Research Report,
ULAP-T28, I.R. Bothwell and B. Grycz, 1974)*

front of the anode. Contours of copper concentration (entrained from the copper
cathode) are shown in Figure 12.23b. The constricted nature of the copper vapour
jet is apparent, particularly close to the cathode derived from the ratio of copper to
nitrogen line intensities. Temperature estimates made from the ratio of two copper
line intensities (Figure 12.23c) show that close to the cathode the copper vapour temperature (1.7×10^5 K) is considerably lower than that of the host discharge plasma
(2.9×10^5 K).

Such test results provide an insight into the manner in which electrical discharges
in high voltage equipment can erode metallic members such as tank walls, switch
contacts, busbars etc.

A form of absorption spectroscopy (Figure 12.13b) is currently being used with
a free space laser beam for identifying SF_6 leakages from high voltage equipment.
The laser wavelength is tuned to one of the SF_6 absorption spectral lines and absorption maps produced which are recorded with electronic imaging. The method is

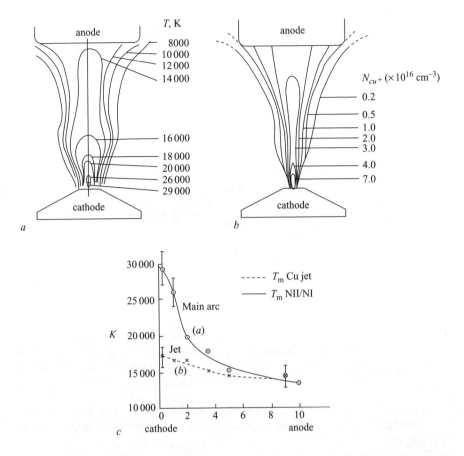

Figure 12.23   Properties of 10 kA free burning air arcs derived from measured spectra

a   gas temperature map
b   copper concentration map
c   comparison of gas and copper jet temperatures (courtesy of University of Liverpool, Arc Research Report, ULAP-T18, D.E. Roberts, 1972)

an important contribution to the quest to reduce greenhouse gas emission into the atmosphere, SF_6 being such a potent greenhouse effect gas.

12.4.3   Coherent scattering results

Two-dimensional images obtained with shadow, Schlieren and interferometric techniques (section 12.3.3.1) can yield details of the size, structure and properties of not only the electrical discharge but also its surroundings. The radial refractivity variations determined from an interferometric image of a 9.7 kA electric arc are shown in Figure 12.24. Various regions of the discharge are identifiable, namely the

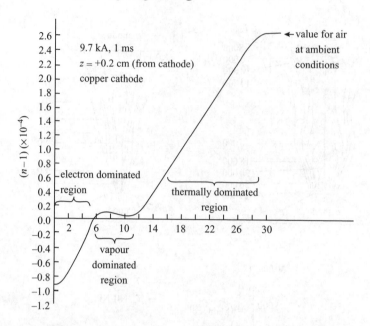

Figure 12.24   *Radial variation of refractivity for a 9.7 kA electric arc discharge determined interferometrically (courtesy of G.R. Jones, 'High pressure arcs in industrial devices: diagnostic and monitoring techniques', published by Cambridge University Press, 1988)*

electrical plasma region (with the negative refractivity due to electron domination), a surrounding copper vapour region, a hot but only low ionisation region and the ambient air.

The time evolution of radial temperature profiles surrounding an electrical discharge may be determined from interferometer refractivity profiles of the form shown in Figure 12.24 to yield the results in Figure 12.25. These results are for a decaying 3 kA electric arc and illustrate the severe gradient of temperature at the boundary of the discharge (α indicates the five per cent and ten per cent accuracy limits).

Although shadowgram results do not provide detailed profile information of the kind available from interferometric measurement, they may be used to determine the location of the outer thermal boundary (i.e. the locations of the steep temperature gradients of Figure 12.25) with reasonable accuracy.

Test results of this form are useful for understanding gas compression and decompression effects which can be produced by transient, high current discharges in high voltage equipment enclosures and which can lead to catastrophic fracture of the containers.

12.4.4   Incoherent scattering results

Incoherent light scattering (section 12.2.4.1) from a discharge plasma may be measured with systems of the form described in section 12.3.3.2, plasma light scattering,

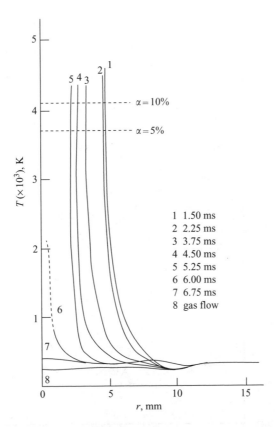

Figure 12.25   *Change in the radial boundary temperatures of a decaying 3 kA elec-*
tric arc determined from refractivity values yielded by interferograms
(courtesy of G.R. Jones, 'High pressure arcs in industrial devices:
diagnostic and monitoring techniques', published by Cambridge
University Press, 1988)

using a high power laser source. Light intensity and the change in spectral width of
the laser line can yield values of the electron density, the electron temperature and
the ion temperature. As an example, Figure 12.26 shows the radial variation of the
concentration of various ion species in a 9.7 kA electric arc in air at an axial locality
1 cm above the cathode and 1 ms after discharge initiation [32].

An example of results from incoherent light scattering from micron-sized particulates (section 12.2.4.1) for measuring the flow of plasma and gases around electrical
discharges (laser Doppler technique (section 12.3.3.2, micron particles scattering))
is given in Figure 12.27 [33]. The dashed curve shows the position of the discharge
heated gas boundary determined shadowgraphically (section 12.3.3.1). The significance of these types of measurement is that they show how the electrical discharge
reacts fluid-dynamically with its surroundings. In this particular case, gas entrained
into the discharge heated region is accelerated to high velocities, while a flow wake

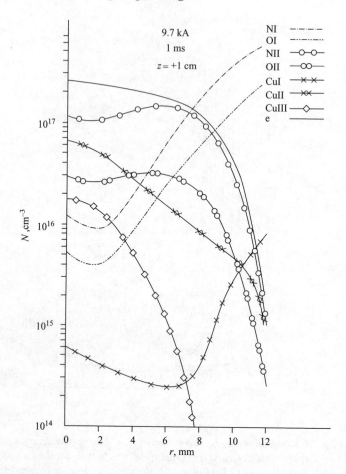

Figure 12.26   Radial profiles of species concentration derived from light scattering experiments (courtesy of F.A. Rodriguez, PhD thesis, University of Liverpool, 1974)

is formed in front of the discharge electrode (cathode). Typically, velocities in excess of 2×10^3 ms^{-1} can occur within the plasma of a 3 kA electric arc discharge.

A knowledge of such fluid-dynamic behaviour is important in designing the geometry and structure of high voltage circuit breakers so that their dielectric recovery function is not impeded by a redistribution of plasma and heated gases after fault current interruption.

12.4.5   Optical fibre transducer results

Because of the inherent electrical insulating nature of optical fibre sensors coupled to their EMI immunity, the results of tests with such sensors yield online information about high voltage equipment which was previously difficult if not impossible to

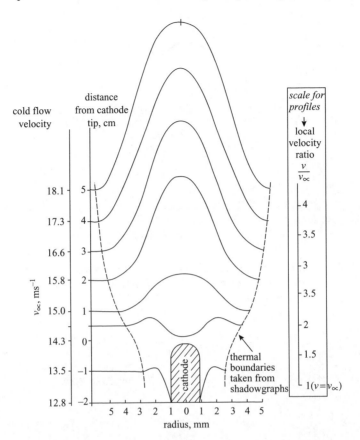

Figure 12.27   Flow velocity profiles under arcing condition; 30 A arc in accelerating flow (courtesy of University of Liverpool, Arc Research Report, ULAP-T13, Collings et al., 1973)

obtain. Some examples of such results obtained with a range of different optical fibre sensors are described.

Figure 12.28 shows the fault current variation through a high voltage circuit breaker using a hybrid optical current transformer (HOCT) [27], (section 12.3.2). Comparison of the results with those obtained with a resistive current shunt indicates a maximum deviation of only 0.4 per cent.

Figure 12.29 [29] shows the time variation of temperature with the calibration curve (Figure 12.20*b*) derived from the Fabry–Perot temperature probe (section 12.3.4.1, electrode temperature) embedded in the contact stalk of a high voltage circuit breaker, 13 cm from the contact tip. The results show temperature variations following the interruption of a number of different currents in the range 17–60 kA peak. Temperature changes of up to about 11°C occur. Two half cycles of operation at 60 kA peak produces a prolonged transient which only returns to ambient in about 0.5 s.

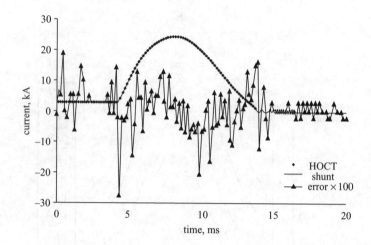

Figure 12.28   High voltage circuit breaker current measured with a hybrid optical current transformer (HOCT) [27] (differences between HOCT and shunt measured currents shown by triangles)

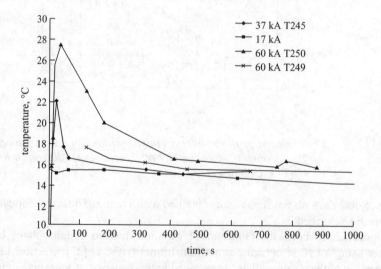

Figure 12.29   The variation with time temperature within the contact of a high voltage circuit breaker for various peak arc current and measured with a Fabry–Perot optical fibre probe [29] (60 kA, T250 is for two half cycles of current)

Figure 12.30*a* [18] shows the time variation of dominant wavelength of optical emissions from a number of fault arc currents in a high voltage circuit breaker ranging from 2–15 kA peak, using the probe of Figure 12.21*a*. The arc emissions are superimposed upon a continuous optical fibre signal from an external polychromatic source

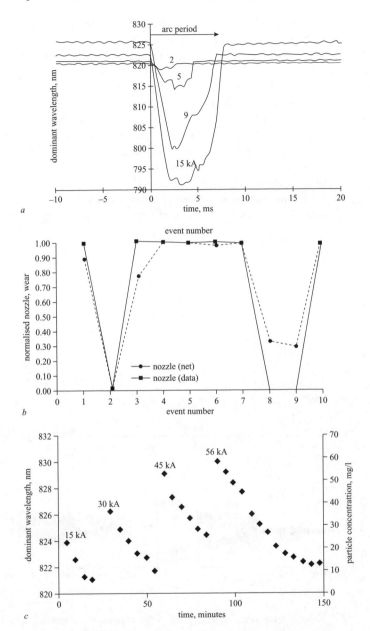

Figure 12.30   *Results from discharge emission/microparticles probe (Figure 12.20).*
a   time variation of dominant wavelength (section 12.2.5);
b   neutral network prediction of HV circuit breaker nozzle wear
from optical emission monitoring; *c*   microparticles concentration
as a function of HV circuit breaker fault currents and time determined
from dominant wavelength of scattered light

leading to a convenient referenced method for plasma intensity monitoring. One implication of such optical emission results is that they may be cross correlated with circuit breaker nozzle ablation indicating the extent of nozzle wear. Figure 12.30*b* [34] shows the prediction of nozzle wear due to radiative effects from such measurements compared with directly measured values of the mass reduction in the nozzle.

The shift in dominant wavelength pre and post arcing on Figure 12.30*a* is due to Mie scattering (section 12.2.4.1) of polychromatic light from micron-size particulates formed by the arc discharge. By tracking changes in the dominant wavelength following a series of high current arcs (15–56 kA) and using the calibration curve of Figure 12.21*b*, the variation in particulates concentration with time and duty of a high voltage circuit breaker can be determined (Figure 12.30*c*) [35].

Examples of optomechanical fibre sensor results are shown in Figures 12.31*a* and *b* [36]. Figure 12.31*a* shows the time variation of pressure in the compression

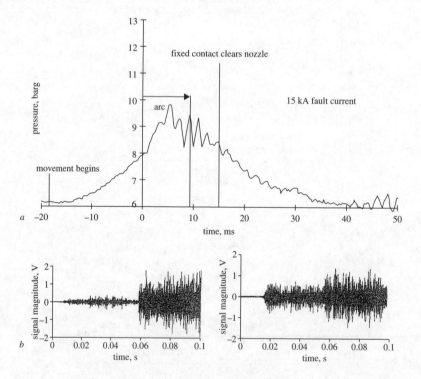

Figure 12.31   Results from optical fibre probes for measuring mechanical parameters on high voltage circuit breaker [30, 38]

    a   time variation of piston pressure in a high voltage circuit breaker under 15 kA fault current condition measured with a Fabry–Perot pressure sensor (Figure 12.20*a*) (Issac, 1997)

    b   optoacoustic fibre monitoring of mechanical vibrations from an SF_6 puffer circuit breaker under zero and 31.1 kA fault current conditions (Cosgrave *et al.*, 1997) cylinder of a puffer circuit breaker during and after the separation of contacts and arcing at 15 kA fault current; Figure 12.31*b* shows the difference in the optoacoustic signals produced by a high voltage circuit breaker when operating with no fault current and with a fault current of 31 kA peak. The results were obtained with the distributed optical fibre vibration sensor of section 12.3.4.1.

12.4.6 Time and wavelength response of optical fibre and free space techniques

Although optical-fibre-based techniques offer attractive advantages over the free space alternative (e.g. access into enclosures, reduced interference from external effects etc.), care is required in ensuring that measurements are not compromised by properties of the fibres themselves. An example of such a fibre limitation is in monitoring electrical breakdown between the contacts of a disconnector switch in an SF_6 filled gas-insulated system (GIS).

Figure 12.32 shows a simplified geometry of such a GIS mounted disconnector with a viewing port in the tank wall and the busbar lying along the tank axis. The rise time of the optical emission from the breakdown discharges is extremely rapid (lower part of Figure 12.32) being of the order of 5–6 ns. Reflections of electric field perturbations from the tank ends occur on timescales of 7 and 35 ns, and optical emissions radially from the tank wall occur in about 2 ns. These timescales are comparable with response limiting dispersion effect in optical fibres (section 12.2.5) which are in the range 3 ns (waveguide dispersion), 4 ns (chromatic dispersion) and 7.5 ns (intermodal dispersion). Discharge emission rise time measurements can, therefore, be prejudiced by the transmission dispersion effects of optical fibres.

With free space techniques, time response limitations are governed by the opto electronic detectors. A photomultiplier can provide a rise time of 2–3 ns, and that for a vacuum photodiode may be as short as 0.3 ns. Consequently, free space techniques can, where useable, provide better transient responses than optical-fibre-based systems when extremely rapid events are involved.

It should also be recognised that conventional silica optical fibres have transmission properties which are severely wavelength dependent in the range 300 nm to 2 μm (section 12.2.5). Consequently, such wavelength degrading effects need to be evaluated before application to a particular discharge monitoring situation. Special purpose optical fibres may therefore be needed (e.g. quartz cored fibres for ultra-violet transmission) with additional cost penalties.

12.5 Conclusions

The range of measurement and monitoring possibilities of optically-based techniques for use under high voltage conditions has been described. Utilisation and deployment of such approaches in the future are likely to be driven not only by the need of new and conventional high voltage applications but also by the substantial developments in the optoelectronic sector itself. The evolution of high specification, lower costs digital imaging systems, economic white quantum sources (without thermal power wastage)

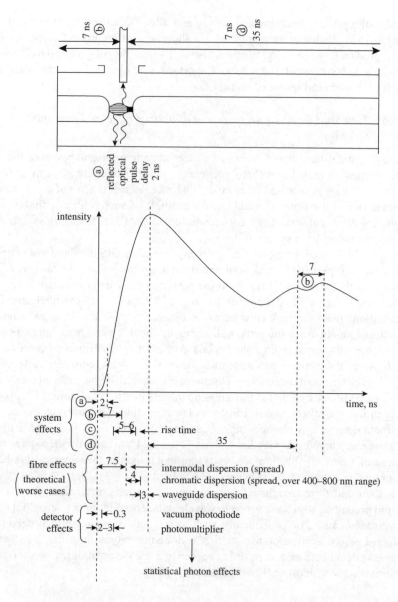

Figure 12.32   Transient response of optical fibres and photodetectors for recording electrical discharge formation in GIS

and short wavelength transmitting optical fibres (polymeric fibres etc.) are all set to drive further the usage of optical monitoring for a considerable range of industrial applications including those involving high voltage technology. Coupled with developments in intelligent information extraction systems [40], these approaches have considerable potential for future deployment.

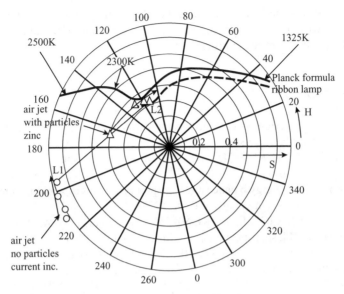

O without particles: (140, 180, 220, 250 A, Z = 30 mm)
△ with particles: (particle injection rate = 0.63 gs⁻¹
240 A, Z = 60 – 140 mm)

Figure 12.33   Distinction between plasma and particulates optical emission using chromatic mapping [37]

As an example of the evolving possibilities, Figure 12.33 shows some test results obtained from monitoring a high voltage plasma jet for producing microparticles for surface spraying of superconducting coatings [37]. The results are in the form of a dominant wavelength:effective bandwidth (H-S) map (section 12.2.5) of emissions monitored with a tristimulus chromatic system. It shows how the conditions of plasma and micro particles in the jet can be separated in chromatic space without recourse to data overload or expensive and cumbersome conventional spectroscopy and provide outputs easily assimilable into the operating control system (e.g. power, gas, particulates flow etc.) of the plasma torch system as a whole via modern information extraction means. This is but one of several examples which are emerging.

12.6 Acknowledgements

The authors appreciate the efforts of Miss M. Burns in the preparation of this chapter.

12.7 Note

[1] Line broadening mechanisms include broadening due to finite lifetime of an excited state ($\nabla\lambda \sim 10^{-5}$ nm), strong magnetic fields interacting with electron spins (Zeeman shift, $\nabla\lambda \sim 0.03$ nm), strong electron fields influencing the radiating atoms

(Stark broadening), the random movement of the emitting atoms (Doppler broadening, $\nabla\lambda \sim 0.01$ nm), pressure broadening ($\nabla\lambda \sim 0.02$ nm), ($\nabla\lambda$ values are typical for a high pressure electric arc discharge).

12.8   References

1   JONES, G.R.: 'High pressure arcs in industrial devices' (Cambridge University Press, Cambridge, 1988)

2   LOCHTE-HOLTGREVEN, W.: 'Plasma diagnostics' (North-Holland, Amsterdam, 1968)

3   AIREY, D.R.: 'Energy balance and transport properties of very high current SF_6 arcs'. PhD thesis, University of Bath, 1977

4   STANISLAVJEVIC, M., and KONJEVIC, N.: *Fizika*, 1972, **4**, (13)

5   GRIEM, H.: 'Plasma spectroscopy' (McGraw-Hill, 1964)

6   WIESE, W.L.: *in* HUDDLESTONE, R., and LEONARD, S.L. (Eds): 'Plasma diagnostics techniques' (Academic Press, New York, 1965)

7   TRAVING, G.: *in* LOCHTE-HOLTGREVEN, W. (Ed.): 'Plasma diagnostics' (North Holland, Amsterdam, 1968)

8   THEIMER, O.: *Phys. Lett.*, 1966, **20**, p. 639

9   PLATISA, M.: PhD thesis, University of Liverpool, 1970

10   KATO, M.: 'Fluctuation spectrum of plasma with few electrons in Debye volume observed by collective light-scattering', *Phys. Fluids*, 1972, **15**, p. 460

11   GEORGE, T.V., GOLDSTEIN, L., SLAMA, L., and YOKOYAMA, M.: 'Molecular scattering of ruby laser light', *Phys. Rev.*, 1965, **137**, p. 369

12   KERKER, M.: 'The scattering of light'. Academic Press, Library of Congress Cat. Card 69-2664, 1969

13   MAZUNDER, M.K.: NASA, CR-2031, 1970

14   DITCHBURN, R.W.: 'Light' (Blackie and Son, 1957)

15   DYSON, J., WILLIAMS, R.V., and YOUNG, K.M.: 'Interferometric measurement of electron density in high current discharge', *Plasma Phys.*, 1964, **6**, p. 105

16   DHAR, P.K., BARRAULT, M.R., and JONES, G.R.: 'A multi ring radio frequency technique for measuring arc boundary variations both at high currents and close to current zero', University of Liverpool, Arc research report, ULAP-T64, 1979

17   JONES, G.R., JONES, R.E., and JONES, R.: 'Multimode optical fibre sensors', *in* GRATTAN and MEGGITT (Eds): 'Optical fibre sensor technology' (Kluwer Academic Publishers, 2000), chap. 1

18   JONES, G.R.: 'Optical fibre based monitoring of high voltage power equipment', *in* RYAN, H.M. (Ed.): 'High voltage engineering and testing' (The Institution of Electrical Engineers, London, 2001, 2nd edn.), chap. 21

19   JONES, G.R., and RUSSELL, P.C.: 'Chromatic modulation based metrology', *Pure Appl. Opt.*, 1993, **2**, pp. 87–110

20   ROBERTS, D.E.: 'Spectroscopic investigation of the 10 kA free burning arc in air', University of Liverpool, Arc research report, ULAP-T8, 1972

21   KOGELSCHATZ, U., and SCHNEIDER, W.R.: 'Quantitative Schlieren techniques applied to high current arc investigations', *Appl. Opt.*, 1972, **11**, pp. 1822–1832

22   TIEMANN, W.: *IEEE Trans. Plasma Sci.*, 1980, **PS-8**, pp. 3368–3375

23   BLACKBURN, T.R., and JONES, G.R.: 'Measurement of density and temperature in an ORIFICE arc', *J. Phys. D, Appl. Phys.*, 1977, **10**, pp. 2189–2200

24   WALMSLEY, H., JONES, G.R., and BARRAULT, M.R.: 'Properties of the thermal region of a gas blast a.c. arc particularly near current zero', University of Liverpool, Arc research report, ULAP-T56, 1978

25   TODOROVIC, P.S., and JONES, G.R.: *IEEE Trans. Plasma Sci.*, 1985, **PS-13**, pp. 153–162

26   IEEE Working Group: 'Optical current transducers for power systems: a review', *IEEE Trans. Power Deliv.*, 1994, (4), pp. 1778–1788

27   PATE, A., HUMPHRIES, J.E., GIBSON, J.R., and JONES, G.R.: 'The measurement of fault arc currents using a hydrid opto electronic current transformer (HOCT)'. Proceedings of XIV international conference on *Gas discharges and their applications*, Liverpool, 2002, pp. 180–183

28   JONES, G.R., LI, G., SPENCER, J.W., ASPEY, R.A., and KONG, M.G.: 'Faraday current sensing employing chromatic modulation', *Opt. Commun.*, 1998, **145**, pp. 203–212

29   MESSENT, D.N., SINGH, P.T., and HUMPHRIES, J.E., *et al.*: 'Optical fibre measurement of contact stalk temperature in a SF_6 circuit-breaker following fault current arcing'. Proceedings of 12th international conference on *Gas discharges and their applications*, Greiffswald, 1998, **1**, pp. 543–546

30   ISAAC, L.T.: 'Puffer circuit breaker diagnostics using novel optical fibre sensors'. PhD thesis, University of Liverpool, 1997

31   ISAAC, L.T., SPENCER, J.W., JONES, G.R., HALL, W.B., and TAYLOR, B.: 'Live monitoring of contact travel on EHV circuit breakers using a novel optical fibre technique'. Proceedings of 11th international conference on *Gas discharges and their applications*, Tokyo, 1995, **1**, pp. 238–241

32   RODRIGUEZ, F.A.: PhD thesis, University of Liverpool, 1974

33   COLLINGS, N., BLACKBURN, M.R., JONES, G.R., and BARRAULT, T.R.: 'Laser Doppler velocimetry of flow fields surrounding low and high current arcs', University of Liverpool, Arc research report, ULAP-T13, 1973

34   COSGRAVE, J.: Private communication, 1998

35   ISAAC, L.T., JONES, G.R., HUMPHRIES, J.E., SPENCER, J.W., and HALL, W.B.: 'Monitoring particle concentrations produced by arcing in SF_6 circuit-breaker using chromatic modulation probe', *IEE Proc., Sci. Meas. Technol.*, 1999, **146**, (4), pp. 199–204

36   RUSSELL, P.C., COSGROVE, J., TOMPTIS, D., VOURDAS, A., STERGIOULAS, L., and JONES, G.R.: 'Extraction of information from

acoustic vibration signals using Gabor transform type devices', *Meas. Sci. Technol.*, 1998, pp. 1282–1290

37   RUSSELL, P.C., DJAKOV, B.E., ENIKOV, R., OLIVER, D.H., WEN, Y., and JONES, G.R.: 'Monitoring plasma jets containing micro particles with chromatic techniques', *Sens. Rev.*, 2003, **23**, (1), pp. 60–65 (Emerald Group Publishing Limited)

38   COSGRAVE, J., HUMPHRIES, J.E., and SPENCER, J.W., *et al.*: 'Chromatic characterisation of optoacoustic signals from fault current arcs in high voltage circuit-breakers'. Proceedings of XIIth international conference on *Gas discharges and their applications*, Greifswald, 1997

39   IBUKI, K.: 'Spectroscopic study of the high current vacuum arc'. PhD thesis, University of Liverpool, 1979

40   JONES, G.R., RUSSELL, P.C., VOURDAS, A., COSGROVE, J., STERGIOULAS, L., and HABER, R.: 'The gabor transform basis of chromatic monitoring', *Meas. Sci. Technol.*, 2000, **II**, pp. 489–498

Chapter 13

Pulsed power – principles and applications

J.E. Dolan

13.1   Introduction

Pulsed power deals with the generation of extremely high power, short duration impulses. Peak powers typically range from megawatt (MW) to terawatt (TW) levels, and pulse durations from nanoseconds to milliseconds. The aim of this review is to indicate the development of pulsed power as a discipline, and to discuss some of the key elements in pulsed power systems. The range of applications of pulsed power is also outlined.

Pulsed power began in the 1920s when Erwin Marx at the Technical University of Braunschweig devised a novel form of high voltage impulse generator for lightning testing of high voltage power transmission equipment. In the Marx generator (Figure 13.1) a set of n capacitors are charged in parallel at moderate voltages (typically 10–100 kV). The capacitors are then switched into series connection to achieve an output impulse voltage which is a multiple n of the charging voltage. The series connection arises due to the sequential over-volting and breakdown of the spark gaps once the first gap breaks down. Output voltages up to 2 MV are readily achieved. The spark gap closure maintains the capacitors in series for the duration of the ensuing current discharge. In more recent years, the development of high energy and fast rise-time Marx banks has been driven by the requirements of flash radiography, ion beam generation, and plasma fusion drivers for ~MJ energy levels to be delivered in sub-microsecond timescales.

The first pulsed power systems were probably the power modulators used to drive radar magnetrons at ~100 kW level in the Second World War era. The magnetron requires a repetitive pulse driver developing typically 10–50 kV at 1–10 A, and running at 1–10 kHz repetition rates. The necessary pulse width is of the order of microseconds. The problem lay in the peak pulse power and the fact that this was at a high repetition rate. Spark gaps could not conduct the required current levels, and still recover in the available timescale.

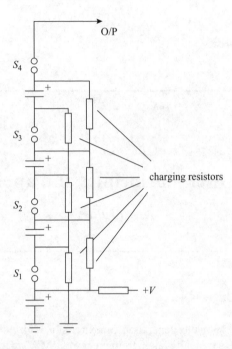

O/P

S_4

S_3

S_2

S_1

charging resistors

+V

Figure 13.1   Four-stage Marx generator with positive polarity output

The solution was found in magnetic pulse compression [1]. A series of saturable magnetic stages are driven by the primary closing switch and energy is transferred between stages in progressively shorter timescales. The primary switch both benefits from turn-on snubber action and does not handle the peak magnetron power and current. Although rotary spark gaps were often used to drive the magnetic pulse compressors, these would only handle a moderate current and so the recovery time would be adequate. Magnetic pulse compression fell into disuse post war as the power handling capabilities of various tubes and the hydrogen thyratron in particular were raised enormously by development. However, the technique has enjoyed a steady renaissance from the early 1980s through to the present day due to novel requirements in high power lasers, ultrawideband radars, corona reactors etc.

Since the 1940s, pulsed power has gone through a series of distinct phases as applications have developed. The modern day pulsed power collective is very much centred on the US, and this is reflected in the preeminence of the biannual IEEE International Pulsed Power conference. The shape of pulsed power has been moulded largely by the requirements of nuclear weapons, inertial confinement fusion (ICF) and directed energy weapons (DEW) programmes. All of these have benefited from extensive support in the United States and former Soviet Union and, to a lesser extent, in Japan and Western Europe. Nonetheless, industrial applications have always been an important aspect of pulsed power. By the 1970s, moderately high power pulse techniques were becoming well established for a range of technical/industrial applications – flash photography, metal forming, pulsed lasers, electric fences, pulsed electrostatic precipitators. This is covered in detail in the series of books by Früngel [2]. In something of a paradigm shift, the spread of applications and pulse techniques that Früngel describes is in present day terminology largely subsumed under 'industrial electronics'. Meanwhile pulsed power has shifted strongly towards the requirements of high energy physics.

The first high energy pulsed power area to develop was that of flash radiography. Much pioneering work was carried out in the UK at the Atomic Weapons Research Establishment, largely due to the individual genius of J.C. 'Charlie' Martin. Flash radiography is the production of intense submicrosecond duration x-ray beams of extremely short duration – typically of the order of 10–100 nanoseconds. The particular application was and remains today the x-ray imaging of nuclear warhead test detonations from outside the metal casing. The pulsed power requirement is for submicrosecond pulses of 1–10 MV and $\sim$1 MA levels to drive the electron beam which produces the x-rays by the Bremsstrahlung mechanism. Rise times of 50–100 ns or less are also required. This led to the development both of fast rise time high energy Marx banks in oil tanks, and of transmission line techniques for shaping of pulses and channelling the energy to the e-beam load. J.C. Martin established many of the ground rules which distinguish pulsed power from general high voltage technology. In particular, the realisation that dielectrics can be operated considerably above their DC breakdown strength for very short impulse durations. This led to the much used 'large area breakdown formulae' for water and oil as transmission line dielectric liquids. The Martin equation for coaxial geometry is:

$$V_b = kA^{-1/10}dt^{-1/3} \text{ (MV)} \tag{13.1}$$

where V_b is the applied voltage in MV, k is a value in the range 0.3–0.6 depending on whether oil or water is used and on the polarity, d is the radial separation of the plates in cm and t is the time to breakdown in μs.

Martin also established formulae for spark gap behaviour which are still in use today, and developed many other pulsed power techniques. A highly readable account of this work may be found in 'Charlie Martin on pulsed power' [3]. As a consequence of the Aldermaston's capabilities, very strong links developed between the Aldermaston group and US groups including Los Alamos, Sandia National Laboratories and the Naval Research Laboratories.

The second major area to develop related to inertial confinement fusion (ICF) programmes based on laser and on particle beams, respectively. The pulsed power requirement is to energise either intense krypton fluoride (KrF) lasers or to produce particle beams, which in turn compress and heat the fusion pellet target. Useful outlines may be found for the Titania KrF scheme [4], the Sandia National Laboratories PBFA-II (Particle Beam Fusion Accelerator-II) scheme [5], and the ongoing Megajoule project in France [6].

Directed energy weapons (DEW) programmes have run continuously since the 'Star Wars' Strategic Defense Initiative (SDI) under President Reagan in the mid 1980s. The term directed energy weapons refers to a whole range of weapons with a variety of functions. High power lasers are envisaged as a means of disabling or directly destroying items such as ballistic missiles or sensor components, in particular by interception in space or near space. High power microwave devices and other radio frequency sources would principally be intended to disrupt enemy communications, computing or control systems.

The decline in funding for nuclear weapons programmes since the end of the Cold War has led to a certain degree of realignment of pulsed power activities towards commercially viable industrial applications. The principal areas of industrial pulsed power application today can largely be grouped under the following headings: environmental cleanup, materials treatment, biotreatments, food treatment, rock breaking and particle generation. The general rationale for the application of high power pulsed techniques rather than forms of continuous DC or AC excitation is that it enables operating regions to be accessed which are not accessible in the steady state. For example, higher values of pulsed fields may be applied to a corona-generating assembly without electrical breakdown occurring than for steady state fields. If electric fields are to be applied in water, the conductivity of the water means that only pulsed fields may realistically be applied at the kV/cm level and above. Some systems may have a non-linear response which can only be driven effectively in a pulsed mode.

Environmental cleanup is applied principally in the areas of flue and exhaust gas cleanup, and of destruction of volatile organic compounds (VOCs). The chemistry can be broadly understood as the stimulation of reactions by means of ionised particles or free radicals. The injection of ionised particles can be by means of an electrical discharge or by ion beam. Controlled electrical discharges can be achieved effectively by means of pulsed corona, which results in the generation of microstreamers. Various other plasma techniques may also be used. Principal among these is the use of cold (non-thermal) plasmas established using the dielectric barrier (DBD) technique. The big issue is not so much whether gas cleanup can be achieved, but whether schemes can be devised which give cost effective cleanup in terms of the joules per cubic centimetre of treated substance (J/cc). Which approaches are likely to be cost effective remains a very open question. For example, Penetrante *et al.* [7] present experimental data showing that for a given rate of NO_x removal, ion beam injection requires only 20 per cent of the input energy required by pulsed corona. However, the capital cost of an ion beam source compared with a pulsed corona source is likely to be high. A general problem with commercial applications of pulsed power is that solutions based upon submicrosecond pulses inherently tend to require relatively complex, costly, systems, as the simplest spark-gap-based approaches are not capable of long time operation between parts replacement. This is typified by the abandonment by the 1950s of static and rotating spark gaps in favour of more stable and controllable gas discharge tubes for radar modulators. Perhaps some of the most promising future applications areas for pulsed power are in fact in very high technology areas with large payback, such as extreme ultra-violet/soft x-ray lithography for the next generation of digital semiconductors.

13.2   Pulsers and topologies

Any formal definition of pulsed power will include the concepts of pulse compression and pulse forming. Pulse compression involves the discharge of energy from a storage

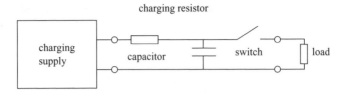

Figure 13.2   Capacitor discharge circuit

device at a faster rate than it has been charged. This is exemplified by capacitor discharge systems, although there are various other forms. Pulse forming is the use of circuit, switching and transmission line techniques to achieve the pulse shape required by the load. A discussion of generic pulse compression and shaping techniques is given by Pai and Qi Zhang [8]. A good up-to-date review of circuit topologies and analysis techniques is given by Smith [9].

13.2.1   Capacitive discharge

Capacitor discharge circuits are certainly the most widely used pulsed power topology. Figure 13.2 indicates a generalised discharge circuit. A suitable charging system is used to charge a capacitor to the required voltage and energy level. The charging process may take minutes for a set of μF capacitors being charged to 50 or 100 kV in a high energy, single-shot system. On the other hand, it can be as short as milliseconds for a 1 nF capacitor in a high repetition rate system. When the capacitor is charged to the required level, a closing switch connects the capacitor to the load. The resulting discharge has a pulse width defined by the load time constant $\tau = C R_{load}$. In pulsed power applications, the discharge time is typically of the order ns–μs. The pulse rise time into the load may be of more significance than the pulse width. The rise time is likely to be affected principally by the closing switch characteristics and by the inductance of the load circuit. One means of obtaining a more rapid voltage rise onto the load is by use of a sharpening gap, as indicated in Figure 13.3. The circuit has a second capacitor intermediate between the source capacitor and the load. This 'peaking' capacitor is of much lower value than the source capacitor. When the primary switch is closed, this intermediate capacitor charges at a rate determined by the closing switch circuit. The sharpening gap is designed to self-break as the intermediate capacitor reaches a suitably high voltage, and thereby connects the intermediate capacitor across the load. The load voltage rise time may then be very rapid since the inductance of the peaking capacitor load loop can be made very low. Elements of the capacitive circuit topology are briefly outlined below.

13.2.2   Charging supplies

Resistive charging has the benefit of simplicity and low cost. The DC source could typically be any of the following: mains transformer with rectifier, Cockcroft–Walton multiplier, switched-mode high voltage DC supply. However, an inherent disadvantage is that when a capacitor is charged from a fixed voltage source, as indicated in

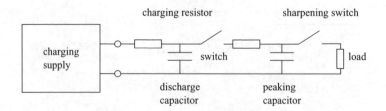

Figure 13.3   Capacitor discharge circuit with peaking capacitor and sharpening switch

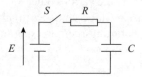

Figure 13.4   Resistive charging circuit for capacitor

Figure 13.4, an energy equal to that stored in the capacitor is dissipated in the charging resistance. For this reason, the use of switched-mode power electronic charging units, generally termed capacitor chargers, is becoming increasingly common. This is particularly the case in rep-rate applications. Charger units range from the miniature systems found in camera flash systems to multi-kW systems with outputs of up to 50–100 kV. The benefit of specialised capacitor charging systems is that the instantaneous output voltage during charging is generally controlled to maintain a constant charging current into the capacitor, giving very high efficiency. Companies producing capacitor chargers include General Atomics, Lambda EMI and ALE.

Although mains supplies will generally be used, battery storage is often employed for mobile systems, for systems with extremely high current demands, or where EMI/EMC concerns can otherwise cause problems. Low voltage ultracapacitors (typically rated at several farads and a few volts) may provide an interface between batteries and faster pulsed circuits, as has been done in some electric vehicle applications. However, these techniques are in the realm of power systems operation and essentially fall outside the pulsed power area. Railgun and coil gun systems will require MW–GW of prime power for several milliseconds pulse duration, and this generally implies the use of specialised pulsed alternators such as compulsators [10, 11].

In passing, 'Electrical interference and protection' by E. Thornton [12] should be mentioned as a near unique, invaluable practical guide to the design of pulsed circuits in terms of the necessary shielding, filtering, screening and grounding.

13.2.3   Capacitors

In order of pulse discharge speed, the main types of capacitor employed in pulsed power work are electrolytic, wound paper and foil, metallised plastic film, and ceramic dielectric. Despite their ubiquity in power electronics rectifier and smoothing applications, electrolytic types are only suitable for millisecond timescale discharges due to their high equivalent series resistance (ESR).

Capacitors based on wound plastic and aluminium foil layers or metallised plastic are suitable for nF–μF ranges and have been developed extensively for high voltage (i.e. 1–100 kV) and fast pulse discharge characteristics. Foil-wound capacitors can be used for discharges with timescales as short as a few hundred nanoseconds. The 1960s' technologies of Kraft paper impregnated with oil were effective for μs timescale discharges, but have been largely superseded by the more recent plastics technologies. Ceramic capacitors based on ferroelectrics, e.g. BaTiO₃, are capable of nanosecond timescale discharges. Ceramic capacitor types are commercially available up to 50 kV in values between ~100 pF and a few nF. A general feature of all types of capacitor used in pulse work is that the lifetime is highly dependent on the degree of voltage reversal imposed as well as on the actual voltage. The highest voltage rating can be achieved if it is possible to ensure no voltage reversal during pulse operation. The greater the voltage reversal, the lower the voltage rating. Reversal characteristics are therefore standard data sheet items for pulse capacitors.

13.2.4   Voltage multiplication: the Marx bank

Limitations imposed by DC corona, DC insulation, power electronics and transformer design mean that generally 50–100 kV is the maximum feasible static capacitor design voltage. Many pulsed power requirements are for voltages in the range 100 kV–2 MV and higher. One solution to this is provided by the Marx generator; this was shown in Figure 13.1. A set of capacitors is charged in parallel to a voltage V via a set of charging resistors. Closure of the switches – typically spark gaps – reconfigures the capacitors in series and a high voltage impulse of the order of microseconds duration can be delivered to the load. Very often, the load will be a transmission line which is resonantly charged by the Marx bank. Generally, the lowest (first) switch in the Marx bank will be actively triggered, and subsequent switches will close due to a combination of ultra-violet (UV) illumination and overvolting. Many high energy Marx banks will use additional spark gaps with mid-plane gaps which can be triggered, either directly or by use of interstage coupling. Reliable operation of Marx banks with high numbers of stages (i.e. 20–30) generally depends upon the use of some triggering linkages between stages and careful optimisation. Ideally, the voltage produced by an n-stage Marx will be nV. Loading by the resistive charging stacks and voltage drops across the switches will reduce this somewhat. Nonetheless, voltages within a few per cent of the ideal can be achieved in practice. Variations on the Marx topology include bipolar charging, which enables doubling of the stage voltage within the ±100 kV charging voltage constraint; use of more than one charging stack in parallel to give faster charge time constants; use of inductive charging for low dissipation in rep-rated applications, etc. Note that a typical high energy Marx bank may have a series $\sqrt{LC}$ time constant of a few μs. This is determined by the total series inductance, due to the combination of the geometrical layout and the internal inductance of the capacitors and switching elements. Pai and Qi Zhang [13] provide a detailed review of Marx bank technology.

13.2.5   Compact, fast rise time Marx banks

The development of compact high voltage sources is of self-evident importance for applications of all types outside the laboratory.

The present day development of Marx bank techniques therefore focuses on compact and high voltage systems. For high voltage, low energy systems (500 kV–1 MV and 10–100 J/shot) the use of ceramic capacitors from manufacturers such as TDK and Murata has become the de facto standard [14]. Such devices enable 50 kV or 100 kV charge voltages, as with high energy Marx banks. Up to MV outputs can therefore be achieved with reasonable numbers of stages (i.e. 4–30). The capacitors have low series inductance contributing to fast load rise times of 10–100 ns. The spark gaps may also be designed for low inductance. Hydrogen is typically used as the switching gas as it has both high voltage hold-off capability and fast recovery rate, so that repetition rates of 1 kHz are readily achieved. Pressurised SF_6 is commonly used as an insulant dielectric gas around the Marx components [15].

Interest in the use of ultrafast Marx generators to provide nanosecond rise time pulses relates to potential applications in ultrawideband (UWB) radar or directed energy (DEW) applications [16, 17]. Achieving this level of rise time typically relies on the Marx being given a layout which enables effective wave propagation between input and output. As with avalanche semiconductor switching generators and DBD lines, a degree of pulse sharpening can occur as the output pulse propagates and builds up between successive spark gaps along the Marx stack.

13.2.6   Pulse compression

Energy may be transferred resonantly between two capacitors by a circuit with a suitable closing switch and inductor. If a second LC loop is added, energy can then be transferred into the third capacitor via the second closing switch. If the $\sqrt{LC}$ resonant period of the second loop is made shorter, the energy transfer occurs in a shorter period and with a higher associated peak current. Figures 13.5 and 13.6 show the generalised LC circuit and waveforms, respectively. With reference to Figure 13.5, the time constant of the first loop is given by:

$$\tau_1 = \pi \sqrt{L_1 \frac{C_0 C_1}{C_0 + C_1}} \qquad (13.2)$$

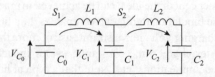

Figure 13.5   Single-stage LC pulse compression circuit

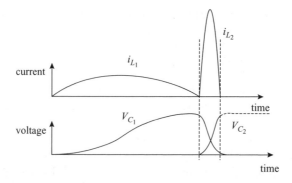

Figure 13.6   Idealised current and voltage waveforms in the pulse compression circuit of Figure 13.5

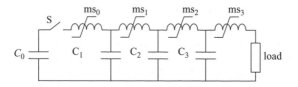

Figure 13.7   Three-stage magnetic pulse compressor circuit

and the time constant of the second loop is given by:

$$\tau_2 = \pi \sqrt{L_2 \frac{C_1 C_2}{C_1 + C_2}} \tag{13.3}$$

This process is termed pulse compression. The switching of the second switch will usually be made to depend upon the state of the circuit. For example, a self-breaking spark gap could be used. An inherent problem with the second and further switches in pulse compressors is that pulse compression inherently means higher power levels and a more arduous switching duty.

13.2.7   The Melville line – magnetic pulse compressor

One way to make the pulse compression switching process occur automatically was devised by Melville in the 1940s [1]. The inductors in the magnetic pulse compression circuit (Figure 13.7) have saturating soft magnetic cores, typically of NiFe ferrite or amorphous metal. These give the inductors a magnetic switching characteristic. While the core is unsaturated ($\mu_0 \mu_r$), the inductor presents a high impedance to current flow. When the core is driven into saturation, the inductor presents a low impedance (μ_0), and current is able to flow according to the value of L_{sat}. The transition between unsaturated and saturated states can occur in 10–100 ns timescales, which means that the inductor exhibits a quite rapid magnetic switching characteristic. The two main parameters for the magnetic switch are the volt–second hold-off and the saturated inductance. The volt–second hold-off is equivalent to the flux swing in the inductor when it is driven from its initial bias point to saturation:

$$(B_s - B_r)AN = \int V \, dt \qquad (13.4)$$

where B_s is the value of saturated magnetic core flux density, B_r is the value of reset magnetic core flux density, A is the magnetic core cross-sectional area and N is the number of turns.

The operation of the magnetic pulse compression circuit of Figure 13.7 is as follows. MS_0 is initially in a high inductance state. After the primary switch S – typically a thyristor – is closed, MS_0 is driven into saturation and the same process repeats through MS_2 and MS_3. C_0 resonantly transfers its charge to C_1. Inductor MS_1 is designed to remain in a high inductance until the point when C_1 is fully charged. MS_1 then saturates, and resonant current flows between C_1 and C_2. Meanwhile, reverse current between C_1 and C_0 is blocked by the fact that MS_0 requires to be driven into negative saturation before it is closed for current in the reverse direction. Only magnetisation current flows during the magnetisation reversal phase. Further pulse compression stages may be added. Between one and four compression stages are usual.

The magnetic pulse compressor (MPC) has gained considerable application in repetitively pulsed applications [18]. These include driver circuits for TEA CO_2 gas lasers [19], excimer lasers, copper vapour lasers [20], corona discharge reactors and ultra-wideband (UWB) radar systems.

13.2.8   Transmission line circuits

The high voltage aspect of most pulsed power circuits imposes generally quite large dimensions and spacings on circuit elements in order to achieve workable electric stress levels. When electrical energy is required to be delivered in the form of short pulses, the physical separation between circuit elements often makes it necessary for these pulses to be delivered along some form of transmission line. The transmission lines will be two-conductor lines e.g. coaxial or parallel-plate, supporting TEM-mode waveforms as the principal mode. Helical slow wave lines are also used. The transmission line transmits power according to the telegraphers' wave equations. It can support forward and backward travelling waves.

13.2.9   Charge lines

Of particular significance in pulsed power applications is the use of transmission line cables as energy stores and pulse sources. A single charge line circuit is shown in Figure 13.8 and a coaxial geometry in Figure 13.9. A length of cable or transmission line is charged to a DC voltage level V by a suitable arrangement. When the switch connecting the transmission line to the load is closed, the load begins to discharge the transmission line. The load pulse rise time is equivalent to the switching time. The values of voltage and current in the load may be found by solving the transmission line equations. In the case of a matched load, the load voltage is equal to half the charging voltage. The load pulse then also has duration of $2T$, where T is the cable

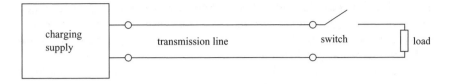

Figure 13.8 Single charge line circuit

Figure 13.9 Schematic arrangement of coaxial charge line with spark-gap switched load

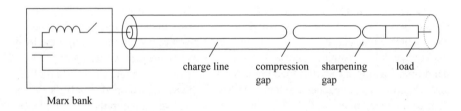

Figure 13.10 Transmission line pulse compression and pulse sharpening arrangement

transit time, given by:

$$T = \frac{l\sqrt{\varepsilon_r}}{c_0} \tag{13.5}$$

where l is the cable length, $c_0 = 3 \times 10^8$ m/s is the free-space velocity of EM waves and ε_r is the relative dielectric constant of the cable.

In the case of the matched load, $2T$ is the time required for the transmission line to become fully discharged. The load pulse waveform is essentially rectangular, with rise and fall times defined by the turn-on time of the switch and the inductance of the line-load connection. Note that the length of transmission line required for a given pulse width depends upon the dielectric constant of the medium. The relative dielectric constants for air/vacuum, oil and water are 1.0, 2.1 and 81, respectively. A 1 m line length therefore corresponds to a double transit time $2T$ (i.e. pulse width) of 6.6 ns in air/vacuum, 10 ns in oil and 90 ns in water.

The charge line configuration is ideally suited to pulse compression and pulse sharpening, as were outlined previously in the circuit context. Figure 13.10 shows the general arrangement. The charge line is resonantly charged by a Marx generator in the order of 1–2 μs. The compression gap initiates a transmission line pulse discharge with a pulse width of typically 100–400 ns. This represents the pulse compression phase. The sharpening gap holds off the initial voltage rise until near peak voltage, then closes. The rise time through the sharpening switch is typically 10–20 ns, and this then propagates forward to the load.

In a variation of the charge line configuration, the charge line is switched to ground by the closing switch. In principle, this applies the same waveform to the load, but has the benefit that the closing switch is not floating. One disadvantage is that the charge line outer conductor floats instead. A second disadvantage is that the switching impulse travels along the charge line before it reaches the load, so that the rise time may be degraded (lengthened) by transmission line losses. This type of system is more suited to relatively low voltage, power and energy levels.

13.2.10   The Blumlein circuit

A variation on the charge line circuit is the Blumlein circuit, invented by A.D. Blumlein. This uses two charge lines with the load interposed, as shown in Figure 13.11. For matched operation, the impedance of each of the charge lines should be one-half of the load impedance. The two lines are initially charged by some suitable charger; note that the right-hand line charging path is through the load impedance. The closing switch is between the two plates (conductors) of the left-hand line. Closing the switch then launches a travelling wave into the left-hand line, and when this wave reaches the load, the load voltage is developed. Analysis shows that, for matched operation, the magnitude of the load voltage impulse is equal to the charge voltage V. The duration of the load impulse is $2T$, where T is the transit time of each of the two charge lines. The benefit of the Blumlein is that the load pulse voltage is equal to the full charge voltage, rather than one-half of the charge voltage as in the single charge line configuration discussed above. A coaxial implementation of the Blumlein is shown in Figure 13.12.

13.2.11   Inductive voltage adders

Inductive voltage adders for pulsed power arose as a means of achieving multi-MV pulses for radiographic and gamma-ray generators. A prime example is the 20 MV Hermes system [21]. The basic principle is that of a series-coupled set of 1:1 transformers (Figure 13.13). Each primary is fed by a separate pulse generator, which may

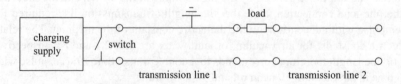

Figure 13.11   Blumlein charge line circuit

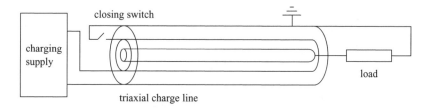

Figure 13.12 *Coaxial Blumlein arrangement; the Blumlein outer conductor is usually grounded*

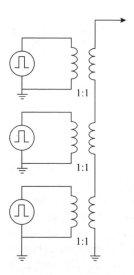

Figure 13.13 *Three-stage inductive adder circuit using 1:1 transformers*

be referenced to ground, and all of the secondaries are coupled in series. The output voltage is the summation of the secondary outputs. In the transmission line embodiment of this method (Figure 13.14), the secondary forms the centre conductor (stalk) of a coaxial line which leads to the anode or cathode of the load, typically an e-beam generator. The magnetic cores are located around the outside of the coaxial line and provide the inductive coupling between the individual primaries with the central coaxial stalk. In the Hermes system, there are twenty 1 MV primaries driven by 1 MV Marx banks, giving the 19–20 MV output voltage. However, systems are not necessarily this large. In MOSFET-switched pulsers manufactured by Kentech Instruments, the outputs of 25 boards, each producing 400 V pulses, are added inductively to form a 10 kV output into 50 Ω. The output rise time is near equivalent to that of a single switching MOSFET element at 2–3 ns. Two-stage inductive adders are used in the RHEPP systems at Sandia. Inductive adders are currently being developed for the next generation radiographic machine (LINX) at AWE Aldermaston in the UK [22, 23].

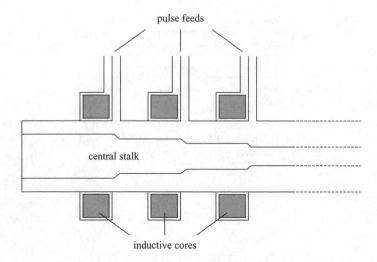

Figure 13.14   Showing first three stages of a transmission line inductive adder

13.2.12   Inductive energy storage

Although capacitive storage has the merit of simplicity, inductive storage potentially offers energy densities around 100–1000 higher. This is based, respectively, on an assumed electric field strength in the capacitive store of $\sim$25 MV/m and magnetic flux density in the inductive store of $10T$ [24]. The caveat lies in the technology: for inductive energy to be extracted, it is necessary to open a switch in order to divert the inductor current into the load. Current interruption is inherently extremely difficult. In the pulsed power context, the two major opening switch types are the plasma erosion opening switch (PEOS) and, increasingly, semiconductor switches. The PEOS addresses the MA current range, whereas semiconductor device types such as MOS-FETs and GTOs will open against $\sim$0.1–5 kA. A new semiconductor approach due to the Ioffe Institute, St Petersburg, Russia is based on exploiting the reverse recovery spike in pn diodes. When a diode current reverses, the diode initially continues to carry current while the pn junction is still in a conductive state. However, as the pn junction recovers into a blocking state, the high impedance produces a reverse recovery voltage spike due to the current which is still flowing. If not properly controlled, this can cause serious damage and device failure in conventional circuits. However, if the spike voltage can be connected across a load, it will give a useful pulse. Work in Russia has focused on optimising pn diodes to maximise this effect. The diode is driven by a relatively short ($\sim$100–200 ns) resonant current waveform. Reverse recovery pulse rise times of a few ns can be obtained, at voltages of several kV, and devices may be stacked to obtain up to $\sim$100 kV total pulse voltage. The device type is referred to as a drift step recovery diode (DSRD) or silicon opening switch (SOS). Most work and exploitation to date has been carried out in Russia, and pulser units are commercially available from Russian companies such as Megapulse and FID Technologies. Some amount of technology transfer out of Russia has also

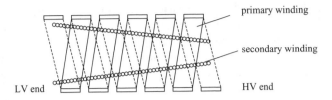

Figure 13.15   *Concentric primary and secondary spiral windings for the Tesla transformer*

occurred. Teramoto *et al.* [25] report a system using pulse compression from 60 Hz mains via magnetic pulse compression and voltage transformation circuit, with final pulse sharpening by the SOS resulting in an output pulse with 140 kV amplitude and 70 ns FWHM.

13.2.13   The Tesla transformer

The open-cored Tesla transformer is generally seen as an exotic curiosity due to its unsuitability for utility power distribution applications. However, amateur enthusiasts construct versions with extreme turns ratios and low coupling coefficients. When driven by primary spark gaps, these can produce radio frequency outputs with amplitudes of up to several MV and capable of producing lengthy ionised channels in air. The former Soviet Union (FSU) has treated the Tesla transformer more seriously and developed the technology considerably [26]. Although the magnetisation current is high, the coupling coefficient is nonetheless generally in the range 0.5–0.8 and pulsed output voltages up to 1 MV can readily be achieved. Figure 13.15 shows a typical Tesla transformer coil arrangement. Primary windings of a few turns are used, which may be axially spaced coils or simply several turns of copper foil wound concentrically. Turns ratios of the order of 1:1200 are typical. The principle of operation is that of a pulse transformer: a thyristor or other primary switch is triggered in the primary circuit and this applies a capacitor voltage to the transformer. Rise times of the order of μs are produced on the secondary winding. Typically, a transmission line or charge line is capacitively charged by the Tesla coil output. Closure of a self-breaking or triggered high pressure spark gap then initiates pulse discharge of the transmission line into the load with a pulse width defined by the charge line length, and subnanosecond rise times can be achieved. Typical loads include UWB emitters and high power X-band oscillators (10 GHz at 150 MW pulse power) with impedances in the range 20–150 Ω.

13.3   Semiconductor switching

13.3.1   Introduction

The prospect of long lifetime, low jitter electronic switches in pulsed power has been a tantalising one over the past decades. The take-up of commercial power devices has been limited, because available device characteristics have only mapped onto certain quite specific and limited areas. Nonetheless, as semiconductor power switching devices have generally developed toward multi-kV, multi-kA voltage and current ratings, combined with submicrosecond switching speeds, the ability to match power devices with applications has steadily broadened. The thyristor has long been applied to microsecond and longer timescale pulse requirements in the 1–100 kA range. The ability of the thyristor to carry large pulse currents is due to the large die areas of up to tens of square cm. Gate turn-off (GTO) and similar thyristor types with heavily interdigitated gate structures are used, due to their inherently rapid switching times, which are of the order of 0.1–1 μs. The bipolar transistor when operated in avalanche mode is capable of subnanosecond switching times, for which there are certain specialised applications. As voltage and current ratings have increased, power MOSFETs and IGBTs are increasingly being used for rapid switching applications. These devices are generally driven within their standard operating envelope. In terms of circuit topology, the two configurations of principal interest are the series/parallel stack and the semiconductor Marx bank. Finally, a range of novel pulsed power semiconductor devices operating at thyristor power levels combined with nanosecond switching speeds has emerged from work at the Ioffe Institute in St Petersburg.

13.3.2   Thyristor

The thyristor is an ideal single shot switch in that as a minority carrier device it is inherently capable of carrying very large impulse currents relative to its continuous current rating. The trigger requirements are also very modest relative to the switched power. Note that the active turn-off capability will generally not be exploited in pulsed power, so that the heavy reverse gate currents required by GTO thyristor devices when switching off large currents are not usually an issue. The Marx bank principle is readily applied to thyristors. Gregory *et al.* [27] describe a four-stage thyristor Marx. The general circuit is shown in Figure 13.16. Diodes are used to clamp the individual thyristor voltages, and each thyristor is actively triggered by resistive coupling to the gate. The use of active triggering to all devices is a general feature of semiconductor Marx circuits, and contrasts with the classic Marx, which relies on sequential overvolting of the spark gaps in the stack. The system described by Gregory *et al.* is relatively modest, producing 2.2 kV, 250 A, 2.5 μs pulses into a 10 Ω load, but it illustrates the thyristor Marx concept very well. Development work has shown that 10–15 stage thyristor Marxes are feasible, using capacitive links between each stage and the next thyristor gate. Repetitive operation is practical except for the efficiency problems posed by resistive charging. Repetitive Marxes have been applied for application in corona flue gas treatment. Again, parallel diodes provide voltage-limiting control on the thyristor devices during the Marx erection phase [28].

In Japan, there is also strong interest in the use of static induction thyristors [29, 30]. These are relatively novel devices with MOSFET-type gate characteristics, with a rapid turn-on time of ∼100 ns, and capable of carrying 30 kA at 15 V forward drop. Pulse widths of 100–500 ns can be switched efficiently. It is envisaged that

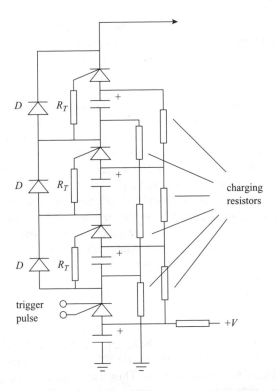

Figure 13.16   Four-stage thyristor Marx with resistive triggering of upper stages

this switching capability could be exploited effectively in rapid resonant charging of corona discharge capacitance for exhaust gas processing. To date, a three-device stack switching 10 kV, 10 kA repetitively has been developed and demonstrated.

13.3.3   Bipolar transistor avalanche mode switching

Baker provides a useful outline of avalanche switching in bipolar junction transistors (BJTs) [31]. A distinction is drawn between slow avalanche processes and current mode second breakdown. The latter is characterised by much higher currents and more complete voltage collapse, but also device destruction if the pulse energy is not curtailed. The subnanosecond switching of BJTs is due to current mode second breakdown, although it is generally referred to as avalanche mode. The successful application of devices in this mode, therefore, depends upon switching only a modest energy per pulse.

Transistors operated in avalanche mode have proved to be reliable and stable switching elements for operation in various forms of Marx bank circuit, switching typically 1–2 kV in <1 ns. Applications are typically micro channel plate systems, pockels cells, and providing the beam sweep in streak cameras. Streak cameras are used to image x-ray emissions, typically over 1–2 ns timescales. The x-rays are directed onto a photo-emissive plate, and the intensity of the emitted photo-electron beam varies over time with the x-ray intensity. The photo-electron beam is then directed between the time axis sweep plates onto photographic film. The requirement on the sweep plate drive voltage is typically for a pulse rise time of 1–2 kV in the order of 1 ns. Circuits for streak camera application are given by a number of authors [32, 33].

Grouping the avalanche transistors into series stacks can give benefits in reducing the effect of the transistor junction capacitance, which acts as a parasitic element on the Marx circuit. Rai *et al.* [34] use five stacks with three series transistors per stack. Each transistor stack may be treated as a single switching element.

Transmission line layouts for avalanche mode switching circuits have been developed at the Lawrence Livermore National Laboratories in the US [35, 36]. Since the inherent switching time of transistors in avalanche mode can be 50–100 ps, the limitation on circuit rise time is generally due to the circuit inductance capacitance and physical size. Design of the Marx layout and ground plane on transmission line principles enables the Marx voltage to propagate from stage to stage in the form of a travelling wave as it develops. Rise times as low as 100 ps have been achieved, with output voltages of up to 8 kV.

13.3.4   MOSFET Marx

The Marx bank configuration has also been applied to power MOSFETs [37]. The achievable rise time is limited by the MOSFET switching speed, which at minimum is typically of the order of 3 ns. This limitation is essentially due to the minimum feasible gate capacitance charging times. As with all semiconductors, it is imperative to trigger the gates of all devices. This may be done by providing each device with its own gate drive, or alternatively by providing suitable coupling capacitances to the gate of each device. The second approach can give equally fast switching times and may give a circuit with a much lower total element count. A disadvantage of the second approach is that it requires a more complex design analysis.

13.3.5   MOSFET switching stacks

MOSFETs can readily be stacked in series to form high voltage stacks, just as thyristors are used in HVDC conversion schemes. Conventional drive arrangements to each floating gate gives the most reliable arrangement. A basic system is described by Continetti *et al.* [38] which was for applying potential difference ion beam experiments. However, the stack switching time in this work was very slow at around 250 ns. Baker and Johnson [39] also describe the design of MOSFET stacks. They reiterate the need for careful gate drive design to ensure that the maximum MOSFET V_{gs} rating of typically 20 V is not exceeded, as this will result in oxide layer punch through. They use passive coupling to drive the gates of devices higher in the stack. This is generally highly effective for stacks at turn-on, as each gate is driven by the switching on of the device beneath it. Clearly, passive coupling arrangements to each device gate generally have cost and simplicity benefits. However, passive coupling is less effective when stacks are required to turn off fast, and this results in an asymmetry between turn-on times of a few nanoseconds, and turn-off times of hundreds of nanoseconds.

Stacked MOSFET techniques are increasingly being adopted by industry. Fully packaged MOSFET stack modules with integral triggering circuits handle voltages up to 60 kV and currents up to 800 A [40]. IGBTs are employed for higher currents of up to 4800 A, albeit at lower switched voltages of 0–15 kV. Applications for such switches are wide ranging and include EMC test equipment, klystron drivers, EMC burst generators, nanosecond pulse generators, pockels cell drivers, thyratron triggering, deflection grid drivers etc. However, when passive coupling triggering techniques are used, this results in slow turn-off characteristics. The mean power ratings for these devices are also relatively low.

A more recent approach now being adopted widely is series connection of commercial MOSFET devices to form 20–50 kV stacks capable of handling 100–200 A [41]. Each MOSFET has a fully active gate driver circuit, and so the stack can be switched off as well as on in the order of 10–20 ns. The series stack essentially functions as a direct replacement for hard tube modulators. This is one reason for engineering a series stack rather than a paralleled switch system driving a pulse transformer. Applications include magnetron modulators, x-ray drivers and very high duty modulators with burst mode pulse repetition frequencies of up to 750 kHz. A typical example of a series/paralleled stack of MOSFETs achieves 55 kV, 200 A switching at repetition rates up to 1333 Hz [42]. The stack volume is 0.1 m^3 and it weighs 91 kg. Forced oil cooling is employed, and the on time may be varied between 0.5–6 μs at duty cycles up to 0.1 per cent.

Kicker pulsers are required for precise ion beam steering and marker applications in linear accelerator installations. A μs length beam can be cleaved into sections by using kicker magnets to deflect segments of the beam as it passes by. It is necessary to energise the kicker magnets with very square pulse waveforms. The pulse specifications are of the order of 10 ns rise and fall times, and 20–200 ns pulse width at 10–15 kV and 200–300 A, i.e. the kicker magnet coils are designed to present a 50 Ω impedance. The requirement has previously been addressed with either planar triode or thyratron switching, but is now increasingly achievable using solid state MOSFET stack approaches which can readily achieve the rise and fall time specifications [43].

13.3.6   MOSFETs with inductive coupling

One approach to high voltage pulse generation is the use of a pulse transformer with a single low voltage primary winding and single high voltage secondary winding. The design of this type of pulse transformer is in fact quite difficult due to the large voltage ratio, high numbers of turns and high voltage insulation requirements. In particular, achieving fast rise times becomes difficult. An alternative approach is to couple the secondaries of a number of 1:1 transformers in series. Each primary can be driven by a grounded pulse circuit, as indicated in Figure 13.13. In the form of a set of primary boards coupling to a transmission line inductive adder it lends itself to highly modular design. Rise times below 3 ns can be achieved.

13.3.7   General semiconductor switching design issues

The attraction of semiconductor switching is the promise of very good lifetimes and high controllability compared with most gas discharge switching devices. However, as with all electronic circuit design, in order to achieve reliable circuit operation it is necessary to ensure that the ratings for device voltage, current and dissipation are adhered to absolutely. In the presence of very large switching currents, this is not necessarily a trivial task. A particular problem with semiconductor switching is the relative sensitivity of low voltage trigger circuits to noise. EMC design and screening of the trigger circuits takes on at least as much significance as it does in measurements and diagnostics. Circuits need careful engineering to ensure that false or self-induced triggering does not occur with consequent runaway failures.

13.3.8   Novel semiconductor devices

Over the past two decades, the group led by Kardo–Sysoev and Grekhov at the Ioffe Institute in St Petersburg has devised at least four novel device types. The group has approached the pulsed power semiconductors issue from a rather different perspective than the US and other western nations. Essentially, they have set out to develop novel devices suitable for pulsed power applications, i.e. having both rapid switching and high power capabilities. This contrasts with the approach taken in the west, which is to apply standard commercial semiconductor devices. The principal limitation of the western approach is the power–speed product. Although thyristor kA and kV ratings are high, the switching times even for high speed GTO devices are broadly in the μs range. Most systems switched by thyristor will therefore require pulse compression or pulse sharpening elements. Semiconductors optimised for general power electronic application may well also have internal features different to what would be ideal for pulsed applications. The approach taken by Kardo–Sysoev and Grekhov has been to revisit generic device types including the pn diode, thyristor, and bipolar transistor and reengineer both device topologies and the junction design specifically to enhance high speed capabilities. Most work has been carried out using silicon as the semiconductor material. The four device types that have arisen are the reverse switching dynistor (RSD), field ionisation dynistor (FID), drift step recovery diode (DSRD) and the delayed breakdown diode (DBD). The RSD and FID devices are essentially variants of the thyristor structure, but with switching times of the order of 0.5–5 ns [44]. The RSD is a two-terminal device and is switched by a reverse pumping current followed by a relatively rapidly applied forward voltage. The FID is a three-terminal device and is more conventionally triggered by a gate pulse of the order of 50–100 V. Both FID and RSD device types are inherently thyristor-like, large die area devices, and can be designed to carry up to several kA and kV.

The drift step recovery diode (DSRD) is a pn junction diode designed for a very abrupt reverse recovery characteristic. Figure 13.17 shows a drive circuit.

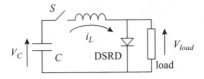

Figure 13.17   Drift step recovery diode (DSRD) pulse generating circuit

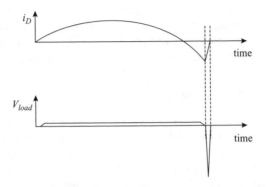

Figure 13.18   Diode current (i_D) and load voltage (V_{load}) waveforms in the DSRD circuit

Figure 13.18 shows device waveforms. The device is driven by a resonant current pulse, initially in the forward direction. As the resonant current reverses, the junction initially conducts, but as the junction regains its blocking state, the current is forced to flow into the high impedance load [45]. Voltages of several kV per device may be generated, and devices may be stacked. This device is often referred to in the literature as the silicon opening switch (SOS).

Finally, the delayed breakdown device (DBD) is a small pn junction device which is driven into reverse breakdown by a suitably high voltage and fast rise time pulse in the 1–10 kV range [46]. If the overvoltage can be applied fast enough, i.e. in a nanosecond timescale, the avalanche breakdown across the pn junction takes the form of a plasma shock wave, and the current rise time through the device is much faster than that of the applied voltage. The device acts as a pulse sharpener.

13.3.9   Applications of novel semiconductors

A typical example of the application of reverse switch-on dynistor (RSD) is driving a pulsed electrostatic precipitator [47]. In this case, the RSD circuit produces 26 kV, 2.5 μs pulses into a 50 pF load. The pulse voltage is superimposed onto a steady 50 kV DC component on the precipitator. The pulsed voltage improves particle charging effectiveness and inhibits the breakdown that would occur for higher values of DC voltage. Energy recuperation is a key capability since most of the load capacitance energy is recoverable. This is done via magnetic pulse compression circuits. The dynistor is a 1.5 kV operating voltage, 6 kA device, i.e. ratings equivalent to a typical GTO thyristor. The key features are that the RSD device is triggered by applied circuit voltages and current reversal. From a conventional viewpoint, the dynistor circuit operation may appear relatively complicated. However, the benefits are that the rapid switching characteristics of the RSD enable good energy transfer efficiency between source and load. Also, a fully solid state solution between the command triggering thyristor input stage and the load is achieved.

Rukin *et al.* [48] describe a system using SOS devices to drive a pulsed corona streamer. A thyristor-triggered magnetic pulse compressor (MPC) is used to drive a series stack of ten paralleled pairs of SOS devices. The output pulse parameters are voltage 1.2 MV, current 4 kA, rise time 10 ns and duration 40–60 ns.

13.3.10   *Electro-optic switching*

Photoconductive switching (PCSS) involves placing a semiconductor slab into a conducting state by means of laser light illumination. Primary applications of the PCSS include ground-probing radar, precision gas switch triggering, short pulse ion accelerators and pockels cell drivers. Pockels cells typically require 0.3–2 kV and 20 A to switch polarisation state. Benefits of PCSS are the low jitter of ∼10 ps, rapid turn-on down to 50–100 ps and low switch inductance. Essentially, there are three operating modes for photoconductive semiconductor switching (PCSS): linear mode, non-linear mode and lock-on. The linear mode requires a high ratio of laser power to the switched electrical power, since one photon drives one electron into conduction mode. The non-linear mode requires high on-state fields ∼10–30 kV/cm, and hence electrical switch losses are high. In the lock-on mode, avalanching occurs which is promising for low on-state loss. In principle, using avalanche mode photoconductive switching in bulk semiconductors, it is possible to control gigawatt levels of peak power with <1 ns rise time [49]. However, current filamentation may lead to low device lifetimes.

The generic issues for efficient optical switching are therefore lock-on and current filamentation. Lock-on is the required condition where the device remains switched on, i.e. in a conductive state, until the current falls below a holding value. This is analogous to the latching effect when a thyristor is triggered. However, lock-on is generally associated with filamentation. Filamentation leads to hot spots and early failure. Resolving these issues is proving to be a major technological challenge.

13.3.11   *Conclusions on semiconductor switching*

It is clear that successful semiconductor applications in pulsed power require extremely careful circuit engineering. This is in order to limit and control fault modes on the one hand, and to ensure that devices are not stressed above their voltage, current and dissipation ratings on the other. However, as with the progressive application of power electronics to industrial drives and traction applications, it is clear that the developing range of device capabilities available is resulting in increased penetration of the pulsed power applications sector. Various applications match the principal device types; MOSFETs, IGBTs and thyristors (GTO, asymmetric and field-controlled). IGBT and MOSFET stacks are being developed as drop-in replacements for thyratrons. Stacked thyristors have been used for some time in kicker magnet drivers, for laser pulsers and also as coil and rail gun switches. The less well known, fast-switching Russian devices are making inroads into novel repetitive applications including corona discharge reactors and ultrawideband sources. Electrooptic switches are being developed for subnanoseconds applications.

13.4 Non-linear transmission lines

The use of non-linear transmission lines (NLTLs) as a means to pulse sharpening was first outlined in detail by Katayev in 1966 [50]. Electromagnetic shock waves may develop in non-linear transmission lines where the point-on-wave velocity varies [51]. The small-signal phase velocity is generally defined as:

$$c_{ph} = \frac{c_0}{\sqrt{\varepsilon_r \mu_r}} \tag{13.6}$$

where c_0 is the phase velocity of electromagnetic waves in free space, and ε_r and μ_r are the relative permittivity and permeability, respectively. If the permeability varies with the applied H-field:

$$\mu_r = f(H) \tag{13.7}$$

then the phase velocity clearly also varies with H. A decreasing or saturating characteristic for μ_r with increasing H results in steepening of the waveform rising edge, and stretching of the falling edge. The effect is analogous to generation of shock waves in air. Clearly, the same mechanism operates for non-linear dielectric, and non-linear transmission lines may be based on both saturable magnetic and non-linear capacitance effects. Equivalent circuits for discrete element NLTLs are indicated in Figures 13.19 and 13.20. Because of the varying point-on-wave velocity, the leading edge rise time of the waveform in a continuous transmission line (Figure 13.21) becomes steeper until it is limited by the magnetic or dielectric frequency response. Shock lines based on ferrite cores are the most widespread type of high power NLTL. These are generally operated in magnetisation reversal mode [52, 53]. In reversal mode the rate of change of magnetisation is governed by the expression [54]:

$$\frac{dM}{dt} = \mu_0 H M_s \gamma \frac{\alpha}{1 + \alpha^2} \left(1 - \frac{M^2}{M_s^2}\right) \tag{13.8}$$

where M is the ferrite magnetisation, M_s is the saturation magnetisation, H is the pulse field in the core in A/m, $\gamma = 1.6 \times 10^{11}$ (rads/s/Tesla) is the gyromagnetic ratio and α is the damping factor, typically in the range 0.1–1.0.

Impedances are typically 50 Ω, but lower values have also been demonstrated [55]. It is possible to combine several shock line sections to obtain very high repetition rates in a burst mode [56]. Since the early 1990s, application of an axial magnetic

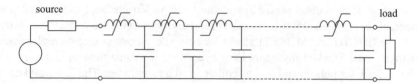

Figure 13.19   Discrete-element NLTL based on non-linear magnetic elements

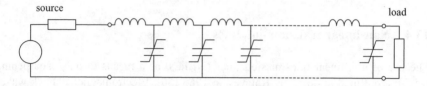

Figure 13.20   Discrete element NLTL based upon non-linear capacitances

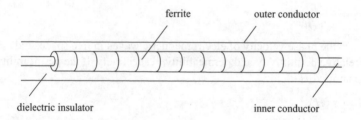

Figure 13.21   Ferrite-loaded coaxial NLTL

field bias has been used to increase switching speed by nearly an order of magnitude. This type of shock line has been demonstrated to be capable of producing sub-100 ps rise times at up to 100 kV into 50 Ω systems [57–59]. Analysis and design of the non-linear transmission line is difficult [60, 61], and lines need to be operated at relatively high electric stress levels in order to achieve high enough non-linearity in the pulsed fields. Nonetheless, the ability to achieve subnanosecond rise times at 10–100 kV levels and at high repetition rates is near-unique and creates a niche for non-linear pulse sharpening lines.

Non-linear capacitance has also received considerable attention as a means for obtaining pulse sharpening [62, 63]. However, the degree of non-linearity (i.e. the ratio of initial to final permittivity at peak field) is far lower for paraelectric or ferroelectric materials such as barium titanate than for ferrites. The high permittivity of most non-linear dielectrics also makes it difficult to design systems for reasonably high impedance. The typical relative dielectric constant for barium titanate of 1000–2000 gives rise to fractional ohm systems, whereas most applications at nanosecond rise time level require 25–50 Ω.

In a discrete element transmission line with periodic non-linear elements such as varactor diodes or capacitors [64, 65], the rise time may be limited by the non-linear

Figure 13.22 *Parallel plate line with periodic dielectric slab loading for soliton generation*

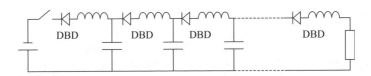

Figure 13.23 *DBD-based pulse sharpening transmission line circuit*

relaxation characteristic. However, it may also be limited by the $\sqrt{LC}$ time constant of each section if this is slower. In the latter case, it is also possible to obtain soliton-like oscillatory waveforms. Interest in solitons as a means of obtaining fast rise time pulses has enjoyed recurrent interest. Soliton formation requires a medium which is both non-linear and dispersive. The non-linear material should also be low loss. Slabs of non-linear dielectric material may be used to form a periodic structure (Figure 13.22), resulting in the generation of recurrent waveforms at 100s MHz to a few GHz. The low impedance typical of these systems enables GW power levels to be achieved [66].

Another form of pulse sharpening is an extension of the sharpening gap approach whereby a switching device transiently holds off an applied rising voltage waveform before switching in a shorter timescale. A series of delayed breakdown devices (DBD) may be used in a variation of the shock line circuit to provide pulse sharpening (Figure 13.23). Output pulses into 50 Ω of 3.8 kV with 100 ps rise time are measured [67].

13.5 Pulsed power applications

13.5.1 Introduction

Industrial applications are driven by a varying combination of technology push and technology pull. The push is exerted by organisations with established pulsed power capabilities searching for new applications or markets. This is exemplified by the Atomic Research Centre (Kernforschungszentrum Karlsruhe) at Karlsruhe, Germany. In the early 1980s, an inertial confinement fusion (ICF) programme based on ion beams was established. The pulsed power requirement was for a 1.7 MV, 1.5 TW, 50 ns beam. However, the ICF programme was then dropped, and only industrial applications of pulsed power are now being pursued. In practice, this means finding uses for the two existing ion beam sources. The types of application being addressed include surface treatments and materials properties characterisation by ion beam impact driven shock wave [68]. The pull is due to applications requirements demanding solutions. Typical examples are in the exhaust gas cleanup and food processing areas. In this section, a variety of industrial and other applications areas for pulsed power techniques will be discussed.

13.5.2   Ion beam materials treatment

The two electron beam facilities at Karlsruhe are capable of 150 keV and 400 keV, respectively. The impact power of these intense beams is of the order of MW/cm^2 over tens of square cm, high enough to melt the surface adiabatically to 20–30 μm and 100 μm, respectively. The rapid melting and subsequent quenching of a thin surface layer is an effective means of achieving required heat treatment effects, e.g. surface hardening of gas turbine blades. A further issue in this application is that the high inlet temperatures that are required for high efficiency operation of gas turbines mean that the turbine blades require a ceramic thermal barrier coating on top of the conventional metallic oxidation protection layer. Typical difficulties result from the grain structure of the protection layer and adhesion of the ceramic layer. Remelting of the surface with an electron beam is found to lead to much improved performance. Lifetimes of 10 000 hours at 950°C have been demonstrated.

In the US, ion beam systems are also being developed for surface treatments. One repetitive system at Sandia National Laboratories is driven by a magnetic pulse compressor [69, 70]. The pulsed power system is a thyratron-controlled five-stage MPC feeding into a two-stage linear induction adder to generate 400 kV 150 ns pulses at a rate of 10 pulses per second. At the time of reporting, the system had been commercially operational for 18 months with only two four-hour down times due to minor component failures.

13.5.3   Air treatment and pollution control

There is a general requirement for cost-effective means of reducing NO_x and SO_x emissions from vehicles and power stations, as well as reducing solid particulate emissions. The cleanup of diesel fumes emitted from vehicle engines is a particular issue. The problem lies in burning off the diesel particulates, and this may rely partially on some type of ceramic reactor complex. The general chemistry for diesel particulate cleanup is well established. Active corona reactors, surface discharges and beds of high dielectric constant beads (which create partial discharges in the interstices) all provide means to burn diesel particulates. However, the energy efficiency of such processes is not well established. The issues appear to be very much the technical questions of how to provide effective insulation, given that carbon is a moderately good conductor, and how to provide the free radicals or ozone at a reasonable cost in energy/power.

Gas discharge reactors are being investigated widely for use in diesel exhaust remediation [71] and other applications. The principal problem lies in achieving a cost-effective treatment process, i.e. achieving adequate design robustness, reliability and lifetime at a reasonable cost and efficiency [72]. A general difficulty with exhaust remediation is that the pulsed corona scheme alone is not necessarily energy efficient [73]. This is attributed to thermalisation of the plasma at > 10 ns timescales. It is likely that a combination of corona reactors with catalytic techniques will be required [74].

Given the need for particle collection, e.g. in diesel exhaust remediation, the idea of combining corona precipitators with a corona reactor is attractive. A brief outline of the corona precipitator follows.

13.5.4 Pulsed corona precipitators

The basic principle in electrostatic precipitators is that dust particles become charged by the corona charge cloud around the high voltage electrode. The electric field then provides a drift force, which results in the particle being attracted to the ground electrode.

Once a particle has acquired charge, the electric field required to produce an adequate drift field is quite moderate. However, the magnitude of electric field required to ionise the region around the high voltage electrode effectively and to charge the particles needs to be much higher. If applied continuously, this will generally result in a transition from partial discharge and corona to arcing and breakdown. This problem is avoided by providing a separate DC bias and pulsed corona energisation fields. To date, pulsed fields have generally been of the order of 100–500 μs width. This is effective for larger particles such as are found in coal burning power station flues, and pulsed voltages may be switched by thyristor technology. However, it appears likely that for sub 2.5 μm particles effective corona discharge precipitators will require faster pulses. This will be an area for investigation, e.g. in diesel exhaust cleanup.

One of the classic problems with pulsed corona reactors in which the reaction plates are charged as capacitors is the need for a circuit to remove the energy from the circuit. This is inherently difficult. Resonant pulser circuits could be constructed in order to enable reverse energy flow, for example based on MOSFET stacks with antiparallel diodes. Alternatively, the magnetic pulse compressor (MPC) is inherently a bidirectional circuit and may also be used to transfer energy back towards a supply.

The general view seems to be that an electrical scheme alone is not efficient. Thermalisation of pulsed corona plasmas for pulse durations > 10 ns is a major problem. The combination of a suitable catalytic converter with the pulsed corona approach is likely to provide a solution.

13.5.5 Biological applications

Biological applications cover a number of areas ranging from sterilisation to the processing of foodstuffs. There is particular interest in the application of pulsed electric fields to kill microorganisms such as *eschericia coli* by electroporation. The bacterial cell wall is perforated by values of electric field of the order of 5 kV/mm [75]. For adequate pulse magnitude, duration and repetition rate, this causes cell death. The pulse repetition rate is found to be of particular significance since there are cell repair or recovery mechanisms which operate over ms–s timescales. If a pulse repetition rate of 100 Hz or above can be sustained, then the treatment effectiveness is much enhanced [76]. It is possible that different values of field level may be required with radio frequency exposure, but this is not yet a well understood or well quantified area. Alternatively, irradiation with gamma rays also has potential for the sterilisation of food surfaces. The benefit of a pulsed power approach to gamma-ray generation based on e-beam technology is that no radioactive materials would be required.

13.5.6   Biofouling and ballast water treatment

Discharge of ships' ballast water can lead to non-indigenous species being spread around the globe with undesirable consequences. The decontamination of ballast waters of large ships is of increasing importance given present-day environmental concerns [77]. On the basis of practical demonstrations, it is claimed that a suitable spark discharge would allow decontamination by five to six orders of magnitude at an energy cost of 27 kWh per 1000 litres of water [78]. This could provide a viable alternative to chemical dosing or direct thermal (heating) approaches.

Biological fouling is generally a major current issue. The use of tributyl-tin (TBT) as an antifouling coating on ships is now banned in the US with the consequence that most US vessels are simply treated in overseas waters, i.e. the problem is shifted to countries with less stringent regulations. There is therefore a genuine need for a non-polluting replacement.

Experimental work shows that biological fouling of oceanographic sensors (e.g. for salinity) can be remedied by application of $\sim$7 kV/mm, 100 ns duration pulses to water as it is drawn through the sensor [79].

13.5.7   Food processing

Other areas in which technology push is exemplified are the food processing industries. In the production of citrus juices, Fishback and Nubbe [80] have recently found that the application of pulsed discharges can lead to the inactivation of enzymes which initiate degradation and limit shelf life. Typical pulse widths are 100 ns–5 μs, and only moderate pulse energies are required per unit volume of juice (0.1–0.4 J/ml). This is in contrast to the much higher energies required to have biocidal (sterilising) effects. The system is driven by an AC mains-fed magnetic pulse compressor with thyristor switching of the first stage. The system is straightforward and reliability is good.

Juice extraction from diverse fruits and vegetables can also be enhanced by electrically-driven breakdown of the cell walls. A benefit of this method is that the temperature rise is extremely low. For example, in sugar processing, the application of Marx-driven 300 kV $\sim$20 A discharges at 20 Hz p.r.f. is being found to make juice extraction from beet more effective than alternative means [81]. Specific energy consumption is 3 kJ/kg. The electroporation pretreatment then allows mechanical extraction of juice to take place at reduced temperatures, which saves energy and facilitates purification of the juice. Demonstrator plants to date have operated at a few 100 kg/hour. This compares with typical throughput in large plants of 1000 tonnes/hour.

The sterilisation of liquid foods such as juices and milks by means of pulsed electric field (PEF) is an area of investigation [82]. Advantages over thermal techniques are claimed to be that the taste is not adversely affected.

13.5.8   Water purification

Water purification relates to both bacterial decontamination and the removal of heavy metals, organic and inorganic materials such as PCBs, hydrocarbons etc. This is generally a multifactor process with synergies at work. Application of an electric field followed by a pulsed discharge gives rise to direct electric field effects (electroporation, cell rupture, death), which are followed by production of ultra-violet, ozone production, other active particles or radicals and possibly shock waves in the fluid. These can give rise to chemistry effects on pollutants as well as on biological contaminants. The combined effect of all these is far greater than that due to electric field alone [83]. Electrical breakdown can be achieved at moderate applied fields by sparging of the liquid with air, nitrogen or other gases. The electric breakdown occurs in the gas bubbles due to field enhancement effects. Another approach is the use of a packed-bed reactor containing particles with high dielectric constant [84].

Due to industrial pollution, the need for techniques for water purification for potable use is extreme in western Siberia [85]. The principal requirement is for reduction of heavy metals, hydrocarbons and phenols. Pulsed dielectric barrier discharge (DBD) systems are currently being used at 40 installations [86]. This type of system principally produces photochemical oxidation via ultra-violet, H_2O_2 hydrogen peroxide and ozone (O_3), and the water is sprayed to maximise surface area. The DBD process also has a bactericidal effect. The DBD is driven by a thyristor-triggered magnetic pulse compressor, and after the DBD process, filtering is employed to remove resulting particulates.

13.5.9   Mechanical applications of spark discharges

The shock waves produced by electric spark discharges can be exploited for a range of applications. Much of the work to date has taken place in Russia, although there is increasing interest in its application in Germany [87] and Japan [88]. Spark discharges can be used for the direct fragmentation of reinforced concrete in recycling applications. Concrete is well suited to shock-wave-induced fragmentation since it is weak in tension and shear stress conditions. This potentially offers considerable environmental benefits, given that the production and use of concrete contributes ten per cent of greenhouse gas emissions [89]. Spark discharge techniques are also being applied at pilot plant level at the ash treatment plant in Bielefeld–Herford, Germany. A seven-stage Marx produces 350 kV with a rise time of $\sim$150 ns. This is rapid enough that the water around the ash acts as an insulator and effectively breakdown is channelled through the solid components. Spark discharge enables heavy metal concentrations – Zn and Pb in particular – to be reduced below statutory levels.

However, the materials are still in solution and have to be removed from the water. Energy consumption is ~3 kW/h 2.5 tonnes/hour and 0.5 m^3 water per tonnes of treated ash; the treated ash can be used in road construction, for example. Drilling, boring and fragmentation of rocks by pulsed power spark discharge are also feasible. A fluid is required in order to transfer the shock energy; from an electrical efficiency perspective the fluid should be insulating oil, but from an environmental perspective water is preferable. Livitsyn *et al.* [90] review the processes of electric pulse internal breakdown, electrohydraulics, surface flashover, plasma blasting and the electric explosion of wires. Their experimental work exploits the lower bulk breakdown and surface flashover strength of rock relative to water. A 100–200 kV Marx generator is used in single-shot, both with direct connection of Marx to load, and in a pulse compression mode with inductive storage of Marx output with opening foil switch.

Granulation and crushing of ores are claimed to be areas in which spark discharge (electroshock) technologies can outperform mechanical crushing. This can enable valuable materials such as gold or diamond to be more effectively recovered from ores in which the concentration may be less than one per cent. It is claimed by a number of authors that the mechanical damage to the recovered ore and contamination of the recovered product is very low [91].

Treatment of sewage sludge is an area in which use of electroshock techniques accelerates the decomposition of organisms. In one test plant [92], a 10 kV voltage is applied to the electrodes. After breakdown, a 600 A peak pulse current flows. The repetition rate is 100 Hz. Although the treatment requires significant amounts of energy at 1000 kJ/kg, nonetheless this compares with 10 kJ/kg for ultrasonics or high pressure homogeneous crushers.

The principle exploding is being investigated as a means of substitution for high explosive TNT in rock blasting applications. A high current discharge of the order of 100 kA through fine aluminium wire contained in a cartridge can be used to produce 0.5–1 kg high explosive equivalent [93]. Electrical explosion of fine wires can also be used in the production of nanopowders, i.e. powders with particle sizes in the range 50 nm–5 μm.

Spark discharges can also be used as a means of providing intense ultra-violet illumination for surface treatment applications. Voronov *et al.* [94] describe the use of a 300 kA 200 μs wide pulse with 50 MW arc power to double the hardness of a steel surface layer over a depth of 40 μm.

13.5.10   Medical applications

One medical application of pulsed power lies in the use of pulsed magnetic fields to provide electric field stimulation within the human body, for example for the stimulation of kidney peristalsis following lithotripsy. A 15 kV, 15 kA coil design is described by the Pulsed Power Group at Loughborough [95] which produces a pulsed electric field at the kidney of ~350 V/m with a FWHM of 1 ms. The possibilities of using pulsed electric fields to induce drug delivery via cell wall electroporation effects are currently being investigated [96]. Use of low intensity millimetre microwaves

(MMW) as a general therapeutic technique has been developed since the early 1980s in Eastern Europe, China and Russia [97].

13.5.11 Ultrawideband and HPM applications

There is considerable interest in the effects of high power RF radiation on electronic and electrooptic systems of all types. On the one hand, armed forces are interested in extending their electronic warfare capabilities. On the other, there is a need to guard both civil and military targets against potential terrorist threats based on RF techniques. Since the development of the atomic bomb, nuclear electromagnetic pulse (NEMP) hardening has been mandatory for military equipment of all types. The NEMP waveform has double exponential form and is essentially similar to a lightning waveform but with a much faster rise time of a few nanoseconds and a fall of a few microseconds. However, the high power RF threat is much broader. It encompasses a variety of waveforms including not only double exponential but also impulse and short pulse sinusoidal waveforms with a range of envelope characteristics and frequencies between a few MHz and tens of GHz. High power short pulse waveforms radiated above 1 GHz are generally referred to as high power microwaves (HPM). Impulse waveforms with rise times of the order 0.1–1 ns are generally referred to as ultra-wideband (UWB) due to their very broad frequency spectrum. A generally accepted definition of an ultrawideband signal is when the half-power bandwidth $f_2 - f_1$ is greater than 25 per cent of the centre frequency f_c of the signal spectrum i.e.:

$$\frac{f_2 - f_1}{f_c} \geq 25\% \tag{13.9}$$

The reality is that effective application of RF weapons is a sophisticated task requiring detailed knowledge of the target system, coupling mechanisms, generator technologies and so on. Although there has been military and defence industry activity in the area for a number of years, programmes of university research into pulsed RF techniques are also taking shape. In the US, multidisciplinary university research initiative (MURI) programmes provide for collaboration between the Department of Defense and the university sector. The types of effect being investigated include the effects on electronic circuits and systems of pulse shape, modulation, timing, clock states, pulse repetition rates, multiple failure mode synergies and induced internal arcing and breakdown. Thus, many individual aspects of the RF circuit threat are being brought together that have been studied separately over the years.

Published ultrawideband research focuses largely on source technologies. These include subnanosecond spark gap switching [98, 99], solid state switching and light-activated switching systems driving into parabolic antennae or specialised types such as TEM horns which can radiate impulse waveforms particularly effectively [100]. Source pulsed powers may be of the order of 10 MW to 1 GW and pulse repetition rates up to several hundred Hz. The requirement is probably to induce fields of several kV/m at a target location.

13.5.12   X-ray simulators

Ware *et al.* [101] provide an excellent review of the historical development of x-ray simulators in the US, under the aegis of the present Defense Threat Reduction Agency (DTRA), between the 1960s and the late 1990s. The main cost of modern simulators in fact lies in the power conditioning and flow arrangements between the capacitor Marx bank and the actual plasma or ion beam load. Early systems made a transition from oil to water dielectric for transmission lines, enabling much lower impedance and higher power flow. All such systems exploit the discovery by J.C. Martin that for short, single-shot pulses, water dielectric can be operated at well above its breakdown strength, because the breakdown streamer channel impedance is high enough that relatively little energy is lost. The 1970s saw much basic physics established: diode physics beam transport, pinch and plasma radiation sources. The 1980s and 1990s saw stressing of all components to minimise size. The late 1990s have seen the discovery of implosion of radiative plasmas in wire arrays (the 'Z' machine) as a means of producing intense x-rays. The size of the 'Z' machine at Sandia is awesome. It contains around 1.7 million litres of Diala insulating oil and 2.7 million litres of deionised water. For compactness, future directions for x-ray generators are likely to involve ultrafast Marxes, plasma opening switches and non-water systems.

Most x-ray simulator systems are modest in comparison with machines such as 'Z'. A good description of a modern day, compact, moderate energy x-ray generator for simulator studies is provided by Miller *et al.* [102].

The current direction of ICF fusion research is summarised by Cook [103]. In 1998–1999, the Sandia 'Z' machine was built up to achieve 1.8–1.9 MJ of x-rays in a 6 ns burst. The energy conversion efficiency from Marx bank to x-ray yield is surprisingly high at $\sim$15 per cent.

13.6   Conclusions

Pulsed power originated with the invention of the Marx bank in the early 1920s, enabling the simulation of lightning strikes and switching operations on power system components. Between the 1960s and the beginning of the twenty-first century, the demands of high energy pulsed power applications have stretched Marx bank techniques enormously. Refinements of triggering and switching techniques, and control and exploitation of the stray capacitances in the systems, enable 100–200 ns rise times to MV levels and above with MJ of energy being stored and delivered. The requirements of exotic radiographic, e-beam and plasma pinch loads have demanded the use of pulse forming networks or lines interposed between the Marx and the load to provide the requisite pulse shaping and power levels. Meanwhile the rise time capability of low energy Marxes has been reduced to nanosecond levels as a result of developments in capacitor and switching technologies. The latest direction is now for the development of ultrafast high energy Marxes with adequately low inductance for direct feed of compact e-beam loads.

Although the largest and most spectacular pulsed power systems at present remain the single shot radiographic and fusion research facilities, there is an ever increasing requirement for repetitive systems. These range from ultrawideband radar pulsers through corona reactors to ion beam material processing facilities, and typically may have mean power levels in the kW–MW range. The mean power and lifetime requirements give rise to a whole new set of design problems and constraints. Magnetic pulse compressors were first developed in the 1940s for radar magnetron modulators. The notable high power pulse compression qualities of these systems have been revived and developed enormously for a range of new applications, including high power gas laser drivers, e-beam drivers, UWB pulsers and pulsed corona reactors and precipitators. The benefits of magnetic pulse compression include high energy efficiency, rep-ratability and the ability to recover energy from reactive loads. The development of magnetic switch technology also led to the development of inductive adders, which use the same form of magnetic core. Based on amorphous alloy materials, these enable multi-MV pulse generation by summing the output voltages of a number of Marx or other modules.

Pulsed power has principally developed in response to high energy physics and weapons programme requirements; plasma drivers, x-ray drivers, magnetron drivers, laser drivers, ion beam steering and acceleration. However, the industrial application of pulsed power is increasing. Examples include the use of high energy electron beams in plasma reactors for flue gas cleanup, and in ion beam surface treatments, typically of high technology turbine blades for operation under high temperatures. High current (≥ 100 kA) electric discharges may be used to fragment valuable mineral ores. Fragmentation of structural materials in recycling situations is also being developed. A general feature in pulsed power is the increasing insertion of power electronic devices into areas which even ten years ago were reserved for spark gaps and other gas discharge devices. This is due first to the increasing power and switching capabilities of well established commercial device types such as MOSFETs, IGBTs and thyristors. Second, there has been the recent discovery and development of nanosecond timescale switching capabilities in a range of non-standard device types, including the field ionisation dynistor (FID), the drift step recovery diode/silicon opening switch (DSRD/SOS) and the delayed breakdown device (DBD). The repetitive and long lifetime switching of voltage and power levels in the 10–100 kV and 2–200 MW range using semiconductors will therefore increasingly become the norm.

For further reading see References 2, 3, 8, 9, 12 and 104.

13.7   References

1   MELVILLE, W.S.: 'The use of saturable reactors as discharge devices for pulse generators', *IEE Proc.*, Part 3 (Radio and Communication), 1951, **98**, pp. 185–207

2   FRÜNGEL, F.B.A.: 'High speed pulse technology', (Academic Press), vol. I: 'Capacitor discharges, magnetohydrodynamics, x rays, ultrasonics', 1965; vol. II: 'Optical pulses, laser measurement techniques', 1965; vol. III: 'Capacitor discharge engineering', 1976; vol. IV: 'Sparks and laser pulses', 1980

3   MARTIN, T.H., GUENTHER, A.H., and KRISTIANSEN, M. (Eds): 'J.C. Martin on pulsed power, series: advances in pulsed power technology' (Plenum Press, 1996)

4   KIDD, A.K., ANGOOD, S.M., BAILLY-SALINS, R., CARR, P.S., HIRST, G.J., and SHAW, M.J.: 'Pulsed power system for the Titania KrF laser module'. Proceedings of 9th IEEE *Pulsed power* conference, 1993, pp. 718–722

5   MARTIN, T.H., TURMAN, B.N., and GOLDSTEIN, S.A., *et al.*: 'The pulsed power characterisation phase'. Proceedings of IEEE *Pulsed power* conference, 1987, pp. 225–232

6   RUBIN DE CERVENS, D., and MARRET, J.P.: 'Energy bank of laser mega-joule CEA-Project'. Proceedings of IEEE *Pulsed power* conference, 1999, pp. 356–358

7   PENETRANTE, B.M., HSIAO, M.C., and BARDSLEY, J.N., *et al.*: 'Electron beam and pulsed corona processing of volatile organic compounds and nitrogen oxides'. Proceedings of IEEE *Pulsed power* conference, 1995, pp. 144–149

8   PAI, S.T., and QI ZHANG: 'Introduction to high power pulse technology', advanced series in electrical and computer engineering – vol. 10' (World Scientific, 1995), pp. 1–5

9   SMITH, P.W.: 'Transient electronics' (Pulsed Circuit Technology, Wiley, 2002)

10 DRIGA, M.D.: 'Compulsators: advanced pulsed power supplies for hyper-velocity accelerators'. Proceedings of 12th IEEE *Pulsed power* conference, 1999, pp. 324–328

11 SCOTT, D.J., CALFO, R.M., and SCHWENK, H.E.: 'Development of high power density pulsed ac generators'. Proceedings of 8th IEEE *Pulsed power* conference, 1991, pp. 549–552

12 THORNTON, E.: 'Electrical interference and protection' (Ellis Horwood, London, 1991)

13 PAI, S.T., and QI ZHANG: 'Introduction to high power pulse technology, advanced series in electrical and computer engineering – vol. 10' (World Scientific, 1995), pp. 12–28

14 GOERZ, D.A., FERREIRA, T.J., NELSON, D.H., SPEER, R.D., and WILSON, M.J.: 'An ultra-compact Marx HV generator'. Proceedings of 13th IEEE *Pulsed power* conference, 2001, pp. 628–631

15 DRAGT, A.J., and ELIZONDO, J.M.: 'Compact, battery powered, 400 kV 40 joule portable Marx generator'. Proceedings of 13th IEEE *Pulsed power* conference, 2001, pp. 1555–1558

16 MAYES, J.R., CAREY, W.J., NUNNALLY, W.C., and ALTGILBERS, L.: 'The Marx generator as an ultrawideband source'. Proceedings of 13th *Pulsed power* conference 2001, pp. 1665–1668

17 MAYES, J.R., CAREY, W.J., NUNNALLY, W.C., and ALTGILBERS, L.: 'The Gatling Marx generator system'. Proceedings of 13th *Pulsed power* conference, 2001, pp. 504–507

18 HARJES, H.C., PENN, K.J., and REED, K.W., *et al.*: 'Status of the repetitive high energy pulsed power project'. Proceedings of

8th IEEE *Pulsed power* conference, San Diego, 16–19 June 1991, pp. 543–548

19 SWART, P.H., VON BERGMANN, H.M., and PRETORIUS, J.H.C.: 'All solid state switched pulser for multikilowatt, multi-kilohertz excimer and CO2 TEA lasers'. Proceedings of 8th IEEE *Pulsed power* conference, 17–19 June 1991, pp. 743–749

20 COOK, E.G., BALL, D.G., and BIRX, D.L., *et al.*: 'High average power magnetic modulator for copper lasers'. Proceedings of 8th IEEE *Pulsed power* conference, 1991, pp. 537–542

21 RAMIREZ, J.J., PRESTWICH, K.R., and BURGESS, E.L., *et al.*: 'The Hermes-III program'. Proceedings of 6th IEEE *Pulsed power* conference, 1987, pp. 294–303

22 THOMAS, K.J., WILLIAMSON, M.C., and PHILLIPS, M.J.: 'LINX and future IVA machines'. IEE *Pulsed power* symposium, 1999, digest 99/030, pp. 2/1–2/4

23 THOMAS, K.J., WILLIAMSON, M.C., CLOUGH, S.G., and Phillips, M.J.: 'Prototype IVA module (PIM)'. Proceedings of 13th IEEE *Pulsed power* conference, 2001, pp. 306–309

24 PAI, S.T., and QI ZHANG: 'Introduction to High Power Pulse Technology', advanced series in electrical and computer engineering – vol. 10' (World Scientific, 1995), pp. 28–30

25 TERAMOTO, Y., DEGUCHI, D., KATSUKI, S., NAMIHIRA, T., AKIYAMA, H., and LISITSYN, I.V.: 'All solid state trigger-less repet-itive pulsed power generator utilising semiconductor opening switch'. Proceedings of 13th *Pulsed power* conference, Las Vegas, 2001, pp. 540–543

26 GUBANOV, V.P., KOROVIN, S.D., PEGEL, I.V., ROITMAN, A.M., ROSTOV, V.V., and STEPCHENKO, A.S.: 'Compact 1000pps high volt-age nanosecond pulse generator', *IEEE Trans. Plasma Sci.*, 1997, **25**, (2), pp. 258–265

27 GREGORY, K., STEVENSON, P., and BURKE, R.: 'Four stage Marx generator using thyristors', *Rev. Sci. Instrum.*, 1998, **69**, (11), pp. 3996–3997

28 MAEYAMA, M., and YOSHIDA, M.: 'Improvement of a high repetitive rate static impulse voltage generator'. Proceedings of 13th IEEE *Pulsed power* conference, 2001, pp. 1264–1267

29 NISHIKAWA, K., OKINO, A., WATANABE, M., HOTTA, E., KO, K-C., and SHIMIZU, N.: 'High rep rate operation of pulsed power modulator using high voltage static induction thyristors'. Proceedings of 13th IEEE *Pulsed power* conference, 2001, pp. 1571–1574

30 TSUNODA, K., MAEYAMA, M., HOTTAM, E., and SHIMIZU, N.: 'Switch-ing properties of series connected static induction thyristor stack'. Proceedings of 13th IEEE *Pulsed power* conference, 2001, pp. 1786–1789

31 BAKER, R.J.: 'High voltage pulse generation using current mode second breakdown in a bipolar junction transistor', *Rev. Sci. Instrum.*, 1991, **62**, (4), pp. 1031–1036

32   BAKER, R.J., and JOHNSON, B.P.: 'Sweep circuit design for a picosecond streak camera', *Meas. Sci. Technol.*, 1994, **5**, pp. 408–411

33   LIU, J.Y., SHAN, B., and CHANG, Z.H.: 'High voltage fast ramp pulse generation using avalanche transistor', *Rev. Sci. Instrum.*, 1998, **69**, (8), pp. 3066–3067

34   RAI, V.N., SHUKLA, M., and KHARDEKAR, R.K.: 'A transistorised Marx bank circuit providing sub-nanosecond high voltage pulses', *Meas. Sci. Technol.*, 1994, **5**, pp. 447–449

35   FULKERSON, E.S., and BOOTH, R.: 'Design of reliable high voltage avalanche transistor pulsers'. Proceedings of IEEE *Power modulator* symposium, June 1994

36   FULKERSON, E.S., NORMAN, D.C., and BOOTH, R.: 'Driving pockels cells using avalanche transistor pulsers'. Proceedings of IEEE *Power modulator* symposium, 1997

37   BAKER, R.J., and JOHNSON, B.P.: 'Applying the Marx bank circuit configuration to power MOSFETs', *Electron. Lett.*, 1993, **29**, (1), pp. 56–57

38   CONTINETTI, R.E., CYR, D.R., and NEUMARK, D.M.: 'Fast 8 kV metal-oxide semiconductor field-effect transistor switch', *Rev. Sci. Instrum.*, 1992, **63**, (2), pp. 1840–1841

39   BAKER, R.J., and JOHNSON, B.P.: 'Stacking power MOSFETs for use in high speed instrumentation', *Rev. Sci. Instrum.*, 1992, **63**, (12), pp. 5799–5801

40   Behlke Instruments GmbH catalogue, Germany

41   RICHARDSON, R., RUSH, R., ISKANDER, M., GOOCH, P., and DEAL, K.: 'High voltage direct switching solid state MOSFET modulators for driving RF sources'. Proceedings of IEE *Pulsed power* symposium, April 1999, paper 13, pp. 14–15

42   RICHARDSON, R., RUSH, R.J., and ISKANDER, S.M., *et al.*: 'Compact 12.5 MW, 55 kV solid state modulator'. Proceedings of 13th IEEE *Pulsed power* conference, 2001, pp. 636–639

43   DE HOPE, W.J., CHEN, Y.J., COOK, E.G., DAVIS, B.A., and YEN, B.: 'Recent advances in kicker pulser technology for linear induction accelerators'. Proceedings of 12th IEEE *Pulsed power* conference, 1999, pp. 416–419

44   EFANOV, V.M., and KRIKLENKO, A.V.: 'Small size solid state nano- and picosecond pulsers on the basis of fast ionization devices'. Proceedings of 13th *Pulsed power* conference, Las Vegas, 17–22 June 2001, pp. 479–481

45   GREKHOV, I.V., and MESYATS, G.A.: 'Physical basis for high power semiconductor nanosecond opening switches'. Proceedings of 12th IEEE *Pulsed power* conference, 1999, pp. 1158–1161

46   KARDO-SYSOEV, A.F., EFANOV, V.M., and CHASHNIKOV, I.G.: 'Fast power switches from picosecond to nanosecond timescale and their application to pulsed power'. Proceedings of 10th IEEE *Pulsed power* conference, 1995, pp. 342–347

47   GREKHOV, I.V., KOROTKOV, S.V., KOZLOV, A.K., and STEPANYANTS, A.L.: 'A high-power high-voltage pulse generator based

on reverse switch-on dynistors for an electrostatic precipitator power supply', *Instrum. Exp. Tech.*, 1997, **40**, (5), pp. 702–704

48 RUKIN, S.N., MESYATS, G.A., PONOMAREV, A.V., SLOVIKOVSKY, B.G., TIMOSHENKOV, S.P., and BUSHLYAKOV, A.I.: 'Megavolt repetitive SOS based generator'. Proceedings of 13th IEEE *Pulsed power* conference, 2001, pp. 1272–1275

49 ZUTAVERN, F.J., LOUBRIEL, G.M., and MAR, A., *et al.*: 'Photoconductive semiconductor switch technology'. Proceedings of 12th IEEE *Pulsed power* conference, 1999, pp. 295–298

50 KATAYEV, I.G.: 'Electromagnetic shock waves' (Iliffe Books, 1966)

51 OSTROVSKII, L.A.: 'Formation and development of electromagnetic shock waves in transmission lines containing unsaturated ferrite', *Sov. Phys.-Tech. Phys.*, 1964, **8**, (9), pp. 805–813

52 WEINER, M., and SILBER, L.: 'Pulse sharpening effects in ferrites', *IEEE Trans. Magn.*, 1981, **MAG-17**, pp. 1472–1477

53 SEDDON, N., and THORNTON, E.: 'A high-voltage, short rise-time pulse generator based on a ferrite pulse sharpener', *Rev. Sci. Instrum.*, 1988, **59**, pp. 2497–2498

54 GYORGY, E.M.: 'Rotational model of flux reversal in square loop ferrites', *J. Appl. Phys.*, 1957, **28**, (9), pp. 1011–1015

55 SHVETS, V.A.: 'Nonlinear multichannel pulse-sharpening line containing ferrite rings with nonrectangular hysteresis loop', *Instrum. Exp. Tech.*, 1982, pp. 908–912 (*Pribory i Tekhnika Eksperimenta*, 1982, (4), pp. 116–119)

56 IKEZI, H., DE GRASSIE, J.S., LIN-LIU, R., and DRAKE, J.: 'High power pulse burst generation by magnetically segmented transmission lines', *Rev. Sci. Instrum.*, 1991, **62**, (12), pp. 2916–2922

57 BENSON, T.M., POULADIAN-KARI, R., and SHAPLAND, A.J.: 'Novel operation of ferrite loaded coaxial lines for pulse sharpening applications', *Electron. Lett.*, 1991, **27**, (10), pp. 861–863

58 POULADIAN-KARI, R., SHAPLAND, A.J., and BENSON, T.M.: 'Development of ferrite line pulse sharpeners for repetitive high power applications', *IEE Proc. H, Microw. Antennas Propag.*, 1991, **138**, (6), pp. 504–512

59 DOLAN, J.E.: 'Effect of transient demagnetisation fields on coherent magnetic switching in ferrites', *IEE Proc. A, Sci. Meas. Technol.*, 1993, **140**, (4), pp. 294–298

60 DOLAN, J.E.: 'Simulation of ferrite loaded coaxial lines', *Electron. Lett.*, 1993, **29**, (9), pp. 762–763

61 DOLAN, J.E., and BOLTON, H.R.: 'A length equation for ferrite-loaded high voltage pulse sharpening lines', *Electron. Lett.*, 1998, **34**, (13), pp. 1299–1300

62 BAKER, R.J., HODDER, D.J., JOHNSON, B.P., SUBEDI, P.C., and WILLIAMS, D.C.: 'Generation of kilovolt subnanosecond pulses using a nonlinear transmission line', *Meas. Sci. Technol.*, 1993, **4**, pp. 193–895

63 WILSON, C.R., TURNER, M.M., and SMITH, P.W.: 'Electromagnetic shock-wave generation in a lumped element delay line containing nonlinear ferroelectric capacitors', *Appl. Phys. Lett.*, 1990, **56**, (24), pp. 2471–2473

64   RODWELL, M.J.W., *et al.*: 'Active and nonlinear wave propagation devices in ultrafast electronics and optoelectronics', *Proc. IEEE*, 1994, **82**, (7), pp. 1037–1059

65   BAKER, R.J., HODDER, D.J., JOHNSON, B.P., SUBEDI, P.C., and WILLIAMS, D.C.: 'Generation of kilovolt-subnanosecond pulses using a nonlinear transmission line', *Meas. Sci. Technol.*, 1993, **4**, pp. 893–895

66   IKEZI, H., deGRASSIE, J.S., and DRAKE, J.: 'Soliton generation at 10 MW level in the very high frequency band', *Appl. Phys. Lett.*, 1988, **58**, (9), pp. 986–987

67   FROST, C.A., MARTIN, T.H., FOCIA, R.J., and SCHOENBERG, J.S.H.: 'Ultra low jitter repetitive solid state picosecond switching'. Proceedings of 12th IEEE *Pulsed power* conference, 1999, pp. 291–294

68   BLUHM, H., BAUMUNG, K., and BOEHME, R., *et al.*: 'Pulsed power science and technology at Forschungszentrum Karlsruhe'. Paper A.04, proceedings of international conference on *Pulsed power applications*, University of Applied Sciences, Gelsenkirchen, 27–29 March 2001

69   BARRETT, D.M., COCKREHAM, B.D., and PRAGT, A.J., *et al.*: 'A pulsed power modulator system for commercial high power ion beam surface treatment applications'. Proceedings of IEEE *Pulsed power* conference, 1999, pp. 173–176

70   STINNETT, R.W., BUCHHEIT, R.G., and NEAU, E.L., *et al.*: 'Ion beam surface treatment: a new technique for thermally modifying surfaces using intense, pulsed ion beams'. Proceedings of IEEE *Pulsed power* conference, 1995, pp. 46–55

71   PUCHKAREV, V., KHARKOV, A., GUNDERSEN, M., and ROTH, G.: 'Application of pulsed corona discharge to diesel exhaust remediation'. Proceedings of 12th IEEE *Pulsed power* conference, 1999, pp. 511–514

72   WEKHOF, A.: 'Multi-parameter non-thermal corona laboratory system with a build-in process reactor'. Paper B.09, proceedings of international conference on *Pulsed power applications*, University of Applied Sciences, Gelsenkirchen, 27–29 March 2001

73   TAKAKI, K., JANI, M.A., and FUJIWARA, T.: 'Oxidation and reduction of NO_x in diesel engine exhaust by dielectric barrier discharge'. Proceedings of IEEE 12th *Pulsed power* conference, 1999, pp. 1480–1483

74   GROTHAUS, M.G., KHAIR, M.K., PAUL, P., FANICK, E.R., and BANNON, D.R.: 'A synergistic approach for the removal of NO_x and PM from diesel engine exhaust'. Proceedings of IEEE *Pulsed power* conference, 1999, pp. 506–510

75   HEINZ, V., ANGERSBACH, A., and KNORR, D.: 'Membrane permeabilization in biological materials in response to pulsed electric fields – a highly efficient process for biotechnology'. Paper B.03, proceedings of international conference on *Pulsed power applications*, University of Applied Sciences, Gelsenkirchen, 27–29 March 2001

76   ALY, R.E., JOSHI, R.P., STARK, R.H., SCHOENBACH, K.H., and BEEBE, S.J.: 'The effect of multiple, microsecond electrical pulses

on bacteria'. Proceedings of IEEE *Pulsed power* conference, 2001, pp. 1114–1117

77 RUIZ, G.M., RAWLINGS, T.K., and DOBBS, F.C., *et al.*: 'Global spread of microorganisms by ships', *Nature*, 2000, **408**, p. 49

78 ABOU-GHAZALA, A., KATSUKI, S., SCHOENBACH, K.H., DOBBS, F.C., and MOREIRA, K.R.: 'Bacterial decontamination of water by means of pulsed corona discharges'. Proceedings of 13th IEEE *Pulsed power* conference, 2001, pp. 612–615

79 BLOCK, R., LEIPOLD, F., and LEBAHN, K., *et al.*: 'Pulsed electric field based antifouling method for salinometers'. Proceedings of 13th IEEE *Pulsed power* conference, 2001, pp. 1146–1149

80 FISHBACK, J.R., and NUBBE, M.E.: 'Preliminary investigation of enzyme inactivation in citrus juices by submerged electric discharge'. Paper B.08, proceedings of international conference on *Pulsed power applications*, University of Applied Sciences, Gelsenkirchen, 27–29 March 2001

81 BLUHM, H., BAUMUNG, K., and BOEHME, R., *et al.*: 'Pulsed power science and technology at Forschungszentrum Karlsruhe'. Paper A.04, proceedings of international conference on *Pulsed power applications*, University of Applied Sciences, Gelsenkirchen, 27–29 March 2001

82 GAUDREAU, M.P.J., HAWKEY, T., PETRY, J., and KEMPKES, M.A.: 'A solid state pulsed power system for food processing'. Proceedings of 13th IEEE *Pulsed power* conference, 2001, pp. 1174–1177

83 ESPIE, S., MARSILI, L., MACGREGOR, S.J., and ANDERSON, J.G.: 'Pulsed power inactivation of liquid-borne microorganisms'. Paper B.07, proceedings of international conference on *Pulsed power applications*, University of Applied Sciences, Gelsenkirchen, 27–29 March 2001

84 BYSTRITSKII, V, YANKELEVICH, Y., and WOOD, T., *et al.*: 'Pulsed discharge in the fluidized packed bed reactor for toxic water remediation'. Proceedings of IEEE *Pulsed power* conference, 1999, pp. 464–467

85 YAVOROVSKY, N.A., and BOEV, S.: 'Electropulse water treatment'. Proceedings of IEEE *Pulsed power* conference, 1999, pp. 181–184

86 YAVOROVSKY, N.A., PELTSMANN, S.S., KHASKELBERG, M.B., and KORNEV, J.I.: 'Pulsed barrier discharge application for water treatment and disinfection'. Paper B.02, proceedings of international conference on *Pulsed power applications*, University of Applied Sciences, Gelsenkirchen, 27–29 March 2001

87 HOFMANN, J., and WEISE, Th. H.G.G.: 'Material disintegration and material breakdown with electrically generated shock waves'. Paper E.01, proceedings of international conference on *Pulsed power applications*, University of Applied Sciences, Gelsenkirchen, 27–29 March 2001

88 USHAKOV, V.Ya., and DULZON, A.A.: 'Performance capability of technological installations using the electrical discharge of conductor electrical explosion energy'. Paper A.03, proceedings of international conference on *Pulsed power*

applications, University of Applied Sciences, Gelsenkirchen, 27–29 March 2001

89   BLUHM, H., BAUMUNG, K., and BOEHME, R., *et al.*: 'Pulsed power science and technology at Forschungszentrum Karlsruhe'. Paper A.04, proceedings of international conference on *Pulsed power applications*, University of Applied Sciences, Gelsenkirchen, 27–29 March 2001

90   LIVITSYN, I.V., INOUE, H., KATSUKI, S., AKIYAMA, H., and NISHIZAWA, I.: 'Drilling and demolition of rocks by pulsed power'. Proceedings of 12th IEEE *Pulsed power* conference, pp. 169–172

91   DULZON, A.A., KURETZ, V.I., and TARAKANOVSKI, E.N., *et al.*: 'Electric pulse selective disintegration of mountain breeds, of technical rubber of workpieces and extraction of substances from vegetative raw material'. Paper E.05, proceedings of international conference on *Pulsed power applications*, University of Applied Sciences, Gelsenkirchen, 27–29 March 2001

92   JUNG, M., WEISE, T.H.G.G., WITTENBERG, M., MUELLER, J.A., and DICHTL, N.: 'Sludge disintegration by electrical shock waves'. Paper B.01, proceedings of international conference on *Pulsed power applications*, University of Applied Sciences, Gelsenkirchen, 27–29 March 2001

93   VORONOV, A., WEISE TH., and BERG, M.: 'Pulsed power in explosions driven electrically'. Paper E.03, proceedings of international conference on *Pulsed power applications*, University of Applied Sciences, Gelsenkirchen, 27–29 March 2001

94   VORONOV, A., WEISE TH., and BERG, M.: 'Pulsed power discharge for producing an ultra-thin hardened layer'. Paper E.04, proceedings of international conference on *Pulsed power applications*, University of Applied Sciences, Gelsenkirchen, 27–29 March 2001

95   YOUNG, A.J., NOVAC, B.M., SMITH, I.R., LYNN, B., and MILLER, R.A.: 'A pulsed-power system for producing high-intensity magnetic and electric fields for medical applications'. Paper C.01, proceedings of international conference on *Pulsed power applications*, University of Applied Sciences, Gelsenkirchen, 27–29 March 2001

96   SCHOENBACH, K.H., BARKER, R.J., and LIU, S.: 'Nonthermal medical/biological treatments using electromagnetic fields and ionized gases'. Proceedings of 12th IEEE *Pulsed power* conference, 1999, pp. 497–501

97   PAKHOMOV, A., and MURPHY, M.: 'Low intensity millimetre microwaves as a novel therapeutic modality'. Proceedings of 12th IEEE *Pulsed power* conference, 1999, pp. 23–28

98   PRATHER, W.D., BAUM, C.E., and LEHR, J.M., *et al.*: 'Ultrawideband source research'. Proceedings of 12th IEEE *Pulsed power* conference, 1999, pp. 185–188

99   ANDREEV, Y.A., BYANOV, Y.I., and EFREMOV, A.M., *et al.*: 'Gigawatt power level ultrawideband radiation generator'. Proceedings of 12th IEEE *Pulsed power* conference, 1999, pp. 1337–1340

100 JUNG, M., WEISE, TH. H.G.G., BRAUNSBERGER, U., and SABATH, F.: 'High power compact UWB systems'. Paper G.01, proceedings of international conference on *Pulsed power applications*, University of Applied Sciences, Gelsenkirchen, 27–29 March 2001

101 WARE, K., GULLICKSON, R., PIERRE, J., SCHNEIDER, R., and VITKOVITSKY, I.: 'Technologies for development of more affordable large X-ray simulators'. Proceedings of 12th IEEE *Pulsed power* conference, 1999, pp. 264–268

102 MILLER, A.R., GILBERT, C., RAUCH, J., RIX, W., PALKUTI, L., and WARE, K.: 'Pulsed power design for a compact X-ray simulator'. Proceedings of 12th IEEE *Pulsed power* conference, 1999, pp. 271–274

103 COOK, D.: 'Z, ZX, and X-1: a realistic path to high fusion yield'. Proceedings of 12th IEEE *Pulsed power* conference, 1999, pp. 33–37

104 SARJEANT, W.J., and DOLLINGER, R.E.:'High power electronics' (TAB Books, 1989)

Index